NEUTRINO FACTORIES AND SUPERBEAMS

Related Titles from AIP Conference Proceedings

698 Intersections of Particle and Nuclear Physics, 8th Conference; CIPANP 2003
Edited by Zohreh Parsa, February 2004, CD-ROM included, 0-7354-0169-1

689 Neutrinos, Flavor Physics, and Precision Cosmology: Fourth Tropical Workshop on Particle Physics and Cosmology
Edited by José F. Nieves and Raymond R. Volkas, October 2003, 0-7354-0160-8

672 Short Distance Behavior of Fundamental Interactions: 31st Coral Gables Conference on High Energy Physics and Cosmology
Edited by Behram N. Kursunoglu, Metin Camcigil, Stephan L. Mintz, and Arnold Perlmutter, June 2003, 0-7354-0139-X

670 Particles and Fields: Tenth Mexican School
Edited by U. Cotti, M. Mondragón, G. Tavares-Velasco, June 2003, 0-7354-0135-7

655 Particle Physics and Cosmology: Third Tropical Workshop on Particle Physics and Cosmology – Neutrinos, Branes, and Cosmology
Edited by José F. Nieves and Chung N. Leung, February 2003, 0-7354-0112-8

624 Cosmology and Elementary Particle Physics: Coral Gables Conference on Cosmology and Elementary Particle Physics
Edited by B. N. Kursunoglu, S. L. Mintz, and A. Perlmutter, July 2002, 0-7354-0073-3

623 Particles and Fields: Eighth Mexican Workshop
Edited by J. L. Diaz-Cruz, J. Engelfried, M. Kirchbach, and M. Mondragon, July 2002, 0-7354-0072-5

549 Intersections of Particle and Nuclear Physics: 7th Conference, CIPANP2000
Edited by Zohreh Parsa and William J. Marciano, December 2000, 1-56396-978-5

542 Physics Potential and Development of Muon Colliders and Neutrino Factories: Fifth International Conference
Edited by David B. Cline, November 2000, 1-56396-970-X

540 Particle Physics and Cosmology: Second Tropical Workshop
Edited by José F. Nieves, October 2000, 1-56396-965-3

530 Colliders and Collider Physics at the Highest Energies: Muon Colliders at 10 TeV to 100 TeV, HEMC'99 Workshop
Edited by Bruce J. King, August 2000, 1-56396-953-X

To learn more about these titles, or the AIP Conference Proceedings Series, please visit the webpage
http://proceedings/aip.org/proceedings

NEUTRINO FACTORIES AND SUPERBEAMS

5th International Workshop on
Neutrino Factories and Superbeams

NuFact 03

New York, New York   5 – 11 June 2003

EDITOR
Adam Para
Fermi National Accelerator Laboratory
Batavia, Illinois

SPONSORING ORGANIZATIONS
National Science Foundation
Columbia University
Brookhaven National Laboratory

Melville, New York, 2004
AIP CONFERENCE PROCEEDINGS ■ VOLUME 721

Editor:

Adam Para
Fermi National Accelerator Laboratory
MS 220
Pine Street
Batavia, IL 60510
USA

E-mail: para@fnal.gov

Authorization to photocopy items for internal or personal use, beyond the free copying permitted under the 1978 U.S. Copyright Law (see statement below), is granted by the American Institute of Physics for users registered with the Copyright Clearance Center (CCC) Transactional Reporting Service, provided that the base fee of $22.00 per copy is paid directly to CCC, 222 Rosewood Drive, Danvers, MA 01923. For those organizations that have been granted a photocopy license by CCC, a separate system of payment has been arranged. The fee code for users of the Transactional Reporting Service is: 0-7354-0201-9/04/$22.00.

© 2004 American Institute of Physics

Individual readers of this volume and nonprofit libraries, acting for them, are permitted to make fair use of the material in it, such as copying an article for use in teaching or research. Permission is granted to quote from this volume in scientific work with the customary acknowledgment of the source. To reprint a figure, table, or other excerpt requires the consent of one of the original authors and notification to AIP. Republication or systematic or multiple reproduction of any material in this volume is permitted only under license from AIP. Address inquiries to Office of Rights and Permissions, Suite 1NO1, 2 Huntington Quadrangle, Melville, N.Y. 11747-4502; phone: 516-576-2268; fax: 516-576-2450; e-mail: rights@aip.org.

L.C. Catalog Card No. 2004110102
ISBN 0-7354-0201-9
ISSN 0094-243X
Printed in the United States of America

CONTENTS

Preface..xi

PLENARY

Physics of Massive Neutrinos 3
  V. Barger
Solar Neutrino Results, KamLAND and Prospects 12
  K. Nakamura
Near Future Accelerator-Based Experiments 20
  W. C. Louis
Neutrino Scattering Physics at Superbeams and Neutrino Factories 29
  S. Kumano
Beta-Beams: Present Design and Expected Performances 37
  J. Bouchez, M. Lindroos, and M. Mezzetto
Neutrino Factory R&D in Europe 48
  H. D. Haseroth, B. Autin, R. Bennett, F. Méot, S. Gilardoni, C. Prior,
  P. Sievers, and A. Verdier
Neutrino Factory R&D in the U.S. 60
  M. S. Zisman for the Neutrino Factory and Muon Collider Collaboration
Particle Physics with Intense Muon Beams.................................. 68
  A. Baldini
Non-Particle Physics with Intense Muon Beams 77
  K. Ishida
The Future of Neutrino Experiments at Nuclear Reactors.................... 83
  J. M. Link
Overview of Neutrino Factory Simulations.................................. 90
  R. C. Fernow
MuCool Results and Plans 97
  A. D. Bross
MICE Status 106
  Y. Torun
The MuScat Experiment—Status and Plans................................ 114
  R. Edgecock
Impact of Neutrino Oscillation Measurements on Theory 122
  H. Murayama
Neutrinos in Cosmology... 130
  G. G. Raffelt
Summary of Working Group 2—Muon Physics............................. 136
  M. Aoki and A. Baldini
NuFact'03 Machine Working Group Summary............................. 144
  T. R. Edgecock, S. Machida, and R. A. Rimmer

WORKING GROUP 1

Global Fits to Neutrino Properties.................................... 165
    M. Maltoni
Double Beta Decay and the Absolute Neutrino Mass Scale 170
    C. Giunti
Natural Expectations for the Value of $|U_{e3}|$? 175
    A. de Gouvêa
MINOS Status and Physics Goals 179
    G. S. Tzanakos
SNO II: Salt Strikes Back. Update from the Sudbury
Neutrino Observatory 183
    J. A. Formaggio for the SNO Collaboration
Measurement of θ_{13} by Reactor Experiments............................ 190
    O. Yasuda
The Race for θ_{13}—Superbeams versus Reactors? 194
    P. Huber
Precision Measurement of Oscillation Parameters with Reactors............ 198
    S. Choubey
Scenarios for an Entry-Level Neutrino Factory........................... 202
    M. Campanelli
Overview of Degeneracies 206
    H. Minakata
Resolving JHF Degeneracy 211
    H. Sugiyama
Combining Superbeams... 215
    K. Whisnant
NUFACT'03: The Fate of the Clones................................... 219
    A. Donini
The Synergy of the Golden and Silver Channels at the
Neutrino Factory... 223
    P. Migliozzi
Resolving Degeneracies for Different Values of θ_{13} 227
    W. Winter
CNGS, OPERA and ICARUS Status................................... 231
    K. Kodama for the OPERA Collaboration
JHF Sensitivity and the 2km Intermediate Detector 235
    J. Burguet-Castell and D. Casper
BNL Very Long Baseline Experiment with a Super Neutrino Beam.......... 239
    S. Kahn
India-based Neutrino Observatory...................................... 243
    G. Rajasekaran
Low Energy Neutrino-Nucleus Interactions 247
    C. W. Walter, K. McConnel, and M. Sakuda
Lepton Flavor Violation in a Long-Baseline Experiment 252
    T. Ota and J. Sato

The Cosmological Energy Density of Neutrinos from
Oscillation Measurements 256
    K. Abazajian
Neutrino Masses in Theories with Dynamical Symmetry Breaking 261
    T. Appelquist and R. Shrock
Extrinsic CPT Violation in Neutrino Oscillations 265
    T. Ohlsson
An Overview of Neutrino Masses and Mixing in SO(10) Models 269
    M.-C. Chen and K. T. Mahanthappa

WORKING GROUP 2

Theoretical Aspects of Charged-Lepton Flavor Violation 275
    A. de Gouvêa
A Future Muon $(g-2)$ Experiment to $<\pm$ 0.1 ppm at a High Flux
Muon Facility 281
    B. L. Roberts
Sensitive Measurement of the EDM of the Muon? 285
    W. Morse
Status of the MEG Experiment 289
    A. Baldini
Precise Measurement of the Positive Muon Lifetime at the
RIKEN-RAL Muon Facility 293
    D. Tomono, S. N. Nakamura, Y. Matsuda, M. Iwasaki, G. Mason,
    K. Ishida, T. Matsuzaki, I. Watanabe, S. Sakamoto, and K. Nagamine
Towards an Improved Determination of the Fermi Coupling Constant
from the μLan Experiment.. 297
    C. J. G. Onderwater for the μLan Collaboration
A Precision Measurement of μ^+ Decay 301
    P. Kitching for the *TWIST* Collaboration
Measurement of the Transverse Polarization of the e^+ from the Decay
of Polarized μ^+ 305
    W. Fetscher, K. Bodek, A. Budzanowski, N. Danneberg, C. Hilbes,
    L. Jarczyk, K. Kirch, S. Kirstryn, K. Köhler, J. Lang, A. Kozela,
    J. Smyrski, E. Stephan, A. Strzałkowski, A. Von Allmen, and J. Zejma
Radioactive Muonic Atom Studies with Intense Muon Beams................ 309
    P. Strasser, K. Nagamine, T. Matsuzaki, K. Ishida, Y. Matsuda, K. Itahashi,
    and M. Iwasaki
Recent Development of a Point Positive Muon Source at the
RIKEN-RAL Muon Facility 313
    Y. Matsuda, P. Bakule, P. Strasser, K. Ishida, T. Matsuzaki, M. Iwasaki,
    Y. Miyake, K. Shimomura, S. Makimura, and K. Nagamine
A Review of Target Developments in Europe............................. 317
    J. R. J. Bennett
MECO Production Target Development 321
    J. L. Popp

Targetry R&D for PRISM Project 325
    K. Yoshimura, H. Ohnishi, T. Nakamoto, A. Yamamoto, Y. Ajima,
    M. Aoki, N. Fukasawa, K. Ishibashi, Y. Kuno, T. Miura, K. Nakahara,
    N. Nosaka, M. Numajiri, T. Ogitsu, A. Sato, A. Yamanoi, B. Autin,
    and P. Sievers

Preliminary Optimization of the Pion Capture and Decay Channel 329
    K. Paul and C. Johnstone

Horn R&D for 2002–2003 334
    S. Gilardoni, G. Grawer, G. Maire, J.-M. Maugain, S. Rangod,
    and F. Voelker

Optimization of the Transmission in an Alternating Gradient Muon Collection Channel 338
    B. Autin, F. Lemuet, F. Méot, and A. Verdier

High Intensity Surface Muon Beam Using a Large Acceptance Axial Focusing Channel 342
    H. Miyadera, K. Nagamine, K. Shimomura, K. Nishiyama, H. Tanaka,
    Y. Ikedo, and K. Ishida

Super Omega—New Concepts of Super Intense Surface Muon Beam 346
    K. Shimomura, K. Ishida, H. Miyadera, and K. Nagamine

Looking for Strangeness with Neutrino-Nucleon Scattering 350
    W. M. Alberico, S. M. Bilenky, and C. Maieron

Neutrino Scattering on the Nucleon and Determining Parton Distribution Functions 354
    Y. Miyachi

Unified Approach for Modelling Neutrino and Electron Nucleon Scattering Cross Sections from Very High Q^2 to $Q^2 = 0$ 358
    A. Bodek and U.-k. Yang

Near Liquid Argon TPC Detectors for Near Future 363
    F. Sergiampietri

First Results from E158; Measuring Parity Violation in Moller Scattering 367
    I. Younus

DIS-Parity: Measuring $\sin^2(\theta_W)$ with Parity Violating Deep Inelastic Scattering 371
    P. E. Reimer

Low Energy Neutrino Cross Sections 375
    G. P. Zeller

Working Group 2: Neutrino Scattering Physics 379
    B. T. Fleming

WORKING GROUP 3

Tetra Muon Cooling Ring 387
    S. Kahn, R. Fernow, V. Balbekov, R. Raja, and Z. Usubov

RFOFO Cooling Ring: Simulation Results 391
    J. S. Berg, R. C. Fernow, J. C. Gallardo, and R. B. Palmer

A Muon Ring Cooler with Lithium Lenses 395
    Y. Fukui, D. Cline, A. Garren, and H. Kirk
Features of a Muon Cooling Ring for a Neutrino Factory 399
    S. J. Brooks, M. R. Harold, C. R. Prior, and G. H. Rees
Frictional Cooling of Protons 403
    R. Galea
"High Frequency" Buncher and Phase Rotation 407
    D. Neuffer
Propagation of a Large-Emittance Muon Beam through a Straight,
Quadrupole-Based Precooling Channel 413
    M. Berz, K. Makino, and C. J. Johnstone
Tetra Cooler Ring Simulation in COSY INFINITY 418
    K. Makino and M. Berz
ISIS as a Proton Driver for a Neutrino Factory 422
    C. R. Prior
Bunch Production for a Muon Collider 428
    V. Balbekov
Novel Ideas for Beam Profiling in a Muon Cooling Channel 432
    K. D. Hoffman
The MuCool/MICE LH_2 Absorber Program 436
    M. A. Cummings
Plans for MICE at RAL 441
    P. Drumm for the MICE Collaboration
200MHz Superconducting RF Cavity Development for RLAs 445
    R. L. Geng, H. Padamsee, D. Hartill, P. Barnes, J. Sears, R. Losito,
    E. Chiaveri, H. Preis, and S. Calatroni
Optimized Beam Optics for Muon Acceleration 449
    S. A. Bogacz
Time-Energy Densities in $\pi \to \mu$ Decay 455
    B. Autin and F. Méot
A Pulsed Synchrotron for Muon Acceleration at a Neutrino Factory 463
    D. J. Summers, A. A. Garen J. S. Berg, and R. B. Palmer
Nonlinear Acceleration Modes in FFAGs with Fixed RF 467
    S. Koscielniak and C. J. Johnstone
Lattice Design and Particle Tracking of FFAG 475
    S. Machida
FFAG Construction for PRISM 479
    A. Sato for the PRISM-Working Group

Author Index 483

Preface

NUFACT03, the fifth annual international workshop in the NUFACT series, was held from 5-11 June 2003 at Columbia University, New York. Like its predecessors, NUFACT03 provided a unique forum in which neutrino physicists and accelerator scientists could discuss the motivation for future neutrino facilities, the progress in understanding how to design and build these facilities, and the associated R&D plans for the future. Indeed, the NUFACT meetings provide an important mechanism for ensuring a healthy dialog between the accelerator R&D community and the neutrino physics community. One result of this dialog is that in NUFACT03 the organizers felt it appropriate to extend the scope of the meetings to explicitly include discussions about neutrino Superbeams, in addition to Neutrino Factories.

In the last few years Neutrino Factory R&D has become increasingly international, and the NUFACT series has offered an essential mechanism for the international R&D community to review its progress and plan its future. The international cooling experiment (MICE) collaboration arose from discussions in NUFACT01. In NUFACT03 discussions took place that we hope will prepare the way for other international Neutrino Factory related projects, modeled very much after the MICE example.

The NUFACT03 meeting included both plenary and working group sessions. There were three working groups, covering (i) Neutrino Oscillation Physics, (ii) Neutrino Scattering and Muon Physics, and (iii) Machine Design and R&D. It is a pleasure to thank all of the plenary and working group speakers, the working group organizers, the International Advisory Committee, and my colleagues on the Scientific Program Committee. Together all of these many contributions resulted in an interesting and very productive meeting. It is also a pleasure to thank all of the approximately 170 participants. Finally, the local organizers, and the NUFACT03 Chairpersons, well deserve the standing ovation they received at the end of the meeting.

Each year the NUFACT organizing team discusses the next workshop: when and where ? At the present time it is clear that there is a need to continue holding NUFACT workshops annually, both because of the rapid R&D progress, and because of the evolving experimental neutrino oscillation results. The next meeting, NUFACT04, will be held at the University of Osaka in Japan, July 26-August 1. We hope to see you there.

Steve Geer
Chair, NUFACT03 Scientific Program Committee

PLENARY

Physics of Massive Neutrinos

V. Barger

Physics Department, University of Wisconsin, Madison, WI 53706

Abstract. The recent revolutionary accomplishments in the study of neutrino mass are reviewed. Key outstanding neutrino issues and how they are being or can be solved are then discussed.

THE NEUTRINO REVOLUTION (1998–2003) AND BEYOND

After decades of intensive searches for evidence of neutrino mass that gave increasingly restrictive upper limits, experiments have definitely established that neutrino flavors oscillate and hence that neutrinos have finite masses. An exciting era of further discovery and precision lies before us. The fundamental properties of neutrinos are finally within our reach. A variety of experimental pathways are falling into place, including reactors, off-axis beams, superbeams, and new detector technologies. Ultimately and inevitably this progression will lead to neutrino factories. The goal is to unravel the enigma of flavor physics.

It is not possible in this short review to do justice to the extensive literature on this science. A recent bibliography can be found in Ref. [1].

NEUTRINO OSCILLATIONS

The flavor states ν_α ($\alpha = e, \mu, \tau$) of the charged-current weak interaction are related to the mass states ν_i ($i = 1, 2, 3$) by the MNSP mixing matrix $V_{\alpha i}$. The vacuum oscillation probabilities are given by

$$P(\nu_\alpha \to \nu_\beta) = \left| \sum_{j=1}^{n} V_{\beta j} \, e^{-i\frac{m_j^2 L}{2E\nu}} V_{\alpha j}^* \right|^2. \tag{1}$$

These probabilities depend on neutrino mass-squared differences. For three neutrinos, the mixing matrix is specified by three mixing angles ($\theta_a, \theta_s, \theta_x$) and three complex CP-violating phases (δ, ϕ_2, ϕ_3) as

$$V = \begin{bmatrix} 1 & 0 & 0 \\ 0 & c_a & s_a \\ 0 & -s_a & c_a \end{bmatrix} \begin{bmatrix} c_x & 0 & s_x e^{-i\delta} \\ 0 & 1 & 0 \\ -s_x e^{i\delta} & 0 & c_x \end{bmatrix} \begin{bmatrix} c_s & s_s & 0 \\ -s_s & c_s & 0 \\ 0 & 0 & 1 \end{bmatrix} \begin{bmatrix} 1 & 0 & 0 \\ 0 & e^{i(\frac{1}{2}\phi_2)} & 0 \\ 0 & 0 & e^{i(\frac{1}{2}\phi_3+\delta)} \end{bmatrix}, \tag{2}$$

where $c_i = \cos\theta_i$ and $s_i = \sin\theta_i$. The first matrix on the right-hand side is relevant to atmospheric neutrinos, the second matrix is presently unknown, the third matrix pertains to solar neutrinos, and the fourth matrix involves the Majorana phases ϕ_2, ϕ_3 that do not enter in the oscillation probabilities but are important in neutrinoless double-beta decay.

The observed atmospheric and solar neutrino oscillations have very different δm^2 scales and are nearly decoupled (θ_x is small). Using an effective 2-neutrino approximation where one δm^2 is dominant, the vacuum oscillation probabilities are of the form

$$P(\nu_\alpha \to \nu_\beta) \simeq \sin^2 2\theta \sin^2 \Delta \qquad (3)$$

$$P(\nu_\alpha \to \nu_\alpha) \simeq 1 - \sin^2 2\theta \sin^2 \Delta, \qquad (4)$$

with θ the mixing angle and

$$\Delta \equiv \frac{\delta m^2 L}{4E}. \qquad (5)$$

In the propagation of electron neutrinos through matter, the scattering of ν_e on the electrons modifies both the amplitude and the wavelength. The amplitude modification has the resonant form

$$\sin^2 2\theta^m = \frac{\sin^2 2\theta}{\left(\frac{2\sqrt{2} G_F N_e E_\nu}{\delta m^2} - \cos 2\theta\right)^2 + \sin^2 2\theta} \qquad (6)$$

where N_e is the electron density. The oscillation amplitude is enhanced for $\delta m^2 > 0$ and suppressed for $\delta m^2 < 0$. The matter modification is crucial for solar ν_e propagation through the dense core of the Sun (N_e varies) and for accelerator neutrinos traversing long baselines through the Earth. There are analogous effects on oscillations from active to sterile neutrinos (for which N_e in the above formula is replaced by $N_e - \frac{1}{2}N_n$ for $\nu_e \leftrightarrow \nu_s$ and $-\frac{1}{2}N_n$ for $\nu_\mu, \nu_\tau \leftrightarrow \nu_s$).

PRESENT STATE OF KNOWLEDGE

Atmospheric neutrinos. The SuperKamiokande, Macro and Soudan experiments have found definitive evidence for the oscillations of muon-neutrinos produced in the Earth's atmosphere by cosmic rays. The ν_μ and $\bar\nu_\mu$ partially disappear (with oscillation parameters $\delta m_a^2 \sim 2 \times 10^{-3} \text{eV}^2$, $\theta_a \sim 45°$), while the ν_e and $\bar\nu_e$ do not (θ_x small).

Solar Neutrinos. The SNO, SuperKamiokande, Gallium and Chlorine experiments have established that ν_e from the sun partially disappear. The measurements of the neutral current flux by SNO have confirmed that the ν_e deficit corresponds to a ν_μ, ν_τ flux appearance. The oscillation solution is Large Mixing Angle (LMA) with $\delta m_s^2 \sim 6 \times 10^{-5} \text{eV}^2$ and $\theta_s \sim 33°$. There is an enhancement due to matter effects ($\delta m_s^2 > 0$) but the solution is not resonant. Maximal solar mixing, $\theta_s = \pi/4$, is excluded at 5.4σ by a global analysis including the SNO neutral current measurements on salt.

Reactor Antineutrinos. The KamLAND experiment, at an average distance $L \sim 175$ km from surrounding reactions, confirms the LMA solution with a best fit value $\delta m_s^2 \sim 7 \times 10^{-5} \text{eV}^2$. The combined data from the KamLAND and solar neutrino experiments further constrain the allowed δm_s^2 range. Figure 1 shows the allowed regions

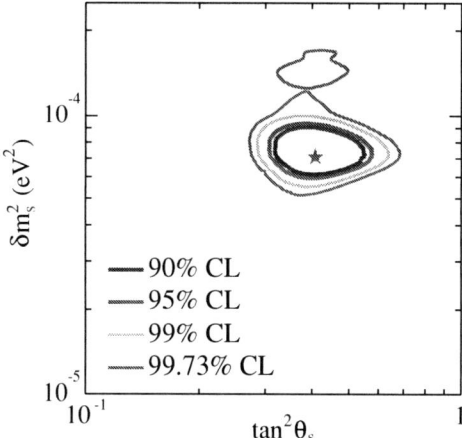

FIGURE 1. 90%, 95%, 99% and 3σ C. L. allowed regions from a combined fit to KamLAND and solar neutrino data. The best-fit point is at $\delta m_s^2 = 7.1 \times 10^{-5}\,\text{eV}^2$ and $\tan^2\theta_s = 0.41$.

from a global fit; the narrow horizontal ellipse represents expectations for 3 years of KamLAND data. The rejections of all solar solutions but LMA has been called "the KamLAND massacre". The earlier CHOOZ reactor experiment at a distance $L \approx 1$ km from the source found that the $\bar{\nu}_e$ did not disappear. This places a 90% C.L. of $\theta_x \lesssim 13°$ for $\delta m_a^2 = 2 \times 10^{-3}\,\text{eV}^2$.

Accelerator Antineutrinos. The LSND experiment found evidence at 3.3σ significance for $\bar{\nu}_\mu \to \bar{\nu}_e$ appearance. The effect would correspond to oscillations with $\delta m^2_{\text{LSND}} \sim 0.2\text{--}1\,\text{eV}^2$ and $\theta_{\text{LSND}} \sim 1.5°\text{--}5°$. The null result from the similar KARMEN experiment constrains the allowed oscillation parameters to a narrow band of oscillation parameters.

Since three neutrinos give only two independent δm^2 values, an oscillation interpretation of the LSND data requires the introduction of one or more sterile neutrinos. However, there are other stringent constraints on oscillations to sterile neutrinos: $\nu_\mu \to \nu_s$ from atmospheric data, $\nu_e \to \nu_s$ from solar data, $\nu_\mu \to \nu_\mu$ from accelerator data at short baselines, and $\bar{\nu}_e \to \bar{\nu}_e$ from reactor data at short baselines. Because of these constraints, global fits to the data in models with sterile neutrinos are barely acceptable; there is a tension between the atmospheric/solar data and the short-baseline data. The final resolution of the LSND neutrino awaits the MiniBooNE experiment at Fermilab.

Neutrino Counting in the Early Universe. Photons and light neutrinos dominate the energy density of the Universe at very early epochs and extra neutrinos would speed up the expansion of the Universe. The Cosmic Microwave Background, which originated at an age of about 380,000 years, and Big Bang Nucleosynthesis, which occurred at an age of a few minutes, constrain the total number of neutrinos. From the precision WMAP data, the constraint is $N_\nu = 0.9\text{--}8.3$ (2σ range). From the primordial abundances of D and ^{4}He, the constraint is $N_\nu = 1.3\text{--}3.0$ (2σ range). The LSND sterile neutrino would be fully thermalized by the BBN era and thus Standard BBN rejects it.

Houdini's Escape from the BBN Constraint. A large asymmetry between the numbers of ν_e and $\bar{\nu}_e$ in the early Universe still allows extra neutrinos to be compatible with BBN. The asymmetry L_e in the number densities n of ν_e and $\bar{\nu}_e$ is expressed in terms of a degeneracy parameter $\xi_e = \mu_e/T$, where μ_e is the chemical potential and T is the temperature, as $L_e = (n_{\nu_e} - n_{\bar{\nu}_e})/T \simeq 0.7\xi_e$. The equilibrium ratio of neutrons to protons is modified to $(n/p)_{\text{equil}} = \exp(-\Delta m_{np}/T - \xi_e)$, where Δm_{np} is the neutron-proton mass difference. Then $\xi_e \sim 0.1$ can reconcile the LSND sterile neutrino with BBN prior to BBN by modification of the reaction rates to suppress its production. The existence of the LSND sterile neutrino implies $L_e \sim 0.01$–0.1, which is huge compared to the baryon asymmetry $n_B/n_\gamma \sim 10^{-9}$.

Neutrino Mass and Large Scale Structure in the Universe. Even small Σm_ν, the sum of neutrino masses, influences the power spectrum of galaxy correlations. Heavier neutrinos suppress power more on small scales. Analyses of the 2dF, CMB, and Lyman alpha forest data have obtained the following bounds: $\Sigma m_\nu \leq 2.2$ eV (2dF), $\Sigma m_\nu \leq 0.7$ eV (2dF + Lyα Forest + WMAP), $\Sigma m_\nu \leq 0.63$ eV (2dF + WMAP). The LSND neutrino would imply $\Sigma m_\nu \geq \sqrt{\delta m_{\text{LSND}}^2} \geq 0.45$ eV if it is thermalized.

Neutrino Parameters. A summary of our present knowledge of neutrino parameters is given in Table 1 along with the future experiments that can improve this knowledge. This summary brings into focus the critical needs for future experimentation that we next discuss.

TABLE 1. Present knowledge of neutrino parameters and future ways of improving this knowledge.

3-neutrino observables	Present knowledge ($\sim 95\%$ C. L.)	Near future
θ_a	$45° \pm 10°$	$P(\nu_\mu \to \nu_\mu)$ MINOS, CNGS
θ_s	$32.5° \pm 3.6°$	SNO NC, KamLAND
θ_x	$\leq 13°$ (for $|\delta m_a^2| = 2.0 \times 10^{-3}$ eV^2)	$P(\bar{\nu}_e \to \bar{\nu}_e)$ Reactor, $P(\nu_\mu \to \nu_e)$ LBL
$|\delta m_a^2|$	$(2.0^{+1.2}_{-0.8}) \times 10^{-3}$ eV^2	$P(\nu_\mu \to \nu_\mu)$ MINOS, CNGS
sgn(δm_a^2)	unknown	$P(\nu_\mu \to \nu_e), P(\bar{\nu}_\mu \to \bar{\nu}_e)$ LBL
$|\delta m_s^2|$	$(7.1^{+1.8}_{-1.1}) \times 10^{-5}$ eV^2	$P(\bar{\nu}_e \to \bar{\nu}_e)$ KamLAND
sgn(δm_s^2)	+ (MSW)	done
δ	unknown	$P(\nu_\mu \to \nu_e), P(\bar{\nu}_\mu \to \bar{\nu}_e)$ LBL
Majorana	unknown	$0\nu\beta\beta$
ϕ_2	unknown	$0\nu\beta\beta$ (if $\simeq 0, \pi$)
ϕ_3	unknown	hopeless
m_ν	$\Sigma m_\nu < 1$ eV	LSS, $0\nu\beta\beta$, β-decay

KEY NEUTRINO ISSUES AND HOW THEY ARE BEING / CAN BE SOLVED

Key Issue #1: Verify Oscillations/Precision. The immediate challenge is to "see" the oscillation wiggles versus energy and thereby more precisely determine the associated δm^2 scales. Present experiments determine only the averaged suppressions of the oscillation probabilities. The energy dependence of $P(\bar{\nu}_e \to \bar{\nu}_e)$ at the solar δm_s^2 scale should be measurable in the KamLAND experiment. The energy dependence of $P(\nu_\mu \to \nu_\mu)$ at the δm_a^2 scale will be measureable in the K2K (250 km), MINOS (730 km) and CNGS (730 km) experiments. In addition, the CNGS experiments OPERA and ICARUS will observe ν_τ appearance, $P(\nu_\mu \to \nu_\tau)$.

Key Issue #2: How Small is θ_x? Reactor experiments are under consideration with two detectors at short L ($<$ a few km) to measure θ_x from the wiggles in $P(\bar{\nu}_e \to \bar{\nu}_e)$ versus energy. The distances of the detectors in these proposals are

Site	L_1 (km)	L_2 (km)
Krasnoyarsk	0.1	1.0
Kashiwazaki	0.3	1.7
Diablo Canyon	0.15	1.2

The sensitivity limit of the reactor experiments is $\sin^2 2\theta_x \approx 0.01$.

Future accelerator experiments will measure θ_x via the appearance probabilities

$$P(\nu_\mu \to \nu_e) \text{ or } P(\nu_e \to \nu_\mu) \approx \sin^2 2\theta_x \sin^2 \Delta_a. \tag{7}$$

The "magic" of off-axis beams ($\theta_{OA} \approx 1\text{--}2°$) is to give approximately monochromatic neutrino energy and lower backgrounds. JPARC (295 km) and FNAL (730 km) have proposed off-axis beam facilities.

The next stage of long-baseline experiments with accelerators will be Superbeams with intensity upgrades of 4 to 5 times that of present beams; they may be either off-axis or wide-band (BNL proposal). The binning of quasi-elastic events from a wide-band beam gives the equivalent of many narrow-band beams.

The ultimate accelerator facility for neutrino oscillation experiments is a neutrino factory, with the decays of stored muons along straight sections of an oval ring giving ν_e and $\bar{\nu}_\mu$ (or $\bar{\nu}_e$ and ν_μ) beams of the highest achievable intensity. A neutrino factory will provide the first ν_e beam. The "golden" oscillation channel at a neutrino factory is $\nu_e \to \nu_\mu$.

Along with the foregoing accelerator developments, new detector technologies are being developed for future detectors of 50 to 500 kton sizes. These studies include low-Z calorimeters, liquid Argon, water Cherenkov, and iron scintillator.

The approximate discovery reaches in $\sin^2 2\theta_x$ in future experiments are listed in Table 2. How low in $\sin^2 2\theta_x$ will we need to go before a signal is detected? This is presently an unanswerable question.

Key Issue #3: Mass Hierarchy? The present oscillation data allow two possible orderings of the neutrino mass eigenstates as shown in Fig. 2. The normal and inverted

TABLE 2. Approximate 3σ reaches in $\sin^2 2\theta_x$ of future neutrino oscillation experiments.

Experiment	Reach in $\sin^2 2\theta_x$		
	Discovery	sgn(δm_a^2)	CP violation
Reactor	0.01	–	–
Conventional ν beam	0.01	–	–
Superbeam	0.003	0.01	0.03
Entry-level NuFact	0.0005	0.001	0.002
High-performance NuFact	0.00005	0.0001	0.0005

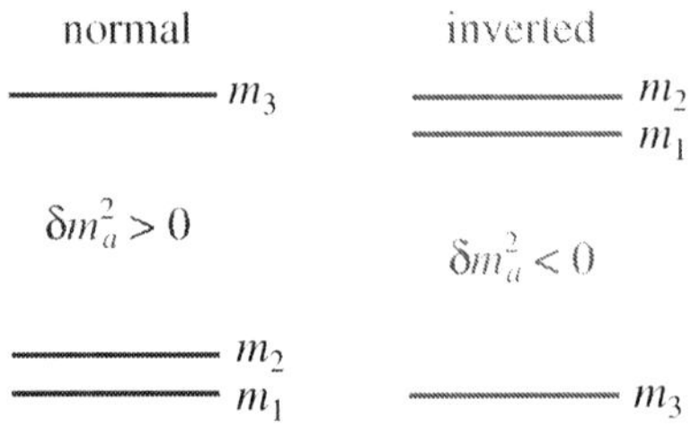

```
        normal                      inverted
      _____ m_3               _____ m_2
                                 _____ m_1

    δm_a^2 > 0                   δm_a^2 < 0

      _____ m_2
      _____ m_1               _____ m_3
```

FIGURE 2. The patterns of relative mass differences in normal and inverted neutrino mass hierarchies.

orderings can be distinguished by Earth matter effects in long-baseline oscillation experiments, which enhance $P(\nu_\mu \to \nu_e)$ and suppress $P(\bar{\nu}_\mu \to \bar{\nu}_e)$, or vice versa, depending on the sign of δm_a^2. The matter effects increase with the baseline. Long baselines ($L > 900$ km) are needed to determine the mass hierarchy. The dependence of appearance probability rates on the sign of δm_a^2 is illustrated in Fig. 3 for a neutrino factory. The effects of a CP-violating phase are also indicated in the figure.

Key Issue #4: CP Violation? If there is intrinsic CP violation in the neutrino sector then $P(\nu_\mu \to \nu_e) \neq P(\bar{\nu}_\mu \to \bar{\nu}_e)$. The difference in these CP-conjugate vacuum probabilities has the parameter dependences

$$\Delta P \propto \left(\frac{\delta m_s^2}{\delta m_a^2}\right) \sin\delta \sin^2\theta_x, \quad \frac{\Delta P}{P} \propto \frac{\Delta_s \sin\delta}{\theta_x}. \quad (8)$$

Thus, the feasibility of a measurement of the CP phase δ depends on the value of θ_x (the phase enters the MNS matrix V only through the combination $\sin\theta_x e^{-i\delta}$). Moreover, both δm_s^2 and δm_a^2 oscillations must contribute to have $\Delta P \neq 0$. A cautionary note is that intrinsic CP violation must be distinguished from the fake CP violation of matter effects. The magic baselines for CP-violation considerations in $P(\nu_\mu \to \nu_e)$ are:

- $L/E_\nu \approx 620$ km/GeV, for which $P(\nu_\mu \to \mu_e)$ depends only on $\sin\delta$ (not $\cos\delta$);
- $L \approx 7600$ km, for which there are no CP-violation effects at the matter oscillation wavelength.

Approximate discovery reaches in $\sin^2 2\theta_x$ for CP-violation are given in Table 2.

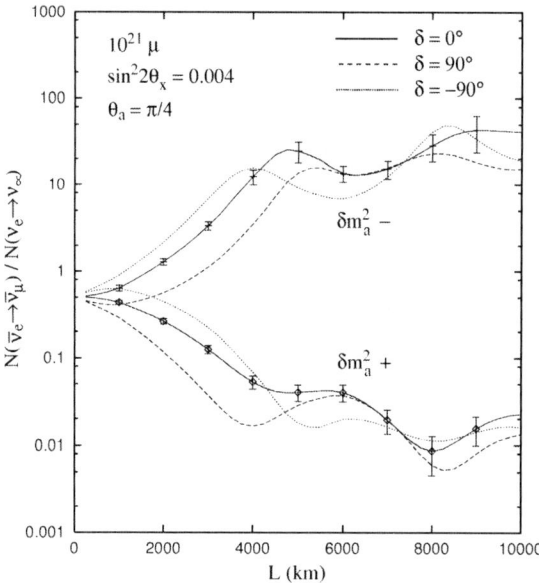

FIGURE 3. Ratio of antineutrino to neutrino appearance events versus baseline in a neutrino factory for $\sin^2 2\theta_x = 0.004$ and several values of δ. Both $\delta m_a^2 > 0$ and $\delta m_a^2 < 0$ cases are shown.

To establish CP violation in long-baseline experiments, we must be able to resolve degeneracies that can confuse CP-violating and CP-conserving fits to the data. Different parameter sets can give the same values of $P(\nu_\mu \to \nu_e)$ and $P(\bar{\nu}_\mu \to \bar{\nu}_e)$ at one L. There is a possible eight-fold degeneracy in the parameters, consisting of the following three two-fold degeneracies: (δ, θ_x), $\text{sgn}(\delta m_a^2) = \pm 1$, $(\theta_a, \frac{\pi}{2} - \theta_a)$. The best strategies for resolving the degeneracies are (1), to have one detector at the first appearance peak, (2) to have at least one detector at a long baseline, and (3) to have detectors at two or more baselines.

Key Issue #5: 3 × 3 Mixing Matrix Unitarity? To test unitarity of the neutrino mixing matrix, and thereby determine if there is "leakage" to sterile neutrinos, we need to measure all its elements. For this purpose ν_e beams are required, and these exist only at a neutrino factory. The six oscillation probabilities ($\nu_\mu \to \nu_\mu$, $\nu_\mu \to \nu_e$, $\nu_\mu \to \nu_\tau$, $\bar{\nu}_e \to \bar{\nu}_e$, $\bar{\nu}_e \to \bar{\nu}_\mu$, $\bar{\nu}_e \to \bar{\nu}_\tau$) can all be determined through the detection of charged leptons μ^-, e^-, τ^-, e^+, μ^+, τ^+). Neutral currents also test unitarity, since oscillations to sterile neutrinos would deplete the observable flux.

With ν_e beams, tests can also be made of time reversal violation, for which

$$P(\nu_e \to \nu_\mu) \neq P(\nu_\mu \to \nu_e) \tag{9}$$

Key Issue #6: Dirac or Majorana? The only way to answer this fundamental question about the nature of neutrinos is by measurement of neutrinoless double-β decay, which occurs *only if* neutrinos are Majorana. The matrix element for this process is given to an excellent approximation by

$$M_{ee} = \left(2(\Sigma m_\nu) - \sqrt{(\Sigma m_\nu)^2 + 3\delta m_a^2}\right) \left|c_s^2 + s_s^2 e^{i\phi_2}\right|/3, \tag{10}$$

where ϕ_2 is a Majorana phase. The predicted bands of Σm_ν versus M_{ee}, given our present knowledge of θ_s, is shown in Fig. 4. The solid band represents a normal mass hierarchy and the dotted band an inverted hierarchy. For the latter there is a higher minimum value of $|M_{ee}|$. A non-zero measurement of neutrinoless double-β decay can constrain Σm_ν, giving both upper and lower bounds. The present bounds on Σm_ν from single-β decay and from cosmology (large scale structure) are shown by the horizontal dotted lines in Fig. 4.

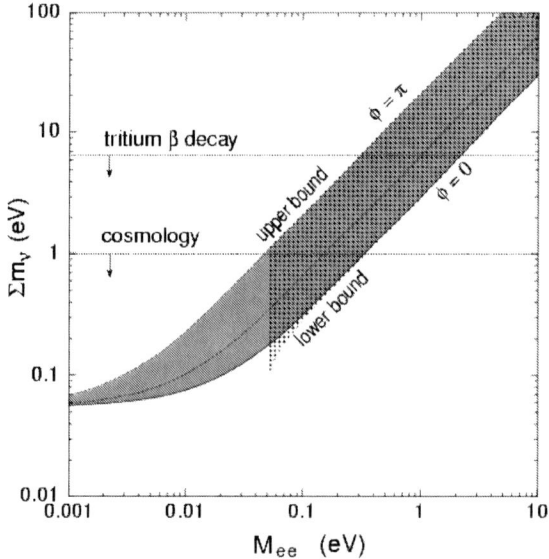

FIGURE 4. Σm_ν vs. M_{ee} for the normal (shaded) and inverted (dotted) hierarchies. For the inverted hierarchy, $M_{ee} \geq \sqrt{|\delta m_a^2|}$. (Here, $|\delta m_a^2|$ was taken to be 3×10^{-3} eV^2). The 95% C. L. bounds from tritium β decay and cosmology are shown.

Key Issue #7: What Theory? The see-saw is the favored theoretical mechanism since it naturally explains the smallness of neutrino masses as the consequence of heavy right-handed neutrinos N_R, $m_\nu \sim m_D^2/M_N$. The N_R occur as singlets in SO(10) Grand Unified Theory (GUT) representations of the fermions. The masses of the N_R are unprotected by symmetries and are thus naturally expected to be comparable to the GUT scale, while the Dirac mass m_D is at the electroweak scale. GUT models can accommodate all quarks and lepton data. Their predictions for neutrino masses and mixings are somewhat flexible.

Key Issue #8: Leptogenesis? The matter-antimatter asymmetry is due to processes that violate CP in the early Universe. A necessary condition is baryon number violation, which could originate through lepton number violation. A lepton asymmetry in the early Universe could arise from the CP-violating decays of the heavy right-handed neutrinos. In some models there is a connection beteween the CP violation in N decays and the CP violation measurable in long-baseline experiments. These models make other testable predictions at low energies.

SUMMARY

Neutrino mass is the first discovery of physics beyond the Standard Model. Oscillation experiments "on the table" have great potential for another breakthrough in measuring θ_x. The future of oscillation physics is very bright, with long baselines and Superbeams as the next horizon. Whatever experiments accomplish over the next decade, Neutrino Factories will be essential to reconstruct all neutrino mixings with high precision. The combination of Superbeam and Neutrino Factory data will further improve this knowledge.

If theoretical prejudices for Grand Unified Theories are correct, neutrino masses owe their origin to right-handed neutrinos with masses comparable to the GUT scale. The sign of the baryon asymmetry may be related via leptogenesis to the CP phase in neutrino oscillations and thereby be subject to experimental investigation. These and other fundamental ideas may soon be "put to the test", at least in the context of models, by measurements of θ_x, $\text{sgn}(\delta m_a^2)$ and δ.

Neutrino physics has always been full of surprises. There very likely are more to come!

ACKNOWLEDGMENTS

This work was supported in part by the U.S. Department of Energy under Grant no. DE-FG02-95ER40896 and in part by the Wisconsin Alumni Research Foundation.

REFERENCES

1. V. Barger, D. Marfatia, and K. Whisnant, hep-ph/0308123, to be published in Int. Jour. Mod. Phys. E (2003).

Solar neutrino results, KamLAND and prospects

K. Nakamura[1]

TUNL, Department of Physics, Duke University, Durham, NC 27708, USA

Abstract. The solar neutrino problem has puzzled physicists for 30 years. There were two beautiful results announced in 2002 that solved this problem. The SNO experiment measured the solar neutrino flux via both charged current and neutral current interactions which gave strong evidence for flavor transitions of neutrinos on their way between the sun and the earth. In late 2002, KamLAND provided confirmation for this solar neutrino solution independent of any solar neutrino results. It measured the flux and the energy spectrum of anti-neutrinos coming from nuclear reactors and provided the first evidence for anti-neutrino disappearance. Assuming CPT invariance, the LMA region is the only solar neutrino solution consistent with KamLAND.

I focus my discussion here on KamLAND results and prospects.

INTRODUCTION

Starting in the 1970's, the Homestake experiment measured the solar neutrino (v_e) flux. The measured flux was about 1/3 of the Standard Solar Model prediction [1]. This solar neutrino measurement was followed by several experiments such as SAGE, GALLEX, Kamiokande and SuperKamiokande [2]. However, the measured solar neutrino fluxes were still less than predicted [3]. The interpretation was either an unknown phenomenon related to neutrino propagation or a limitation of the Standard Solar Model.

The common explanation of this unknown phenomenon is neutrino oscillations [4][5]. Using the two flavor mixing scenario, the survival probability of v_e after propagating for a distance L in vacuum is given by

$$P(v_e \to v_e) = 1 - \sin^2 2\theta \sin^2 \left(\frac{1.27 \Delta m^2 [\text{eV}^2] \text{L[m]}}{\text{E[MeV]}} \right). \quad (1)$$

Here, the mixing angle, $\sin^2 2\theta$, and the mass difference between two neutrino flavors, Δm^2, are the neutrino oscillation parameters. Taking the electron density in the sun into account and combining the experimental results, there are four possible allowed regions for the neutrino oscillation parameters: LMA, SMA, LOW, VAC [6][7].

In 2002, the SNO experiment [8] released a result for the solar neutrino flux via both charged current and neutral current interactions on deuterium. The charged current reaction, $v_e + d \to p + p + e^-$, is only sensitive to v_e, while the neutral current reaction, $v_x + d \to v_x + n + p$ ($x = e, \mu, \tau$), is sensitive to all types of active neutrinos with equal cross section and it measures the flux of all active neutrinos coming from the sun. The

---

[1] This work is presented on behalf of KamLAND collaboration

SNO result is

$$\phi_{SNO}(CC) = \phi_{SNO}(\nu_e) = 1.76^{+0.06+0.09}_{-0.05-0.09} \times 10^6/\text{cm}^2/\text{s} \qquad (2)$$

$$\phi_{SNO}(NC) = \phi_{SNO}(\nu_{e\mu\tau}) = 5.09^{+0.44+0.46}_{-0.43-0.43} \times 10^6/\text{cm}^2/\text{s}. \qquad (3)$$

While the neutral current measurement is in good agreement with the Standard Solar Model prediction ($\phi_{SSM} = 5.05^{+1.01}_{-0.81} \times 10^6/\text{cm}^2/\text{s}$) [9], the measured ν_e flux is less than the predicted flux. This is strong evidence for neutrino flavor transition between the sun and the earth. By combining the results of all solar neutrino experiments, the LMA solution is the most likely solution, although others are still allowed at lower confidence level.

Although these results are consistent with neutrino oscillations, it is important to verify this interpretation in an experiment not based on solar neutrinos, and to measure the neutrino energy distortion predicted by equation (1). KamLAND uses man-made anti-neutrinos from nuclear reactors and it measures the anti-neutrino energy spectrum.

KAMLAND

The primary motivation of KamLAND (Kamioka Liquid scintillator Anti-Neutrino Detector) is to measure $\bar{\nu}_e$ disappearance using commercial power reactors as the neutrino sources. It was built in the same cavity as the Kamiokande experiment in the Kamioka zinc mine (2700 mwe). Construction was completed at the end of 2001 and data taking started in early 2002. The first results [10] correspond to the period from 4th of March 2002 to 6th of October 2002 (life-time=145.1 days).

Nuclear reactors are well known anti-neutrino sources. Reactor neutrino experiments have a ~30 year history and their anti-neutrino intensity and energy spectrum are well understood. Since the average neutrino energy is ~4 MeV and the average baseline is 180 km (Fig. 1 (left)), the sensitivity of KamLAND just covers the LMA region ($\Delta m^2 \sim 6 \times 10^{-6} \text{ eV}^2$, $\sin^2 2\theta > 0.1$). Assuming CPT invariance (ν_e has the same properties as $\bar{\nu}_e$), KamLAND can confirm the LMA solution independent of solar neutrino results.

The anti-neutrino detection is via inverse beta decay which can occur for anti-neutrinos with eneries above 1.8 MeV.

$$\bar{\nu}_e + p \rightarrow e^+ + n. \qquad (4)$$

The final state positron produces scintillation light and subsequently annihilates with an electron emitting two γ-rays. The minimum detection energy is 1.0 MeV. The neutron thermalizes quickly and is captures on a proton emitting a mono-energetic γ-ray (2.2 MeV). The detection of this prompt positron + γ's and the delayed (~200 μsec) 2.2 MeV γ which is emitted from the neutron capture is the anti-neutrino signal. This delayed coincidence technique reduces the background considerably. A total mass of 1 kton of liquid scintillator is used as target and detector.

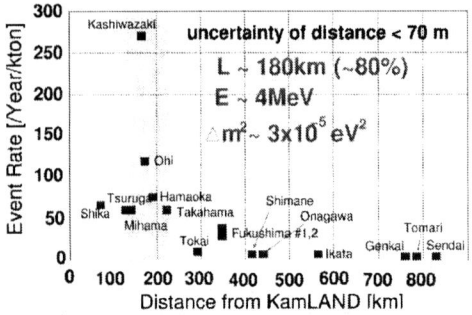 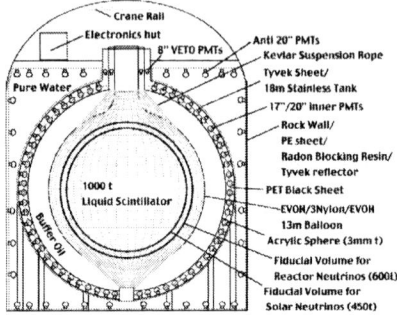

FIGURE 1. (left) Expected event rate at KamLAND from each reactor in Japan. (right) KamLAND schematic view.

KamLAND detector

Figure 1 (right) shows the schematic view of KamLAND. The liquid scintillator is contained in a 13 m diameter spherical balloon which is suspended by Kevlar ropes. The region between the balloon and the 18 m diameter stainless steel spherical tank is filled with buffer oil which does not scintillate. The scintillation light is detected by 1325 newly developed 17-inch diameter PMTs and 554 old Kamiokande 20-inch PMTs mounted in the spherical tank facing the liquid scintillator. Only 17-inch PMTs were used for the first analysis and the cooresponding photocathode coverage is 22%.

The stainless steel tank is surrounded by a water Cherenkov outer detector with 225 20-inch PMTs. It vetos any events in the liquid scintillator when a muon is detected in the outer detector. Since KamLAND observes the reaction (4), neutrons coming from outside can create fake signals that look like the anti-neutrino signal. In addition, the outer detector absorbs γ-rays and neutrons produced by cosmic-ray muons in the surrounding rock.

Detector performance

The energy scale between 0.5 MeV and 7.5 MeV is calibrated with γ-ray sources (^{68}Ge, ^{65}Zn, ^{60}Co, AmBe). Gamma-rays from the capture of cosmic-ray muon induced neutrons on hydrogen and carbon in the liquid scintillator are also used for energy calibration. The PMT gain, solid angle, density of PMTs, shadowing by suspension ropes and transparencies of the liquid scintillator and buffer oil are corrected for in the calibration procedure. Figure 2 (left) (a) shows the fractional deviation of the reconstructed energies from the known source energies. The energy resolution is $7.5\%/\sqrt{E(MeV)}$. Since the AmBe source also emits neutrons, it can be used for studying the delayed coincidence efficiency. A value of 95.3±0.5% was estimated.

The fiducial volume uncertainty was estimated from the vertex distribution of capture events of cosmic-ray muon produced spallation neutrons. Figure 2 (left) (b) shows the vertex positions normalized to the balloon radius (6.5 m). The ratio of total number of

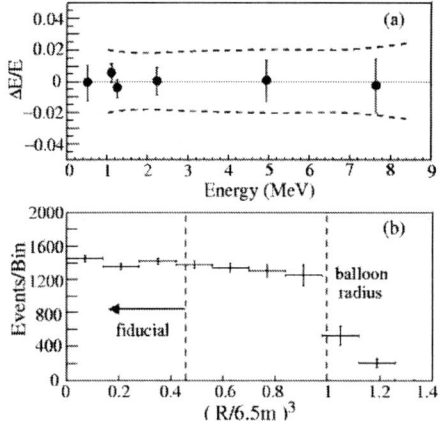
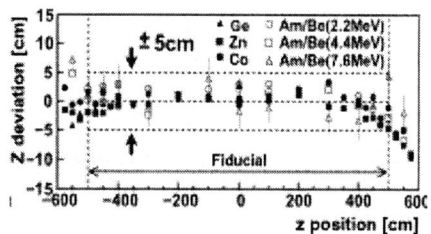

FIGURE 2. (left) (a) The fractional difference of the reconstructed average energies and known γ-source energies. The dashed lines show the adopted systematic error. (left) (b) The R^3 vertex distribution of 2.2 MeV neutron capture γ's. The level of uniformity over the fiducial volume is used in the estimate of the fiducial volume uncertainty. (right) Vertex reconstruction for γ sources deployed at several positions along Z axis.

TABLE 1. Systematic uncertainties

Total LS mass	2.1	Reactor power	2.0
Fiducial mass ratio	4.1	Fuel composition	1.0
Energy threshold	2.1	Time lag	0.28
Efficiency of cuts	2.1	$\bar{\nu}$ spectra	2.5
Life-time	0.07	Cross section	0.2

events and the number in the fiducial volume (R<5 m) agree with the geometric fiducial fraction to within 4.1%. Since we measured the total mass of the liquid scintillator with 2.1% precision, we estimated the fiducial mass volume uncertainty to be 4.6%.

The vertex reconstruction uncertainty was estimated by deploying γ-ray sources at different positions along the Z axis. The vertices are reconstructed within an accuracy of 5 cm (Fig. 2 (right)).

Table 1 provides a summary of the systematic uncertainties.

Background

The radioactive background in the liquid scintillator was studied using real data. ^{238}U and ^{232}Th concentrations were estimated from Bi-Po sequential decays. Figure 3 (left) shows the observed time difference between ^{214}Bi decay and ^{214}Po decay in the ^{238}U chain in KamLAND. The time constant of 232±7 μsec was obtained and it is consistent with the ^{214}Po decay time (237 μsec). ^{238}U and ^{232}Th concentrations were estimated to be $(5.2\pm0.8)\times 10^{-17}$ g/g and $(3.5\pm0.5)\times 10^{-18}$ g/g, respectively. The number of background events coming from ^{238}U and ^{232}Th is negligible in the data sample.

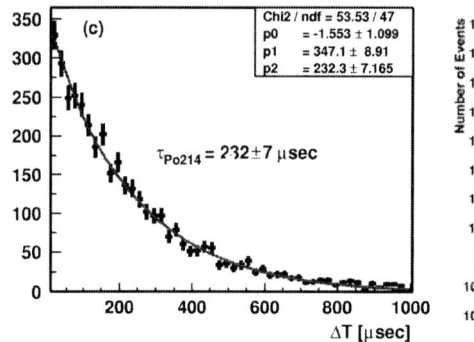 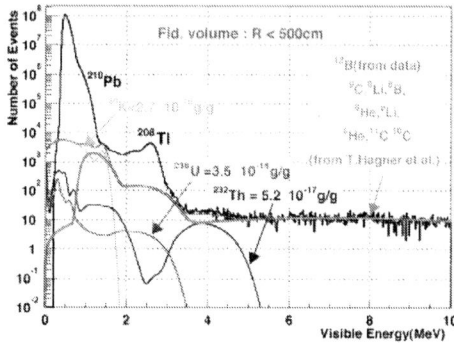

FIGURE 3. (left) Observed time difference between ^{214}Bi and ^{214}Po decay signals. It is consistent with the ^{214}Po decay time. (right) Summary of the background in our liquid scintillator [11].

TABLE 2. Background summary

Background	Number of events
Accidental	0.0086±0.0005
^{9}Li, ^{8}He	0.94±0.85
Fast neutrons	<0.5

Radioactive spallation products, such as ^{8}He ($T_{1/2}$ = 119 msec) and ^{9}Li ($T_{1/2}$ = 178 msec), which emit both e^- and n can be a background for the delayed coincidence. These were eliminated by requiring two time/geometry cuts: (a) 2 sec veto in the entire fiducial volume following a muon which deposits a large amount of energy. (b) for other muons, delayed events within 2 sec and 3 m from the muon track were rejected. The background coming from these spallation products are estimated to be 0.94±0.85 in our data sample.

Figure 3 (right) is the summary of the radioactive background in our liquid scintillator and Table 2 provides the background summary.

KAMLAND RESULT

The event selection criteria are

- No outer detector signal.
- Time difference between the prompt and delayed signal is 0.5-660 μsec.
- Delayed energy is between 1.8 and 2.6 MeV.
- Fiducial volume (R<5 m).
- Distance between prompt and delayed vertex is within 1.6 m.
- Spallation cuts
  - Veto for 2 msec after muons

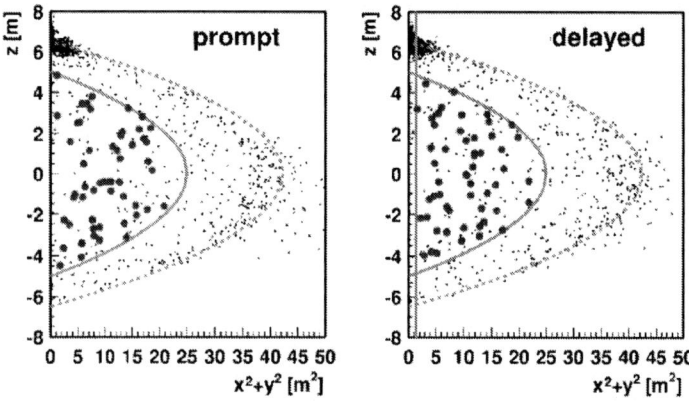

FIGURE 4. Vertex position distribution for prompt and delayed signals before fiducial volume cut (dots) and after (circles). There is a higher activity around the chimney region (z=6-7 m) because there is less water for shielding.

- Veto for 2 sec after muons with E>10^6 photo electrons (p.e.) or cylindrical (R=3 m) 2 sec veto around muons with E<10^6 p.e.

The total tagging efficiency is 78.3 % and finally 54 events were obtained for E_{prompt}>2.6 MeV. The vertex distribution of the 54 events is shown in Fig. 4. The solid curve corresponds to the fiducial volume.

Figure 5 shows the distribution of $\bar{\nu}_e$ candidates after fiducial volume, vertex correlation and spallation cuts. The prompt energy stands for positron energy and the delayed energy stands for neutron capture on proton. The event with ~5 MeV of delayed energy is considered to be due to neutron capture on ^{12}C. One event is consistent with this capture cross section.

In the absence of $\bar{\nu}_e$ disappearance, the expected number of $\bar{\nu}_e$ events is 86.8±5.6 which is calculated from thermal power, burn-up and fuel exchange records provided by the electric power companies operating the nuclear reactors. Uncertainties in this calculation are listed in Table 1.

The ratio of observed and expected $\bar{\nu}_e$ events is

$$\frac{N_{obs} - N_{BG}}{N_{expected}} = 0.611 \pm 0.085(stat) \pm 0.041(syst). \quad (5)$$

A clear deficit of events is observed. The probability that the KamLAND result is consistent with the no disappearance hypothesis is less than 0.05%.

Figure 6 (left) shows the prompt energy distribution. The dots in the lower frame are data and the upper histogram represents the spectrum expected from the no-oscillation scenario. Contributions from geo-anti-neutrinos are expected below 2.6 MeV [12]. The lower histogram above 2.6 MeV is the best fit spectrum with the neutrino oscillation parameters: $\sin^2 2\theta = 1$ and $\Delta m^2 = 6.9 \times 10^{-5}$ eV^2. This should be compared to the LMA best fit values of $\sin^2 2\theta = 0.83$ and $\Delta m^2 = 5.5 \times 10^{-5}$ eV^2.

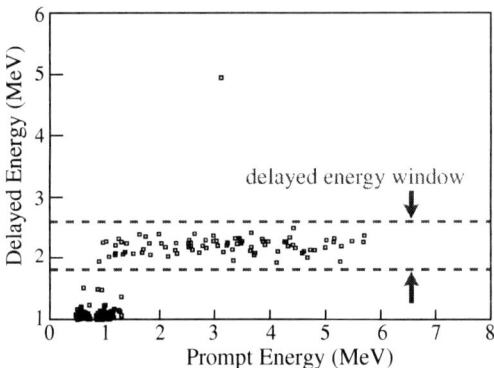

FIGURE 5. Delayed energy vs prompt energy

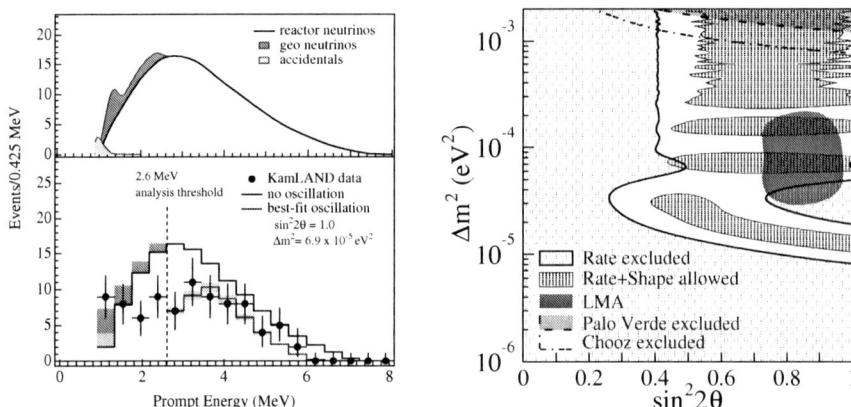

FIGURE 6. (left) Upper: Expected reactor $\bar{\nu}_e$ prompt energy spectrum along with $\bar{\nu}_{geo}$ and background. Lower: Energy spectrum of the observed prompt events (solid circles), along with the expected no oscillation spectrum (upper histogram) and best fit including neutrino oscillations (lower histogram). (right) Excluded regions of neutrino oscillation parameters for the rate analysis and allowed regions for the combined rate and shape analysis at 95% C.L. The 95% allowed region of the LMA solution is also shown.

The data are consistent with the distorted spectrum shape expected from neutrino oscillations at 93% C.L.

The neutrino oscillation parameter region for two neutrino mixing is shown in Fig. 6 (right). The KamLAND result is consistent with the LMA solution, assuming CPT invariance. In addition, KamLAND has limited the allowed region of the LMA solution to two regions through its energy-shape analysis: LMA-I and LMA-II.

PROSPECTS

Although KamLAND observed a clear deficit of events, the energy spectrum is still consistent with a scaled no-oscillation shape at 53% C.L. A new reactor Shika-II will be in operation starting in 2006. It is located 80 km away from KamLAND and the thermal power will be 3926 MW. It will be useful for further study of the LMA-I and LMA-II solutions. For these reasons, it is important to keep the data taking stable to precisely determine the neutrino oscillation parameters. In fact, KamLAND has been running very stable up to now.

In addition, since there is no direct measurement of the ^{7}Be neutrinos yet, the solar neutrino phase of KamLAND is being planned. The expected precision of a ^{7}Be neutrino flux measurment would be 5% after three years of running.

ACKNOWLEDGMENTS

KamLAND is supported by the Center of Excellence program of the Japanese Ministry of Education, Culture, Sports, Science and Technology, and funding program of the United States Department of Energy. The reactor data are provided by courtesy of the Electric Power Companies in Japan.

REFERENCES

1. J.N. Bahcall and M.H. Pinsonneault, *Rev. Mod. Phys.*, **67**, 781 (1989).
2. J.N. Bahcall, *http://www.sns.ias.edu/jnb/* (2002).
3. J.N. Bahcall and R. Davis, *Science*, **91**, 264 (1976).
4. Z. Maki et al., *Prog. Theor. Phys.*, **28**, 870 (1962).
5. B. Pontecorvo, *Sov. Phys. JETP*, **6**, 429 (1957).
6. L. Wolfenstein, *Phys. Rev. D*, **17**, 2369 (1978).
7. S.P. Mikheyev and A.Yu. Smirnov, *Sov. J. Nucl. Phys.*, **42**, 1441 (1985).
8. SNO collaboration, *Phys. Rev. Lett.*, **89**, 011301 (2002).
9. J.N. Bahcall, M.H. Pinsonneault and Sarbani Basu, *Astrophys. J.*, **555**, 990 (2001).
10. KamLAND collaboration, *Phys. Rev. Lett.*, **90**, 021802 (2002).
11. T. Hagner et al., *Astropart. Phys.*, **14**, 33 (2000).
12. R.S. Raghavan et al., *Phys. Rev. Lett.*, **80**, 635 (1998).
13. Y. Kishimoto, *talk at LowNu 2003* (2003).

Near Future Accelerator-Based Experiments

W. C. Louis

Los Alamos National Laboratory, Los Alamos, NM 87545

Abstract. This report covers accelerator-based neutrino oscillation experiments that will be conducted in the near future: the MiniBooNE short-baseline experiment at FNAL, the MINOS long-baseline experiment at FNAL/SOUDAN, and the OPERA and ICARUS long-baseline experiments at CERN/Gran Sasso. The results from these experiments will help provide answers to fundamental questions in neutrino physics.

INTRODUCTION

This report covers accelerator-based neutrino oscillation experiments that will be conducted in the near future. The first section discusses the current state of neutrino oscillation evidence coming from solar experiments [1, 2, 3, 4, 5, 6], atmospheric experiments [7, 8, 9], and the LSND experiment [10], and shows that, taken at face value, these experiments imply physics beyond the Standard Model, such as lepton number violating muon decay [11], light, sterile neutrinos [12], or CPT violation [13]. The next sections describe, in chronological order, the MiniBooNE short-baseline experiment at FNAL, the MINOS long-baseline experiment at FNAL/SOUDAN, and the OPERA and ICARUS long-baseline experiments at CERN/Gran Sasso. The results from these experiments will help provide answers to fundamental questions, such as: (i) What is the resolution of the $3 - \Delta m^2$ paradox? (ii) What are the neutrino masses and hierarchy? (iii) What are the neutrino mixings? (iv) Do light, sterile neutrinos exist? (v) Is CP conserved in the neutrino sector? (vi) Is CPT conserved in the neutrino sector? and (vii) Are neutrinos Dirac or Majorana?

CURRENT STATE OF NEUTRINO OSCILLATION EVIDENCE

Fig. 1 displays the current state of neutrino oscillation evidence coming from the solar experiments [1, 2, 3, 4, 5, 6], the atmospheric experiments [7, 8, 9], and the LSND experiment [10]. As shown in Table 1, the solar and atmospheric experiments observe large mixing at relatively low Δm^2 values, while the LSND experiment observes small mixing at relatively high Δm^2 values. If all of these experiments are correct, then, taken at face value, they imply physics beyond the Standard Model because it is not possible to explain such disparate Δm^2 regions with only three neutrinos. Examples of such beyond the Standard Model physics include: (i) lepton number violating muon decay ($\mu^+ \to e^+ \bar{\nu}_\mu \bar{\nu}_i$), which will be tested by the TWIST experiment at TRIUMF [11]; (ii) light, sterile neutrinos, which could have a huge impact on astrophysics in terms of BBN,

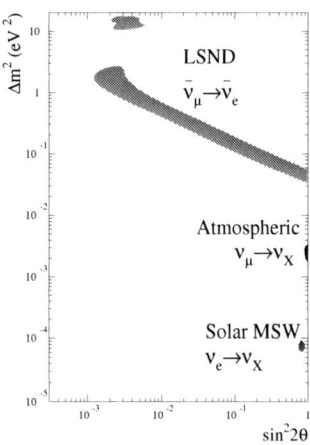

FIGURE 1. The current state of neutrino oscillation evidence coming from the solar experiments, the atmospheric experiments, and the LSND experiment.

TABLE 1. The current state of neutrino oscillation evidence coming from the solar experiments, the atmospheric experiments, and the LSND experiment.

Experiment	Type	Δm^2 (eV^2)	$\sin^2 2\theta$
Solar	$\nu_e \to \nu_{\mu,\tau}$	$\sim 7 \times 10^{-5}$	~ 0.8
Atmospheric	$\nu_e \to \nu_\tau$	$\sim 2 \times 10^{-3}$	~ 1
LSND	$\bar\nu_\mu \to \bar\nu_e$	~ 1	$\sim 3 \times 10^{-3}$

the r-process in supernovae neutrino bursts, cold dark matter, and possibly even dark energy [12]; and (iii) CPT violation, which is motivated by theories of extra dimensions and has the potential to explain the baryon asymmetry of the universe [13].

THE MINIBOONE EXPERIMENT AT FNAL

A schematic drawing of the MiniBooNE experiment is shown in Fig. 2. MiniBooNE is fed by the 8-GeV protons from the Booster, which at full intensity will deliver approximately 5×10^{20} protons on target (POT) per year. The protons interact in a 71-cm long Be target located at the upstream end of a magnetic focusing horn. The horn pulses at 5 Hz (10^8 pulses per year), operates at a current of 170 kA and a voltage of 2.5 kV, and focuses the pions and kaons from the proton-Be interactions. (The horn can be operated at either positive polarity for neutrino running or negative polarity for antineutrino running.) The pions and kaons decay into neutrinos in a 50-m decay pipe located just downstream of the horn, and the neutrinos then pass through the detector tank positioned 500 m downstream of the decay pipe. The average neutrino energy is approximately 1 GeV, and the intrinsic ν_e background is about 0.4% of the primary

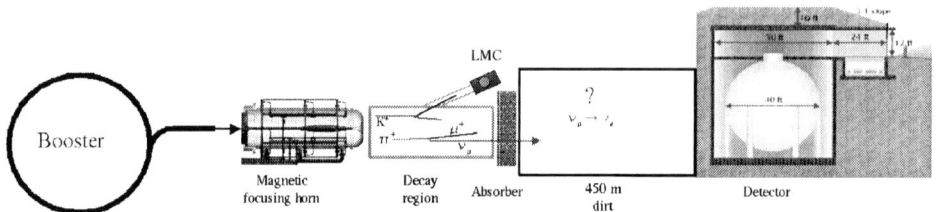

FIGURE 2. A schematic drawing of the MiniBooNE experiment.

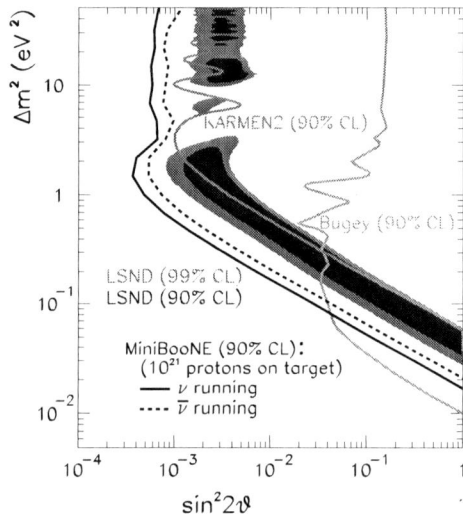

FIGURE 3. The expected MiniBooNE sensitivity after two full years of data collection.

ν_μ flux. The detector consists of a 40-ft diameter spherical tank filled with 800 tons of mineral oil (450 ton fiducial volume) and covered on the inside by 1520 8-inch phototubes (1280 detector phototubes and 240 veto phototubes). After 2 full years of data collection (10^{21} POT), MiniBooNE will be able to fully cover the entire LSND allowed region, as shown in Fig. 3.

After the first year of operation, MiniBooNE has collected about 162 K neutrino events, corresponding to 1.5×10^{20} POT (or about 30% of the yearly goal of 5×10^{20} POT.) The detector is working well with 99% of the phototube channels operational and a data acquisition livetime of about 99%. The time, charge, position, and angular resolutions, as well as the neutrino event rate, are consistent with expectations. Finally, the detector is clearly reconstructing charge-current μ^- events, neutral-current elastic events, and neutral-current π^0 events. Fig. 4 shows the π^0 mass distribution after requiring > 40 MeV per ring. The mass and width of the π^0 peak are consistent with expectations. Neutrino oscillation results are expected by the summer of 2005.

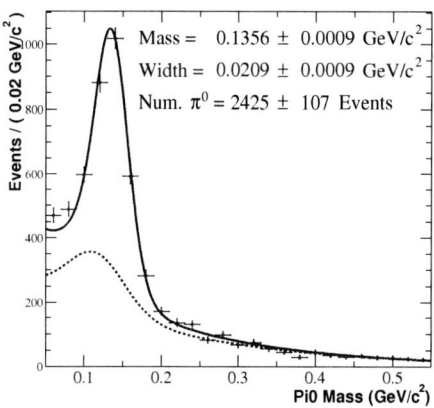

FIGURE 4. The π^0 mass distribution after requiring > 40 MeV per ring. The mass and width of the π^0 peak are consistent with expectations.

THE MINOS EXPERIMENT AT FNAL/SOUDAN

The MINOS experiment will be a definitive test of the atmospheric neutrino oscillation results and will be capable of making a precision measurement of the oscillation parameters, including a possible determination of θ_{13}. Neutrinos are produced by the 120-GeV protons from the Main Injector interacting in a 1-m long segmented graphite target, followed by a two horn focusing system and a 675 m long decay tunnel. The average neutrino energy can be varied from 3 GeV to 18 GeV by adjusting the locations of the two horns, and MINOS will start with the low-energy horn configuration. The intrinsic v_e component in the neutrino beam is < 1%. MINOS consists of two detectors, both consisting of a magnetized toroid and plastic scintillator strips. The near detector is located on site at FNAL at a distance of $\sim$ 1 km from the neutrino source and has a mass of 1 kton, while the far detector, shown in Fig. 5, is located in the Soudan mine in northern Minnesota at a distance of 735 km and has a mass of 5.4 ktons. Both detectors have an energy resolution of $\sim 60\%/\sqrt{E}$ for hadronic energy and $\sim 25\%/\sqrt{E}$ for electromagnetic energy. Fig. 6 shows the measurement of oscillations in MINOS for $v_\mu \to v_\tau$ oscillation parameters of $\Delta m^2 = 0.0025$ eV^2 and $\sin^2 2\theta = 1.0$. Finally, MINOS will have a θ_{13} sensitivity of < 7.1 degrees.

Construction of the MINOS far detector is complete, and the detector has begun making the world's first measurement of atmospheric v_μ and $\bar{v}_\mu$ interactions separately. Construction of the NuMI beamline is nearing completion. The tunnel excavation is complete, and the outfitting and final civil construction is on schedule. First protons on target are planned for December 2004, and MINOS should be fully operational by September 2005.

FIGURE 5. A schematic drawing of the MINOS far detector, located in the Soudan mine in northern Minnesota.

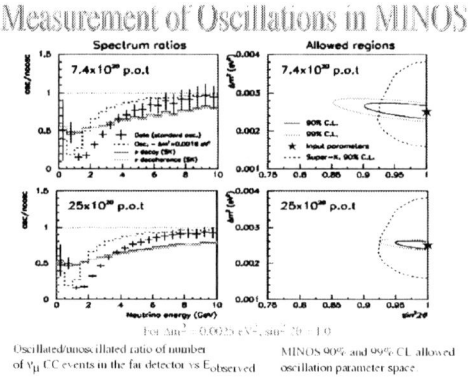

FIGURE 6. The measurement of oscillations in MINOS for $\nu_\mu \to \nu_\tau$ oscillation parameters of $\Delta m^2 = 0.0025$ eV^2 and $\sin^2 2\theta = 1.0$.

OPERA AND ICARUS AT CERN/GRAN SASSO

The OPERA and ICARUS experiments are both designed to be a definitive test of atmospheric $\nu_\mu \to \nu_\tau$ oscillations and ν_τ appearance. The experiments are located in the Gran Sasso underground laboratory and will detect neutrinos produced by the CNGS beam at CERN. The CNGS beamline, shown in Fig. 7, is fed by the 400 GeV protons from the SPS, which should be able to supply 4.5×10^{19} POT/y. The protons interact in a target made of graphite rods, and downstream of the target are magnetic focusing horns and a decay volume. The average neutrino energy is 17 GeV, the distance from CERN to Gran Sasso is 732 km, and the intrinsic ν_e background is $\sim 8 \times 10^{-3}$. First beam to Gran Sasso is scheduled for May 2006.

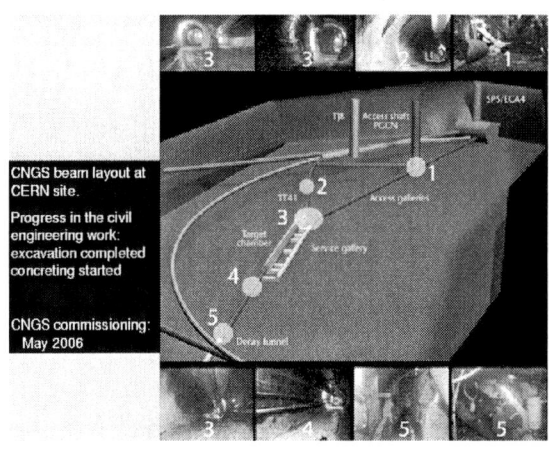

FIGURE 7. A schematic drawing of the CNGS beamline.

The OPERA experiment, shown in Fig. 8, consists of 1.8 ktons of emulsion interleaved with Pb plates. There are a total of 206,336 bricks, where each brick consists of 56 emulsion and PB layers. The detector lies in a 1.6 T magnetic field transverse to the neutrino beam, and an electronic target tracker determines the locations of neutrino events, so that candidate bricks can be removed for subsequent analysis. The energy resolution of the detector is $\sim 40\%/\sqrt{E}$, and the efficiency for neutrino oscillation events is $\sim 9.1\%$. After 5 years of data collection, OPERA should be able to observe a signal of 19.8 clean ν_τ events (for $\nu_\mu \to \nu_\tau$ oscillation parameters of $\Delta m^2 = 2.5 \times 10^{-3}$ eV^2 and $\sin^2 2\theta = 1.0$) on top of a background of only 0.67 events. Furthermore, OPERA will have a θ_{13} sensitivity of < 7.1 degrees, which is similar to MINOS.

A schematic drawing of the ICARUS detector is shown in Fig. 9. ICARUS is a 3 kton liquid Ar TPC, consisting of ten 300-ton half-modules. Each half module is a $4 \times 4 \times 20$ m^3 cryostat. The expected momentum resolution is $\sim 20\%$ at 10 GeV, and the expected efficiency for oscillation events is about 5.9%. After 5 years of data collection, ICARUS should observe an oscillation signal of 11.9 events (for oscillation parameters of $\Delta m^2 = 2.5 \times 10^{-3}$ eV^2 and $\sin^2 2\theta = 1.0$) on top of a background of only 0.7 events. In addition, ICARUS will have an excellent θ_{13} sensitivity of < 5.8 degrees.

OPERA

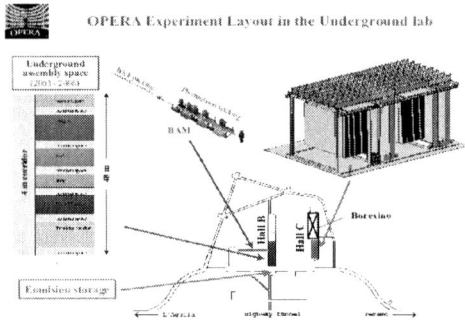

FIGURE 8. A schematic drawing of the OPERA detector, consisting of 1.8 ktons of emulsion interleaved with Pb plates.

CONCLUSIONS

This report summarizes the accelerator-based neutrino experiments that are planned for the near future: MiniBooNE, MINOS, OPERA, and ICARUS. The MiniBooNE experiment at Fermilab will be a definitive test of the LSND oscillation evidence, and if it confirms LSND, it will be able to probe new physics beyond the Standard Model. The MINOS experiment at Fermilab and Soudan will be a definitive test of the atmospheric oscillation evidence and will be capable of making precision measurements of the oscillation parameters. Finally, the OPERA and ICARUS experiments at CERN/Gran Sasso will be a definitive test of atmospheric $v_\mu \to v_\tau$ oscillations and will be capable of measuring v_τ appearance. We can confidently conclude that the results from these experiments will lead to a rich program at a future Neutrino Factory.

ICARUS

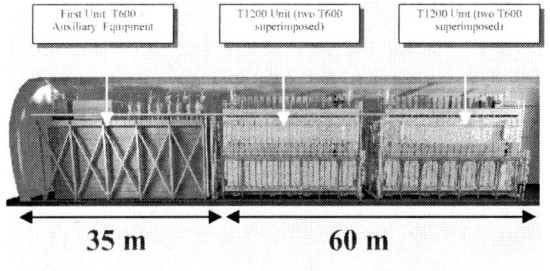

FIGURE 9. A schematic drawing of the ICARUS detector, consisting of a 3 kton liquid Ar TPC.

ACKNOWLEDGMENTS

We thanks the organizers of NuFact03 for a stimulating and productive workshop.

REFERENCES

1. R. Davis, Prog. Part. Nucl. Phys. **32**, 13 (1994).
2. P. Anselmann et al., Phys. Lett. B **328**, 377 (1994).
3. J. N. Abdurashitov et al., Phys. Lett. B **328**, 234 (1994).
4. J. Yoo et al. [Super-Kamiokande Collaboration], Phys. Rev. D **68**, 092002 (2003) [arXiv:hep-ex/0307070]; S. Fukuda et al. [Super-Kamiokande Collaboration], Phys. Lett. B **539**, 179 (2002) [arXiv:hep-ex/0205075].
5. Q. R. Ahmad et al. [SNO Collaboration], Phys. Rev. Lett. **89**, 011302 (2002) [arXiv:nucl-ex/0204009].
6. K. Eguchi et al. [KamLAND Collaboration], Phys. Rev. Lett. **90**, 021802 (2003) [arXiv:hep-ex/0212021].
7. Y. Hirata et al., Phys. Lett. B**335**, 237 (1994).
8. Y. Totsuka, Nucl. Phys. A**663**, 218 (2000); Y. Fukuda, et al. Phys. Rev. Lett. **81**, 1562 (1998); T. Toshito [Super-Kamiokande Collaboration], arXiv:hep-ex/0105023.

9. M. H. Ahn *et al.* [K2K Collaboration], Phys. Rev. Lett. **90**, 041801 (2003) [arXiv:hep-ex/0212007].
10. A. Aguilar *et al.*, Phys. Rev. D **64** (2001) 112007.
11. K.S. Babu and Sandip Pakvasa, hep-ph/0204236.
12. M. Sorel, J. Conrad, and M. Shaevitz, hep-ph/0305255.
13. G. Barenboim, L. Borissov, J. Lykken, and A. Y. Smirnov, JHEP **0210**, 001 (2002) [arXiv:hep-ph/0108199]; H. Murayama and T. Yanagida, hep-ph/0010178.

Neutrino Scattering Physics at Superbeams and Neutrino Factories

S. Kumano

Department of Physics, Saga University, Saga, 840-8502, Japan

Abstract. Neutrino scattering physics is discussed for investigating internal structure of the nucleon and nuclei at future neutrino facilities. We explain structure functions in neutrino scattering. In particular, there are new polarized functions g_3, g_4, and g_5, and they should provide us important information for determining internal nucleon spin structure. Next, nuclear structure functions are discussed. From F_3 structure function measurements, valence-quark shadowing should be clarified. Nuclear effects on the NuTeV $sin^2\theta_W$ anomaly are explained. We also comment on low-energy neutrino scattering, which is relevant to current long-baseline neutrino oscillation experiments.

INTRODUCTION

Nucleon structure has been investigated experimentally by various scattering experiments. Now, the perturbative QCD is well understood. The nonperturbative part is studied by theoretical models and lattice calculations, and they are tested experimentally. Because of these efforts, many aspects of the nucleon substructure are understood. However, there are still missing points. For example, spin is a fundamental quantity, and yet nucleon spin is poorly understood. We still do not know how the spin is constituted in terms of quarks and gluons. Future neutrino facilities should be able to provide important information on the internal hadron structure including the spin.

High-energy neutrino reactions have been already used for investigating the nucleon structure and determining fundamental constants such as the running coupling constant α_s and weak-mixing angle $sin^2\theta_W$. From accurate neutrino deep inelastic scattering (DIS) data, the structure functions, F_1, F_2, and F_3, have been extracted. Future neutrino facilities, superbeams [1] and neutrino factories [2], will provide new insight into the hadron substructure. Specialized talks are presented in the working group 2 (WG2) of this workshop, so that the details should be found in its summary [3] and presentations [4, 5, 6, 7]. Neutrino beams are strong enough to allow proton and polarized targets at the considered neutrino factories. Therefore, the nucleon structure functions and the fundamental constants are obtained without worrying about nuclear corrections. In addition, it is important that polarized structure functions, especially new functions g_3, g_4, and g_5, could be measured. Using these polarized structure functions, we expect that the internal nucleon spin structure will be precisely understood.

The future neutrino facilities are supposed to contribute also to nuclear physics. In the present neutrino DIS, accurate measurements have been done mainly for the nuclear target, iron, so that neutrino-nucleus scattering data already exist. However, there is no accurate deuteron or proton data for investigating nuclear corrections in neutrino reactions by taking the ratio $\sigma_{\nu A}/\sigma_{\nu D}$. Because the proton and deuteron cross sections should be accurately measured at the future facilities, we could shed light on the nuclear corrections. In particular, measurements of the function F_3 will clarify the valence shadowing phenomenon. On the other hand, we could investigate nuclear effects such as Pauli exclusion and nucleon-nucleon correlation in the low-energy scattering.

This paper consists of the following. The unpolarized and polarized neutrino-nucleon scattering processes are explained, and then nuclear structure functions are discussed. We also comment on low-energy neutrino scattering. Finally, the discussions are summarized.

UNPOLARIZED NEUTRINO-NUCLEON SCATTERING

The cross section for unpolarized neutrino-nucleon DIS is calculated by assuming a one-boson exchange process, and then the charged-current (CC) cross section is expressed in terms of three structure functions, F_1, F_2, and F_3:

$$\frac{d^2\sigma_{CC}}{dx\,dy} = \frac{G_F^2 s}{2\pi(1+Q^2/M_W^2)^2}\left[xy^2 F_1 + (1-y)F_2 \pm y(1-y/2)x F_3\right]. \quad (1)$$

Here, $+$ and $-$ of $\pm$ indicate neutrino and antineutrino reactions, respectively, G_F is the Fermi coupling constant, s is the center-of-mass squared energy, Q^2 is defined by the momentum transfer q: $Q^2 = -q^2$, and M_W is the W boson mass. The kinematical variables x and y are defined by $x = Q^2/(2Mq^0)$ and $y = q^0/E$ with the nucleon mass M and the initial neutrino energy E. There are sum rules for these structure functions:

$$S_A = \int_0^1 \frac{dx}{x}\left[F_2^{\bar{\nu}p}(x,Q^2) - F_2^{\nu p}(x,Q^2)\right] = 2,$$

$$S_{Bj} = \int_0^1 dx\left[F_1^{\nu n}(x,Q^2) - F_1^{\nu p}(x,Q^2)\right] = 1 - \frac{2}{3}\frac{\alpha_s(Q^2)}{\pi} + \cdots + O(1/Q^2), \quad (2)$$

$$S_{GRS} = \frac{1}{2}\int_0^1 dx\left[F_3^{\bar{\nu}p}(x,Q^2) + F_3^{\nu p}(x,Q^2)\right] = 3\left[1 - \frac{\alpha_s(Q^2)}{\pi} + \cdots\right] + O(1/Q^2).$$

These are called Adler, unpolarized Bjorken, and Gross-Llewellyn Smith sum rules. There are perturbative QCD corrections to the last two sum rules, and they have been investigated up to the α_s^4 level [8]. Therefore, sum-rule measurements will provide valuable information for an accurate determination of α_s. Possible ambiguities come from the higher-twist corrections $O(1/Q^2)$. Therefore, it is important to understand twist-four corrections theoretically, and such studies should be tested experimentally in the small-Q^2 region at the future neutrino facilities. The details of these points are summarized in the previous workshop [8].

The structure functions are expressed in terms of parton distribution functions (PDFs). The CC cross section is calculated in the parton model by using the current

$$J_{CC}^\mu = \bar{u}\gamma^\mu(1-\gamma_5)[d\,cos\theta_c + s\,sin\theta_c] + \bar{c}\gamma^\mu(1-\gamma_5)[s\,cos\theta_c - d\,sin\theta_c], \quad (3)$$

where θ_c is the Cabbibo angle. Comparing the obtained cross section with Eq. (1), we have the leading-order (LO) expressions for the structure functions in terms of the PDFs:

$$2x(F_1^{\nu p})_{CC} = (F_2^{\nu p})_{CC} = 2x(\bar{u} + d + s + \bar{c}),$$
$$2x(F_1^{\bar{\nu} p})_{CC} = (F_2^{\bar{\nu} p})_{CC} = 2x(u + \bar{d} + \bar{s} + c), \qquad (4)$$
$$x(F_3^{\nu p})_{CC} = 2x(-\bar{u} + d + s - \bar{c}), \quad x(F_3^{\bar{\nu} p})_{CC} = 2x(u - \bar{d} - \bar{s} + c).$$

Neutron structure functions are obtained by using the isospin symmetry for the PDFs. Parton-model expressions for neutral current (NC) structure functions are not shown here, but they are found, for example, in Refs. [9, 10].

Using the neutrino DIS data together with other lepton and hadron scattering data, we obtain the PDFs in the nucleon. The present situation is illustrated in Fig. 1 [11], where the MRST02 distributions are shown at $Q^2 = 10$ GeV^2 as an example. Because these distributions are rather well determined, we had better focus on other aspects such as polarized and nuclear PDFs at future neutrino facilities.

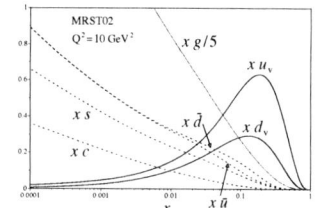

FIGURE 1. Parton distribution functions at $Q^2 = 10$ GeV^2.

POLARIZED NEUTRINO-NUCLEON SCATTERING

Polarized structure functions have been investigated by electron and muon DIS. Current polarized PDFs are determined by analyzing these data. Inclusive data are listed by the spin asymmetry A_1, which is expressed $A_1 \cong 2x(1+R)g_1/F_2$, where R is the longitudinal-transverse structure function ratio and g_1 is a polarized structure function. The g_1 is given by the polarized PDFs which are expressed by a number of parameters. These parameters are determined by a χ^2 analysis with the spin asymmetry data.

Recent analysis results are illustrated in Fig. 2 [12], where the polarized PDFs and their errors by Blümlein and Böttcher are shown as an example. The polarized valence-quark, antiquark, and gluon distributions are

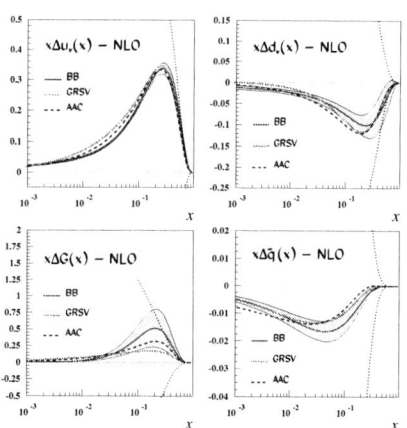

FIGURE 2. Recent polarized PDFs [12].

shown. Three different parametrization results are compared, and they agree each other except for the gluon distribution. The error bands for the valence-quark distributions are small; however, the error is large especially for the gluon distribution. It indicates that the polarized gluon distribution cannot be fixed at this stage.

The results may seem to indicate that the polarized PDFs are rather well determined except for the gluon. However, there are important points to be investigated. The overall magnitudes of the polarized valence-quark distributions are fixed by low-energy seimi-leptonic decay data with a flavor symmetric assumption for the antiquark distributions ($\Delta \bar{u} = \Delta \bar{d} = \Delta \bar{s}$). Furthermore, the quark spin content $\Delta\Sigma$ cannot be determined from the current electron and muon DIS experiments although the analyses indicate a small fraction $\Delta\Sigma = 10-30\%$. These issues could be clarified by future neutrino DIS studies as explained in the following.

In addition to g_1 and g_2, there exist extra functions g_3, g_4, and g_5 in neutrino reactions. There are various definitions for g_3, g_4, and g_5 depending on researchers, so that one should be careful in reading related papers. In the following, we use the convention in Refs. [2, 13, 14]. The asymmetry $\Delta\sigma$ is the difference between polarized cross sections: $\Delta\sigma = \sigma_{\lambda_p=-1} - \sigma_{\lambda_p=+1}$, where λ_p is the proton helicity, and it is expressed as

$$\frac{d\Delta\sigma^{\lambda_\ell}}{dx\,dy} = \frac{G_F^2}{\pi(1+Q^2/M_W^2)^2}\frac{Q^2}{xy}\left[-\lambda_\ell\,x\,y\,(2-y)\,g_1 - (1-y)\,g_4 - x\,y^2\,g_5\right], \quad (5)$$

for the CC process by neglecting M^2/Q^2 correction terms. Here, λ_ℓ is the lepton helicity. In the parton model, the leading-twist structure functions g_1, g_4, and g_5 are expressed in terms of the polarized PDFs. The g_4 and g_5 are related by the Callan-Gross type relation $g_4 = 2xg_5$ in the LO, and the CC structure functions g_1 and g_5 are expressed:

$$\begin{aligned}
g_1^{\nu p} &= \Delta\bar{u} + \Delta d + \Delta s + \Delta\bar{c}, & g_1^{\bar\nu p} &= \Delta u + \Delta\bar{d} + \Delta\bar{s} + \Delta c, \\
g_5^{\nu p} &= \Delta\bar{u} - \Delta d - \Delta s + \Delta\bar{c}, & g_5^{\bar\nu p} &= -\Delta u + \Delta\bar{d} + \Delta\bar{s} - \Delta c.
\end{aligned} \quad (6)$$

It is important that the g_1 structure functions directly probe the flavor singlet distribution: $\Delta\Sigma(x) = g_1^{(\nu+\bar\nu)p} = \Delta u + \Delta\bar{u} + \Delta d + \Delta\bar{d} + \Delta s + \Delta\bar{s} + \Delta c + \Delta\bar{c}$. Therefore, the quark spin content issue could be clarified by the neutrino scattering although the measured x range is limited. In addition, combining the g_5 functions for the proton, we obtain $g_5^{\nu p} + g_5^{\bar\nu p} = -(\Delta u_v + \Delta d_v) - (\Delta s - \Delta\bar{s}) - (\Delta c - \Delta\bar{c})$. The g_5 functions are important for determining the polarized valence-quark distributions.

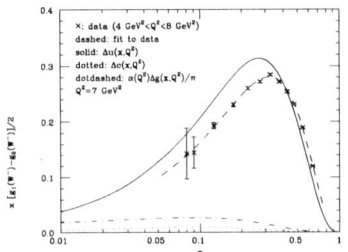 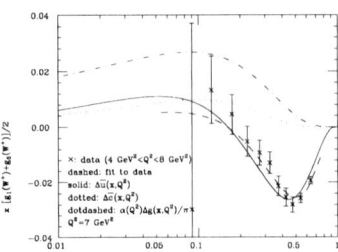

FIGURE 3. Feasibility studies for a neutrino factory [14]. Expected $x(g_1^{\bar\nu p} - g_5^{\bar\nu p})/2$ and $x(g_1^{\nu p} + g_5^{\nu p})/2$ are shown.

Feasibility is studied for the European neutrino factory in Ref. [14], and some results are shown in Fig. 3. Eight-year running of the neutrino factory with the butanol target is assumed for estimating the errors. As shown in the figure, Δu and $\Delta \bar{u}$ are the dominant contributions to the combinations $g_1^{\bar{\nu}p} - g_5^{\bar{\nu}p}$ and $g_1^{\nu p} + g_5^{\nu p}$, respectively, and they should be determined by the polarized reactions. However, luminosity has be increased as much as possible for accurate measurements.

A recent HERMES analysis indicates a slightly positive $\Delta s(x)$ at small x [4] in contrast to the parametrization results in Fig. 2. On the other hand, the polarized strangeness Δs could be investigated by other neutrino reactions [4]. In elastic neutrino scattering, the axial vector form factor $G_A(Q^2)$ can be measured. If non-strange contributions are known, the strange part is extracted: $G_A^s(Q^2 \to 0) = \Delta s$. At this stage, the analysis of BNL734 data indicates that G_A^s is consistent with zero. However, there is a proposal to measure it at Fermilab by the FINeSE project [4]. We expect that the strange spin will be clarified by G_A as well as the DIS experiments.

NEUTRINO-NUCLEUS SCATTERING

Nuclear modification of the PDFs is investigated in lepton DIS and high-energy hadron reactions. There are two major parametrizations, EKRS [15] and HKM [16], for nuclear PDFs. Current situation of the HKM studies is shown for the ^{40}Ca nucleus in Fig. 4, where $w(Ca,x)$ indicates nuclear modification. A χ^2 analysis has been made by using the data on the structure-function ratios $F_2^A/F_2^{A'}$ and Drell-Yan cross-section ratios $\sigma_{DY}^{pA}/\sigma_{DY}^{pA'}$. The nuclear PDFs are expressed by a number of parameters, which are then determined

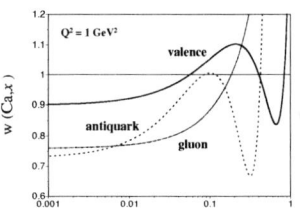

FIGURE 4. Nuclear modification of the PDFs [16].

by the χ^2 analysis with the data. The valence-quark distributions are determined well in the medium-x region. Because the nuclear modification is negative in this region as shown in Fig. 4, it is cancelled by the positive one at $x \approx 0.2$ so as to satisfy the charge and baryon-number conservations. However, these conservations do not pose a strong constraint at small x, so that the small-x modification is not obvious for the valence-quark distributions, and it should be tested by future neutrino DIS measurements. The antiquark distributions are fixed by the observed F_2 shadowing at small x and the Drell-Yan data at $x \sim 0.1$; however, the medium-x behavior is not obvious unless new data are obtained. It is difficult to determine the nuclear gluon distributions at this stage.

Because the iron target has been used in neutrino scattering, there are already neutrino-nucleus scattering data. However, it is not possible to investigate nuclear modification due to the lack of accurate deuteron data. At future neutrino facilities, proton and deuteron measurements will become possible, so that the nuclear modification could be investigated. In particular, the F_3 structure function is specific in the neutrino reaction. Although the F_2 shadowing is investigated well in the electron and muon scattering, F_3 shadowing has not been studied at all. The function F_3 provides information on the valence-quark distribution: $(F_3^{\nu N} + F_3^{\bar{\nu}N})_{CC}/2 \cong u_v + d_v$. The NuMI project [17] and ultimately the neutrino factories will provide data for the difference between the F_2 and

F_3 shadowing modifications, so that this issue will become clear in future.

We comment on a possible nuclear modification of the longitudinal-transverse structure function ratio R. It is sometimes called "HERMES effect". The effect was suggested by the HERMES collaboration in 2000 [18]; however, it was not observed in a CCFR/NuTeV analysis of neutrino data [19] and also in a subsequent HERMES reanalysis with careful radiative corrections [18]. Nonetheless, there could be a nuclear modification at large x with small Q^2, which is not the observed region by these experiments. A physics origin is the admixture of longitudinal and transverse nucleon structure functions in a nucleus due to nucleon Fermi motion [20]. Such an effect could be investigated by JLab experiments [21] and possibly by future neutrino reactions.

$sin^2\theta_W$ anomaly from a nuclear physicist's point of view

The NuTeV collaboration reported anomalously large weak mixing angle: $sin^2\theta_W = 0.2277 \pm 0.0013 \text{(stat)} \pm 0.0009 \text{(syst)}$ [6, 22] in comparison with a global analysis result $sin^2\theta_W = 0.2227 \pm 0.0004$ without neutrino-nucleus scattering data [23]. In the WG2 of this workshop, there are discussions on the NuTeV result and future experimental studies by parity-violating DIS and Møller scattering [6]. Although there may be new physics [24] behind this difference, it is more natural to seek a mechanism in nuclear corrections of the target iron.

The Paschos-Wolfenstein relation, $R^- = (\sigma^{\nu N}_{NC} - \sigma^{\bar{\nu} N}_{NC})/(\sigma^{\nu N}_{CC} - \sigma^{\bar{\nu} N}_{CC}) = \frac{1}{2} - sin^2\theta_W$, plays an important role for extracting $sin^2\theta_W$ from neutrino and antineutrino scattering data. This relation is valid for the isoscalar nucleon, but there are four correction factors in a nucleus: $R^-_A = \frac{1}{2} - sin^2\theta_W + O(\varepsilon_v) + O(\varepsilon_n) + O(\varepsilon_s) + O(\varepsilon_c)$. Here, $O(\varepsilon)$ indicates a correction of the order of ε, and the detailed expressions should be found in Refs. [25, 26, 27]. The correction factors are defined by

$$\varepsilon_n(x) = \frac{N-Z}{A} \frac{u_v(x) - d_v(x)}{u_v(x) + d_v(x)}, \qquad \varepsilon_v(x) = \frac{w_{d_v}(x,A) - w_{u_v}(x,A)}{w_{d_v}(x,A) + w_{u_v}(x,A)},$$

$$\varepsilon_s(x) = \frac{s^A(x) - \bar{s}^A(x)}{w_v(x,A)[u_v(x) + d_v(x)]}, \qquad \varepsilon_c(x) = \frac{c^A(x) - \bar{c}^A(x)}{w_v(x,A)[u_v(x) + d_v(x)]}, \qquad (7)$$

where w_{u_v} and w_{d_v} indicate the nuclear modifications of up- and down-valence quark distributions, and w_v is defined by $w_v = (w_{u_v} + w_{d_v})/2$. The function ε_n comes from the non-isoscalar nature of the nucleus, ε_v is related to the nuclear modification difference between u_v and d_v, and ε_s (ε_c) is proportional to the difference between s and $\bar{s}$ (c and $\bar{c}$). The charm correction is expected to be small. The strange correction is also found to be a small effect according to a NuTeV estimate, and it tends to increase the deviation [27]. The valence-quark correction ε_n was found to be small according to model estimates [25] although it should be tested by future experiments. The isovector correction ε_n was included in the NuTeV analysis. It was later investigated in Ref. [26]; however, NuTeV kinematical effects may reduce such a contribution. Therefore, it seems that the nuclear effects [25, 26, 27] are not enough for explaining the whole deviation at this stage.

COMMENTS ON LOW-ENERGY NEUTRINO SCATTERING

We have discussed high-energy neutrino reactions; however, current long baseline neutrino experiments have been done in the low-energy region. In order to understand the neutrino oscillation parameters in a few percent accuracy, the neutrino cross section should be understood accurately as well [7]. There are two important factors. One is to understand the neutrino interaction with the ^{16}O nucleus, another is to describe the cross sections in both DIS and resonance regions. There is a dedicated workshop for this topic, so that the details are found in its web page [28].

First, nuclear corrections should be accurately taken into account [29]. At high energies, they are expressed in terms of the nuclear PDF modifications. At low energies, the corrections include the effects of nuclear binding, Fermi motion, Pauli exclusion, and nucleon-nucleon (NN) correlation. For example, a final-state nucleon in a neutrino reaction suffers from the exclusion effect due to the existence of other nucleons. Such effects modify the small Q^2 part of the cross section significantly. If the cross section is shown as a function of neutrino energy, the exclusion effect is typically 8% as shown in Fig. 5 [29].

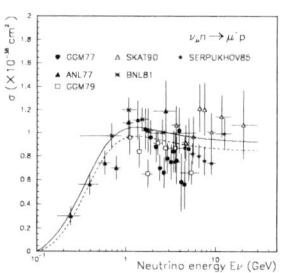

FIGURE 5. Quasi-elastic cross section [29]. The dashed curve includes exclusion effects.

From the figure, it is also obvious that the cross section is not accurately measured, and this fact makes it difficult to determine the oscillation parameters accurately. Furthermore, the NN correlation mechanism gives rise to a large momentum tail beyond the Fermi momentum, and it also modifies the cross section significantly. All of these nuclear corrections should be understood clearly for the precise neutrino-oscillation physics.

Second, an appropriate model should be studied for describing the cross section smoothly from the DIS to the resonance region because the neutrino data could contain both contributions. It is shown in Ref. [30] that a simple change of the scaling variable $[x_w = x(Q^2 + 0.624)/(Q^2 + 1.735x)]$ could describe the measured F_2 structure functions fairly well even in the small Q^2 region (Q^2=0.07, 0.22, and 0.85 GeV^2). Such a simple prescription could be also applied to the neutrino cross sections for the description in both low- and high-energy regions.

SUMMARY

Future superbeam and neutrino factories provide us a unique opportunity for investigating nucleon substructure, which cannot be studied by other lepton and hadron probes. Nucleon spin structure will be clarified by the leading-twist structure function g_1 and g_5. The valence-quark shadowing will be investigated by the F_3 structure functions for nuclei. We pointed out that these studies together with low-energy nuclear structure studies affect the long-baseline experiments as nuclear corrections. The hadron-structure studies are important, for example, for finding a quark-gluon plasma signature and any exotic signature beyond the current physics framework. The future neutrino facilities should play an important role in establishing the hadron-structure physics.

ACKNOWLEDGMENTS

S.K. would like to thank Y. Kuno for motivating him to hadron-structure studies at the neutrino factories. He thanks the Elsevier Science for permitting him to quote Figs. 2, 3, and 5 directly from its publications [12, 14, 29]. He was supported by the Grant-in-Aid for Scientific Research from the Japanese Ministry of Education, Culture, Sports, Science, and Technology.

REFERENCES

1. D. Michael, talk at this workshop.
2. C. Albright *et al.*, hep-ex/0008064; M. L. Mangano *et al.*, hep-ph/0105155; Y. Kuno *et al.*, NuFact-J studies, version 1.0 (unpublished).
3. M. Aoki, A. Baldini, and B. Fleming, summary of working group 2 at this workshop. See also http://www.cap.bnl.gov/nufact03/agenda-wg2.xhtml.
4. W. M. Alberico, R. Tayloe, and Y. Miyachi, talks on strangeness at this workshop.
5. U. K. Yang, F. Sergiampietri, and R. Bernstein, talks on neutrino DIS at this workshop.
6. I. Younus, P. E. Reimer, and J. Yu, talks on $sin^2\theta_W$ at this workshop.
7. G. P. Zeller, C. W. Walter, and K. S. McFarland, talks on low-energy neutrino scattering at this workshop.
8. A. L. Kataev and S. Kumano, J. Phys. G **29**, 1925 (2003).
9. R. G. Roberts, *The Structure of the Proton*, Cambridge University press (1990).
10. *Neutrino Physics*, edited by K. Winter, Cambridge University press (2000).
11. PDF codes are taken from http://durpdg.dur.ac.uk/hepdata/pdf.html.
12. J. Blümlein and H. Böttcher, Nucl. Phys. B **636**, 225 (2002). Other curves in Fig. 2 are taken from M. Glück *et. al.*, Phys. Rev. D **63**, 094005 (2001); Y. Goto *et. al.*, Phys. Rev. D **62**, 034017 (2000).
13. J. Blümlein and N. Kochelev, Nucl. Phys. B **498**, 285 (1997).
14. S. Forte, M. L. Mangano, and G. Ridolfi, Nucl. Phys. B **602**, 585 (2001).
15. K. J. Eskola *et. al.*, Nucl. Phys. B **535**, 351 (1998); Eur. Phys. J. C **9**, 61 (1999).
16. M. Hirai, S. Kumano, and M. Miyama, Phys. Rev. D **64**, 034003 (2001); in preparation.
17. J. G. Morfin, J. Phys. G **29**, 1935 (2003); S. A. Kulagin, hep-ph/9812532.
18. K. Ackerstaff *et al.*, Phys. Lett. B **475**, 386 (2000); A. Airapetian *et al.*, hep-ex/0210067 & 0210068.
19. U. K. Yang *et al.*, Phys. Rev. Lett. **87**, 251802 (2001).
20. M. Ericson and S. Kumano, Phys. Rev. C **67**, 022201 (2003). See also G. A. Miller, S. J. Brodsky, and M. Karliner, Phys. Lett. B **481**, 245 (2000); G. A. Miller, Phys. Rev. C **64**, 022201 (2001).
21. A. Brüll *et al.*, http://www.jlab.org/exp_prog/proposals/99/PR99-118.pdf.
22. G. P. Zeller *et. al.*, Phys. Rev. Lett. **88**, 091802 (2002).
23. D. Abbaneo *et. al.*, hep-ex/0112021.
24. S. Davidson *et. al.*, J. High Energy Phys. **02**, 037 (2002); W. Loinaz *et. al.*, Phys. Rev. D **67**, 073012 (2003) and references therein.
25. S. Kumano, Phys. Rev. D **66**, 111301 (2002).
26. S. A. Kulagin, Phys. Rev. D **67**, 091301 (2003).
27. K. S. McFarland *et. al.*, Nucl. Phys. B **112**, 226 (2002); S. Kovalenko, I. Schmidt, and J.-J. Yang, Phys. Lett. B **546**, 68 (2002); W. Melnitchouk and A. W. Thomas, Phys. Rev. C **67**, 038201 (2003).
28. http://neutrino.kek.jp/nuint01/; http://nuint.ps.uci.edu/.
29. M. Sakuda, Nucl. Phys. B **112**, 109 (2002); E. A. Paschos, J.-Y. Yu, and M. Sakuda, hep-ph/0308130.
30. A. Bodek and U. K. Yang, Nucl. Phys. B **112**, 70 (2002).

Beta-Beams: present design and expected performances

Jacques Bouchez*, Mats Lindroos† and Mauro Mezzetto**

*DAPNIA, CEA Saclay, France
†CERN, Geneva, Switzerland
**Istituto Nazionale Fisica Nucleare, Sezione di Padova, Italy.

Abstract. We give the present status of the beta-beam study, which aims at producing intense ν_e and $\bar{\nu}_e$ beams from the decay of relativistic radioactive ions. The emphasis is put on recent technical progress and new ideas. The expected performances in terms of neutrino mixing parameters (θ_{13} and CP violating phase δ) using a megaton water Cerenkov detector installed in the Fréjus underground laboratory are shown to be excellent, and the synergy with a a companion SuperBeam is underlined.

1. MOTIVATIONS

Super-Kamiokande has given strong evidence for a maximal oscillation between ν_μ and ν_τ [1], and several projects with accelerators have been designed to check this result. The first results of the K2K experiment [2] confirm the oscillation, and future projects (MINOS in the USA, OPERA and ICARUS at Gran Sasso) should refine the oscillation parameters by 2010.

More recently, after the results from SNO [3] and Kamland [4], a solid proof for solar neutrino flavour oscillations governed by the so-called LMA solution has been established. We can no longer escape the fact that neutrinos have indeed a mass, although the absolute scale is not yet known. Furthermore, the large mixing angles of the two above-mentioned oscillations and their relative frequencies open the possibility to test CP violation in the neutrino sector if the third mixing angle, θ_{13}, is not vanishingly small (we presently have only an upper limit set at 10 degrees on θ_{13}, provided by the CHOOZ experiment [5]). Such a violation could have far reaching consequences, since it is a crucial ingredient of leptogenesis, one of the presently preferred explanations for the matter dominance in our Universe.

The ideal tool for these studies is thought to be the so-called neutrino factory, which would produce through muon decay intense neutrino beams aimed at magnetic detectors placed several thousand kilometers away from the neutrino source.

However, such projects would not be launched unless one is sure that the mixing angle θ_{13}, governing the oscillation between ν_μ and ν_e at the higher frequency, is such that this oscillation is indeed observable. This is why physicists have considered the possibility of producing new conventional neutrino beams of unprecedented intensity, made possible by recent progress on the conception of proton drivers with a factor 10 increase in power (4 MW compared to the present 0.4 MW of the FNAL beam) The present limit on θ_{13} is 10 degrees, these new neutrino "superbeams" would explore θ_{13} down to 1 degree (i.e a factor 100 improvement on the ν_μ- ν_e oscillation amplitude).

European working groups have studied a neutrino factory at CERN for some years, based on a new proton driver of 4 MW, the SPL. Along the lines described above, a subgroup on neutrino oscillations has studied the potentialities of a neutrino SuperBeam produced by the SPL. The energy of produced neutrinos is around 270 MeV, so that the ideal distance to study ν_μ to ν_e oscillations happens to be 130 km, that is exactly the distance between CERN and the existing Fréjus laboratory. The present laboratory cannot house a detector of the size needed to study neutrino oscillations, which is around 1 million cubic meters. But the recent decision to dig a second gallery, parallel to the present tunnel, offers a unique opportunity to complete the needed extension in 2012 for a reasonable price.

Due to the schedule of the new gallery, a European project would be competitive only if the detector at Fréjus reaches a sensitivity on θ_{13} around 1 degree, since other projects in Japan (JHF phase 1) and USA (NuMI off-axis) will have reached 2.5 degrees by 2013. The working group has then decided to study directly a water Čerenkov detector with a mass around 1 megaton, necessary to reach the needed sensitivity. It has benefited from a similar study by our American colleagues, the so-called UNO detector [6] with a total mass of 660 kilotons. Simulations have shown that the sensitivity on θ_{13} at a level of 1 degree could indeed be fulfilled. However, the study of CP violation requires the SPL to be run sequentially with neutrinos and antineutrinos, and due to the fact that less antineutrinos are produced (less π^- than π^+ are produced) and that the $\bar{\nu}$ cross section is 5 times lower at the considered energies, 10 years of running should be shared roughly in 2 years with ν and 8 years with $\bar{\nu}$. This would be a strong limitation on CP sensitivity.

This is where the beta beam concept, initially proposed by Piero Zucchelli [7], comes into play. The idea is to produce well collimated and intense $\nu_e(\bar{\nu}_e)$ beams by producing, collecting, accelerating to energies with γ factor around 100 and storing in a final decay ring radioactive ions chosen for their ability to be copiously produced and with a lifetime around 1 second. The best candidates happen to be ^{18}Ne for ν_e and ^6He for $\bar{\nu}_e$. A baseline study for such a BetaBeam complex has been produced at CERN [8], where there is a strong expertise on ion beams, both for nuclear physics through ISOLDE and for high energy experiments.

The initial goal was to produce a ν_e beam which could be run simultaneously with the ν_μ SPL SuperBeam, so that 10 years of data could be accumulated for the each of the 2 time reversed oscillations, $\nu_e \to \nu_\mu$ and $\nu_\mu \to \nu_e$. This project was already presented at NuFact02 [9, 10].

A workshop took place in march 2003 at Les Arcs [11], where nuclear physicists (mainly those concerned with the EURISOL project), neutrino physicists and machine scientists have met to discuss BetaBeam issues. The aim of this workshop was to get updated on recent progress on the BetaBeam project, explore the synergies between beta beams and EURISOL [12], and identify common studies which could benefit to both communities.

Apart from new ideas which simplify the overall BetaBeam design, the major innovation was the proposal to run simultaneously with both types of ions (β^+ and β^- emitters) stored in the ring. This opens up the exciting possibility of performing efficiently CP violation studies with beta beams alone, and get very useful redundancies by comparing SuperBeam and BetaBeam data.

The section 2 describes the machine aspects of beta beams, the section 3 gives the expected performances on the measurement of the mixing angle θ_{13} and the CP violating δ phase.

2. THE BETABEAM COMPLEX

The beta-beam complex is shown schematically on figure 1. Technical details and recent progress on this project can be found at http://beta-beam.web.cern.ch/beta-beam/

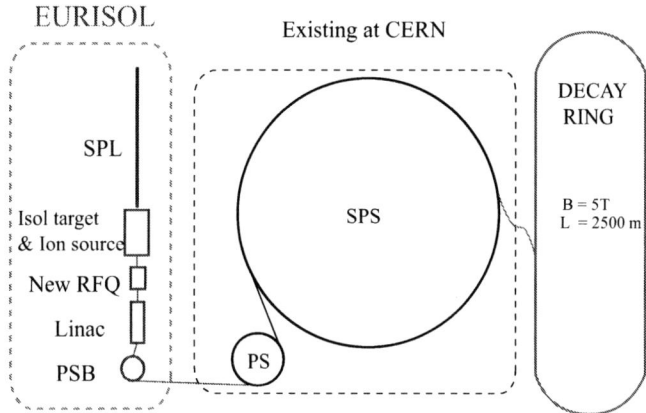

FIGURE 1. Schematic layout of the beta-beam complex. At left, the low energy part is largely similar to the EURISOL project. The central part (PS and SPS) uses existing facilities. At right, the decay ring has to be built.

The protons are delivered by the Super Proton Linac (SPL) [13], which is being studied at CERN in the framework of the neutrino factory [14]. Such an intense proton driver would deliver 2mA of 2.2 GeV (kinetic energy) protons hopefully by 2012. An ISOL target would need only 100 μA, that is 5 % only of the total proton intensity.

2.1. The target and the ion source

The targets are similar to the ones envisioned by EURISOL [12]: for ^{6}He, it consists either of a water cooled tungsten core or of a liquid lead core which works as a proton to neutron converter surrounded by beryllium oxide [15] ,aiming for 10^{15} fissions per second. ^{18}Ne can be produced by spallation reactions, in this case protons will directly hit a magnesium oxide target. The collection and ionization of the ions is performed using the ECR technique. The pulsed "ECR-duoplasmatron" under development at Grenoble, using very dense plasmas (10^{14}/cm^3) and high magnetic fields (2 to 3 Teslas) submitted to high frequencies (60 to 90 GHz) is aimed to produce 10^{12} to 10^{13} ions in very short bunches (20 to 100 μs) at 100 keV with repetition rates reaching 16 Hz [16].

The advantage over the standard ECR technique is that the downstream complex can be simplified, due to the achieved bunching and hopefully to ions which are totally ionized.

2.2. First acceleration and storing

Then the first acceleration process can be achieved using a LINAC rather than a cyclotron or a FFAG as initially considered. Ions would be accelerated to 20-100 MeV/u in 16 batches per second. Then comes the first storage ring, a rapid cycling synchrotron using multiturn injection (40 turns), delivering a single 150 ns bunch at 300 MeV/u.

2.3. Final acceleration

16 bunches (consisting of $2.5 \; 10^{12}$ ions each in the case of He) are then accumulated into the PS, and reduced to 8 bunches during their acceleration to intermediate energies. Due to the fact that the PS is a slow machine, this is the place where radiation levels due to ion decays is the most severe (the replacement of the PS by a more rapid machine would ease this problem and many others in the CERN complex of accelerators). Furthermore, the space charge bottleneck at SPS injection will require a transverse emittance blow-up. The SPS will finally accelerate the 8 bunches to the desired energy ($\gamma \simeq 100$) using a new 40 MHz RF system and the existing 200 MHz RF system, before ejecting them in batches of four 10 ns bunches into the decay ring.

2.4. The decay ring

This ring has the shape of an hippodrome, with a total length of 6880 m (matching the SPS) and straight sections of 2500 m each (36%). Due to the relativistic time dilatation, the ion lifetimes reach several minutes, so that stacking the ions in the decay ring is mandatory to get enough decays and hence high neutrino fluxes. The challenge is then to inject ions in the decay ring and merge them with existing high density bunches. As conventional techniques with fast elements are excluded, a new scheme (asymmetric merging) was specifically conceived for this task [17]. It schematically consists in injecting an off-momentum ion bunch on a matched dispersion trajectory, then rotate this fresh bunch in longitudinal phase space by a 1/4 turn into a starting configuration for bunch merging. This technique had been proven to work on computer simulations [17], but it received very recently a first experimental confirmation [18].

2.5. Neutrino fluxes

One of the ideas presented at the Moriond meeting was that it should be possible to run together Neon and Helium ions in the decay ring (of course, in different bunches). Due to their different rigidities, these ions would have relativistic γ factors in the 5 to 3 ratio, which is quite acceptable for the physics program. This will impose constraints on the lattice design for the decay ring, but no impossibility has been identified.

An ECR source coupled to an EURISOL target would produce $2 \cdot 10^{13}$ ^{6}He ions per second. Taking into account all decay losses along the accelerator complex, and estimating an overall transfer efficiency of 50%, one estimates that $4 \cdot 10^{13}$ ions would permanently reside in the final decay ring for $\gamma = 60$. That would give an antineutrino flux aimed at the Fréjus underground laboratory of $2.1 \cdot 10^{18}$ per standard year (10^7 s).

For ^{18}Ne, the yield is expected to be only $8 \cdot 10^{11}$ ions per second. Due to this smaller yield, which could be certainly improved with some R&D, it was then proposed to use 3 EURISOL targets in sequence connected to the same ECR source. Again taking into account decay losses plus a 50% efficiency, this means that $2 \cdot 10^{13}$ such ions would reside in the decay ring for $\gamma = 100$, giving rise to a neutrino flux of $0.35 \cdot 10^{18}$ per standard year.

All these numbers are preliminary and need to be refined. They are however based on the present state of the art for the technology, and suppose using the present PS, while the SPS cycle is set at 16 s; a shorter cycle for the SPS would improve the accumulation factor substantially, while a faster PS would increase the intensity of ions making it to the decay ring.

In the following study, it was supposed that the neutrino flux from ^{18}Ne could be increased by a factor 3 over the present conservative estimate, having room for improvements both in the cycle duration of PS and SPS and in the ^{18}Ne production at the targets with a dedicated R&D, while only a 40 % improvement was put on antineutrino fluxes.

2.6. Radiation issues

The main losses are due to decays of He ions, and reach 1.2 W/m in the PS and 9 W/m in the decay ring. This seems manageable, although the use of superconducting bending magnets in the decay ring requires further studies. Activation issues have been recently addressed [19], and show that the dose rate on magnets in the arcs is limited to 2.5 mSv/h at contact after 30 days operation and 1 day cooling. Furthermore, the induced radioactivity on ground water will have no impact on public safety.

3. PHYSICS REACH

The following study is based on the hypothesis that a UNO-like water Cerenkov detector (440 kt fiducial mass) will be installed in the underground Fréjus laboratory and receive neutrino beams produced at CERN, 130 km away.

3.1. Signal and backgrounds

The neutrino beam energy depends on the γ of the parent ions in the decay ring. As discussed in ref. [10], the optimization of this energy, is a compromise between the advantages of the higher γ, as a better focusing, higher cross sections and higher signal efficiency; and the advantages of the lower γ values as the reduced background rates (see the following) and the better match with the probability functions. Given the decay ring constraint (see sect. 2.4): $\gamma(^6\text{He})/\gamma(^{18}\text{Ne}) = 3/5$ the optimal γ values result to be $\gamma(^6\text{He}) = 60$ and $\gamma(^{18}\text{Ne}) = 100$. A flux of $2.9 \cdot 10^{18}$ ^{6}He decays/year and $1.1 \cdot 10^{18}$ ^{18}Ne decays/year, as discussed in sect. 2.5, will be assumed. Fig. 2 shows the BetaBeam neutrino fluxes computed at the 130 Km baseline, together with the SPL Super Beam (SPL-SB).

The mean neutrino energies of the $\overline{\nu}_e$, ν_e beams are 0.24 GeV and 0.36 GeV respectively. They are well matched with the CERN-Frejus 130 km baseline. On the other hand energy resolution is very poor at these energies, given the influence of Fermi motion and other nuclear effects and in the following all the sensitivities are computed for a counting experiment with no energy cuts.

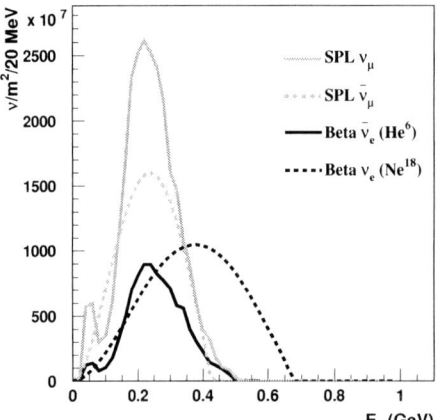

	Fluxes $\nu/m^2/yr$	$\langle E_\nu \rangle$ (GeV)
$\overline{\nu}_e(\gamma = 60)$	$1.97 \cdot 10^{11}$	0.24
$\nu_e(\gamma = 100)$	$1.88 \cdot 10^{11}$	0.36
ν_μ	$4.78 \cdot 10^{11}$	0.27
$\overline{\nu}_\mu$	$3.33 \cdot 10^{11}$	0.25

FIGURE 2. Beta Beam fluxes at the Frejus location (130 km baseline). Also the SPL Super Beam ν_μ and $\overline{\nu}_\mu$ fluxes are shown in the plot.

The signal in a Beta Beam looking for $\nu_e \rightarrow \nu_\mu$ oscillations would be the appearance of ν_μ charged-current events, mainly via quasi-elastic interactions. These events are selected by requiring a single-ring event, the track identified as a muon using the standard Super-Kamiokande identification algorithms (tightening the cut on the pid likelihood value), and the detection of the muon decay into an electron. Background rates and signal efficiency have been studied in a full simulation, using the NUANCE code [20], reconstructing events in a Super-Kamiokande-like detector.

The Beta Beam is intrinsically free from contamination by any different flavor of neutrino. However, background can be generated by inefficiencies in particle identification, such as mis-identification of pions produced in neutral current single-pion resonant interactions, electrons (positrons) mis-identified as muons, or by external sources such as atmospheric neutrino interactions.

The pion background has a threshold at neutrino energies of about 450 MeV, and is highly suppressed at the Beta Beam energies. The electron background is almost com-

TABLE 1. Event rates for a 4400 kt-y exposure. The signals are computed for $\theta_{13} = 3°$, $\delta = 90°$ sign$(\Delta m^2) = +1$. "δ-oscillated" events indicates the difference between the oscillated events computed with $\delta = 90°$ and with $\delta = 0$. "Oscillated at the Chooz limit" events are computed for $\sin^2 2\theta_{13} = 0.12$, $\delta = 0$.

	Beta Beam		SPL-SB	
	$^6He(\gamma=60)$	$^{18}Ne(\gamma=100)$	ν_μ(2 yrs)	$\overline{\nu}_\mu$(8 yrs)
CC events (no osc, no cut)	19710	144784	36698	23320
Oscillated at the Chooz limit	612	5130	1279	774
Total oscillated ($\delta = 90°$, $\theta_{13} = 3°$)	44	529	93	82
δ oscillated	-9	57	-20	12
Beam background	0	0	140	101
Detector backgrounds	1	397	37	50

pletely suppressed by the request of the detection of a delayed Michel electron following the muon track. The atmospheric neutrino background can be reduced mainly by timing the parent ion bunches. For a decay ring straight sections of 2.5 km and a bunch length of 10 ns, which seems feasible [8], this background becomes negligible [10]. Moreover, out-of-spill neutrino interactions can be used to normalize this background to the 1% accuracy level.

Signal and background rates for a 4400 kt-yr exposure to ^{6}He and ^{18}Ne beams, together with the SPL SuperBeam (SPL-SB) fluxes [9], are reported in table 1

3.2. Systematic errors

A facility where the neutrino fluxes are known with great precision is the ideal place where to measure neutrino cross sections. In the Beta Beam the neutrino fluxes are completely defined by the parent ions beta decay properties and by the number of ions in the decay ring. A close detector of ~ 1 kton placed at a distance of about 1 km from the decay ring could then measure the relevant neutrino cross sections. Furthermore the γ factor of the accelerated ions can be varied. In particular a scan can be initiated below the background production threshold, allowing a precise measurement of the cross sections for resonant processes. It is estimated that a residual systematic error of 2% will be the final precision with which both the signal and the backgrounds can be evaluated.

The θ_{13} and δ sensitivities are computed taking into account a 10% error on the solar δm^2 and $\sin^2 2\theta$, already reached after the recent SNO-salt results [3] and a 5% and 1% error on δm^2_{23} and $\sin^2 2\theta_{23}$ respectively, as expected from the J-Parc neutrino experiment [21]. Only the diagonal contributions of these errors are considered. In the following the default values for the oscillation parameters will be $\sin^2 2\theta_{23} = 1$, $\delta m^2_{23} = 2.5 \cdot 10^{-3}$eV^2, $\sin^2 2\theta_{12} = 0.8$, $\delta m^2_{12} = 7.1 \cdot 10^{-5}$eV^2, sign$(\Delta m^2)$=+1.

3.3. Parameter correlations and degeneracies

Correlations between θ_{13} and δ are fully accounted for, and indeed they are negligible as can be seen in the fits to θ_{13} and δ shown in Fig. 3.

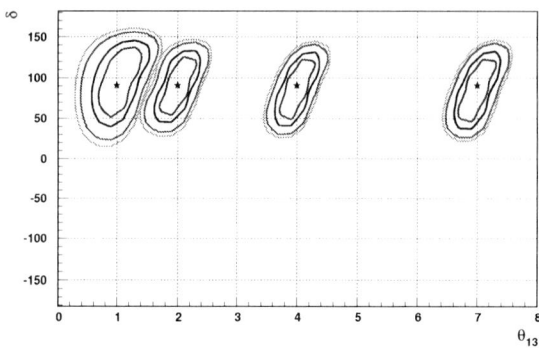

FIGURE 3. Fits to θ_{13} and δ after a 10 yrs BetaBeam run. Plots are shown for $\delta = 1°, 2°, 4°, 7°$. For the other neutrino oscillation parameters see the text. Lines show 1σ, 90%, 99% and 3σ confidence levels.

The net effect of the $\text{sign}(\Delta m^2)$ ambiguity is to make undetectable $\text{sign}(\delta \cdot \text{sign}(\Delta m^2))$. This derives by the negligible matter effects at the 130 km, so that a change of $\text{sign}(\Delta m^2)$ is equivalent to a change of the sign of δ. The performances of the BetaBeam to the two opposite values of $\text{sign}(\delta \cdot \text{sign}(\Delta m^2))$ are different because the neutrino and antineutrino runs have different statistics and backgrounds. This effect will be illustrated in fig. 5 and fig. 6.

Finally the $\theta_{23}/(\pi/2 - \theta_{23})$ ambiguity is formally taken into account, but no effect is found because the BetaBeam performances are computed for the central value of SuperKamiokande: $\theta_{23} = 45°$. A study of the performances of the BetaBeam for different values of θ_{23} is beyond the purpose of this article.

We stress the fact that an experiment working at very short baselines has the smallest possible parameter degeneracies and ambiguities and it is the cleanest possible environment where to look for genuine leptonic CP violation effects.

3.4. θ_{13}/δ sensitivities

The θ_{13} angle can be independently explored both with ν_e and $\overline{\nu}_e$ disappearance measurements. We note that the comparison of the ν_e and $\overline{\nu}_e$ disappearance experiments could set limits to CPT violation effects. Sensitivities to θ_{13}, computed for a 5 yr run and for systematic errors equal to 2%, 1% and 0.5% are shown Fig. 4left). For comparison sake, shown in the same plot are the sensitivities reachable with the appearance channels, computed for $\delta = 0$.

Indeed θ_{13} and δ are so tightly coupled in the appearance channels that the sensitivity expressed for $\delta = 0$ is purely indicative. A better understanding of the sensitivity of the BetaBeam is expressed in the (θ_{13}, δ) plane, having fixed all the other parameters $(\delta m_{23}^2 = 2.5 \cdot 10^{-3} \text{ eV}^2)$, as shown in Fig. 4right). In the same plot the sensitivity of the SPL-SB computed for a 5 yrs ν_μ run is displayed. It can be noted the very large variation of the SPL-SB sensitivity for the different values of δ, characteristic of the single flavour run. The BetaBeam, having both CP neutrino states in the same run, exhibits a much more favourable dependence to the CP phase δ.

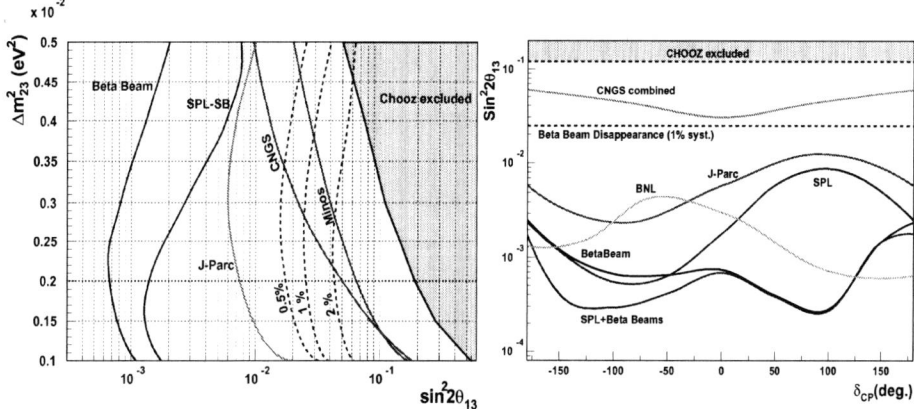

FIGURE 4. LEFT: 90%CL sensitivity of the disappearance channel to θ_{13} in a 5 yrs run drawn as dotted lines. The labels 0.5%, 1% and 2% indicate the systematic errors with which are computed. Also shown are the appearance sensitivities of Beta and SPL beams, computed for $\delta = 0$, sign(Δm^2)=+1. The combined CNGS limit is taken from ref. [22], J-Parc from [21], Minos from ref. [24]. RIGHT: 90%CL sensitivity expressed as function of δ for $\delta m_{23}^2 = 2.5 \cdot 10^{-3} eV^2$. CNGS and J-Parc curves are taken from ref. [22], BNL from ref. [23]. All the appearance sensitivities are computed for sign(Δm^2) = +1.

A search for leptonic CP violation can be performed running the Beta Beam with ^{18}Ne and ^{6}He, and fitting the number of muon-like events to the $p(\nu_e \to \nu_\mu)$ and to the $p(\overline{\nu}_e \to \overline{\nu}_\mu)$ probabilities. The fit can provide the simultaneous determination of θ_{13} and δ, see fig. 3.

Event rates are summarized in Table 1. The region of 99% CL sensitivity to maximal CP violation ($\delta = 90°$) in the δm_{12}^2 and θ_{13} parameter space, following the convention of [25], is plotted in Fig. 5.

The 3σ sensitivity to δ, having fixed $\delta m_{12}^2 = 7.1 \cdot 10^{-5}$ eV^2, is shown in Fig. 6.

3.5. Synergies between the SPL-SuperBeam and the Beta Beam

The Beta Beam needs the SPL as injector, but consumes at most $\sim 10\%$ of the SPL protons. The fact that the average neutrino energies of both the SuperBeam and the Beta Beam are below 0.5 GeV (cfr. fig. 2), with the Beta Beam being tunable, offers the fascinating possibility of exposing the same detector to 2×2 beams (ν_μ and $\overline{\nu}_\mu \times \nu_e$ and $\overline{\nu}_e$) having access to CP, T and CPT searches in the same run.

It is evident that the combination of the two beams would not result only in an increase in the statistics of the experiment, but it would also offer clear advantages in the reduction of the systematic errors, and it would offer the necessary redundancy to firmly establish any effect of violation of CP within the reach of the experiment.

The CP violation sensitivities of the combined BetaBeam and SPL-SB experiments are shown in Fig. 5 and Fig. 6.

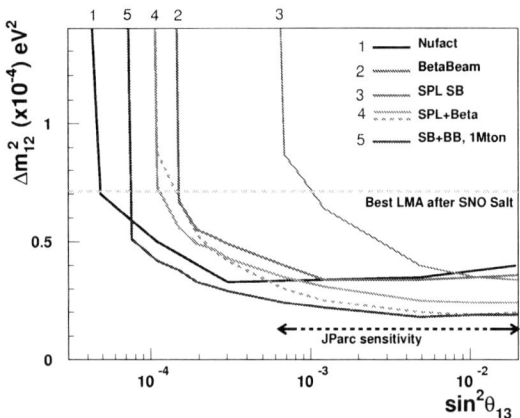

FIGURE 5. 99%CL δ sensitivity of the Beta Beam, of the SPL-SuperBeam, and of their combination, see text. Dotted line is the combined SPL+Beta sensitivity computed for sign(Δm^2)=-1. Sensitivities are compared with a 50 GeV Neutrino Factory producing $2 \cdot 10^{20} \mu$ decays/straight section/year, and two 40 kton detectors at 3000 and 7000 km [25].

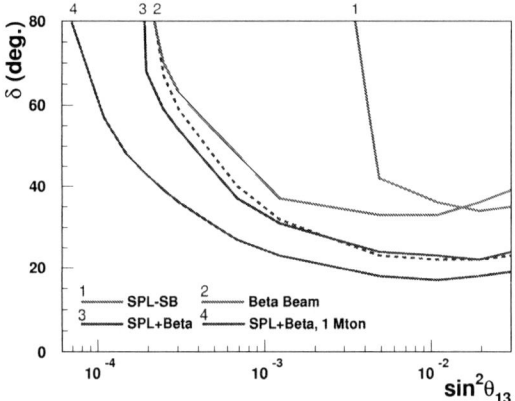

FIGURE 6. δ discovery potential (3σ) as function of θ_{13}. Dotted line are sensitivities computed for sign(Δm^2)=-1

4. CONCLUSIONS

Betabeams are a novel concept which can give very precise insight on the problem of neutrino mixing. Their design has many common features with the EURISOL project aiming at producing high intensity radioactive beams for nuclear physics studies with astrophysical applications. This synergy has been outlined at the Moriond workshop, and common technical studies are already going on. Recently, the potentialities of low energy betabeams (γ factors below 10) has also been emphasized [26].

On the other hand, there have been for some time projects of megaton detectors to study proton decay and detect supernova explosions, but they never got financial support. The fact that they would also be perfect targets for low energy neutrino superbeams or betabeams considerably increases the physical interest of these detectors. Fortunately enough, a possible site exists near CERN at the right distance and could house as soon as 2015 such a detector. CERN has a long lasting expertise on ion production and acceleration, and has announced officially to be site candidate for the Eurisol project.

This offers Europe and CERN a unique opportunity to contribute significantly to megaton physics, neutrino physics, and Eurisol physics by joining efforts of several communities on an ambitious and multidisciplinary project.

REFERENCES

1. The Super-Kamiokande Collaboration, S. Fukuda *et al.*, Phys.Rev.Lett. 85 (2000) 3999-4003
2. The K2K collaboration, M.H. Ahn *et al.*, Phys. Rev. Lett. 90 (2003) 041801
3. S. N. Ahmed *et al.* [SNO Collaboration], arXiv:nucl-ex/0309004.
4. KamLAND collaboration, K. Eguchi *et al.*, Phys.Rev.Lett. 90 (2003) 021802
5. M. Apollonio *et al.*, Phys. Lett. B 466 (1999) 415
6. UNO Collaboration, hep-ex/0005046
7. P. Zucchelli, Phys. Lett. B **532** (2002) 166.
8. B. Autin *et al.*, "The acceleration and storage of radioactive ions for a neutrino factory," arXiv:physics/0306106.
9. M. Mezzetto, J.Phys.G29:1771-1776, 2003; hep-ex/0302005.
10. M. Mezzetto, J.Phys.G29:1781-1784, 2003; hep-ex/0302007.
11. Moriond Workshop on "radioactive beams for nuclear physics and neutrino physics", Les Arcs (France) March 17-22, 2003 http://moriond.in2p3.fr/radio/index.html
12. http://www.ganil.fr/eurisol/
13. B. Autin *et al.*, "Conceptual design of the SPL, a high-power superconducting H- linac at CERN," CERN-2000-012
14. M. Apollonio *et al.*, "Oscillation physics with a neutrino factory." arXiv:hep-ph/0210192.
15. J. Nolen, NPA 701 (2002) 312c
16. P. Sortais, presentations at the Moriond workshop on radioactive beams, Les Arcs (France) 2003 "ECR technology", http://moriond.in2p3.fr/radio/Moriond-Sortais_1.ppt
17. http://beta-beam.web.cern.ch/beta-beam/Presentations/asymmerging.gif
18. M. Benedikt, S. Hancock and J-L. Vallet, "A proof of principle of asymmetric bunch pair merging", CERN note AB-Note-2003-080 MD
19. "Parameters of radiological interest for a beta-beam decay ring", M. Magistris and M. Silari, note CERN-TIS-2003-017-RP-TN
20. D. Casper, Nucl. Phys. Proc. Suppl. **112** (2002) 161 [arXiv:hep-ph/0208030].
21. Y. Itow *et al.*, "The JHF-Kamioka neutrino project,", hep-ex/0106019.
22. P. Migliozzi and F. Terranova, Phys. Lett. B **563** (2003) 73 [arXiv:hep-ph/0302274].
23. M. V. Diwan *et al.*, Phys. Rev. D **68**, 012002 (2003) [arXiv:hep-ph/0303081].
24. G. Tzanakos, proceedings of this conference.
25. J. Burguet-Castell *et al.*, Nucl. Phys. B **608**, 301 (2001) [arXiv:hep-ph/0103258].
26. C. Volpe, "What about a beta beam facility for low energy neutrinos?," arXiv:hep-ph/0303222.

Neutrino Factory R&D In Europe

H. D. Haseroth, B. Autin, R. Bennett*, F. Méot, S. Gilardoni**, C. Prior*, P. Sievers, A. Verdier

CERN, *CCLRC and **Geneva University

Abstract. Until last year the CERN Neutrino Factory Working Group has been quite active together with other European laboratories. The sudden discovery of the bad financial situation at CERN due to the LHC construction put an abrupt halt to these studies. Lots of CERN activities were cut back to allow for the finishing of LHC in time and without any significant global increase in CERN's budget. Nevertheless with a new organisation and several committees and working groups, activity on a European scale is still going on. This paper describes the actual situation and some European ideas in the different areas which may turn out to be very relevant in the future.

ORGANISATION AND ACTIVITIES IN EUROPEAN RESEARCH FOR NEUTRINO FACTORIES

After the LHC problems at CERN became clear and the activity of CERN people in the neutrino factory activities had been reduced to close to zero there was nevertheless a (positive) impact from directors of several big European labs (including CERN) with the intention to contribute towards neutrino R&D in spite of CERN's reduction. They created a committee, called the "European Muon Coordination and Oversight Group" (EMCOG) with participation of CEA (Saclay), CERN, GSI, IN2P3, INFN, PSI, RAL and University of Geneva. The following intentions were defined:

- The long-term goal is to have a Conceptual Design Report for a European Neutrino Factory Complex by the time of LHC start-up, so that, by that date, this would be a valid option for the future of CERN.

- An earlier construction for the proton driver (SPL + accumulator & compressor rings) is conceivable and, of course, highly desirable. The SPL, targetry and horn R&D have therefore to be given the highest priority.

This committee suggested creating a European working group called ENG (European Neutrino Group) with sub-working groups for the proton driver (SPL and proton rings), targets, pion and muon collection, the frontend with phase rotation and cooling, and the muon acceleration including the decay ring.

This group has plenary meetings during the ECFA Muon Weeks, which are being held three times per year at different locations in Europe at participating labs. As the resources at all European labs (manpower and money) are very limited, it was suggested the EU be asked for support. In this context it must be noted that contrary to the past even CERN can now – within the framework of FP6 (Framework Program 6) – apply for support from the European Union. In fact ECFA is encouraged to ask for EU support, which had as a consequence that another meeting of European lab directors (CCLRC, CERN, DAPNIA/CEA, DESY, LNF, Orsay/IN2P3, and PSI in consultation with ECFA) decided to form a European Steering Group on Accelerator R&D (ESGARD) with representatives from these laboratories.

Two Contact Groups (CG) issued from ESGARD have been set up for bidding preparation, one on electron-positron linear colliders and one on super proton accelerators and neutrino beams. These CGs have created working groups with the people of the community involved in the relevant activities. They also maintain close contact with existing committees, boards and teams. The mandate of the CGs is amongst others to act as liaison between ESGARD and the relevant communities and to investigate whether proposals for Design Studies (DS) and/or Construction of New Infrastructures (CNI) as defined by the EU commission are applicable and identify their scope.

The R&D on accelerators for high energy physics is organised around three main future world-wide projects:
1. Electron-positron linear colliders with energies ranging between 500 and 3000 GeV in the centre-of-mass system, using the technology of superconductive high gradient accelerator structures recently developed by the TESLA international collaboration, and aiming at exploiting as well the two-beam technique for obtaining ultra-high gradients at room temperature developed by the CLIC international collaboration.
2. Facilities providing intense neutrino beams using both improvements to the existing methods based on intense proton beams, and the more novel techniques based on radioactive ion or muon beams.
3. Facilities providing proton beams with ultra-high intensities and energies, aiming at very large hadron colliders, and covering as well luminosity and energy upgrades of the LHC at CERN.

ESGARD prepared a proposal to the European Union called "CARE" (Coordinated Accelerator R&D in Europe), which was very much welcomed by ECFA. CARE consists of essentially three networking activities and five so-called Joint Research Areas (JRA). CARE covers many aspects of accelerator R&D; however, one network is dedicated specifically to neutrino beams, one JRA contains part of a possible proton driver and another deals with high field magnets, also relevant for neutrino factories.

SPECIFIC R&D IN EUROPE FOR NEUTRINO FACTORIES

A basic concept for a neutrino factory may consist of the following components [1]:
- Proton driver
- High-power proton beam onto a target

- System for collection of the produced pions and their decay products, the muons.
- Energy spread and transverse emittance may have to be reduced: "phase rotation" and ionisation cooling
- (Fast) acceleration of the muon beam with a linac and "RLAs" (Recirculating Linear Accelerators) or FFAGs
- Muons are injected into a storage ring (decay ring), where they decay in long straight sections in order to deliver the desired neutrino beams.

In this paper we shall mention some of the items which recently have been worked on in Europe.

Proton Driver

Concerning the proton driver there are different ideas, either a Superconducting Proton Linac with 2.2 GeV with accumulator and (bunch-) compressor rings as proposed by CERN, or a driver based on synchrotrons at higher energy as worked out at RAL [2, 3].

At CERN work is continuing, however, mainly on the low energy part of the proton driver [4]. At RAL ideas are presented to upgrade the existing ISIS facilities in several stages which could pave the way to interesting tests essential for a neutrino factory at a very early stage [5].

The ISIS spallation neutron source has been running successfully for more than 15 years and at 160kW of proton beam power remains the most intense source of its kind in the world. Plans are to install a radio frequency quadrupole (RFQ) in the linac and to add a dual harmonic RF system to the synchrotron. This should increase the power to about 240kW. Beyond this, the machine is space-charge limited.

However, a staged upgrade to higher beam powers looks interesting:

Phase 1: Addition Of A New Synchrotron

In phase one of an upgrade to higher beam power an additional synchrotron could be installed. The lattice is based on a racetrack design, described in [6]. With a mean radius of 78m (three times that of the present ring), the two bunches that make up a normal ISIS pulse will be transferred bucket to bucket to the new ring, followed by an (approximately) fourfold increase in energy to the MW level of beam power. The overall dimensions of the lattice are 184m × 107m and it is possible to install it on the RAL site. The cost of these proposals is relatively modest compared with a completely new facility. In the longer term, a combination of two such rings with a new synchrotron booster (replacing the existing ISIS) would give attractive options for both the neutron and neutrino communities.

In this mode of operation, the machine would be used as a source of spallation neutrons. However, the possibility exists of operating at one third of the frequency (i.e. 16.7 Hz), discarding two pulses in three and accelerating the remaining pulse to 8 GeV over 30 ms. No additional RF would be needed. An opportunity would thereby be provided for further experimental tests of bunch compression, for pion target tests and investigations of a prototype pion decay/muon capture channel.

Phase 2: A New Injector

In a second phase, a new booster injector would be built, replacing the ISIS accelerator complex as the feed to the 78m racetrack synchrotron, which would be upgraded to be capable of delivering up to 2.5MW of beam power. The dual role of a neutron facility with applications to neutrino factory development would be maintained.

The proposed injector is based on a 180MeV H⁻ linac feeding two 1.2 GeV, 50Hz booster synchrotrons. Two 1.2 GeV synchrotrons are used for the booster to maintain manageable space charge levels. They have mean radius of 39m (half the radius of the main synchrotron) and their optical parameters are designed for low loss beam accumulation. H⁻ injection is by charge exchange through a carbon, or possibly aluminium oxide, stripping foil at a point in the latttice where the normalised dispersion is $\cong 16$. Simultaneous longitudinal and horizontal transverse phase space painting helps reduce space charge effects and minimise particle loss. Each booster ring would be filled one after the other with three bunches of protons and all six bunches would be transferred together to fill the main ring for final acceleration and, in the case of a neutrino factory, bunch compression.

Phase 3: Addition Of A Second Synchrotron

In the next phase the proton driver would be completed with the addition of a second main racetrack synchrotron, stacked vertically above the ring built in Phase I in order to achieve the 4-5MW level (at 50 Hz).

Target Studies

The neutrino factory target will dissipate about 1 MW of power from a 4 MW proton beam in pulses of a few micro-second duration, at 50 Hz repetition rate. The target, typically 1-2 cm diameter and 20 cm long, will not stop the beam and it will be necessary to dissipate some 3 MW in a beam dump.

Targets for high-power pulsed beams not only have to dissipate the peak and mean powers but have to withstand possible thermal shocks due to the short pulses. In addition the target is located within the confines of a 20 T solenoidal magnet or magnetic horn to capture the pions formed in the target. The design of the target system must incorporate the remote handling for maintenance and replacement, along with the pion capture equipment, the beam dump and other surrounding equipment in this intense radiation environment.

Two parameters determine the lifetime of a solid target: the instantaneous temperature rise and the rise of shock waves that carry a stress that may exceed the elasticity limit of the material. If the particles are delivered to the targets in batches of N particles of energy E and at a repetition frequency f, the beam power is

$$P = NEf$$

If the heat dissipated in the target is mainly due to ionization and the particle energy remains much larger than that of Bragg peak, then the instantaneous temperature rise can be expressed by the relation

$$\delta T = \frac{1}{EfS} \frac{P}{\rho C} \frac{dE}{dx}$$

where S is the cross section, ρ the density, C the heat capacity and dE/dx the energy loss per unit length of the target. The first ratio is energy dependent and, ideally, the beam should have both a high energy and a high repetition frequency. The proton drivers studied for CERN or ISIS, and may be FFAG machines, can provide a high Ef product.

The shock regime arises when the duration of a batch of particles is short with respect to the propagation time of the sound in the target. This is interesting if the target has dimensions of the order of one millimeter in each direction, hence the idea [7, 8] of fragmenting the target into small spheres (Fig. 1).

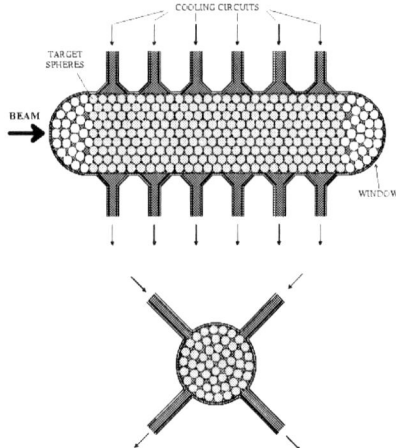

FIGURE 1. Schematic diagram of the granular target.

That structure has the second advantage of an efficient cooling since the cooling medium acts at the core of the target and not at the periphery as in the case of a massive target. It has to be noted that the same idea has been proposed for mercury targets filled with gas bubbles.
Several target schemes look interesting:

Liquid Metal Jet

Historically, due to the experience developed in neutron spallation sources, a mercury target has first been proposed [9, 10]. The production of pions imposes a jet instead of a static target. Although the beam destroys the jet at each pulse, its interaction occurs on a much shorter time scale than the break up of the jet. The jet must reform before the next beam pulse. This target is both robust against shock waves and may handle significantly more power than the 1 MW currently envisaged for the neutrino factory. In addition it does not suffer from radiation damage problems and removal of the highly active target is simply a matter of draining away the mercury. Figure 2 shows one version of the mercury jet utilising a hollow nozzle.

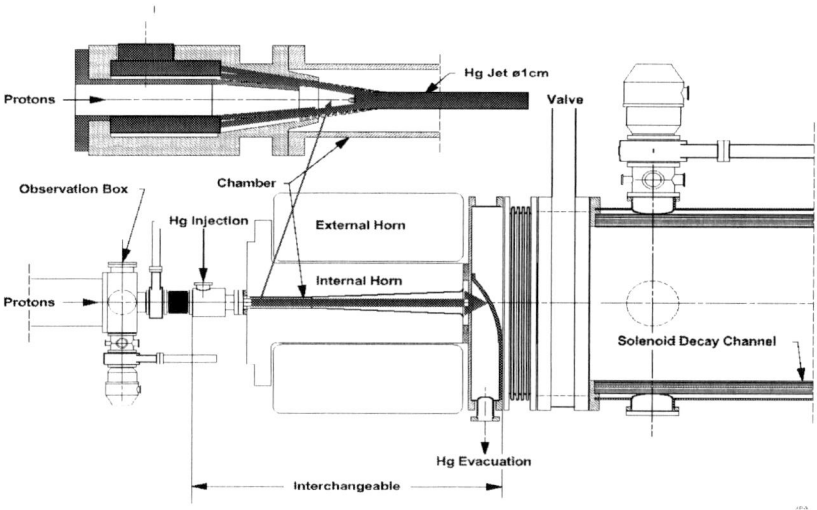

FIGURE 2 Schematic diagram of the mercury jet within a magnetic horn. Note the inset showing the hollow jet nozzle and axial proton beam.

Considerable progress has been made with studies of the interaction of pulsed beams with mercury jets at BNL [11] and with the passage of jets through intense magnetic fields at Grenoble [12]. These tests indicate that the scheme may be viable. However, the mass of mercury involved in the whole circuit is of the order of six tons, the safety problems are very severe and the cost of the target area is very high. Development on solid targets has thus been pursued in parallel.

Granular Targets

A target (see Figure 1) consisting of small spheres, a few mm in diameter, has been suggested to overcome the problem of thermal shocks [13]. The target would be cooled by flowing a coolant such as helium gas over the spheres. The cooling regimes are shown in Figure 3 for a single target and for four targets in parallel. The parameters are summarized in Table 1. When four targets are put in parallel, it is actually four full systems comprising not only the target but also the horn and the cooling systems. The temperatures have to be understood as temperature increments above the input coolant temperature. Although helium seems to be the simplest fluid to operate in a first stage, cooling with water and even a liquid metal in case the target would be pulsed [14] is being contemplated.

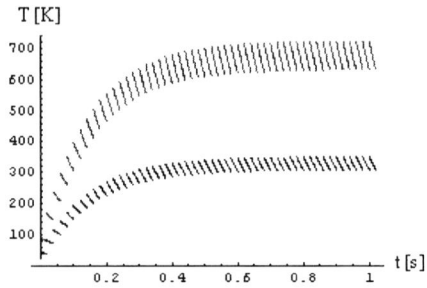

FIGURE 3 Variation of temperature with time in one (upper curve) and four targets in parallel (lower curve).

TABLE 1. Beam and target parameters.

Beam energy [GeV]	2.2	Material	Ta	Cooler	He
Batch length [μs]	3.3	Length [m]	0.33	Flow [kg/s]	0.36
Particles/batch	$2.3 \; 10^{14}$	Cross section [m²]	$2.3 \; 10^{-3}$	$\hat{T}_{He}[K]$	585
Rep. frequency [Hz]	50	Sphere radius [mm]	1	$\hat{T}_{1-target}[K]$	730
Beam power [MW]	4	Weight [kg]	6.15	$\hat{T}_{4-targets}[K]$	225

Rotating Toroidal Target

A rotating solid tantalum toroid (see Figure 4) heated by the beam to ~2000 K will radiate over 10 MW of power depending on the diameter and the speed of rotation of the toroid [15]. It is proposed to levitate, rotate and stabilise the motion of the toroid electromagnetically to avoid any moving parts in the system. The toroid would spin in a vacuum and the walls would be water cooled to remove the power. Experience of other high pulse power targets (for example the pbar target [16] at FNAL) indicate that the toroid would probably not suffer from thermal shock. Currently experiments [17] using 100 kV electron beams on thin tantalum foils indicate that shock is not a problem up to 10^6 pulses [18].

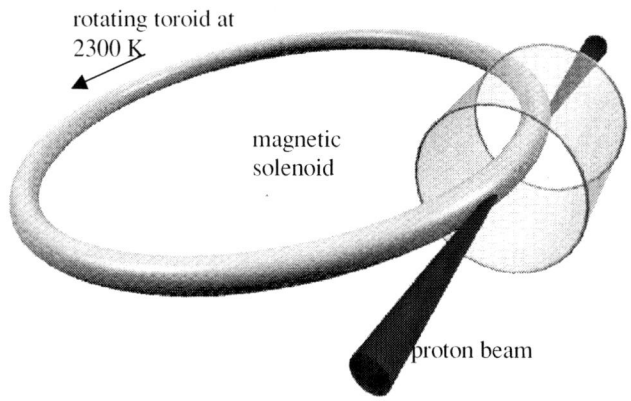

FIGURE 4. view of the thermally radiating rotating toroidal target.

Future Plans

With the current lack of funds, little R&D is currently taking place in Europe. The EU has recently awarded some funds to allow networking to continue. Further funds will be sought from the EU for R&D for jets, granular targets and rotating toroidal targets. The UK is preparing a proposal to its own funding agency for studies of solid targets; this will include the toroid and its levitation, separate targets fired through the beam and re-circulated [19], shock studies in solids and the continuation of the shock studies using electron beams. *Collaboration with the USA and Japan on contained mercury targets in connection with the SNS is being sought as a way forward for target development.*

Finally the problem of in-beam tests must be addressed and suitable facilities at CERN (ISOLDE) and RAL (ISIS) have been identified as providing the required pulse intensity, but it will need considerable expenditure for their use in target tests. *Collaborations between Europe, USA and Japan are being pursued.*

Horn Studies

The horn as focusing device is an essential ingredient in the CERN scheme when using the SPL at 2.2 GeV. This is not only because horns have a long history at CERN combined with a lot of experience. It must be remembered that the first application of the SPL to produce neutrinos might well be by producing a superbeam ("conventional" neutrino beam with higher beam power), where charge selection of either positive or negative pions is wanted. A solenoid would focus both polarities equally well and is hence not suited for this application.

The horn R&D studies at CERN [20] during the last year have been focused on three main topics:
- horn prototype mechanical test;
- horn power supply first prototype construction;
- horn power supply upgrade;

FIGURE 5. Inner horn prototype in the test area. The red cables feed 100 kA at 0.5 Hz. The transparent pipes inject the water inside the cooling circuit.

The final operating condition for the Neutrino Factory horn foresees a current of 300 kA pulsed for 93 µs at a repetition rate of 50 Hz.

Parallel to the construction of the inner horn prototype [Fig. 5], a test power supply to reach 30 kA at 1Hz with a pulse length of ≈100 µs has been realized. These two elements have been used to measure the mechanical vibration eigenfrequencies of the horn, first to assure that the current pulse does not excite any of the horn eigenmodes, and second to check that any vibration is dumped after 20 ms (50 Hz), to avoid a pile up of the mechanical stresses.

The measurements have been taken using a microphone, which records the horn sound resulting from excitation induced by the current. The Fourier analysis of the sounds gives the horn eigenfrequencies (see figure 5 and [21] for a more detailed discussion of the measurements). Two peaks at 193.7 Hz and 549.1 Hz, indicated by the two arrows in figure 6, are the most probable candidates as horn eigenfrequencies. A more precise and expensive method using lasers will be applied to confirm these results, as part of the R&D program for the next year.

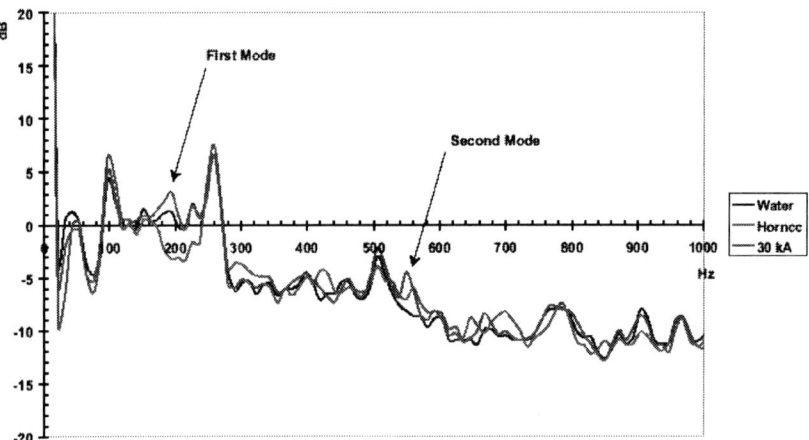

FIGURE 6 Fourier transformation of the horn sound together with the two main background sources, the power supply and the cooling water flowing.

A second step towards the final power supply has also been taken. An upgrade of the power supply prototype reaches 100 kA at 0.5 Hz, fixing the pulse length around 100 μs. The three times more intense current induces ≈10 larger displacements, which simplifies the mechanical vibration measurements and in a condition more similar to the Neutrino Factory operation. Further development is planned at Orsay [22].

Funneling Studies (Target And Decay Channel)

One way to avoid some of the severe problems for the target and for the horn is to use funneling [14], i.e. distributing the beam power among several targets and collection devices and recombining the beams for the subsequent machines. The problems in the CERN scheme are that the target must be of high Z material (because of the low proton energy and the associated low π^- production) and that the horn must be pulsed with the high frequency of the linac (50 Hz). It would be much simpler if instead of 4 MW we had only 1 MW and instead of 50 Hz we had e.g. only 12.5 Hz. It seems this could be achieved without exotic and expensive technology with reasonable lifetime of the components. In addition it could be considered as an evolutionary design starting with a lower beam power at low repetion rate and subsequent upgrading to higher powers with higher repetition rates.

The principle of this scheme is fairly simple. The proton beam is switched to 4 targets in sequence and each of the four pion lines contains an integrated system of target and magnetic horn. The funnel is made of large aperture magnets with quadrupolar and pulsed dipolar coils.

In a staged approach one could start with a single target-horn assembly with a power limited to about 1 MW. Subsequently several could be operated in parallel. The problem here lies in the efficiency of recombination of the beams. The recombination requires quadrupoles and calculations of muon production at the end of the pion decay channel have been compared to the transmission of a single long solenoid. It is true that there is no loss of particles in a solenoid whereas there are significant ones in an alternating gradient beam line. However, the difference in muon collection between the two types of lines disappears as soon as the particles are counted in the transverse and longitudinal acceptances of the downstream machines.

Special care is needed for the choice of the reference particle. In the case of a hybrid beam of pions and muons, the reference particle has been chosen as the center of gravity of the two clouds of particles. A theoretical model supported by simulations determines the variation of momentum of the reference particle along the channel. The adjustment of the quadrupoles can thus be precisely determined. Studies of sensitivity have shown that this adjustment is critical in the funnel. An important output of the calculation is the evaluation of the muon production as a function of the transverse acceptance because it quantifies the interest of muon cooling. In any case, the decay channel must have the greatest possible aperture and, although a preliminary estimate has shown that the funnel quadrupoles can operate at room temperature, the technology of the magnets in the common line is still an open question.

CONCLUSIONS

In spite of severe limitations in funding and manpower useful work has been done in several European institutions. It is hoped that with additional support from government agencies and from the European Union progress can be accelerated.

ACKNOWLEDGEMENTS

Many thanks are due to my European colleagues for their help, in particular to: R. Garoby, R. Edgecock and M. Lindroos.

REFERENCES

H. Haseroth, Neutrino Factory and Cooling Experiment, XXIst Internat. Linear Accelerator Conference, 19-23 August 2002, Kyongiu, Korea

1. Prior, C.R. and Rees, G.H., RAL Proton Driver Studies for a Neutrino Factory. Proceedings of the Neutrino Factory Workshop, NUFACT'00, Monterey, California, June 2000.
2. Prior, C.R. and Rees, G.H., Synchrotron-based Proton Drivers for a Neutrino Factory. Proceedings of the 7th European Particle Accelerator Conference, EPAC'00, Vienna,June 2000.
3. F. Gerigk, M. Vretenar, Design of a 120 MeV H- Linac for CERN High-Intensity Applications, XXIst Internat. Linear Accelerator Conference, 19-23 August 2002, Kyongiu, Korea
4. Prior, C.R. et al, ISIS Megawatt Upgrade Plans, Proceedings of the US Particle Accelerator Conference PAC'03, Portland,Oregon, May 2003.
5. Rees, G.H., Lattices for 8 and 30 GeV Proton Drivers. Proceedings of the Neutrino Factory Workshop, NUFACT'02, London, June 2002.
6. B. Autin. F. Méot, and A. Verdier, Muon collection in an alternating gradient channel, J. Phys. G: Nucl. Part. Phys. 29, 2003, pp. 1777-1780.
7. B. Autin, J. Doornbos, and F. Méot, Particle densities in pion to muon decay process, These proceedings.
8. C. D. Johnson, NuFact99, Lyon. http://cdj.home.cern.ch/cdj/nu-fact/liqu-jet/index.htm
9. H.L. Ravn, the ISOLDE Collaboration and the Neutrino-Factory Working Group, Nucl. Instr. and Meth. B 204 (2003) 197–204.
10. H. Kirk et al., in: P. Lucas, S. Webber (Eds.), Proc. 2001 Particle Accelerator Conf., Chicago, IL, June 2001, Vol. 2, p. 1535.
11. A. Fabich, J. Lettry, Experimental observation of photon induced shocks and magneto-fluid-dynamics in liquid metal, Proc. Int. Workshop NuFACT01 and Muon Storage Rings for a Neutrino Factory, Tsukuba, Japan, May 2001, Nucl. Instr. and Meth. A, 503 (2003) 336.
12. P. Sievers, Nuclear Instruments and Methods A 503 (2003) 344.
13. B Autin, F, Méot, P Sievers, A Verdier, Funnelling Pions and Muons, Proceedings of this Workshop.
14. J R J Bennett, A High Power, Radiation Cooled, Rotating Toroidal Target for a Neutrino Production, Physics Potential and Development of Muon Colliders and Neutrino Factories, 5th International Conference, San Francisco, 1999, Ed. D B Cline, AIP Conf Proc 542, p253,
15. S. O'Day, K. Bieniosek, K. Anderson, "New Target Results from the Antiproton Source", Proceedings of the 1993 IEEE Particle Accelerator Conference, Washington, DC, 17-20 May, 3096-3098, IEEE, 1993.
16. P. Drumm and C. Densham, Material Testing with Electron Beams for Neutrino Factory Targets, Proc. 2001 Particle Accelerator Conference, Chicago, http://accelconf.web.cern.ch/AccelConf/p01/PAPERS/TPAH157.PDF
17. P V Drumm and C J Densham, Private Communication.
18. J. R. J. Bennett, C. J. Densham and P. V. Drumm, A High Power Radiation Cooled Target for a Neutrino Factory, European Particle Accelerator Conference 2000, http://accelconf.web.cern.ch/accelconf/e00/PAPERS/MOP2A06.pdf
19. S. Gilardoni et al., "Horn R&D", in these proceedings
20. S. Gilardoni, "Horn Vibration Acoustic Measurements", CERN-Nufact-Note-126
21. Jean Eric Campagne, Antoine Cazes, The CERN horn prototype revisited, CERN-Nufact-Note-134

Neutrino Factory R&D in the U.S.

Michael S. Zisman[†]
for the Neutrino Factory and Muon Collider Collaboration

Center for Beam Physics
Accelerator & Fusion Research Division
Lawrence Berkeley National Laboratory
Berkeley, CA 94720 U.S.A.

Abstract. We report here on the technical progress and R&D plans of the U.S. Neutrino Factory and Muon Collider Collaboration. Programs in targetry, cooling, acceleration, and simulations are covered. U.S. activities in support of the international Muon Ionization Cooling Experiment (MICE) are also described.

INTRODUCTION

An R&D program aimed at the production, acceleration, and storage of intense muon beams is under way in the U.S. under the auspices of the Neutrino Factory and Muon Collider Collaboration (MC). This program is supported from several sources, including the U.S. Department of Energy, the National Science Foundation, and the State of Illinois Board of Higher Education. The program described here is both complemented and enhanced by the corresponding R&D programs carried out in Europe and Japan.

The MC is attacking the R&D issues associated with intense muon beams on a broad front. This paper describes current activities and plans in the areas of targetry, cooling, acceleration, and simulations, along with an update on the effort aimed at preparing for MICE [1].

R&D PROGRAM PROGRESS

Targetry

There has been a great deal of progress in the targetry program over the last several years. Initial beam tests of both solid and liquid targets were completed at the AGS, using a 24 GeV extracted proton beam. These tests were done at a typical intensity of about 4×10^{12} protons per pulse (compared with a design intensity of 1.6×10^{13} protons per pulse). Results of this work have been reported previously [2] and will not

---

[†]This work supported by the Director, Office of Science, Office of High Energy and Nuclear Physics, Division of High-Energy Physics, of the U.S. Department of Energy under Contract No. DE-AC03-76SF00098.

be covered here. Unfortunately, we are unable to get beam time in 2003 or 2004—a major impediment to rapid progress in this area.

We have continued tests of carbon target sublimation rates at ORNL. Earlier tests in vacuum indicated a target lifetime of the order of one month at a beam power of 1.5 MW [3]. We are preparing the apparatus required to repeat these tests in a helium atmosphere. It is expected that the presence of helium will reduce the sublimation rate, thereby giving a longer target lifetime at a given beam power. Whether this will extend the operating range of a carbon target up to 4 MW remains to be seen.

Other tests to determine the properties of candidate solid-target materials have also been carried out at BNL [4]. Samples of Super-Invar and Inconel were irradiated with a 200 MeV proton beam, after which we looked for changes in mechanical properties (tensile strength and coefficient of thermal expansion, CTE). Comparison of the CTE changes of the two materials is shown in Fig. 1. Although Super-Invar has a low CTE prior to irradiation, the CTE increases by a factor of about five after irradiation. Inconel has an initially higher CTE, but shows little sensitivity to radiation dose.

To improve our knowledge of target behavior with a high intensity beam, we are carrying out bunch-merging tests at the AGS. Figure 2 shows a longitudinal profile of the beam during the bunch-merging process. Two bunches from an $h = 12$ rf system are transferred to an $h = 6$ system. To date, we have demonstrated an extracted bunch intensity of 1×10^{13} protons per pulse. In Fig. 2 there are clearly visible residual oscillations after the bunches merge, so there is work remaining to optimize the process. Nonetheless, the technique is clearly workable and shows good potential for delivering the required intensity of 1.6×10^{13} protons per pulse.

Open questions for the Hg-jet target concept include:

- Injection of the jet into a 20-T solenoidal field
- Nonlinear dynamics of the jet at full proton intensity

To address these issues, the MC has undertaken a number of R&D activities. We have designed a test magnet that can operate at a field level up to about 15 T, we are designing a Hg-jet system capable of providing the required 20–30 m/s jet velocity, and we are continuing our simulation effort to predict and interpret the effects we observe experimentally. An engineering study of the proposed magnet has been completed. The concept we intend to explore is illustrated in Fig. 3. Although

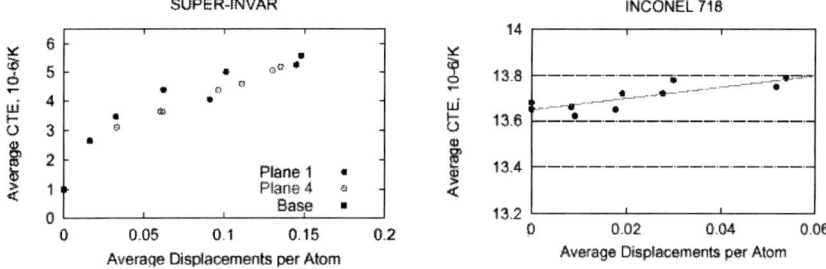

FIGURE 1. CTE changes for Super-Invar (left) and Inconel (right) as a function of radiation dose. Note the very different vertical scales in the two plots.

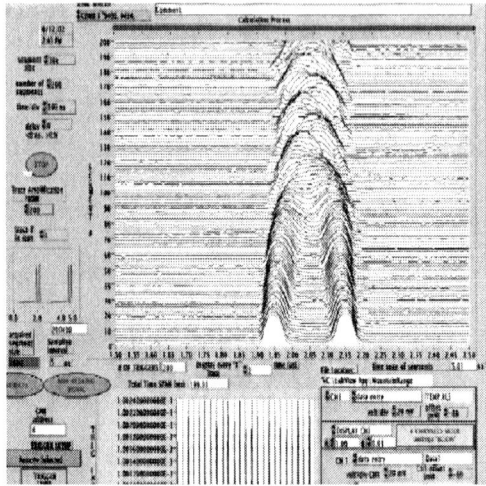

FIGURE 2. Merging of two $h = 12$ bunches into one $h = 6$ bunch at the AGS.

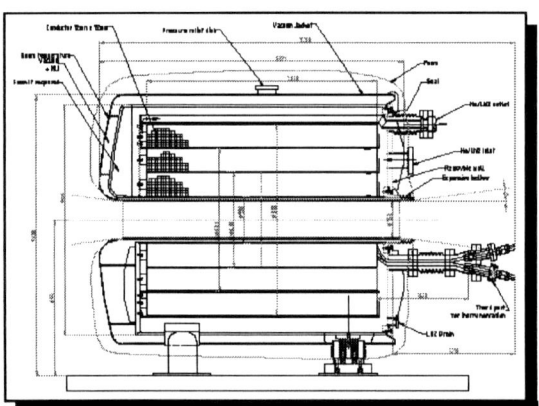 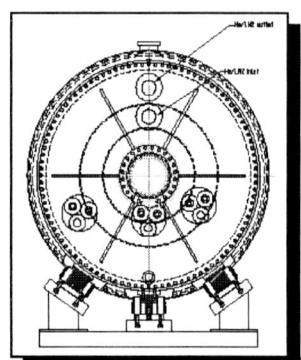

FIGURE 3. Side view (left) and end view (right) of the proposed targetry test magnet. The magnet is designed with three nested coils to provide the required field. It will be operated at cryogenic temperatures for power efficiency reasons, though it will not be superconducting.

consideration was initially given to using hydrogen coolant at the highest field level, the present plan is to use LN_2 and operate with a larger power supply.[1]

Cooling

This activity ("MUCOOL") includes the hardware R&D on rf cavities, absorbers, and solenoids.

---

[1] After evaluating the safety implications of the H_2 system, which show up as additional costs, it appears more cost-effective to avoid hydrogen.

Most of our rf work to date has been done with 805-MHz cavity structures. These serve essentially as quarter-scale models of the cavities we intend to use for a cooling channel. A 201-MHz cavity has been designed [5] and is now under construction. The main question we are addressing is what limits the achievable operating gradient in the cavity. In particular, we are studying breakdown phenomena and dark currents.

Development work on liquid-hydrogen absorbers is being done at several Illinois universities, supported by the Illinois Consortium for Accelerator Research (ICAR), as well as at KEK, supported by U.S.–Japan funds. Much progress has been made in the past few years on the development and testing of large, but very thin, aluminum windows. Due to the proximity of the liquid-hydrogen absorber to the rf cavities and superconducting magnets of a cooling channel, consideration of safety requirements is important. This activity is well along, with solutions being developed to permit testing in the MUCOOL Test Area (MTA) at Fermilab and also to permit operation of a segment of cooling channel beam line for MICE.

Development work for the required superconducting solenoids is also ongoing, with the main goals being to reduce costs and increase reliability.

Cavity R&D

Recent tests have made use of an 805-MHz pillbox cavity [6] having replaceable windows (or possibly grid tubes) to terminate the electric field. Using windows or grids significantly enhances the shunt impedance of the cavity, thereby reducing the power needed to achieve high-gradient operation. These tests are carried out in Lab G at Fermilab, where there exists a 5 T solenoid (provided by the MC) with a bore size sufficient to accommodate the cavity.

In initial tests without a solenoidal field, the cavity (with copper windows) reached an accelerating gradient of 34 MV/m. After the field was turned on, the performance was poorer. The maximum gradient was only 18 MV/m and the radiation levels were much higher. We inferred that the field increased the likelihood of physical damage to the cavity. We did see some evidence for "healing" by reprocessing the cavity without the solenoid, but performance never recovered fully. When the cavity was disassembled in December 2002, we noted significant pitting of the copper window, along with copper dust at the bottom of the cavity.

Next, we installed beryllium windows, coated with a thin layer of TiN to prevent multipactoring. As before, we saw no conditioning problems without a magnetic field. This indicates that the "parallel-plate" geometry is not an intrinsic problem. After conditioning with magnetic field, we again observed a degradation in performance, though not as severe as with the copper windows. Upon opening the cavity for visual inspection, we found no damage to the Be window, but there was sputtered copper on the window surface. The implication is that it is material from the cavity body that is migrating. We will explore the possibility of coatings that might mitigate this effect. Even with the sputtered copper coating, the beryllium windows produced lower backgrounds under comparable conditions, as shown in Fig. 4.

We have completed our design work on a 201-MHz test cavity and begun fabrication. This cavity, which will accommodate either Be windows or an alternative

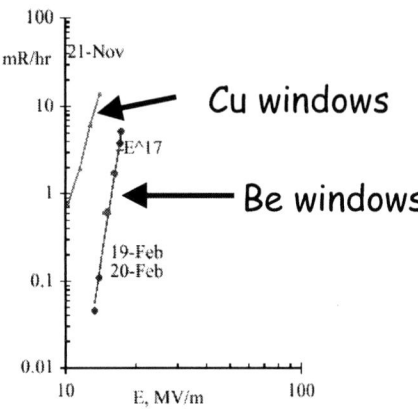

FIGURE 4. Comparison of radiation levels in 805 MHz pillbox cavity produced when using Cu and Be windows.

grid-tube electric field termination, is expected to be available for testing in about one year.

The ideal cavity window would be perfectly conducting, transparent to the muon beam, and mechanically stiff enough to resist changes to the cavity resonant frequency. Our initial concept was to use flat, pre-stressed Be foils. In practice, however, we found it difficult to maintain sufficient flatness to avoid detuning the cavity frequency, even at the smaller dimensions appropriate to the 805-MHz cavity. In addition, the Be window frame had to be very thick to support the pre-stress, making the windows costly.

We have recently adopted a new concept [5], based on a pre-curved window that bows predictably. With proper design, the stresses in the window remain quite low as the foil heats. We are testing various means to manufacture the required shape. One approach, involving clamping in a graphite die and then brazing the window to a frame, is shown in Fig. 5.

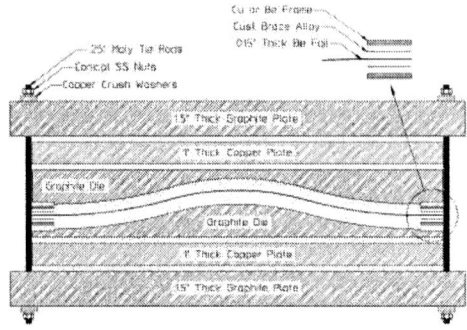

FIGURE 5. Graphite die and clamping arrangement for producing curved Be cavity windows.

Absorber R&D

The absorber program has been very successful in developing thin aluminum windows suitable for absorber use [7]. A new stronger (hence thinner) shape has been developed and will be tested next. The new design involves a double curvature, of the type shown in Fig. 5.[2] The first absorber will be filled with liquid hydrogen later this year. The test will be done in the newly-constructed MUCOOL Test Area (MTA) at Fermilab, which is expected to be ready for beneficial occupancy in October, 2003.

Acceleration R&D

The hardware program in acceleration involves the fabrication and testing of 201-MHz superconducting cavities for a recirculating linear accelerator (RLA). This work, carried out at Cornell under NSF sponsorship, has made good progress. Figure 6 shows the prototype cavity being readied for testing in a vertical dewar. The focus of the effort is to achieve a high design gradient of 16 MV/m. After being returned to CERN once for recleaning, the cavity reached 11 MV/m [8]. This is an encouraging first effort. Still to be done are the design and testing of ancillary items, such as the input coupler, higher-order-mode damper, and tuner.

Simulations

In the past year, the MC has formed a separate group to concentrate on emittance exchange, i.e., on 6D cooling. Cooling rings [9, 10] look particularly attractive, due to their potential for significant cost reduction. While the cooling process itself looks promising, issues remain. First, the ability to inject into (or extract from) a cooling ring appears difficult, and, second, it is not yet clear how to incorporate the beam from a cooling ring into the "standard" Neutrino Factory scenarios.[3]

FIGURE 6. Prototype 201 MHz superconducting cavity being readied for testing in a vertical dewar at Cornell.

---

[2] The idea for the rf cavity window shape referred to earlier was based on this new absorber window concept.
[3] The small circumference of a ring precludes the relatively long bunch train for which present Neutrino Factory scenarios are optimized. However, a cooling ring does lend itself well to a Muon Collider scenario, where only one bunch of muons of each sign is required.

In addition to the more or less standard solenoid-based approaches, we continue to investigate alternatives such as quadrupole-based cooling ring systems [11].

We have also continued to look at the possibilities for utilizing an enhanced version of acceleration hardware. In particular, we are investigating [12] the idea of applying the FFAG (fixed-field, alternating gradient) principle to the design of the RLA arcs. This may permit a simpler and less costly approach to the RLA, in which there is only one FFAG arc on each side rather than a different arc for each pass. Another scheme being looked at is that of very rapidly cycling synchrotrons [13]. To be competitive with the other approaches, the cycle rate must be in the kHz range. This would require very special magnet designs.

Preparations for Study III

In the previous MC-sponsored feasibility studies, we focused mainly on feasibility and performance. In the next such study, we envision developing improved approaches that maintain the performance while giving a more cost-optimized design. Among the ideas we are considering are: improved bunching and phase rotation; 6D cooling (via rings or otherwise); and FFAG acceleration

Our expectation is that Study III will be a "world" study, sponsored by a non-U.S. institution. Participants would come from all three geographical regions—EU, Japan, and U.S. An international organizing group for Study III, now being formed, will become responsible for the organization and execution of the study.

MICE ACTIVITIES

The MICE technical proposal was submitted to RAL in January 2003, and subjected to review by an international panel in February 2003. The review panel subsequently "strongly recommended" approval of the project to RAL management.

The U.S. team for MICE has already submitted a funding proposal to NSF, and hopes to get a response early in 2004. The involvement of the U.S. team will be in the areas of rf cavities and coupling coils, along with contributions to the tracker detector, software development, and experimental simulations.

R&D PLANS

We have developed detailed R&D plans for all of the MC programs. These are briefly mentioned here. The targetry group will fabricate a 15 T magnet to utilize at the AGS or another facility. The cooling group will fabricate and test a 201-MHz cavity at high gradient (≈ 17 MV/m), and will fabricate and test liquid-hydrogen absorbers (first convection cooled, then externally cooled forced-flow type) with all safety aspects. Development of a full prototype of a 201 MHz SCRF cavity module will be pursued by the acceleration group. The simulation group will develop scenarios for use of cooling rings, with the eventual goal of developing a fully engineered cooling ring concept. They will also participate in Study III as part of the "world team."

SUMMARY

The U.S. muon beam program continues to make excellent technical progress on all fronts, despite budgetary problems that make this difficult. We have established good working relationships with our colleagues worldwide, and believe that the muon community serves as a good model for how to work together on large international projects. We are part of a strong international effort for MICE and anticipate obtaining funding for this activity soon. In the meantime, the MUCOOL program of the MC is developing components that serve as prototypes for those needed in MICE. We continue our strong simulation effort, with the intention to participate fully in a World Design Study for a Neutrino Factory in about one year.

ACKNOWLEDGMENTS

The progress reported here is the result of the dedicated work of the 130 members of the MC. I thank them for the high quality of their efforts and the enthusiasm with which they carry them out. Working with such a group is a real privilege.

REFERENCES

1. Gregoire, G., *et al.*, http://hep04.phys.iit.edu/cooldemo/micenotes/public/pdf/MICE0021/MICE0021.pdf.
2. Kirk, H., *et al.*, "Target Studies with BNL E951 at the AGS" in Proc. 2001 Particle Accelerator Conference, Chicago, IL, June 18–22, 2001, ed. P. Lucas, S. Webber, pp. 1535–1537.
3. Haines, J.R., and Tsai, C.C., ORNL/TM-2002/27, Report No. R02-113138, see http://www.ornl.gov/~webworks/cppr/y2002/rpt/113138.pdf.
4. Kirk, H., *et al.*, "Super-Invar as a Target for a Pulsed high-Intensity Proton Beam" in Proc. 2003 Particle Accelerator Conference, Portland, OR, May 12-16, 2003, ed. J. Chew, to be published.
5. Li, D. *et al.*, "A 201 MHz RF Cavity Design with Non-stressed Pre-curved Be windows for Muon Cooling Channels, in Proc. 2003 Particle Accelerator Conference, Portland, OR, May 12-16, 2003, ed. J. Chew, to be published.
6. Li, D. *et al.*, "RF Tests of an 805 MHz Pillbox Cavity at Lab G of Fermilab" in Proc. 2003 Particle Accelerator Conference, Portland, OR, May 12-16, 2003, ed. J. Chew, to be published.
7. Cummings, M.A.C., *et al.*, "Current LH_2-absorber R&D in MUCOOL" in Proc. NuFact02—The 4th International Workshop on Neutrino Factories, ed. K. Long and R. Edgecock, J. Phys. G 29, 1689, 2003.
8. Geng, R.L. *et al.*, "First RF Test at 4.2 K of a 200 MHz Superconducting Nb-Cu Cavity" in Proc. 2003 Particle Accelerator Conference, Portland, OR, May 12-16, 2003, ed. J. Chew, to be published.
9. Palmer, R.B., "Ring Coolers" in Proc. NuFact02—The 4th International Workshop on Neutrino Factories, ed. K. Long and R. Edgecock, J. Phys. G 29, 1577, 2003.
10. Balbekov, V., *et al.*, "Muon Ring Cooler for the MUCOOL Experiment" in Proc. 2001 Particle Accelerator Conference, Chicago, IL, June 18–22, 2001, ed. P. Lucas, S. Webber, pp. 3867-3869.
11. Kirk, H., *et al.*, "Muon Storage Rings for 6D Phase-Space Cooling" in Proc. 2003 Particle Accelerator Conference, Portland, OR, May 12-16, 2003, ed. J. Chew, to be published.
12. Berg, J.S., *et al.*, "FFAGs for Muon Acceleration" in Proc. 2003 Particle Accelerator Conference, Portland, OR, May 12-16, 2003, ed. J. Chew, to be published.
13. Summers, D.J., *et al.*, "Muon Acceleration with a Very Fast Ramping Synchrotron for a Neutrino Factory" in Proc. NuFact02—The 4th International Workshop on Neutrino Factories, ed. K. Long and R. Edgecock, J. Phys. G 29, pp. 1727–1733, 2003.

Particle physics with intense muon beams

Alessandro Baldini

INFN Pisa, Via F. Buonarroti 2, 56127 Pisa

Abstract. The current situation of particle physics experiments which make use of intense muon beams is reviewed with a particular emphasis to experiments looking for lepton flavor violation in rare muon decays. The possible benefits of increasing the muon rate is examined for all the different experimental situations.

INTRODUCTION

Muon physics, which in the past has played a fundamental role in the construction of the standard model of particle physics has not yet exhausted its potential. A great variety of experiments studies muon properties which, in some cases, are very sensitive to theories beyond the standard model. The situation of these experiments is presented in this talk. Neutrino factories could provide muon beams with much greater intensities than the ones presently available. I will examine which experiments could benefit from these future beams. This talk does not claim to be exhaustive: see[1] for a more complete review.

PRECISE DETERMINATION OF THE MUON LIFETIME

The Fermi constant is one of the three free parameters of the standard model bosonic sector

$$\alpha(0.045 ppm), M_Z(23 ppm), G_F(9 ppm)$$

which are currently determined with the relative errors shown in parentheses. G_F can be extracted from the measurement of the muon lifetime with a theoretical uncertainty below 1 ppm[2]. It seems therefore very important to improve the error on the determination of the muon lifetime. We point out however the following two facts:

- V-A is commonly assumed in establishing the relation between G_F and the muon lifetime: in the general case other parameters experimentally determined at the 10^{-3} level appear[1] (see next section);
- electroweak fits depend on the dimensionless product $G_F M_Z^2$ whose uncertainty is dominated by the one of M_Z.

There are three experiments for the muon lifetime measurement currently going on, one at RAL and the others two at PSI.

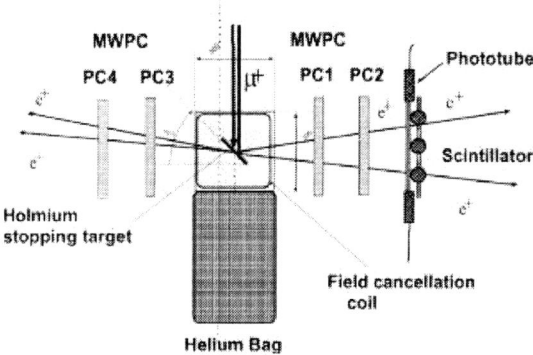

FIGURE 1. A sketch of the experiment for the measurement of the muon lifetime at RAL

The RAL experiment[3](see Figure1) is a segmented device (in order to reduce positrons pile-up)consisting of scintillators and multi wire proportional chambers. Since it has a limited angular coverage it needs to eliminate any dependence on muons polarization by means of a para-magnetic holmium stopping target which quickly depolarizes muons spins. The experiment benefts from the pulsed structure of the beam for the time measurement but it is limited in statistics because of the low accelerator repetition rate (50 Hz). Due to the limited segmentation a maximum of 10^4 events per second can be accepted. In order to achieve a 1 ppm error on the muon lifetime a total of 10^{12} events are needed. This statistics could easily be achieved in a reasonable amount of time if the repetition rate were increased by a factor 10^3. The other two experiments are being performed at PSI. The μLan experiment[4] (Fig.2) detects positrons by means of

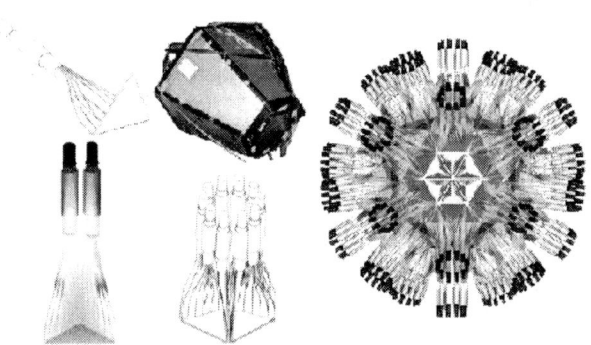

FIGURE 2. A sketch of the μLan experiment

scintillator tiles coupled to photomultipliers via lightguides. The symmetric structure of

TABLE 1. Precisions reachable by the TWIST expriment on the Michel spectrum parameters

	Accepted Value	Standard Model Value	TWIST precision
ρ	0.7518 ± 0.0026	$\frac{3}{4}$	± 0.0001
δ	$0.7486 \pm 0.0026 \pm 0.0028$	$\frac{3}{4}$	± 0.00014
$P_\mu \xi$	$1.0027 \pm 0.0079 \pm 0.0030$	1	± 0.00013
η	-0.007 ± 0.013	0	± 0.003

the detector helps in reducing the muon polarization effects. The pulsed structure of the beam must be artificially created since the PSI beam is continuous. Twenty muons from this beam are used every ten muon lifetimes. A final total statistics of 10^{12} is foreseen.

The FAST experiment uses a pion beam instead of a muon one: it is quoted for the sake of completeness. The detector uses an active target made of scintillating fibers were pions are stopped in the most uniform way by using a suitable plastic degrader. The detection of muon decays, following pions stops, avoids the problems related to muon polarization.

The last two experiments will be taking data in 2004 and expect to reduce the error on the muon lifetime down to 1 ppm. The experiments of this category could gain an additional order of magnitude in sensitivity by an increase of the muon rate if pile-up and detector timing stability could be kept under control.

DETERMINATION OF THE MICHEL PARAMETERS

If the most general form for the muon decay matrix element, compatible with relativitistic invariance, is assumed, muon decay is ruled by a total of 19 parameters (one of wich is G_F) to be determined experimentally. For instance, if terms proportional to m_e/m_μ are neglected, the electron energy distribution in the decay of muons with polarization P_μ can be written as

$$\frac{d\Gamma}{\varepsilon^2 d\varepsilon d\Omega} \propto 3(1-\varepsilon) + \frac{2}{3}\rho(4\varepsilon - 3) \pm P_\mu \xi \cos\theta [1 - \varepsilon + \frac{2}{3}\delta(4\varepsilon - 3)] \qquad (1)$$

where ε is the ratio of the electron energy to its maximum value. The Twist experiment[5] at TRIUMF will measure ρ, δ, $P_\mu \xi$ and the parameter η, which was neglected in 1, with the uncertainties reported in table 1.

If no V-A assumption is made the η parameter appears in the determination of G_F too and it is the error on this parameter that limits the precision reachable on the evaluation of the Fermi constant.

Supersymmetric theories with R parity violation and left-right symmetric gauge theories give new contributions to the ρ and ξ parameters. The TWIST experiment should be competitive with direct searches at high energy colliders in testing such theories.

In another experiment at PSI[6], presenting its results for the first time at this workshop, the transverse polarization of positrons from muon decay is measured. The component (P_{T_1}) of the transverse positron polarization in the plane formed by the muon polarization and the positron momentum may be used to provide a measurement of the η parameter. The second component P_{T_2}, perpendicular to the muon polarization, is non-invariant under time reversal.

The limits of these experiments is not given by muon rates but by systematic effects like positron depolarization in matter.

LEPTON FLAVOUR VIOLATION SEARCHES

Lepton Flavor Violation (LFV) searches have a long history reaching back 1948[7]. The absence of the $\mu \to e\gamma$ decay has played a fundamental role in the construction of the Standard Model of elementary particles physics. In the past 25 years the sensitivity to this decay was raised by two orders of magnitude. The current best limit was given by the MEGA experiment[8] which established a 90% C.L. limit of $1.2\,10^{-11}$ for the branching ratio (BR) of the $\mu \to e\gamma$ decay with respect to the normal muon decay.

Grand unified suspersymmetric (SUSY-GUT) theories, owing to the large top quark mass, predict[9] this decay to happen not much below the current experimental limit. In Figure 3 the SU(5) predictions for the $\mu \to e\gamma$ branching ratio as a function of the right handed selectron mass and for several values of $tan\beta$ are shown. Recent indications from the combined LEP experiments favor values of $tan\beta$ grater than 10. Predictions for SO(10) are even higher (about two orders of magnitude) than SU(5) predictions. In Figure 3 the current experimental limit and the sensitivity that the MEG experiment[10] (see below) at PSI aims at reaching are shown.

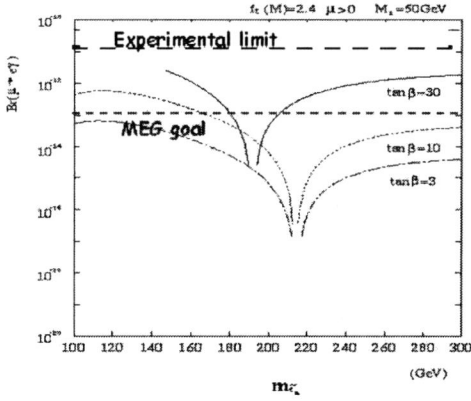

FIGURE 3. SUSY SU(5) predictions for the $\mu \to e\gamma$ decay

Another, independent, source of LFV in SYSY-GUT theories might come from neutrino oscillations. After the recent KAMLAND results the large mixing angle solution, which implies maximum mixing angle and a difference of $10^{-5}eV^2$ between $m_{\nu_e}^2$ and $m_{\nu_\mu}^2$, seems to represent the best solution for the so-called "solar neutrino problem". If a mechanism of the see-saw type is introduced in SUSY-GUT theories to reproduce the pattern of neutrino masses, sizeable contributions (of the same order of magnitude or even higher than the ones discussed above) to the $\mu \to e\gamma$ process take place[11]. These contributions add up to the previous ones, therefore making the $\mu \to e\gamma$ process an extremely sensitive probe of SUSY-GUT theories.

It must also be remarked that the *BR* of the $\mu \to e\gamma$ decay in the standard model due to neutrino oscillations would be completely unobservable ($BR \approx 10^{-54}$). The detection of $\mu \to e\gamma$ events would be a clear, unambiguous sign of physics beyond the standard model.

Experimentally a beam of positive muons is stopped in a thin target and a search is made for a back to back positron-photon couple with the right momenta and timing. The main background in present experiments comes from the accidental coincidence of independent positrons and photons within the resolutions of the used detecors. The best available detectors for low energy positrons and photons must therefore be employed in this kind of experiments.

In the MEG experiment at PSI (see Fig.4) a surface muon beam with an intensity grater than $10^7 \mu/s$ will be stopped in a thin target. A magnetic spectrometer, composed of a superconducting magnet and drift chambers, will be used for the measurement of the positrons trajectories. Positrons timing will be measured by an array of scintillators.

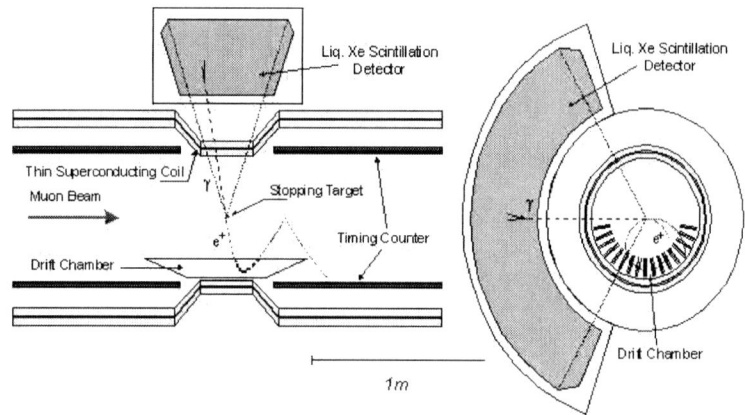

FIGURE 4. A sketch of the MEG detector

Photons will be detected by an innovative electromagnetic calorimeter in which a total of about 800 photomultipliers will detect the light produced by photons initiated showers in about 800 liters of liquid Xenon. Liquid Xenon has optimal scintillation features:

- high light yield: 80% of NaI
- fast emission: three components with exponential decay times of 4, 22 and 45 ns
- short radiation length: 2.77 cm

Two disadvantages are the the operating temperature ($\approx -100°C$) and the ultraviolet light emission spectrum.

Prototypes of all the sub-detectors have being built and have been tested or are currently under test. All the measurements and simulations indicate that a sensitivity down to *BR* of the order of 10^{-13} can be reached by this experiment, with an improvement of two orders of magnitude with repsect to the present best experimental limit. The start of the data taking is foreseen in 2006.

Another interesting channel for LFV investigation is muon to electron (μe) conversion in nuclei. The ratio of the BR for this process with respect to the $\mu \to e\gamma$ one has been calculated by several authors, for various nuclei, under assumptions on the relevant matrix elements which are valid in many SUSY models (see Fig.5[12]).

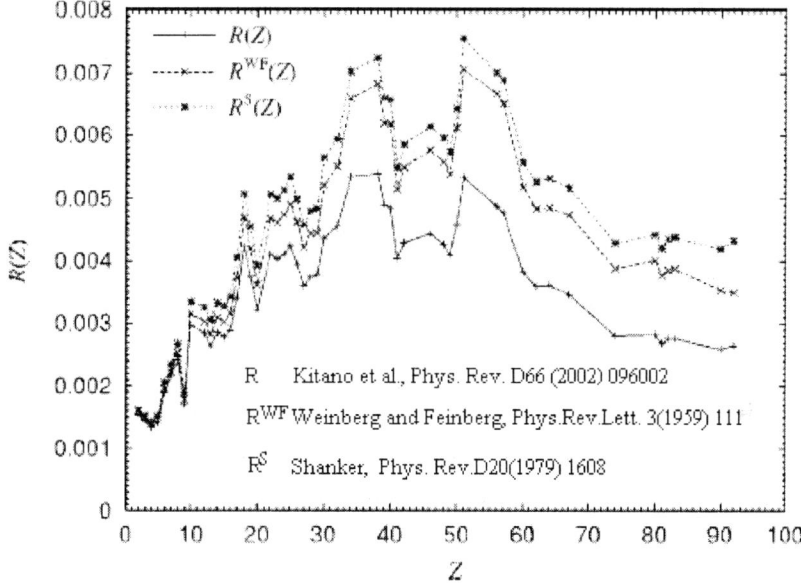

FIGURE 5. Computed ratio of $BR_{\mu e}/BR_{\mu \to e\gamma}$

Experimentally, negative muons are brought to stop in a thin target and are subsequently captured around a nucleus. The energy of a possible converted electron would be equal to the rest muon mass minus the muon binding energy (E_B).

Two main sources of background are:

- a beam correlated background mainly due to radiative pion capture followed by $\gamma \to e^+e^-$ conversions;
- electrons from muon decay in orbit (DIO).

While the first source of background can be reduced the second one in not eliminable. DIO electrons spectrum extends up to the energy region of electrons from μe coversion but with a spectrum proportional to $(E - E_B)^5$. An excellent electron momentum resolution is fundamental in order to keep this background under control.

The best experimental sensitivities to this process were obtained by the SINDRUM experiment at PSI. Pion contamination in the beam was suppressed by means of a moderator which exploited the different ranges of pions and muons. Positrons were detected in cylindrical drift chambers. The final result of the SINDRUM II experiment, which used a gold muon stopping target, is shown in Fig.6. The DIO electrons spectrum

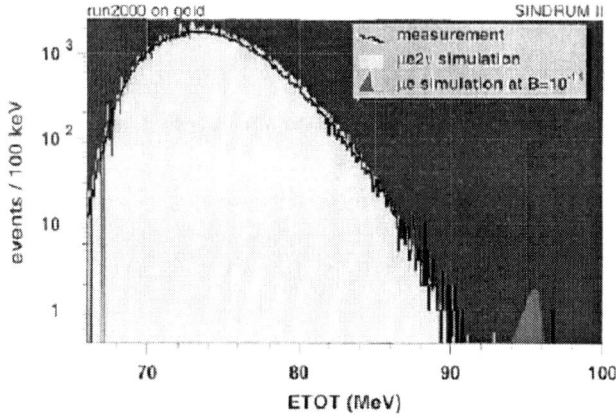

FIGURE 6. Results of the SINDRUM II experiment

is well reproduced by simulations. Also shown is the conversion signal for a $10^{-11} BR$. The momentum resolution at the conversion peak is 2% FWHM. The 90% C.L. limit established by the SINDRUM II experiment is $8 \cdot 10^{-13}$.

The MECO project at BNL[13] plans to use a very intense ($10^{11} \mu/s$) pulsed muon beam for reaching a sensitivity to μe conversion down to $BR \sim 10^{-16}$. The beam will be obtained by capturing most of the lower energy pions produced in a target placed inside a superconducting solenoid magnet. Muons of suitable momentum ($60 - 120 MeV/c$) from pion decays are transported by a curved solenoid to the stopping target and tracking system. This scheme of muons production resembles very much those proposed for ν−factories. The design electron momentum resolution, dominated by interactions in the target is $900 KeV$ FWHM. The pulsed structure of the beam is indispensable to reduce the beam correlated background. A proton extinction factor better than 10^{-9} between two bursts must be obtained in order to reach a sensitivity to $BR \sim 10^{-16}$.

In the PRISM/PRIME project at J-PARC[14] the same muon production scheme as MECO is adopted. After pion production in a solenoid the beam is transported in a circular system of magnets and RF cavities (FFAG ring) which acts as a pion decay section (increasing beam cleaning) and reduces the muon energy spread.

The features of this beam would be the following:

- extremely high intensity ($\approx 10^{12}/s$) of clean muons
- low momentum ($\sim 70 MeV/c$)
- narrow energy spread (few % FWHM)

The last feature is essential to stop enough muons in thin targets. If the electron momentum resolution will be kept below $350 KeV/c$ (FWHM) the experiment will be sensitive to μe conversion down to $BR \sim 10^{-18}$.

MAGNETIC ANOMALY AND ELECTRIC DIPOLE MOMENT

The anomalous magnetic moment of the muon (a_μ) was recently measured at BNL[15] by measuring the difference between the spin precession and the cyclotron frequency in a storage ring. The uncertainty on a_μ is 0.7 ppm. The comparison with theory[16] is not easy. The hadronic vacuum polarization term must be evaluated by using dispersion relations and data from electron-positron annihilation into hadrons or by using hadron τ decay data. These calculations should be improved by better data treatment or by using new experimental data in order to draw firm conclusions about possible discrepancies between the experimental value and the theoretical predictions.

A non zero muon electric dipole moment(*edm*) would violate parity and time-reversal. The best sensitivity to the muon *edm* was reached by the CERN g-2 experiment which measured a value of $3.7 \pm 3.4 \, 10^{-24} \, e \, cm$. A letter of intent for a new experiment[17] to be performed at J-PARC plans to disentangle the *edm* effect from the g-2 precession by means of a radial electric field. A new high intensity beam of 0.5 GeV/c polarized muons woul be needed for this measurement.

CONCLUSIONS

Muons are sensitive probes of physics beyond the standard model. SUSY-GUT Theories predict LFV to be not to far from the present experimental limits. Many of the on-going measurements will benefit from an increase of the muon flux as the one obtainable at neutrino factories. In some cases better experiments should be conceived. This constitutes a challenge for the field of detectors R&D. The effort is however worthwhile since new physics could be just around the corner.

REFERENCES

1. J. Aysto *et al.*, CERN-TH/2001-231
2. T. van Ritbergen and R. G. Stuart , Phys. Rev. Lett. **82** (1999) 488
3. D. Tomono, Precise measurement of the positive muon lifetime at RIKEN-RAL muon facility, these proceedings
4. G. Onderwater, Towards an improved determination of the Fermi coupling constant from the μLan experiment, these proceedings
5. P. Kitching, TWIST. A precision measurement of muon decay at TRIUMF, these proceedings
6. W. Fetscher, Measurement of the polarization vector of the e^+ from the decay of polarized μ^+, these proceedings
7. E.P. Hinks and B. Pontecorvo, Phys. Rev. Lett. **73** (1948) 246
8. M. L. Brooks *et al.* Phys. Rev. Lett. **83** (1999) 1521
9. R. Barbieri and L.J. Hall Phys. Lett. **B338** (1994) 212
10. The MEG experiment: search for the $\mu \to e\gamma$ experiment at PSI, September 2002, available on the web at the address: http://meg.psi.ch/docs/prop_infn/nproposal.pdf
11. J. Hisano and N. Nomura, Phys. Rev. **D59** (1999) 116005
12. R. Kitano *et al.*, Phys. Rev. **D66** (2002) 096002
    M. Koike, Theoretical study on the Lepton Flavour Violating $\mu - e$ conversion in nuclei, these proceedings

13. P. Yamin, The MECO experiment, coherent $\mu \to e$ conversion in the field of a nucleus, these proceedings.
14. A. Sato, PRISM/PRIME, these proceedings
15. L. Roberts, g-2, these proceedings.
16. D. Nomura, Theory of muon g-2 and muon EDM, these proceedings
17. W. Morse, Measuring the electric dipole moment of the muon and deuteron in storage rings, these proceedings.

Non-particle physics with intense muon beams

K. Ishida

Muon Science Laboratory, RIKEN (The Institute of Physical and Chemical Research), Wako, Saitama 351-0198, Japan

Abstract. Typical examples of muon's application, such as for muon catalyzed fusion, nuclear physics, condensed matter physics, surface and nano-scale science, and radiography, are introduced. Also, the importance and the benefit of intense muon beams are stressed.

INTRODUCTION

In matters, a negative muon behaves like a heavy electron while a positive muon behaves like a light proton or a radioactive isotope of hydrogen. A negative muon forms a muonic atom, while a positive muon stays at an interstitial site or makes a bonding like a hydrogen does. These behaviors bring various interesting applications of muons to non-particle physics.

MUON CATALYZED FUSION

When a negative muon is stopped in hydrogen, it forms a small muonic atom with net neutral charge. Because of the reduction of the Coulomb barrier, it is very reactive and easily causes reactions involving muonic atoms and molecules. In a mixture of deuterium and tritium, muon can form a $dt\mu$ molecule. In the $dt\mu$ molecule, the muon binds the two nuclei so closely that the dt fusion can occur very rapidly. After the fusion the muon becomes free again and can be used to catalyse another fusion (figure 1). This muon catalyzed fusion (μCF) cycle[1, 2] is repeated many times until the muon is removed from the cycle either by muon decay or by other muon loss processes. The efficiency of the μCF cycle can be described by two parameters, the cycling rate (λ_c) and the probability of muon loss per cycle (W). The averaged number of fusions per muon is $Y_n = \phi \lambda_c / \lambda_n$, where ϕ is the density of D-T mixture normalized to the liquid hydrogen density since λ_c is usually defined as the rate at liquid hydrogen density, λ_n ($= \lambda_0 + W\phi\lambda_c$) the fusion neutron disappearance rate and λ_0 the free muon decay rate.

Among the loss processes, the probability that the muon sticks to the charged fusion product, α particle, dominates. Theoretically, the final sticking probability ω_s was calculated as $(1-R)\omega_s^0$, where ω_s^0 is the initial sticking probability right after the d-t fusion and R is the muon reactivation probability due to the stripping of muons from the energetic $(\alpha\mu)^+$. The ω_s limits the number of fusion catalysis and the energy-production capability by μCF. Thus it is one of the most important observables in all the μCF studies.

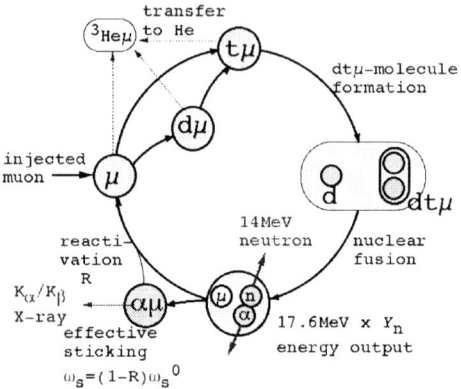

FIGURE 1. Simplified diagram of the muon catalyzed fusion cycle.

So far, λ_c and W have been determined from fusion neutron measurement of λ_n and Y_n ($\lambda_c = \lambda_n Y_n/\phi, W = (\lambda_n - \lambda_0)/\lambda_n Y_n$)[3, 4]. The ω_s is obtained by subtracting from W other contributions such as muon sticking through $dd\mu$ and $tt\mu$ formation and muon transfer to impurity atoms such as ^{3}He which may accumulate by tritium-decay. The effective sticking probability ω_s obtained by neutron measurements gave a value of 0.5 % and has been always smaller than the theoretically predicted values (~0.65 %). The discrepancy has been one of the most important problems of μCF, since the ω_s gives an upper limit on μCF efficiency.

Concerning this problem, characteristic X-rays from $(\alpha\mu)^+$ can give us valuable insights about the atomic process of $(\alpha\mu)^+$ ions. The X-ray measurement had been considered very difficult due to a huge radiation background of bremsstrahlung from tritium β decay. However, with the help of the intense pulsed muon beam at the RIKEN-RAL muon facility[5], the signal to noise ratio was drastically improved. Muons were stopped in the target of solid or liquid D-T mixture of various tritium concentrations. Clear Doppler broadened K_α(n=2 $\rightarrow$ n=1) peaks from recoiling $(\alpha\mu)^+$ were observed at 8.2 keV above bremsstrahlung background. The K_α X-ray yield was not contradicting with the calculated values, while the K_β/K_α intensity ratio was much smaller than the predicted values[6]. These experimental data indicates that muon stripping from $(\alpha\mu)^+$ excited states ($n \geq 3$) could be larger than expected.

Muon transfer to impurities is another important muon loss process. Especially helium-3 is a product of tritium decay. Muon transfer to helium was studied in D-T mixture as well as pure T_2[7, 8]. It was found that ^{3}He stays in solid while it diffuses out quickly from liquid. The transfer mechanism by forming an intermediate tHeμ molecule was confirmed by our observation of the characteristic X-ray from the molecule whose energy is distributed around 6.9 keV.

The cycling rate is another key to improving the efficiency of the μCF. Among the various processes leading to μCF, $dt\mu$ formation rate is usually the bottleneck. Since the $dt\mu$ is formed by resonant mechanism, it is very sensitive to small energy difference and can be controlled by temperature, molecular states etc. Our recent progresses include (1) study of temperature dependence and density effect at low temperatures (20K - 5K)[9], (2) study of $dt\mu$ formation by $t\mu + D_2$ and $t\mu + DT$ by using non-equilibrium liquid mixture of D_2 and T_2, and (3) study of the difference of $dd\mu$ formation rates in normal D_2 and ortho-D_2[10, 11].

In the future, with intense muon beams, further precise measurement of X-rays will become possible by utilizing detectors such as cryogenic calorimeters or X-ray diffraction spectrometers. The intensities of K_α, K_β and K_γ should be measured to further investigate the muonic atom process in the $\alpha\mu$ excited states. Doppler broadening will give us information on the $\alpha\mu$ velocity at the time of X-ray emission.

Extending the study of μCF to much extreme conditions is also worth persueing in order to increase the μCF efficiency. The μCF target condition will be coupled with laser, high density, high temperature, non-equilibrium molecule states etc. Several ideas to enhance the stripping of muons from $(\alpha\mu)^+$, such as the use of a plasma for longer $(\alpha\mu)^+$ mean-free path or an electric field for $(\alpha\mu)^+$ acceleration, are proposed and could be tested.

At present, the average number of neutrons generated per muon at the best condition is around 120, and the energy output (120×17.6 MeV) is still below the calculated energy cost of muon production (~ 5 GeV). However, it is known that the sticking probability is reduced and the cycling rate is increased with increasing density. Practically, we would be able to achieve 2 times the liquid hydrogen density (ϕ_0), and, from the extrapolation of the present data, this would yield a cycling rate $\phi\lambda_c$ of 600 times the muon decay rate and reactivation probability of 0.65, giving neutron yeild of 200 per muon. If a density above 10 ϕ_0 can be realized, we can expect the cycling rate of more than 6000 times the muon decay rate, and the reactivation probability of more than 0.95. This would yield more than 1500 fusions per muon. The investigation would need a special high presuure container, and a muon beam of good quality is necessary to stop muon in the target volume.

MUONIC ATOMS OF RADIOACTIVE NUCLEI

Muonic atom spectroscopy has been used successfully for many years to determine the nuclear charge distribution[12]. With the increased recent interest in unstable nuclei, it is desired to measure the radius of unstable nuclei. There are several scenarios proposed to form and investigate the radioactive antiprotonic and muonic atoms, such as beam merging and cyclotron trap[13].

A project has started to study this by using the fast transfer of muons from hydrogen to heavier nuclei[14]. The concept is to stop both negative muon and nuclear beams simultaneously in a solid hydrogen (H_2/D_2) film. Fast muon transfer reaction from hydrogen to implanted nuclei (Z) to form muonic atoms μZ will occur. At present, studies are being carried out by using the beam of stable nuclei to establish the most efficient way of forming muonic atoms via muon transfer. Clear peaks of muonic x-rays were observed from the argon nuclei implanted at a concentration of 5 ppm. Nearly half of the muons which stopped in the deuterium formed the muonic argon atom. A detail was reported in this conference[15].

CONDENSED MATTER STUDY WITH MUON SPIN

It is well known that a muon is generated from pion decay with its spin polarized along the direction of emission due to the parity violation. In addition, when the muon decays with a lifetime of 2.2 μs, it produces a positron/electron preferentially emitted in the instantaneous direction of the muon's spin. Thus by monitoring the positrons/electrons emitted from a sample following implantation of polarized muons we can study the muon's spin interactions within a sample.

The muon spin relaxation (μSR) technique, especially that with positive muons, is the most commonly used application of muons. Although it is similar to NMR in principle, it has several advantages. The muon is polarized, so the measurement can be performed even without applied magnetic field. The μSR can be applied to a sample where there is no proper nuclei for NMR. The μSR has a unique sensitive time range and time response, which is usually shorter than that of NMR.

The μSR method has been applied to a wide variety of samples. It covers virtually any sample, such as (1) high T_c superconductors (phase diagram, magnetic flux distribution), (2) semiconductors (muon as dilute hydrogen impurity), (3) chemicals (muon simulating reactions of hydrogen), (4) magnetic materials, and (5) polymers and biomolecules (muons probing electron motion in cytochromes and DNA)[16]. Intense muon beams will help us applying μSR to smaller and precious samples.

In contrast to a positive muon, a negative muon loses a large part of its polarization by forming a muonic atom through spin couplings with the orbital angular momentum. The use of negative muons (μ^-SR) was limited by the lower beam intensity and lower residual polarization. In addition, since the muon lifetime depends on the nuclei, the analysis becomes more complicated. However, the μ^- gives us information at the lattice site by its attachment to the nucleous, while the μ^+ tend to stay at interstitial sites. The μ^-SR measurement may be improved by using the tagging with muonic atom x-rays to select the element that the muon is bound. Such a method would need intense muon beams to compensate the lower detection efficeincy.

ULTRA SLOW MUON BEAM

There is a lot of demand of ultra low energy muons with variable energy from a few meV to tens of eV. It can be used for surface and sub-surface layer studies of nanometer scale. It can be used for microstructure studies by making a micro beam by accelerating the ultra low energy muons.

It has not been straightforward to obtain intense ultra low energy muons, since the muon is generated with high energy of MeV order, and simple degradation results in wide spread phase space. There has been two directions taken to open an efficient way for generating ultra low energy muons.

The first method, which was developed at PSI [17], uses a cold moderator of rare gas solid. A 4-MeV surface muon beam passes the moderator and loses its energy. At energies below 50 eV inelastic energy loss is strongly suppressed due to energy gap in the material, so it becomes difficult for muon to lose energy further and the muon may have a chance to leave the solid with a certain energy. Conversion efficiency from MeV μ^+ to slow μ^+ is 10^{-5} to 10^{-4}.

The other method, which was started at KEK[18] and is being developed at RIKEN-RAL[19], uses the resonant laser ionization of thermally emitted muonium from hot tungsten. A surface muon beam whose energy is around 4 MeV is injected into a heated thin tungsten foil. Muonium atoms with thermal energy are evaporated from the surface. Then the electrons of the muonium atoms are removed by irradiation with VUV laser light through a series of transition, 1s → 2p and 2p → unbound. A detailed report of present status was given in this workshop[20]. The efficiency was of the order of 10^{-5} and a further upgrade is expected with increasing the laser power. The time resolution of the slow muon beam is mainly determined by the pulse width of the VUV laser pulse, which is approximately 5 ns. The resonant ionization method is expected to offer better energy spread of slow muon beam, which is limited by the target temperature (~ 0.3 eV).

RADIOGRAPHY WITH MUON BEAMS

Because of the high penetration power of high energy muons, muon is an ideal tool to study inner structure of massive substance. Protons, neutrons and mesons interact strongly with nuclei, gammas and electrons produce showers, neutrinos make only very week interactions. Muon is the only particle that is suited to studying dense materials with the size typically above 1 m.

The first such kind of measurement was done for the pyramid of Giza using cosmic-ray muons[21]. Even higher energy cosmic ray muons of TeV range were used to study volcano mountains, Asama and Iwate[22]. These used the muon's penetration thickness dependence on the initial muon energy. The number of penetrating muons decrease with the density thickness.

Another method with tracking of the scattering angle of incoming and outgoing cosmic-ray muons was recently reported[23]. Since heavy elements cause larger scattering angle, it can be applied to searching heavy elements in dense materials.

The most significant limitation in using cosmic-ray muons is that the cosmic-ray muon has very low intensity (~ 1 /cm^2/min). An intense GeV energy beam of accelerator generated muons will open new area of radiography.

Another important application of muons for radiography is the nondestructive elemental analysis with muonic atom x-rays. A muonic atom x-ray has energy characteristic to the nucleous that the muon is bound after stopping. Basically, every muon emits one muonic atom x-ray in K-series, and other x-rays (in L, M, ... series) may be also observable. Thus this method is applicable to all the elements and the composition of a sample can be studied nondestructively[24].

Since the muonic atom x-ray energies are 200 times higher than those of normal atomic x-rays of the corresponding element, the muonic x-rays have much higher penetration power. It could be the method most suited for studying samples of a few tens of cm, such as human body[25]. The stopping position can be defined by the range-energy distribution in the depth direction and by the beam collimation in the transverse direction. Thus three dimensional mapping of elements is possible.

SUMMARY AND FUTURE PROSPECT

Muon's rather well known and pure interaction with materials has opened several important applications. At present, typical muon beam intensity is of the order of 10^7 /s. For the pulsed muon beam, it is around 10^4 /pulse at RAL. It is expected to increase the pulsed muon beam intensity by one order at J-PARC muon facility[26]. To go beyond, there are several plans using full pion capture solenoid to produce more than 10^{10} muons per second. This will make the present studies much easier, and may open new applications of muons which is so far considered infeasible or even unknown.

REFERENCES

1. Breunlich, W. H., et al., *Ann. Rev. Nucl. Sci.*, **39**, 311 (1989).
2. Nagamine, K., and Kamimura, M., *Advances in Nuclear Physics*, **24**, 151 (1998).
3. Jones, S. E., et al., *Phys. Rev. Lett.*, **56**, 588 (1986).
4. Breunlich, W. H., et al., *Phys. Rev. Lett.*, **58**, 329 (1987).
5. Matsuzaki, T., et al., *Nucl. Inst. and Meth. in Phys. Res. A*, **465**, 365 (2001).
6. Nakamura, S. N., et al., *Phys. Lett. B*, **473**, 226 (2000).
7. Kawamura, N., et al., *Phys. Lett. B*, **465**, 74 (1999).
8. Matsuzaki, T., et al., *Phys. Lett. B*, **527**, 43 (2001).
9. Kawamura, N., et al., *Phys. Rev. Lett.*, **90**, 043401 (2003).
10. Toyoda, A., et al., *Phys. Lett. B*, **509**, 30 (2001).
11. Toyoda, A., et al., *Phys. Rev. Lett.*, **90**, 243401 (2003).
12. Hüfner, J., Scheck, F., and Wu, C., *Muon Physics Vol. 1*, Academic Press, New York, 1975, pp. 201–307.
13. see articles in *Workshop on Radioactive Antiprotonic and Muonic Atoms (RAMA)*, CERN, Geneva, 2001.
14. Strasser, P., et al., *Nucl. Instr. and Meth. A*, **460**, 451 (2001).
15. Strasser, P., et al., in *NuFact03*, AIP Conference Proceedings, American Institute of Physics, New York, 2003.
16. Nagamine, K., et al., *Physica B*, **289/290**, 631 (2000).
17. Morenzoni, E., et al., *Phys. Rev. Lett.*, **72**, 2793 (1994).
18. Nagamine, K., et al., *Phys. Rev. Lett.*, **74**, 4811 (1995).
19. Bakule, P., et al., *Spectrochimica Acta B*, **58**, 1019–1030 (2003).
20. Matsuda, Y., et al., in *NuFact03*, AIP Conference Proceedings, American Institute of Physics, New York, 2003.
21. Alvarez, L., et al., *Science*, **167**, 832 (1970).
22. Nagamine, K., et al., *Nucl. Instr. and Meth. A*, **356** (1995).
23. Borozdin, K. N., et al., *Nature*, **422**, 277 (2003).
24. Daniel, H., et al., *Archaeometry*, **29**, 1 (1987).
25. Hosoi, Y., et al., *Br. J. Radiol.*, **68**, 1325 (1995).
26. Miyake, Y., et al., *Physica*, **326**, 255 (2003).

The Future of Neutrino Experiments at Nuclear Reactors

Jonathan M. Link

Columbia University, Dept. of Physics, New York, NY 10027, USA

Abstract. A next-generation neutrino oscillation experiment using reactor neutrinos could give important information on the size of mixing angle θ_{13}. The motivation and goals for a new reactor measurement are discussed in the context of other measurements using off-axis accelerator neutrino beams. The reactor measurements give a clean measure of the mixing angle without ambiguities. The key question is whether a next-generation experiment can reach the needed sensitivity goals to make a measurement for $\sin^2 2\theta_{13}$ at the 0.01 level. The limiting factors associated with a reactor disappearance measurement are described with some ideas of how sensitivities can be improved. Examples of possible experimental setups are presented.

MOTIVATION FOR A NEXT-GENERATION REACTOR OSCILLATION EXPERIMENT

Information on the masses and mixing angles in the neutrino sector is growing rapidly and the current program of experiments will map out the parameters associated with the solar, atmospheric, and LSND signal. With the recent confirmation by KamLAND [1] and isolation of the Δm_{solar}^2 in the LMA region, the emphasis of many future neutrino oscillation experiments is turning to measuring the last unknown mixing angle, θ_{13}, searching for *CP* violation in the lepton sector, and determining the neutrino mass hierarchy.

Sensitivity to the mass hierarchy and *CP* violation is tightly coupled to the value of θ_{13}. Experiments with v_e appearance, such as the proposed off-axis neutrino beam experiments, can potentially observe *CP* violation and the mass hierarchy as well as $\sin^2 2\theta_{13}$, but they are beset with ambiguities and degeneracies associated with these effects and the uncertainty in the other mixing angles [2]. By making a clean measurement of $\sin^2 2\theta_{13}$, absent these ambiguities, reactor experiments make an important contribution to the understanding these effects.

Reactors are a very high flux source of electron antineutrinos and have been used in the past for several neutrino oscillation searches and measurements [3, 4, 5, 6, 1]. Currently, several groups [7, 8, 9] are considering new reactor oscillation experiments with the primary goal of improved sensitivity to $\sin^2 2\theta_{13}$. To achieve this goal, the new experiments will use a comparison of detectors at both near and far baselines thus minimizing the uncertainties due to the reactor neutrino flux. The current best limit on $\sin^2 2\theta_{13}$ (from the CHOOZ experiment) is $\sin^2 2\theta_13 < 0.18$ at $\Delta m^2 = 2 \times 10^{-3}$ (the preferred value of Δm_{13}^2 from the Super-K atmospheric [10]). Experiments hoping to significantly improve on this limit must carefully control all potential sources systematic errors.

TABLE 1. Systematic errors associated with normalization from CHOOZ [5]

Parameter	Relative Error (%)
Reaction Cross Section	1.9
Number of Protons	0.8
Detection Efficiency	1.5
Reactor Power	0.7
Energy Released per Fission	0.6
Total	2.7

BASIC EXPERIMENT DESIGN

The design of the reactor detectors is similar the CHOOZ and KamLAND detectors: a homogeneous, spherical volume filled with a mineral oil based fluid and viewed by an array of PMTs located on the periphery. The neutrino target is a mixture of mineral oil and scintillator contained in a clear vessel at the center of the tank. It is surrounded by about a meter of pure mineral oil which shields the scintillator from radioactive elements in the PMT glass.

At reactor neutrino energies, the charged current interaction is forbidden for all neutrino flavors except $\bar{\nu}_e$. Therefore the signature of $\bar{\nu}_e$ oscillation is an apparent disappearance of $\bar{\nu}_e$ interactions compared to expectation.

The far defector is located at a baseline of between 1200 and 1800 meters. An identical near detector is located at a distance of between 200 and 400 meters from the reactor core. Both the near and far detectors require significant earth shielding to reduce the cosmic ray rate. A minimum of about 300 meter water equivalent is required at the far detector.

CONTROLLING SYSTEMATIC ERRORS

There are two main sources of systematic error that may limit the sensitivity of a rector measurement. The first comes from the uncertainty in predicting the total event rate. Table 1 shows the sources of systematic error associated with event rate normalization as reported by CHOOZ. The second type of systematic error comes from uncertainty in the background event rate. Methods for dealing with both sources of error are addressed in the following sections.

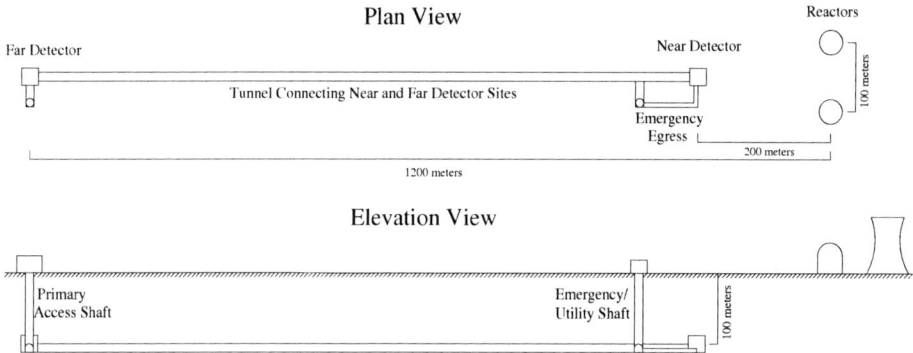

FIGURE 1. A conceptual design of the tunnel facility for the movable detector scenario at a two reactor site.

Event rate uncertainty

Predicting the event rate in a reactor detector requires detailed knowledge of the state of the reactor fuel, the reactor power, the neutrino cross section, the absolute number of target nuclei and the detector efficiency. Previous experiments [4, 5, 6, 1] have demonstrated that the uncertainties on these quantities can be controlled at the level of a few percent. With this level of error it is not possible to achieve the targeted sensitivity of 0.01 in $\sin^2 2\theta_{13}$, which corresponds to a $\bar{\nu}_e$ deficit of less than 0.5%. This problem can be addressed by building a near detector at a distance of about 200 meters from the reactor core. In such a near detector the effects of oscillations are negligible. The uncertainties from the reactor neutrino flux (instantaneous fuel composition and reactor power) and cross section are identical in the two detectors and cancel in the far to near ratio. Further, if the near and far detectors are identical, the target and detector efficiency uncertainties also cancel. While it is reasonable to assume that the target chemistry and total mass can be controlled at the 0.1% level, understanding the relative efficiency of the two detectors at this level is more difficult. One possible solution is to design the experiment so that the far detector can be moved to the near detector site, allowing a direct measurement of the relative efficiency.

In the movable detector scenario, the intense neutrino flux at the near detector site is used to measure the relative efficiency of the two detectors head-to-head. With calibration running of about 10% of the experimental run, the uncertainty on the relative efficiency is smaller than the statistical error on the event rate in the far detector. In other words, the sensitivity to disappearance will be limited statistics and not systematic errors.

Figure 1 is a sketch of a potential layout for a movable detector experiment. The sketch shows two 100 meter deep shafts (one for each detector site) and a tunnel connecting the two sites.

Background uncertainty

The first step in minimizing the systematic error from background events is to reduce the background rate. This is achieved by shielding the detector from cosmic rays with a minimum of 100 meters of rock. In addition, the $\bar{\nu}_e$ charged current interaction on hydrogen, known as inverse β-decay (Eq. 1), provides the means for a significant reduction of background.

$$\bar{\nu}_e + p \rightarrow e^+ + n \qquad (1)$$
$$\hookrightarrow n + p \text{ (Gd)} \rightarrow 2.2 \text{ (8) MeV}$$

Requiring the coincidence in space and time of the positron event and the neutron capture event significantly cuts the rate of background from individuel unrelated events, such as radioactive decays. Adding a small amount of gadolinium[1] to the target reduces the mean neutron capture time from 200 μs to 30 μs, which narrows the window for random coincidences by a factor of seven. The remaining random coincidence events can be understood by measuring the singles rates for positron-like and neutron capture-like events and estimating the rate of random coincidences in the time window allowed for neutron capture.

The more difficult background comes from correlated coincidences, where both parts of the event are generated by the same initial event. Most correlated backgrounds are due to fast neutrons produced by spallation from cosmic ray μ in the materials surrounding the detector. Inside the detector a fast neutron can strike a hydrogen nucleus giving it enough energy to mimic a positron. Alternatively, a neutron-carbon inelastic collision may result in the the excitation of the carbon nucleus which then emits gamma rays decays to get back to the ground state. In both cases the fast neutron will thermalize and capture with nearly the same spatial and time correlation as the neutrons produced in the neutrino interactions.

One possible method of dealing with correlated backgrounds is to build a Muon Veto Neutron Shield (MVNS) system. In the MVNS system (shown in Figure 2), the detector is housed in a bunker of dense material intended to range out neutrons. The outside of this bunker is covered in an array of plastic scintillator. The scintillator array detects muons entering the bunker and vetoes any prompt positron-like events in the detector. Muons entering the bunker material may kick out fast neutrons and cause a correlated background event in the detector. If a muon passes close to the bunker without penetrating it, any neutrons that it generated in the surrounding rock will be ranged out by the bunker. In addition, the events where a veto has been detected can be used to subtract correlated background events that managed to evade the system. This subtraction is achieved by matching the energy distribution for vetoed events to the neutrino candidate events outside the reactor energy range.

---

[1] The ^{157}Gd nucleus has the largest thermal neutron capture cross section of any stable isotope.

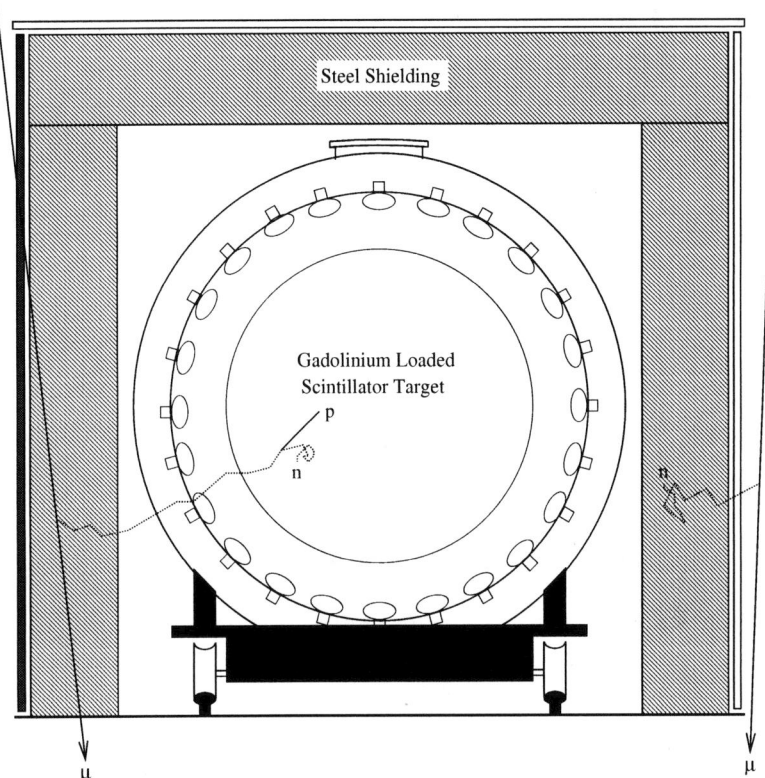

FIGURE 2. Interaction of the Muon Veto Neutron Shield system with muon induced fast neutrons produced both inside and outside the shielding bunker. Fast neutrons produced by muons inside the bunker are vetoed and neutrons produced outside the bunker are ranged out in the bunker material.

POTENTIAL HOST REACTOR SITES IN THE U.S.

Many potential host reactors have been identified around the world including sites in Russia, Japan, Europe, and Brazil. Additionally, many sites are being investigated in the United States. Table 2 lists the top 30 U.S. nuclear reactor facilities ordered by their power performance over the last seven years.

The focus of the U.S. site selection is now centered on two groups of sites. The fist is the Diablo Canyon reactor located on the California coast. Its main asset is the close proximity of hills that could provide significant shielding to a far detector. The second group of reactors are in northeastern Illinois. These reactors, which include Braidwood, Byron and La Salle, are all owned and operated by the Exelon Cooperation and they are all located in close proximity to both Fermilab and Argonne National Lab. To achieve shielding at the Illinois sites requires that shafts be dug to a depth of about 100 meters. The reactor owners for the Illinois and California sites have agreed to cooperate in the early stages of development and are providing the support needed for ongoing feasibility studies.

TABLE 2. List of the top 30 reactor sites in the U.S. by power performance. The neutrino flux is proportional to the reactor thermal power.

Reactor Site	State	Cores	Avg MW_{th}*	Max MW_{th}
Palo Verde	AZ	3	10570	11552
South Texas Project	TX	2	6864	7600
Braidwood	IL	2	6491	7172
Vogtle	GA	2	6456	7130
Byron	IL	2	6442	7172
Browns Ferry	AL	2	6377	6916
Limerick	PA	2	6365	6916
Peach Bottom	PA	2	6290	6916
Sequoyah	TN	2	6209	6822
Oconee	SC	3	6204	7704
Susquehanna	PA	2	6161	6978
Catawba	SC	2	6116	6822
San Onofre	CA	2	6061	6876
Diablo Canyon	CA	2	6043	6749
Comanche Peak	TX	2	5986	6916
McGuire	NC	2	5880	6822
North Anna	VA	2	5129	5786
St. Lucie	FL	2	4925	5400
Edwin Hatch	GA	2	4901	5526
Arkansas Nuclear	AR	2	4844	5383
Calvert Cliffs	MD	2	4813	5400
Joseph Farley	AL	2	4801	5550
Dresden	IL	2	4779	5914
Brunswick	NC	2	4701	5116
Surry	VA	2	4664	5092
Nine Mile Point	NY	2	4500	5317
Quad Cities	IL	2	4481	5914
Indian Point	NY	2	4467	6096
La Salle	IL	2	4323	6978
Salem	DE	2	4281	6918

* Average over the last 7 years [11][12].

CONCLUSIONS

Short baseline reactor neutrino experiments have the potential to make a significant contribution to neutrino oscillation physics. In addition to making a clean measurement of $\sin^2 2\theta_{13}$, reactor experiment help to resolve degeneracies inherent in the ν_e appearance experiments. The prospects for a reactor experiment somewhere in the world to begin taking data in the second half of this decade are very good, and in particular prospects in the U.S. are high. The key to achieving the desired sensitivity is to control the systematic errors.

REFERENCES

1. Eguchi, K., et al., *Phys. Rev. Lett.*, **90**, 021802 (2003).
2. Minakata, H., Nunokawa, H., and Parke, S., *Phys. Rev.*, **D66**, 093012 (2002).
3. Vidyakin, G. S., et al., *JETP Lett.*, **59**, 390–393 (1994).
4. Declais, Y., et al., *Nucl. Phys.*, **B434**, 503–534 (1995).
5. Apollonio, M., et al., *Eur. Phys. J.*, **C27**, 331–374 (2003).
6. Boehm, F., et al., *Phys. Rev.*, **D62**, 072002 (2000).
7. Martemyanov, V., Mikaelyan, L., Sinev, V., Kopeikin, V., and Kozlov, Y., **hep-ex/0211070** (2002).
8. Minakata, H., Sugiyama, H., Yasuda, O., Inoue, K., and Suekane, F., **hep-ph/0211111** (2002).
9. Shaevitz, M. H., and Link, J. M., **hep-ex/0306031** (2003).
10. Nashikawa, K., Slides Presented at Lepton-Photon 2003, http://conferences.fnal.gov/lp2003/program/S10/nishikawa_s10.pdf
11. U.S. Nuclear Regulatory Commission Information Digest, Office of the Chief Financial Officer, U.S. Goverment Printing Office (2002), volume 14.
12. Nucleonics week (1997-2003), published by McGraw-Hill Companies, Inc.

Overview of Neutrino Factory Simulations

Richard C. Fernow

Physics Department, Brookhaven National Laboratory, Upton, NY 11973

Abstract. We give a brief overview of recent simulation activities on the design of neutrino factories. Simulation work is ongoing on many aspects of a potential facility, including proton drivers, pion collection and decay channels, phase rotation, ionization cooling, and muon accelerators.

INTRODUCTION

The neutrino factory is considered by many people [1] to be the ultimate facility for producing large fluxes of well-characterized neutrinos for long baseline oscillation experiments. There are at present three well-developed schemes for neutrino factories based on muon storage rings from the U.S. [2], CERN [3], and Japan [4]. In addition studies have been done on numerous variations for parts of each of the systems. There is in addition a major alternative (beta beam) approach [5], where the neutrinos come from the decay of stored radioactive ions, rather than from muon decays. An overview of machine-related activities in this field in 2002 can be found in the summary of the Machine Working Group at NuFact02 [6].

In this review we will only consider work on the machine designs for neutrino factories. This leaves aside some related ongoing work on "superbeams" and muon colliders. Simulations for targetry and cooling demonstration experiments (MICE) will be omitted since they are covered in other papers in these proceedings. Besides classic Monte Carlo tracking, we generalize the definition of *simulation* to include other computer-aided system designs, such as new lattice designs.

The discussions that follow consider recent simulations on the proton driver, the "front end" systems and the muon accelerators. The front end of a neutrino factory is concerned with all the manipulations required to collect the pions off the target and to prepare a suitable muon bunch for acceleration.

PROTON DRIVER

The baseline CERN proton driver is a 2.2 GeV linac followed by accumulator and bunch compression rings. The baseline U.S. driver is a 16-24 GeV proton synchrotron, while the baseline Japanese driver is the 50 GeV J-PARC synchrotron. Although most recent driver work was done in connection with superbeam facilities, a MW-class proton machine would certainly also be suitable for a neutrino factory if it were built. A new study has been done [7] on upgrading ISIS to 1 MW proton beam power.

A lattice has been designed for a new 8 GeV synchrotron operating at 16 Hz. The machine is about 3 times larger than the present ISIS machine, with a mean radius of 78 m.

PION COLLECTION AND DECAY

The baseline designs for pion collection either make use of a horn (CERN) or a tapered solenoid (U.S. and Japan). This is followed by a pion decay region, consisting of a suitable length of empty magnetic channel. New studies of the CERN channel have been made using ZGOUBI [8]. The simulation used realistic magnetic models for the quadrupoles, dipole and solenoids. A theory of π/μ longitudinal beam transport was developed in order to check the results. The simulation of the longitudinal phase space distribution of the muon bunch 40 m from the target agreed well with a MATHMATICA implementation of the theory. Work is under way to check the transverse phase space simulations and to include the effects of the finite size of the parent bunch.

An alternative CERN collection scheme switches the beam among four target and horn systems. A funneling system consisting of large-aperture quadrupoles and pulsed dipoles has been designed for collecting the four beams into a common decay channel [9].

Another decay channel optimization was done using MARS [10]. The initial beam distribution came from an 80 cm long graphite target. The pion collection was studied as a function of the adiabatic solenoid field taper. A 7.2 m long taper was found to work better than a short 2.4 m long one. The decay channel was 47 m long and used a solenoid field strength of 5 T. This was followed by a second tapered region that brought the final field value down to 1.25 T. Final yields were ~0.19 µ/p for each muon charge. This was an increase of 18% over the muon yields in the U.S. Feasibility Study 1.

BUNCHING AND PHASE ROTATION

The baseline neutrino factory designs make use of phase rotation to decrease the large initial energy spread in the muon beam. Phase rotation is done with induction linacs in the U.S. design and a linear 88 MHz RF channel for the CERN design. No separate phase rotation is required in the Japanese design since the FFAG accelerators have a momentum acceptance of ±50%.

An alternative U.S. scheme first bunches the beam using varying-frequency RF with adiabatically increasing cavity gradient [11]. Fig. 1 shows the bunched beam resulting from a 60 m long buncher. Studies have shown that 10 discrete frequencies are sufficient to get almost the maximum possible yield. This is followed by a short section to do a 90° phase rotation. The RF frequency for the rotator may have a slight "vernier" variation to maximize the throughput. Simulations have taken the phase-rotated beam and injected it into the U.S. Feasibility Study 2 cooling channel. Even though the phase rotated beam was not matched for this channel, the muon yield at the end of cooling was 0.22 µ/p, comparable to the yield in Study 2. These results have generated considerable interest because the system appears to be considerably cheaper than the induction linac solution and the system can in principle collect both signs of muon charges simultaneously.

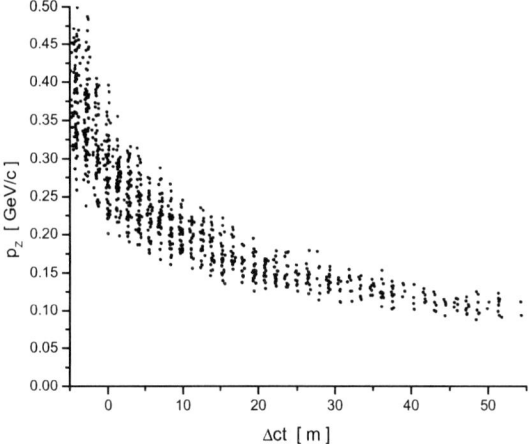

FIGURE 1. Longitudinal phase space following an adiabatic buncher. The initial beam distribution corresponded to 24 GeV protons on mercury. The decay channel length was 76 m and the buncher length was 60 m. The frequencies in the buncher varied from 385 to 216 MHz, while the gradient increased linearly to 6.5 MV/m.

Work is currently looking into producing both shorter and longer bunch trains. A short train would be necessary for injection into a cooling ring, while a longer train would result in a smaller longitudinal emittance in each bunch.

A second study has also considered bunching and phase rotation using high frequency RF [12]. In one configuration a fairly long drift space of 80 m is followed by three 4 m long bunching stations spaced 20 m apart. The frequencies of the cavities in the three stations are ~250, 200, and 165 MHz. This is then followed by another 80 m drift space. When the pulses in the resulting bunch train are overlapped, 73% of the particles lie within a 120° phase interval. Simulations of phase rotation of the bunched beam show that 71% of the particles have an energy spread $\Delta E/E < 10\%$.

IONIZATION COOLING

The baseline U.S. and European schemes use linear cooling channels. The baseline Japanese scheme does not use cooling. There has been a great deal of activity over the past few years examining the performance of cooling rings [13]. Unlike the linear channels in the baseline scenarios, rings also provide emittance exchange with the longitudinal phase space and produce much greater 6-dimensional emittance reduction.

Two common measures of cooling performance are

$$M(s) = \frac{\varepsilon_{6N}(0)}{\varepsilon_{6N}(s)} \frac{N(s)}{N(0)}, D(s) = \frac{N(s)/V}{N(0)/V}$$

where s is the accumulated distance traveled around the ring, ε_{6N} is the normalized 6-dimensional emittance, N is the number of muons and V is a fixed 6-dimensional acceptance volume.

The first successful cooling ring simulation was for the "tetra" ring designed by V. Balbekov [14]. Cooling has been successfully simulated using hard-edge modeling of the magnet fields with ICOOL, GEANT and Balbekov's private design code. Simulations of this ring have produced M-factors as high as 90. Work is in progress on improving the modeling with a more realistic description of the fields. Realistic field maps have been made using TOSCA [15]. As a first step towards more realistic modeling hard-edge simulations have been made leaving gaps in the lattice at the locations where flux returns and field clamps would be located. Extra focusing coils were placed symmetrically at the ends of the solenoid to match into the bending dipoles. The transmission dropped by about 25% when 15 cm gaps were inserted. Other studies are examining the effects of allowing the solenoid fringe field to extend into the dipole region. Recent simulations using COSY have looked at the effects of including fringe fields [16]. Adding fringe fields to the dipoles had a particularly dramatic impact on the performance. Future studies will include the effects of adding damping in the simulations.

A second family of ring coolers uses transverse focusing from quadrupoles or from rotated pole faces on the dipoles [17]. The basic strategy for the design of these rings was to first find a linear optics lattice using SYNCH. The performance was then studied with tracking in hard-edge fields using ICOOL. These rings produce different transverse emittances in the x and y directions since the emittance exchange takes place in the x-s plane. M-factors as high as 1000 have been obtained for some small dipole-only models. One 3.5 m circumference model in particular might be interesting to use for an experimental demonstration of ring cooling. Work has begun on using realistic field models for these rings. Preliminary studies with COSY indicate that the fringe fields could significantly degrade the predicted hard-edge performance. Other work is looking into whether it is possible to include lithium rods in the lattice to greatly reduce the transverse emittance.

The most realistic ring model is currently the RFOFO cooling ring [18]. This ring uses a lattice of alternating polarity solenoids for focusing. The bending field is produced by tipping the solenoid axes by 3°. Modeling of this ring is done with 3-dimensional, Maxwellian fields. The ring has a 33 m circumference divided into 12 identical cells. Energy loss occurs in wedge-shaped liquid hydrogen absorbers. The energy is restored using 201 MHz RF cavities with a field gradient of 12 MV/m. The minimum beta function at the center of the absorbers is 38 cm and the maximum dispersion is 8 cm. Because of the presence of a small radial magnetic field on-axis, the closed orbits follow approximately helical paths. Fig. 2 shows the performance of an "ideal" ring with realistic fields, but before the addition of windows or space for injection. The M-factor of the ideal ring is 112 and the D-factor is 8.9.

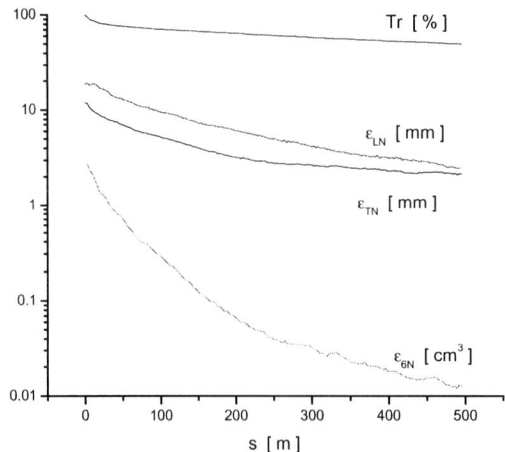

FIGURE 2. Performance of the ideal RFOFO ring cooler; normalized longitudinal, transverse and 6-dimensional emittances and transmission including decay as a function of accumulated distance. One turn is 33 m.

Studies [19] have looked at real world perturbations, such as adding aluminum windows around the absorbers, adding beryllium windows across the aperture of the RF cavities, and leaving two empty cells for the injection/extraction system. This realistic model gives a D-factor of 4.2, still indicating a significant increase in muons suitable for acceleration. Future work will concentrate on including this ring cooler as part of a complete neutrino factory front-end scenario.

The design of a racetrack shaped muon cooling ring has begun at RAL [20]. This design has the virtue of leaving long straight sections for injection and extraction. It uses solenoidal focusing and reaches a minimum beta function of 13 cm at the absorber locations. Tracking of single muons in the ring was done using Opera3D. Future work will include adding absorbers and RF to the ring simulations to study cooling.

Although the cooling performance of these rings looks promising, it is not clear at this time whether they will have a role to play in a neutrino factory[1]. Ring coolers have three serious problems that do not presently have satisfactory solutions. (1) The major uncertainty is whether it is possible to build a kicker with large enough strength and short enough risetime to efficiently inject the muon bunch into the ring. The problem comes from the fact that the muon bunch has a normalized transverse emittance (~12 mm rad) that is very large and that these rings have a small circumference (~35 m). Preliminary designs for a suitable kicker have 3 orders of magnitude more magnetic stored energy than the most powerful kicker built so far [21]. (2) It is uncertain whether the extra power deposited in the liquid hydrogen absorbers by the

---

[1] There is no question that ring coolers will play a vital role for muon colliders.

multiple passes in the ring can be successfully removed. (3) The 50 ns risetime of the assumed kicker and the small ring circumference limit the maximum length of the injected bunch to ~12 m. A suitable buncher and phase rotation system needs to be designed to collect most of the muons into a bunch train this short.

One possible scenario [22] is to take the beam from a bunch compressor ring designed for a muon collider and injected it into the RFOFO cooling ring. The simulated normalized transverse emittance was reduced from 6.3 to 2.5 mm, the normalized longitudinal emittance was reduced from 25 to 3.2 mm, while the yield fell from 0.11 to 0.054 μ/p in this exercise. Presumably a bunch compression ring specifically designed for a neutrino factory could improve the muon throughput.

MUON ACCELERATION

The baseline acceleration system for the U.S. and European designs is a Recirculating Linear Accelerator (RLA), similar to the electron machine at Jefferson Laboratory. The Japanese scheme uses a sequence of Fixed Field Alternating Gradient (FFAG) accelerators. Recent studies [23] have improved on the RLA design used in the U.S. Study 2. The optics was enhanced to give smooth transitions between the arcs and the linacs and in the spreaders and recombiners. Three families of sextupoles were used to restore longitudinal phase space linearity in the arcs. Multi-particle tracking studies showed only minimal emittance dilution with small particle losses of 0.8% from clipping the large amplitude tails.

There has been considerable activity on simulating the performance of FFAGs [24]. The scaling type FFAG is being studied primarily by the Japanese, while non-scaling designs are being examined by other groups. Scaling FFAGs have a betatron tune that is independent of energy. Non-scaling FFAGs have closed orbit variations and time of flight ranges that are small. Studies [25] of the first 0.3-1 GeV/c Japanese FFAG ring show that the accepted transverse emittance is 30 mm, while the longitudinal acceptance is 4.6 eV-s (13 m). Many of the recent simulations have concentrated on the performance of the 10-20 GeV ring [26]. The baseline triplet lattice has been compared with new singlet (FODO) and doublet lattices. The singlet ring has a radius of curvature of 55 m and uses 6 T bending fields. Tracking studies have been performed using the PTC code. The code can read in field maps from TOSCA, for example, then do symplectic kick-drift-kick tracking. Transverse acceptances and coupling between the horizontal and vertical planes has been studied. Non-scaling FFAGs are being studied with FODO, minimum emittance, triplet and racetrack lattices [24,27].

PROSPECTS

Simulation work on neutrino factories is still an active and productive area of research. Much of the current work is devoted to finding ways to improve the performance/cost ratio of future machines. New schemes continue to be proposed and examined. Fortunately code development has progressed in tandem with these new developments. Besides the topics mentioned above, investigations are also continuing on other alternatives, such as cooling channels containing high pressure hydrogen gas [28], quadrupole-focused linear precooling channels, and using FFAG arcs in an RLA [29]. Given continued support, one can anticipate much more simulation progress in the future.

ACKNOWLEDGEMENTS

This work was supported by the U.S. Department of Energy under contract DE-AC02-98CH10886.

REFERENCES

1. Geer, S., Neutrino factories: physics potential, *J. Phys. G: Nucl. Part. Phys.* **29**, 1485-1492 (2003).
2. Alsharo'a, M. et al, Recent progress in neutrino factory and muon collider research within the Muon Collaboration, *Phys. Rev. Special Topics-Accelerators and Beams* **6**, 081001-1-52 (2003).
3. Autin, B. et al, The study of a European neutrino factory complex, CERN NuFact Note 103, 2001.
4. Mori, Y., Neutrino factory in Japan, *J. Phys .G: Nucl. Part. Phys.* **29**, 1527-1536 (2003).
5. Autin, B. et al, The acceleration and storage of radioactive ions for a neutrino factory, *J. Phys .G: Nucl. Part. Phys.* **29**, 1637-1647 (2003).
6. Autin, B. et al, NuFact02 machine working group summary, *J. Phys .G: Nucl. Part. Phys.* **29**, 1785-1795 (2003).
7. Prior, C., ISIS MW upgrade plans, talk at NuFact03 Workshop, New York, 2003.
8. Autin, B., Meot, F., and Verdier, A., Calculated life of π/μ beams in a decay channel (longitudinal motion), talk at NuFact03 Workshop, New York, 2003.
9. Autin, B., Funneling pion and muon beams, talk at NuFact03 Workshop, New York, 2003.
10. Paul, K., Decay channel optimization using MARS, talk at NuFact03 Workshop, New York, 2003.
11. Neuffer, D., New ideas in bunching and phase rotation, talk at NuFact03 Workshop, New York, 2003.
12. Iwashita, Y., Velocity compliant bunching scheme with amplitude modulation, talk at NuFact03 Workshop, New York, 2003.
13. Palmer, R.B., Ring coolers, *J. Phys.G: Nucl. Part. Phys.* **29**, 1577-1583 (2003).
14. Balbekov, V. et al, Muon ring cooler for the Mucool experiment, *Proc. 2001 Part. Accel. Conf.*, Chicago, IL, p. 3867-9.
15. Kahn, S., Tetra cooling ring, talk at NuFact03 Workshop, New York, 2003.
16. Makino, K., Balbekov (tetra) ring simulation results in COSY, talk at NuFact03 Workshop, New York, 2003.
17. Kirk, H., Quad/dipole ring coolers, talk at NuFact03 Workshop, New York, 2003.
18. Berg, J.S., et al, RFOFO cooling ring: simulation results, these proceedings.
19. Fernow, R.C. et al, Muon cooling in the RFOFO ring, MUC-NOTE-COOL-THEORY-273, April 2003. This series of technical notes can be found at (http://www-mucool.fnal.gov/mcnotes/).
20. Rees, G., et al, Muon front end studies at RAL, talk at NuFact03 Workshop, New York, 2003.
21. Palmer, R.B., Ring cooler kickers, talk at NuFact03 Workshop, New York, 2003.
22. Balbekov, V., Cooling of a compressed bunch in the RFOFO ring, MUC-NOTE-COOL-THEORY-276, June 2003.
23. Bogacz, A., Optimized beam optics for muon acceleration, talk at NuFact03 Workshop, New York, 2003.
24. Berg, J.S. et al, FFAGs for muon acceleration, *Proc. 2003 Part. Accel. Conf.*, Portland, OR.
25. Mori, Y., Neutrino factory in Japan, talk at NuFact03 Workshop, New York, 2003.
26. Machida, S., Lattice design and particle tracking, talk at NuFact03 Workshop, New York, 2003.
27. Palmer, R.B. et al, Acceleration costs, talk at NuFact03 Workshop, New York, 2003.
28. Johnson, R., Cooling using gas absorbers, talk at NuFact03 Workshop, New York, 2003.
29. Zisman, M., Neutrino factory R&D in the U.S., talk at NuFact03 Workshop, New York, 2003.

MuCool Results and Plans

Alan D. Bross

Fermi National Accelerator Laboratory, Batavia, IL 60510

For the MuCool Collaboration

Abstract. Recent results from the MuCool Collaboration's R&D program are described along with plans for the coming year. The MuCool collaboration consists of physicists and engineers from 18 institutions in the US, Europe, and Japan. Its mission is to design, prototype, and test all components of a muon ionization cooling channel with a goal to have the technique operationally established by the end of the decade. A detailed engineering design and well-formulated cost of the channel would then be available as input for a construction decision for a Neutrino Factory.

INTRODUCTION

Motivated by experimental evidence for neutrino oscillations and by the potential to reach an energy regime in the 2-4 TeV range with a lepton machine, a great deal of effort has addressed R&D on muon accelerators in recent years [1-4]. Neutrino oscillation data present the first experimental evidence of physics beyond the standard model. Neutrinos from the decay of an intense beam of muons in a storage ring [5], the so-called Neutrino Factory, provide quite possibly the ultimate tool for studying this new physics [6-7]. With a Neutrino Factory, long-baseline detectors can determine oscillation parameters to unprecedented accuracy. In addition experiments at near detectors can explore non-oscillation physics to new levels of precision with detector designs that would not be possible with conventional neutrino beams. A Neutrino Factory would be one of the steps on the road to a muon collider. A muon collider would provide the potential to explore the Higgs (or a Higgs-like object) in much the same way as LEP studied the Z.

Muon beams of sufficient intensity for a Neutrino Factory or a muon collider can only be produced into a large phase space (at least with conventional techniques). In order to utilize such beams, acceleration technologies with large aperture are required or the beam must be cooled. New acceleration techniques [8] do show promise to provide beams of sufficient intensity for a Neutrino Factory without the need for cooling, but a muon collider will require cooling.

The only technique that can cool a muon beam within its lifetime is ionization cooling [9-12]. In a muon ionization-cooling channel, a muon beam repeatedly passes through an absorber (ionization-loss section) followed by an rf cavity (re-acceleration section). Both the absorber and rf reside in a strongly focusing magnetic lattice. As the beam passes through the energy-absorbing medium, ionization loss results in a decrease in all three components of beam momentum. The resultant evolution of transverse beam emittance is given [9-11] approximately by:

$$\frac{d\varepsilon_n}{ds} \approx -\frac{1}{\beta^2}\langle\frac{dE_\mu}{ds}\rangle\frac{\varepsilon_n}{E_\mu} + \frac{1}{\beta^3}\frac{\beta_\perp(0.014)^2}{2E_\mu m_\mu L_R} \qquad (1)$$

Here, angle brackets denote mean value, β is the muon velocity in units of c, muon energy E_μ is in GeV, $\beta_\perp$ is the lattice beta function evaluated at the location of the absorber, m_μ is the muon mass in GeV/c^2, and L_R is the radiation length of the absorber medium.

We see from equation 1 that cooling (first term in the equation) is degraded by heating from multiple Coulomb scattering (second term in the equation). Looking at the heating term, we see that it is minimized by small $\beta_\perp$ and large L_R. The absorbers should therefore be located in a region of strong focusing and they should be made from a low-Z material (hydrogen being the optimum choice). Most design studies have chosen liquid hydrogen for the absorber material and have used superconducting solenoids for the magnetic lattice. As was mentioned above, rf cavities are used to provide the lost longitudinal momentum thus allowing the cooling process to be repeated many times. The acceleration provided by the rf cavities must not only be enough for momentum replacement, but should also allow for "off-crest" operation. "Off-crest" operation gives continual rebunching of the beam, so that even a beam with large momentum spread remains captured in the rf bucket. This implies that accelerating gradients in excess of 10 MV/m will be required. (Most cooling designs require gradients up to 16 MV/m). Even with gradients this large, the cavities dominate the length of a cooling channel. This is clearly evident from Figure 1, which shows the SFOFO cooling lattice design for the Muon Ionization Cooling experiment (see paper by Y. Torun in these proceedings)

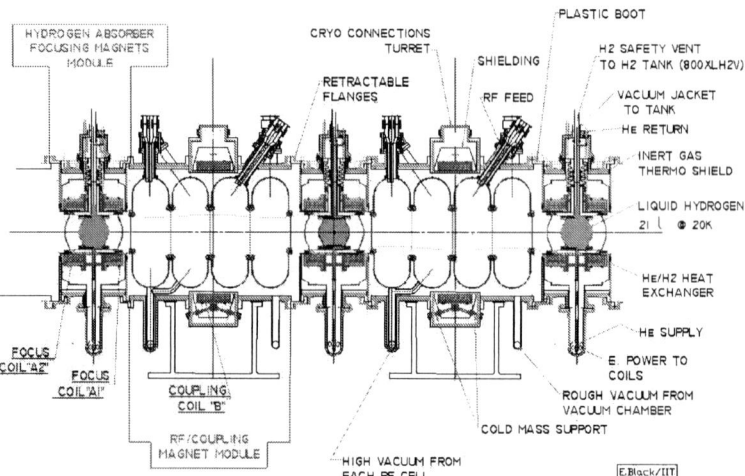

Figure 1. SFOFO Cooling lattice as currently designed for the Muon Ionization Cooling Experiment. Shown are the LH$_2$ absorbers, 201 MHz rf cavities, and the focus and coupling coils.

MUON COOLING COMPONENT DEVELOPMENT

As can be seen from the above discussion, an effective muon ionization cooling channel requires, 1) Low-Z absorbers (preferably hydrogen), 2) RF cavities with the highest possible accelerating gradient, and 3) Multi-Tesla solenoids. The research and development challenge therefore is to address the following questions:

1. Can the heat from dE/dx losses be adequately removed from the absorbers?
2. Can normal conducting rf cavities be built that provide the required accelerating gradient and operate reliably in multi-tesla fields?
3. Can the channel be engineered with an acceptably low thickness of non-absorber material in the aperture (absorber, rf, and safety windows)?
4. Can the channel be designed and engineered to be cost effective?

The MuCool collaboration [13] has ongoing activities in all these areas and plans to develop and test hardware devices that can be sed in a muon ionization-cooling channel.

RF Cavity R&D

Since cooling cell designs require the rf to operate in strong solenoidal magnetic fields, the use of superconducting cavities is precluded. Our work to date has focused on normal conducting 805 MHz cavities. Their higher frequency and thus small size (relative to the 201 MHz cavities which have been chosen for the lattice shown in Figure 1) allows for much less expensive testing. Since the very nature of a muon ionization-cooling channel requires material in the beam, rf cavities with carefully designed windows should be permissible. This closed-cell cavity or "pillbox" configuration approximately doubles the on-axis accelerating field, for a given surface field in the cavity, over that of an open-cell design. However in order to not degrade the cooling performance of the channel, the thickness of the window in radiation lengths must be small compared to that of the absorber. We have used the Lab G rf test facility at Fermilab for cavity tests. This facility consists of an rf "cave", an 805 MHz rf power source (~14 MW peak power, 5 Hz repetition rate with a pulse length up to 50 μs), and a superconducting solenoid with a 44 cm bore. The solenoid was fabricated with two separate coils that can be operated in solenoid mode (coils in series) or in gradient mode (coils in opposition) [14]. In solenoid mode the magnet can produce a 5 T on-axis field. In gradient mode the field varies from +3.5 T to –3.5 T.

6-Cell Open-Cell Cavity Tests

Our first tests were done with a 6-cell open-cell cavity. After approximately 6 weeks of conditioning, the cavity operated at a maximum surface field of 54 MV/m (on-axis ~ 25 MV/m). Large dark currents were observed at high accelerating gradient and they increased with the application of B field (2.5 T maximum). The cavity emission data were fit to the Fowler-Nordheim expression and indicated that the dark current scaled approximately as E^{10} [15] at high accelerating field E. Conditioning without magnetic field was successful, although at the highest gradient the dark currents were quite high. Conditioning with magnetic field was more problematic. Dark currents increased substantially with the application of field and this in combination with focusing effects actually caused two failures of the titanium vacuum windows. Although rf cavities operating at the highest dark current levels observed in these tests would not be acceptable for a cooling channel, nitrogen conditioning and surface treatments can improve this situation significantly. Nitrogen conditioning was applied to this cavity for a period of 2 weeks and resulted in a dark current reduction by 2 orders of magnitude. Cavity geometry will also play a major role with respect to dark currents. In this open cell cavity, the surface field was over two times the accelerating field. As was mentioned above, in a pillbox cavity the surface field is roughly equal to the accelerating field. So with respect to dark currents, a pillbox cavity should offer (for constant accelerating field) a significant advantage over an open-cell cavity.

Pillbox Cavity Tests

Tests with a single-cell pillbox cavity are continuing, but our initial data are somewhat encouraging. With no B field the cavity reached 34 MV/m with little sparking and very low dark currents. The initial tests were first done with thick (0.200") copper windows (taken to the 34 MV/m mentioned above). The cavity was then conditioned to 24 MV/m with thin (0.015") copper windows. Operation in both cases was similar with B=0. As in the case of the open-cell cavity, when the magnetic field was applied the dark current increased substantially. Sparking rates also increased with application of the B field. Figure 2 shows the dark current data for the

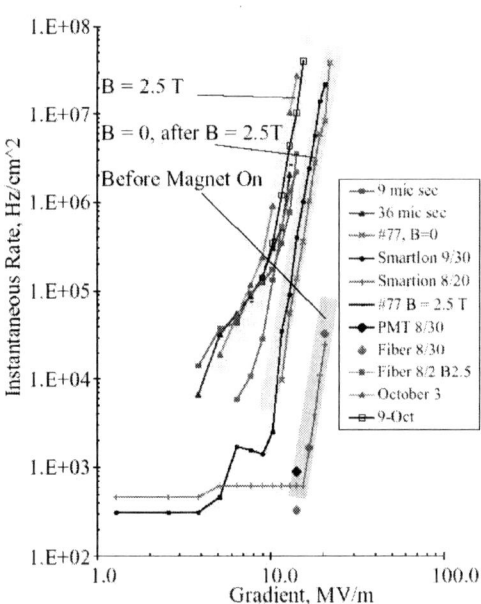

Figure 2. Dark current rates with thin copper windows installed in the pillbox cavity

thin copper window case before, during, and after B field application. It is obvious from these data that some permanent damage did occur when operating with magnetic field. However after extensive additional conditioning over a period of a month, the B=0 dark currents returned to the levels seen before application of the B field. When the copper windows were removed, considerable pitting was observed in addition to numerous areas where molten copper had been deposited.

The next series of tests were performed with TiN-coated Be windows (0.01" thick). The cavity was conditioned in the B-field-off configuration and reached stable operation at 21 MV/m. With B field on, the cavity conditioned to 17 MV/m, but with time the cavity experienced sparking damage, dark currents rose, and stable operation could then be maintained only up to 14 MV/m. The Be windows were then removed for inspection/analysis. What was observed was that the windows appeared to be dusted with copper. There appeared to be little or no damage to the Be or TiN coating, however. This copper contamination came from arcing to the surfaces of the cavity (iris, flange surfaces) which then deposited molten copper onto the Be windows. It is clear from the above discussion that operation of high-gradient rf cavities in high magnetic field can be problematic if one wants routine and stable operation at accelerating gradients greater than 12-14 MV/m. However cavity surface preparation appears to be key in reducing dark current problems and surface coatings such as TiN (applied to both the cavity body and windows) offer the potential for significant improvement.

Although pre-stressed flat windows have been used to date in our studies, other options such as a tube-grid or domed designs are also being considered. The tube-grid window design is being studied using finite element techniques that are being applied to an electromagnetic model of our 805 MHz cavities. Tube-grid designs may provide an attractive alternative to solid rf windows particularly in the case of 201 MHz cavities (see below) which will require windows with approximately 20 cm radius. The critical parameter in the tub-grid design is the maximum surface field enhancement at the tubes. Of the configurations studied to date, a 6 × 6 grid with 1 cm diameter tubes has given the best performance. A surface field enhancement of 1.4 was obtained in this case. Figure 3 shows a section of the 6 × 6 grid design from the simulation.

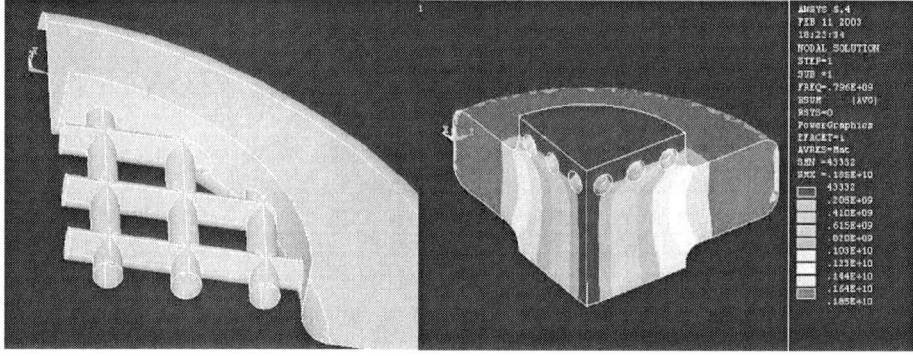

Figure 3. Output from the tube-grid rf window simulation

201 MHz Cavity Design

The electrical and mechanical design for a 201 MHz closed-cell cavity has now been completed [16]. The design allows for detachable windows so that operation with either a solid window or a tube-grid structure can be tested. This prototype will also be a test bed for various surface treatment techniques. Analysis of the design indicates that a peak surface field of 26.5 MV/m should be obtainable (at least in B=0 operation). Figure 4 shows the overall mechanical design.

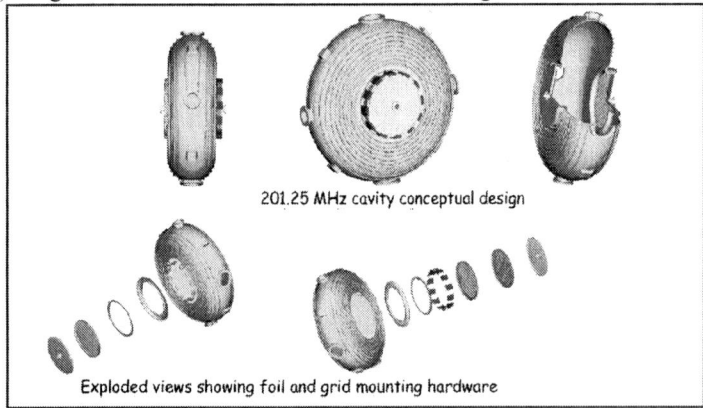

Figure 4. Engineering rendition of 201 MHz cavity

High-Power Liquid-Hydrogen Absorber R&D

The development and test of high-power-handling liquid hydrogen (LH$_2$) absorbers with thin windows are key components of the MuCool R&D program [17]. High-power handling is of paramount importance. For example, in a cooling channel driven with a 4 MW proton driver, the muon beam power dissipation in the LH$_2$ absorber is approximately 1 KW. (In a ring cooler, the power dissipation can be as high as 10 KW). Two possible approaches for removing the heat from the absorbers are being investigated: a "conventional" flow-through design with external heat exchanger, similar to that used for high-power LH$_2$ targets, and a convection-cooled design, with internal heat exchanger built into the absorber vessel. The convection design has desirable mechanical simplicity and minimizes the total hydrogen volume in the cooling channel (a significant safety concern), but is expected to be limited to lower power dissipation than the flow-through design. Figure 5 gives details for the convection-cooled absorber design. To study and optimize the fluid mixing and heat transfer properties of these absorber designs, we have been exploring ways to visualize the flow patterns and temperature distributions within the fluid [18]. This involves Schlieren and Ronchi techniques and has been tested successfully at Argonne National Laboratory using a 20 MeV electron beam and a water target. Two and three dimensional finite element analysis and two dimensional computational fluid dynamics modeling is also being used to understand the performance of both the convective and flow-through LH_2 absorber designs.

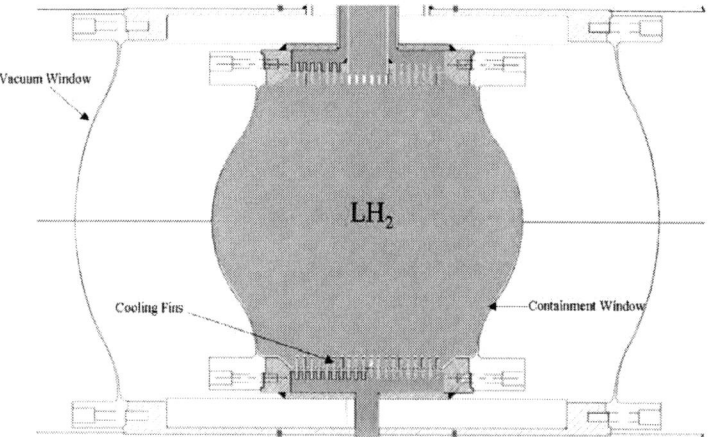

Figure 5. Detail of LH_2 absorber design showing vacuum and absorber windows and the heat exchanger cooling fins.

The windows required in any LH_2 absorber design do degrade the cooling channel performance; thus it is critical that scattering in the window material be minimized. The absorber requires both a set of containment windows and, for safety, vacuum windows. We have tested windows fabricated with 6061-T6 aluminum that utilize a tapered torispherical shape. These windows had an on-axis thickness of approximately 240 μm and passed burst and cryogenic testing. Various techniques have been used to evaluate the shape and performance of fabricated windows and included measurements made with both Coordinate Measuring Machines and with photogrammetry. After our initial window tests, calculations aimed at optimizing the window shape showed that an inflected "thinned-bellows" shape will allow for an on-axis window thickness of only 132 μm (6061-T6 aluminum). The window design shown in Figure 5 incorporates this inflected shape. We are currently fabricating windows with this new shape (in 6061-T6 aluminum) and will perform burst and cryogenic tests on them as was done on the torispherical-design windows. We are also considering the use of a lithium-aluminum alloy such as 2195. This material is much stronger than 6061-T6 so the windows could, in principle, be made even thinner with this material. New fabrication techniques will likely be needed, however, in order to take full advantage of this alloy. Since the radiation hardness/stability of the 2195 alloy is not fully documented, we will be performing radiation tests on this material in order to study its mechanical properties as a function of dose.

MuCool Test Area

In order to effectively test all components of a muon-cooling lattice, a new test facility, the MuCool Test Area (MTA), is being constructed at Fermilab. The MTA is located near the end of the Linac 400 MeV beam line. Figure 6 shows how the MTA will appear when it is completed in the fall of 2003. This location will provide access

Figure 6. MuCool Test Area at Fermilab.

to multi-megawatt rf power (at both 805 and 201 MHz) and eventually to the linac beam. The facility is being designed to accommodate the full linac beam of approximately 2.4×10^{13} p/s. The experimental hall is approximately 20' by 40' and can thus easily accommodate the relatively large components of a cooling cell. The MTA will have cryogenic infrastructure [19] that will allow for the operation of LH_2 absorbers and superconducting magnets. High-power studies of LH_2 absorbers will be possible (full beam intensity corresponds to approximately 600 W power deposition in a 35 cm long LH_2 absorber). High power rf testing under intense beam conditions will also be possible. All components used in a cooling channel will thus be able to be tested at high beam power. Beneficial occupancy is expected in the fall of 2003. The first tests scheduled for the area are operational studies of a convectively-cooled LH_2 absorber.

The MuCool collaboration is also developing beam line instrumentation and infrastructure for use in the MTA. Beam profile monitors using bolometry have been tested and have shown promising results. The concept of using CVD diamond detectors is also being investigated for use as profile monitors. A prototype test of this concept is scheduled for the fall of 2003. Given the harsh radiation environment for tests in the MTA, new rad-hard temperature, pressure, and strain/stress sensors are being evaluated for use in the MTA along with DAQ systems that can operate under strict hydrogen safety guidelines.

PLANS FOR THE COMING YEAR

During the coming year, the MuCool collaboration plans to continue rf studies using 805 MHz test cavities. Our plan is to investigate the effect of surface treatments and window materials on dark current and breakdown. Simulation work on grid-tube designs for rf cavity windows will also continue and prototype windows of the final design will be tested using the 805 MHz pillbox cavity. New tests with inflected "thinned-bellows" windows for LH_2 absorbers will fully study their mechanical and cryogenic properties. In addition to the convective LH_2 absorber tests in the MTA mentioned above, we will begin construction of a LH_2 absorber using the flow-through design that will also be tested in the MTA. Construction of the first 201 MHz RF cavity should be completed by the end of the year and we hope to have it under test in the MTA in the Fall of 2004.

ACKNOWLEDGMENTS

This work was supported in part by the U.S. Department of Energy, the National Science Foundation, the Illinois Board of Higher Education, and the US-Japan Agreement on High Energy Physics.

REFERENCES

1. M. M. Alsharo'a et al., '"Status of Neutrino Factory and Muon Collider Research and Development and Future Plans," FNAL-PUB-02/149-E (July 19, 2002), Phys. Rev. ST Accel. Beams, arXiv:hep-ex/0207031.
2. See e.g. Proceedings of the NuFact Workshop series: NuFact'99, B. Autin, ed., Nucl. Instrum. Meth. **A451** (2000) 1-388; NuFact'00, S. Chattopadhyay, ed., Nucl. Instrum. Meth. **A472** (2001) 323-666; Proc. NuFact'02, K. Long ed., J. Phys. G: Nucl. Part. Phys. **29** 1679-1681.
3. "Feasibility Study on a Neutrino Source Based on a Muon Storage Ring," D. Finley, N. Holtkamp, eds. (2000), http://www.fnal.gov/projects/muon_collider/reports.html.
4. "Feasibility Study-II of a Muon-Based Neutrino Source", S. Ozaki, R.Palmer, M.Zisman, and J.Gallardo, eds., BNL-52623, June 2001, http://www.cap.bnl.gov/mumu/studyii/FS2report.html.
5. S. Geer, Phys. Rev. **D 57**, 6989 (1998); earlier versions of a Neutrino Factory, considered by e.g. D. G. Koshkarev, report CERN/ISR-DI/74-62 (1974), S. Wojicki (unpublished, 1974), D. Cline and D. Neuffer, AIP Conf. Proc. **68**, 846 (1981), and D. Neuffer, IEEE Trans.Nucl. Sci. **28**, 2034 (1981), were based on pion injection into a storage ring and had substantially less sensitivity.
6. C. Albright et. al., Fermilab-FN-692 (May, 2000), hep-ex/0008064; M. Apollonio et al., CERN-TH-2002-208 (Oct. 2002.), hep-ph/021019; M. Lindner, hep-ph/0209083 (2002).
7. See e.g. M. Pluemacher, in Proc. NuFact'02, op. cit., and references therein.
8. "A Feasibility Study of A Neutrino Factory in Japan," Y. Kuno, ed., available from http://www-prism.kek.jp/nufactj/index.html; Y. Mori, "Review of Japanese Neutrino Factory R&D," A. Sato, "Beam dynamics studies of FFAG," D. Neuffer, "Recent FFAG studies," S. Machida, "Muon Acceleration with FFAGs," and C. Johnstone, "FFAG with high frequency RF for rapid acceleration," in Proc. NuFact'02, op. cit.
9. A. N. Skrinsky and V. V. Parkhomchuk, Sov. J. Part. Nucl. **12**, 223 (1981); E. A. Perevedentsev and A. N. Skrinsky, in Proc.12th Int. Conf. on High Energy Accelerators, F. T. Cole and R. Donaldson, eds. (Fermilab, 1984), p 485; R. Palmer et al., Nucl. Phys. Proc. Suppl. **51A**, 61 (1996).
10. D. Neuffer, Part. Acc. **14**, 75 (1983).
11. D. Neuffer, Proceedings of COOL03, to be published
12. Introductory discussions of muon ionization cooling may be found in [11] and D. M. Kaplan, "Introduction to muon cooling," in Proc. APS/DPF/DPB Summer Study on the Future of Particle Physics (Snowmass 2001)}, N. Graf, ed., arXiv:physics/0109061 (2002). More detailed treatments may be found in D. Neuffer, "$\mu^+\mu^-$ Colliders," CERN yellow report CERN-99-12 (1999), K. J. Kim and C. X. Wang, Phys. Rev. Lett. **85**, 760 (2000), and Chun-xi Wang and Kwang-Je Kim, "Linear Theory of Ionization Cooling in 6D Phase Space," Phys. Rev. Lett. **88** 184801 (2002).
13. See http://www.fnal.gov/projects/muon_collider.
14. M.A. Green, J.Y. Chen, S.T. Wang, "A test of a Superconducting Solenoid for the MUCOOL RF Experiment," IEEE Transactions on Applied Superconductivity **11** (2001), 2296
15. J. Norem et al., Phys. Rev. ST Accel. Beams **6**, 7, (2003), 2001.
16. D. Li et al., in Proc. NuFact'02, op. cit.
17. M. A. C. Cummings et al., "Absorber R&D in MUCOOL," in Proc. NuFact'02, op. cit., also see M. A. Cummings et al., these proceedings.
18. J. Norem et al., "Measurement of Beam Driven Hydrodynamic Turbulence," submitted to Proc. 2003 Particle Accelerator Conference, Portland, OR (2003).
19. C. Darve et al., "The Liquid Hydrogen System for the MuCool Test Area", submitted to Proc. Cryogenic Engineering Conference 2003, Anchorage, AK (2003).

MICE Status

Yağmur Torun[†]

Illinois Institute of Technology, Chicago, IL 60616, USA
For the MICE Collaboration

Abstract. The MICE Collaboration has designed an experiment in which a section of an ionization cooling channel is exposed to a muon beam. This channel includes liquid-hydrogen absorbers providing energy loss and high-gradient rf cavities to re-accelerate the particles, packed in a solenoidal magnetic channel. It reduces the beam transverse emittance by >10% for muon momenta between 140 and 240 MeV/c. Spectrometers placed before and after the cooling section perform the measurements of beam transmission and emittance reduction with an absolute precision of ±0.1%.

INTRODUCTION

A Neutrino Factory based on a muon storage ring is the ultimate tool for studies of neutrino oscillations, including possibly the discovery of leptonic CP violation [1, 2, 3]. It is also the first step towards a $\mu^+\mu^-$ collider. Ionization cooling of muons has never been demonstrated in practice but has been shown by end-to-end simulation and design studies to be an important factor for both performance and cost of a Neutrino Factory. This motivates an international program of R&D, including an experimental demonstration. The aims of MICE, the International Muon Ionization Cooling Experiment are:

- To show that it is possible to design, engineer and build a section of cooling channel capable of giving the desired performance for a Neutrino Factory
- To place it in a muon beam and measure its performance in various modes of operation and beam conditions, investigating the limits and practicality of cooling

A proposal [4] has been submitted to Rutherford Appleton Laboratory (RAL) to mount the experiment at ISIS.

EXPERIMENT LAYOUT

The main components of MICE are outlined in Fig. 1. Cooling is provided by one lattice cell from the 201 MHz cooling channel of "Study-II" [5] with some components modified for cost savings and compliance with RAL safety requirements. The incoming muon beam first encounters diffusers to generate a large tuneable input emittance. In this section, a precise time measurement and particle identification are performed. Next comes an input spectrometer consisting of tracking devices within a uniform-field solenoid to measure the phase space coordinates of each particle. This is followed by the cooling section, with hydrogen absorbers, rf cavities and superconducting coils. One

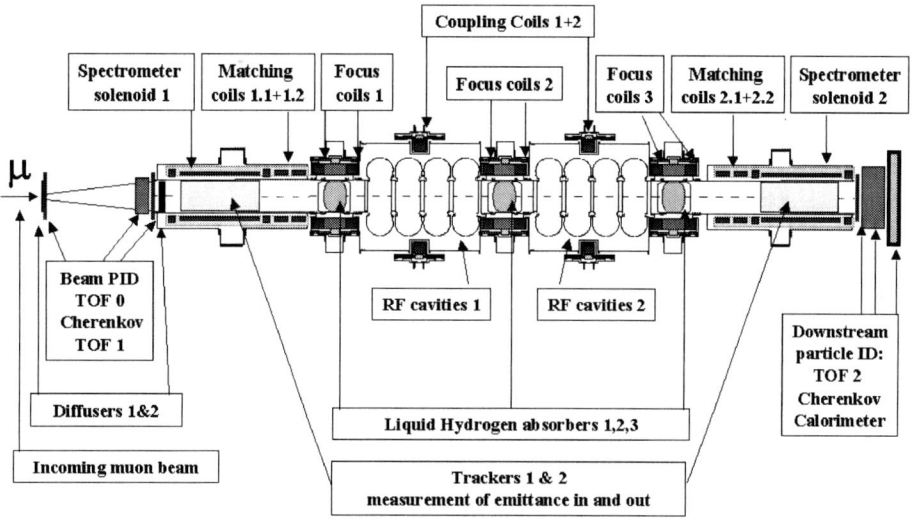

FIGURE 1. MICE layout.

additional absorber finishes the cooling section, both for symmetry and to protect the trackers against dark currents emitted by the rf cavities. Outgoing tracks are measured in a second spectrometer, identical to the first one. At the downstream end of the experiment, another time-of-flight (TOF) measurement is performed, and particle identification by means of a Cherenkov counter and a calorimeter eliminates muons that have decayed in the apparatus. To avoid emittance growth, the magnets in these two cells are matched to the spectrometer solenoids using two sets of matching coils.

MEASUREMENT TECHNIQUE

To allow precision measurement of transmission and emittance, one muon at a time will be tracked through the apparatus and detected using standard particle-physics techniques, which are much more precise than those typically used in beam instrumentation. A "virtual bunch" formed in offline analysis will be used to demonstrate how an actual bunch would have behaved had the beam intensity been orders of magnitude higher. Momentum measurement requires a magnetic spectrometer. Ease of matching into and out of the cooling section and the need to keep a large-emittance beam in a small physical volume has led to the choice of solenoid magnets on each side of the cooling channel. Each detector measures, at given z positions, the coordinates x and y of every incident particle, and the time. Momentum and angles are reconstructed by using several measurement planes. For the experimental resolution not to affect the emittance measurement significantly, the rms resolution of the measurements must be better than about 10% of the rms beam size at the equilibrium emittance in each of the six dimensions. An essential aim of MICE is to measure the equilibrium emittance precisely. For each incident particle it will be possible to determine whether it was lost in the channel or went through successfully. Therefore, losses can be separated clearly from cooling. Except for possible collective effects such as space charge, this technique is equivalent to full-beam measurements, but offers several advantages. Correlations between parameters can be easily measured. The role of each beam parameter (energy, transverse momentum, rf phase, etc.) can be studied using selection cuts in the ensemble of tracks without making changes to the beam parameter settings. Software cuts based on the incoming beam make it possible to derive a variety of results with different input beam conditions from a single data set. Any desired input beam conditions can be reconstructed by appropriate weighting or culling of the observed particles.

POSSIBLE EXPERIMENTS AND TIMELINE

Many different cooling experiments can be performed with the proposed apparatus. First, beam momentum can be varied since the magnets have been designed to allow exploration of momenta as high as 240 MeV/c. The Super-FOFO lattice used here has the property that the beta function at the absorber can be changed by adjusting the currents in the focusing and coupling coils [6]. Different rf voltages and phases can also be used. Another important part of the experimental program will be testing various absorbers. It will be straightforward to replace the liquid hydrogen with liquid helium. The mechanical assembly of the liquid hydrogen absorbers will also allow replacement of one of the absorber windows by a structure supporting solid absorbers.

Since all detectors and parts of the equipment will not be ready at the same time, one can foresee a staged development of the experiment, leading to the scenario shown in Fig. 2. First (step I), the beam can be tuned and characterized using a set of TOF and particle ID detectors. In step II, the first spectrometer solenoid allows a first measurement of 6D emittance with high precision and comparison with the beam simulation. This should allow a systematic study of the tracker performance. In step III, the two spectrometers work together without any cooling device in between which allows the study of systematic errors. Step IV, with one focusing pair between the two spectrometers, should provide experience with operating the absorber and a precise understanding of energy loss and multiple scattering in it. Several experiments with varying beta-functions and momenta can be performed with observation of cooling in normalized emittance. Starting from step V, the real goal of MICE, which is to establish the performance of a realistic cooling channel, will be addressed. Only with step VI will the full power of the experiment be reached.

COOLING CHANNEL

The MICE magnetic channel consists of seven magnet assemblies composed of eighteen superconducting solenoid coils spread over a length of nearly 11.5 m. The baseline MICE channel operates with muons at an average momentum $p=200$ MeV/c and $\beta=42$ cm at the center of the absorber. Eight 201-MHz rf cavities, in two 4-cavity assemblies,

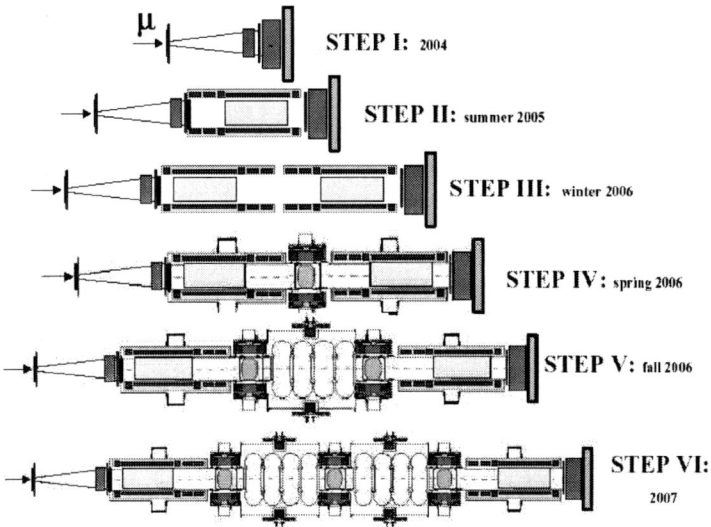

FIGURE 2. Six possible steps in the development of MICE

are needed in the cooling section. Due to the (financial) limitation of having only 8 MW of rf power available, the MICE cavities will operate at a gradient of about 8 MV/m (compared with the 16 MV/m specification for Study-II). The cavity shape chosen is based on a slightly reentrant rounded profile with a large beam aperture and a small nose cone. To achieve high shunt impedance, the beam aperture is terminated electromagnetically using thin beryllium foils or thin-walled Al tubes. Hydrogen was chosen as the most suitable absorber material because of its large ionization energy-loss rate ("cooling") and small probability of multiple scattering ("heating"). For more information on absorber and rf cavity development, see [7].

DETECTORS

The driving design criteria for the MICE detector systems are robustness, in particular of tracking detectors, to potentially severe background conditions in the vicinity of rf cavities and redundancy in particle identification (PID) to keep contamination below 1%. Two technologies (scintillating fiber and TPC) are being vigorously pursued for the tracker design and prototyping and testing is in progress for both. Additional detectors will provide redundant particle identification to eliminate from the sample any residual pions in the incoming beam or muons that decay within the apparatus. These include time-of-flight scintillation counters, Cherenkov detectors and a calorimeter. While these are standard ingredients for particle-physics experiments, measuring an emittance ratio with 0.1% precision has never been done and will require careful design of diagnostics and attention to system integration and calibration.

Tracker

In the absence of severe background, three precise measurement planes can provide adequate resolution in transverse and longitudinal momentum. The tracking detectors are required to have low mass, to avoid photon conversions, and enough redundancy to separate signal hits from a potentially large background of photon hits.

The baseline design is five sets of scintillating fiber planes per tracker, deployed in three stereo views, with the fibers individually read out using cryogenic VLPC photodetectors. Prototypes of this type of detector have already been exposed to rf cavity backgrounds in Lab G at Fermilab, and the performance estimate as a function of background rate, as well as the cost estimate, is reliable. The scintillating fiber tracker can operate effectively at very high background rates, is immune to electromagnetic interference, and poses little risk when operated near liquid hydrogen. The five planes will be supported in a rigid tube so that the assembly can be moved in and out of the spectrometer solenoid as a whole. Each plane consists of three sets of fiber doublets at 120° to each other. This structure gives efficient space-point reconstruction, since hits in any two of the three doublets suffice. Each doublet consists of two overlapping layers of round fibers, giving 100% geometrical coverage over the face of the detector and an approximately uniform thickness of scintillator, independent of where a particle crosses the doublet. This structure also suppresses the effects of background photons since a photon interaction is not likely to produce a triplet hit in a fiber plane. Scintillation light from the fiber will be piped using clear fiber light guides to VLPCs (Visible Light Photon Counter), which were chosen for high quantum efficiency (85%) and gain (50,000). This design is similar to the D0 Fiber Tracker's [8] VLPC system. To keep the occupancy low, the smallest fiber that yields enough light for efficient tracking will be used and 350 μm fiber diameter appears to be a good match. Each plane has an active area defined by a 30-cm-diameter circle. Given a fiber-ribbon pitch to diameter ratio of 1.2, a total of 42900 fibers are required to instrument all planes. If backgrounds can be kept low, the cost and channel count can be reduced by multiplexing the fibers. Seven fibers can be multiplexed into one VLPC pixel with six fibers arranged in a ring around a central one resulting in only about 10% light loss. This would be provided by a 7:1 multiplexing wave guide, to retain the capability of reading out each fiber individually. Should there be a need to cope with higher background situations at a later stage, only wave guides would be changed, not the tracker itself.

An alternative design is also under study in which each spectrometer contains a time projection chamber with GEM amplification (TPG) [9]. This is a promising technique with low material budget and excellent pattern recognition capability. The active gas volume is 30cm diameter by 1m long. Ionization electrons from muon tracks drift along the electric field (parallel to the magnetic field) and are amplified by a factor 10^4 to 10^5 in a set of three gaseous electron multipliers (GEMs) [10]. The readout plane is a hexaboard, made of hexagonal pads arranged to form three sets of strips oriented at 120° to each other with 500 μm pitch. To minimize multiple scattering and photon conversions a low-Z gas is preferred for the TPG and a He-based mixture like the one used in KLOE [11] appears suitable. For a drift velocity of 1.7 cm/μs and 500 ns sampling interval, each muon track would be sampled 120 times.

Particle Identification

Three TOF stations equipped with fast scintillators are foreseen. The first two, upstream of the cooling section and separated by about 10 m, will provide the basic trigger for the experiment, in coincidence with the ISIS clock. These will provide muon identification as well as the phase information (relative to the rf) necessary for the measurement of the input longitudinal emittance. The coincidence with a third station of similar nature, downstream of the second measuring station, will select particles traversing the entire cooling section. The variation of emittance due to losses and decays will thus be distinguishable from cooling. The TOF stations have transverse dimensions of 12×12, 40×40 and 40×40 cm^2 respectively and are 2.5cm thick. A 70ps resolution provides both effective (99%) rejection of beam pions and adequate precision in the measurement of the muon rf phase. The two large stations (TOF1, TOF2) are equipped with 8 scintillator slabs made of Bicron BC-404 scintillator material (with 1.5 ns decay constant and 1.7 m attenuation length). The smallest station (TOF0) could be made of two crossed planes (X-Y), each of two slabs, using BC-420 plastic scintillator, which is faster than BC-404 but has a shorter attenuation length. Each slab is read out at both ends by a fast photomultiplier through a Plexiglas light guide. The fringe field from the spectrometer solenoids can be as high as 1T at the PMT locations without shielding. The choice of PMTs and magnetic shielding for TOF1 and TOF2 are therefore critical and under study.

In a small ($\sim$1%) fraction of events, a muon decays inside the cooling section or one of the spectrometers. Kinematics cuts can reject about 80% of the resulting decay electrons, but this is not sufficient to avoid a bias in the emittance measurement. Dedicated detectors are needed to separate electrons from muons. The strategy is as follows: positively identify muons by requiring low and longitudinally uniform energy deposition in an electromagnetic calorimeter at the very end of the experiment; reject any residual background of electrons in the muon sample defined by the calorimeter by means of a threshold Cherenkov detector immediately upstream of the calorimeter.

A small radiator will be used in the upstream Cherenkov detector to separate beam muons from pions and electrons of the same momentum on the basis of the difference in light yield. This will reduce any background from electrons and pions that might be left after the time-of-flight selection has been made, for example, due to overlaps. A single 125 mm diameter cell of liquid C_6F_{14} with a quartz window is proposed. An air light guide and 45° mirror would bring the light out to a single 125 mm photomultiplier tube. The total length of the device is 30 cm. The index of refraction of C_6F_{14} is 1.25 and it has thresholds of 0.7, 140, and 190 MeV/c for electrons, muons, and pions, respectively. Discrimination is achieved by pulse height analysis [12].

A high-resolution electromagnetic Pb-scintillating fiber calorimeter, of the type built by KLOE [13], is proposed. It is to be built by gluing 1-mm-diameter blue scintillating fibers between 0.5-mm-thick grooved lead plates, with a uniform and symmetric Pb-scintillating fiber structure, with a fiber spacing of 1.35 mm. When layers are superimposed, fibers are located at the vertices of adjacent quasi-equilateral triangles, forming a homogeneous and compact structure with a fiber:lead:glue volume ratio of 48:42:10. The resulting composite, with a density of 5g/cm^3, a Moliere radius of 3.5cm, radiation length of 1.5cm and a sampling fraction of 15% for a minimum ionizing particle, gains considerable stiffness, and can be easily machined to the shape required for the final assembly. This 'spaghetti' design offers the possibility of fine sampling and results in optimal lateral uniformity of the calorimeter. Fibers run mostly transversely to the particle trajectories, reducing sampling fluctuations due to channeling, i.e., showers developing along the fibers' direction, an effect particularly important at the low energies of interest. Finally, the very small lead foil thickness ($< 0.1 X_0$) results in a quasi-homogeneous structure and a high efficiency for minimum ionizing particles and low energy electrons. The lead and fiber planes are perpendicular to the beam axis, structured in four layers.

Further electron rejection will be provided by the downstream Cherenkov detector. In the momentum range of interest to MICE, aerogel ($1.01 < n < 1.06$) appears to be the only adequate radiator from which to build a threshold Cherenkov blind to the passage of muons. The choice of the index of refraction for the radiator is governed by the relative light yields of electrons and muons, and their respective detection efficiencies assuming a fixed detection threshold. The goal is to maximize the response to electrons while minimizing possible contributions from higher energy muons. The proposed aerogel radiator has an index of refraction $n = 1.02$, a total thickness of 10 cm and a transverse size of 90×90 cm^2. A plane mirror, tilted at 45°, reflects light at 90° to the beam axis towards 20-inch-diameter high gain PMTs (same as those used in Super-Kamiokande), well matched to the low light yield of aerogel.

RF BACKGROUND

The layout described above has one major drawback: the detectors will be exposed to a large dark current and x-ray background generated by the nearby high-gradient rf cavities. The understanding of this problem is well underway [14, 15]. The basic mechanism is illustrated in Fig. 3: low energy electrons ('dark current') produced by field emission are accelerated across the cavities and generate bremsstrahlung photons whenever they hit material. Several factors contribute to protect the tracking detectors: i) the rf cavities will be operated at a moderate gradient of 8.3 MV/m, due to the limited availability of rf power; ii) most dark-current electrons are deflected by the field flips; iii) the electrons must also pass through the liquid-hydrogen absorbers, which are thick enough to absorb them completely, letting through x-rays only; iv) the detectors are built of low-Z material and are well able to distinguish muon hits from those generated by x-rays.

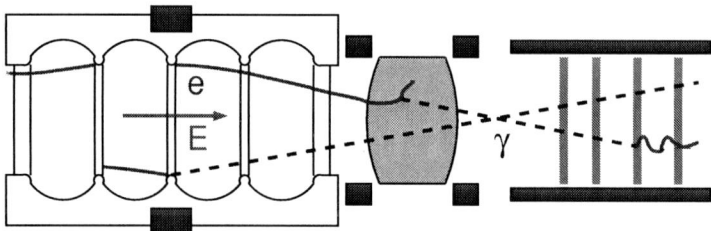

FIGURE 3. Backgrounds from rf cavities

STATUS

The MICE Collaboration has brought together 141 physicists and engineers from the world's accelerator and particle physics communities to tackle the technical challenges of ionization cooling. Together, they have designed an experiment to demonstrate the feasibility of muon cooling and, with enthusiastic support from the UK particle physics community, shown that it can be carried out at Rutherford Appleton Laboratory. By measuring the parameters of each muon individually, MICE will measure cooling in each transverse plane with an absolute precision of 0.1%. The proposed cooling section will be operable with a variety of optical settings and absorber materials, allowing the cooling performance to be mapped out for a range of cooling-channel parameters and beam momenta and compared with the predictions of detailed simulations. By demonstrating that the technology of muon ionization cooling is not only technically feasible, but that its cost and performance are well understood, MICE will pave the way for the start of a Neutrino Factory construction project and will point the way to muon colliders in the longer term. The proposed schedule for the commissioning and operation of MICE will establish the technical feasibility of muon ionization cooling by 2007; we are seeking funding from agencies around the world to realize this schedule.

ACKNOWLEDGMENTS

The author's work was supported by the US Department of Energy and the Illinois Board of Higher Education.

REFERENCES

1. C. Albright *et al.*, arXiv:hep-ex/0008064
2. M. Lindner, arXiv:hep-ph/0209083
3. S. Geer, J. Phys. G 29(2003), 1485
4. Proposal to the Rutherford Appleton Laboratory, An International Muon Ionization Cooling Experiment, MICE-Note 21, http://mice.iit.edu/mnp/MICE0021.pdf
5. Feasibility Study-II of a Muon-Based Neutrino Source, S. Ozaki, R. Palmer, M. Zisman, and J. Gallardo, eds. BNL-52623, June 2001
6. R. B. Palmer and R. C. Fernow, 200 MHz Cooling Experiment Design, MICE-Note 18, http://mice.iit.edu/mnp/MICE0018.pdf
7. A. Bross, MuCool Results and Plans, these proceedings
8. S. Abachi *et al.*, Nucl. Instrum. and Meth. A408(1998)103-109; A. Bross *et al.*, Nucl. Instrum. and Meth. A477(2002)172
9. F. Sauli, Nucl. Instrum. and Meth. A505(2003)195
10. F. Sauli, Nucl. Instrum. and Meth. A386(1997)531
11. M. Adinolfi *et al.*, Nucl. Instrum. and Meth. A482(2002)364
12. D. Bartlett *et al.*, Nucl. Instrum. and Meth. A260(1987)55
13. The KLOE Detector Technical Proposal, LNF-92/019
14. J. Norem *et al.*, J. Phys. G 29(2003) 1697
15. J. Norem *et al.*, Phys. Rev. ST Accel. Beams 6, 072001

The MuScat Experiment - Status and Plans

Rob Edgecock

Rutherford Appleton Laboratory

Abstract. The MuScat experiment is making a precise measurement of the multiple scattering of muons at 180 MeV/c. The data will be compared to a variety of models of multiple scattering and used as input to cooling studies for a Neutrino Factory and Muon Collider and to the MICE experiment.
  The second and last run of MuScat took place at TRIUMF in the summer of 2003. This paper will summarise the experiment and give the status of the analysis of these data.

INTRODUCTION

A vital requirement of the accelerator complex used to supply muons to a future Neutrino Factory or Muon Collider is the ability to cool the muons. In the case of the Neutrino Factory, the aim is about a factor of 10 reduction in the normalised emittance, while for a muon collider the requirement is a 10^6 reduction in the 6-dimensional phase space. Due to the muon lifetime, such cooling needs to be fast and the currently prefered technique is ionisation cooling [1]. In the case of the transverse cooling required for a Neutrino Factory, this involves passing the muons through an absorber in which they lose both longitudinal and transverse momentum. The lost longitudinal momentum is then restored using rf cavities following the absorber.

However, as well as a cooling effect coming from the ionisation energy loss, there is heating coming from multiple scattering and the final cooling achieved is a delicate balance between these. Theory suggests this balance is most favourable for elements with low atomic number, in particular, liquid hydrogen. However, an extensive literature search has failed to find any measurements of the muon scattering distribution in light elements [2]. The most relevant data found comes from the scattering of 2.7 MeV/c electrons on Al, Be and Li [3] (see figure 1). These data show a clear trend: as Z decreases, the agreement with Moliere theory [4] gets worse. If this trend continues to hydrogen, there will be two effects:

1. The level of cooling achieved would be less than expected.
2. Due to the increased scattering in the tails, the fraction of muons scattered out of the cooling channel could be much bigger than expected.

Due to the importance of this to ionisation cooling, the MuScat experiment has been created to measure the scattering of muons of momentum around 180 MeV/c in a number of low atomic number materials, in particular liquid hydrogen. As well as checking these observations, MuScat will compare a range of muon scattering models with the data.

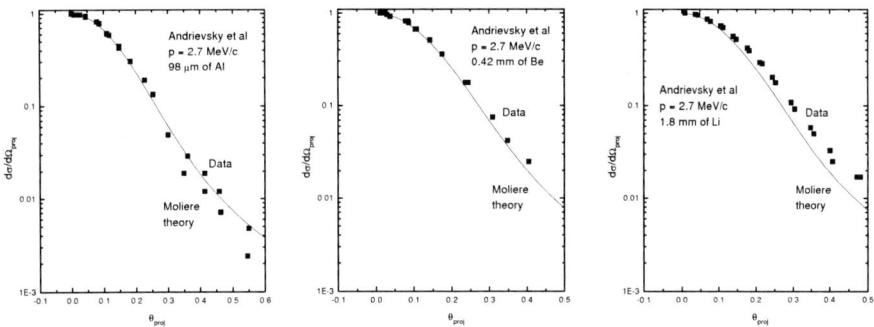

FIGURE 1. Multiple scattering of 2.7 MeV/c electrons measured in aluminium, beryllium and lithium, respectively, from [2]

MuScat had a technical run at the TRIUMF laboratory in the summer of 2000 using the M11 beam. As a result of this, much was learnt about the experiment and modification were made for a second and final data-taking run in April and May 2003 in the M20 beam. This article will describe the version of MuScat used in 2003 and show some first results from the analysis of the data.

THE MUSCAT EXPERIMENT

The Detector

As the aim of the experiment is to make a precise measurement of the multiple scattering of muons, the amount of material that the muons must pass through has to be kept to a minimum. For this reason, it is not possible to do any tracking before the target and a collimation system is employed to reduce the beam dimensions so that the incoming particle position is known accurately enough. In addition, the measurement of the position of the scattered muon relies on the first tracking detector as all subsequent detectors are effected by scattering in the first. Any additional detectors are only used to aid in noise rejection and for checking systematics. To minimise scattering in air, as much of the experiment as possible is mounted in vacuum. Finally, to eliminate particles other than muons, a good time-of-flight system is required.

The detector designed to satisfy these requirements and used in the M20 beam in 2003 is shown in figure 2. The most upstream parts are a veto shield and veto scintillator to eliminate beam halo. These are followed by the first trigger counter, which also acts as the TOF start. This is built from two fingers of scintillator, each 1 mm thick, 28 mm long and 3 mm high. These overlap by 20 mm in length and 3 mm in height. The timing resolution is about 250 ps. The TOF stop comes from the following RF-bucket of the cyclotron. This is almost a square-wave of length 1.9 ns, the smearing of the edges

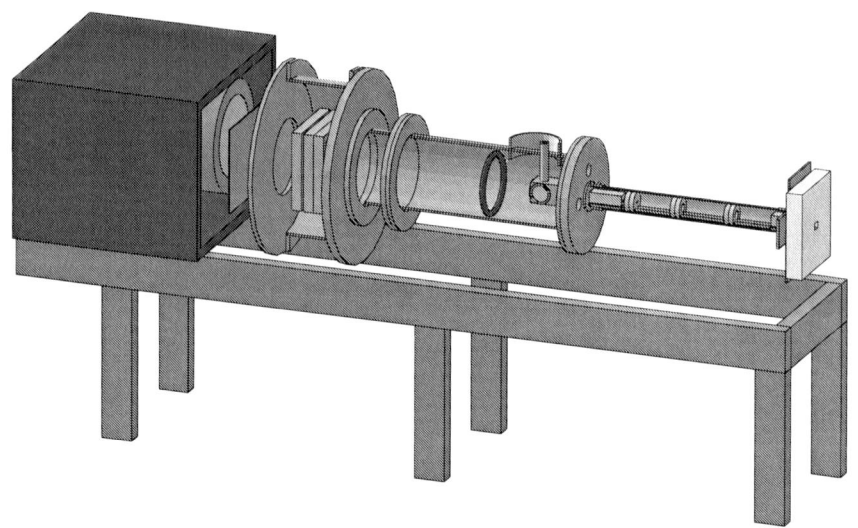

FIGURE 2. The MuScat experiment in 2003.

corresponding to a resolution of about 500 ps.

This trigger scintillator is followed by a 1 m long vacuum tube containing the collimation system. This consists of a 40 mm thick lead block at the front and a 160 mm lead block at the back, with 2 10 mm disks behind the front collimator and another 2 in front of the back collimator. The first block has a slit 20 mm long by 2 mm high cut in it, while the slit in the second block is tapered to prevent large angle scatters off the internal face. With this arrangement, the scattering distribution is measured vertically, in the narrow direction of the slot. The second dimension is longer to increase the particle intensity. There are also 2 pairs intermediate blocks each 10 mm thick with larger slits in them to prevent scattering off the internal faces and an active collimator in front of the back block. The latter consists of a strip of scintillator above and below the slit in the collimator. Finally, the whole collimator tube is wrapped in about 6mm of lead.

The vacuum tube is connected to the main vacuum vessel, which contains the targets. These are:

- Liquid Hydrogen, 100 mm and 150 mm thick
- Lithium, 12.7 mm and 6.3 mm thick
- Beryllium, 3.7 mm and 1.0 mm thick
- Carbon, 2.5 mm thick
- Aluminium, 1.5 mm thick
- CH_2, 4.8 mm thick
- Iron, 0.2 mm and 5.1 mm thick

The "thick" iron sample is used simply to blow the beam up to give a better coverage of the detectors and improve the measurement of the efficiency. The liquid hydrogen targets were built by the Target Group at TRIUMF. The two lengths were achieved in the same structure simply by rotating it through 90 degrees in the horizontal plane.

The solid targets are mounted on a target wheel that can be controlled from outside the vacuum so it is unnecessary to break this each time a target is changed. The wheel has 12 slots, the last of which has no target mounted and is used to measure the intrinsic properties of the beam. These are monitored on a regular basis. In addition, 2 sets of lithium and beryllium targets are used to allow systematic checks.

Three detectors built from scintillating fibres are used for tracking. These consist of two offset planes of 1mm thick fibres in each dimension, to give a uniform efficiency and two dimensional readout. There is a total of 1024 fibres per chamber. The light from the scintillating fibres is transmitted to photomultipliers using clear fibres. The PMTs used are Hamamatsu R5900 L16s [5] and contain 16 anodes, each 16mm long and 0.8mm wide. Bundles of 16 by 16 clear fibres are formed to match these anodes, thus giving a 16-fold multiplexing. To ensure that signals can be de-convoluted, the scintillating fibres are read out at both ends and the PMTs at each end are rotated by 90 degrees with respect to each other. The detectors are mounted inside the vacuum vessel to minimise the amount of material between them and the target. The PMTs, on the other hand, must sit outside and this means the fibre arrays form the vacuum seal. The leak rate from these is sufficiently small, however, that a vacuum of less than 5×10^{-6} Torr was achieved.

A second scintillator for use in the trigger sits behind the tracking detectors, outside the vacuum. The final part of the detector is TINA, a NaI calorimeter of 460 mm diameter and 510 mm depth [6]. It has a measured energy resolution (fwhm) of 3.6% at 90 MeV for electrons, with an energy dependence of $E^{-0.55}$. As it is not big enough to cover the full area of the tracking detectors after extrapolation of the tracks, it is used offset from the centre of the experiment. Nevertheless, it is valuable for both a muon energy measurement and to check the beam composition.

The M20 Beamline

The M20 beamline is a quadrupole muon decay channel, employing two bending magnets with quadrupole transport between them. This allows a dramatic reduction in the background to muons by selecting different momenta for the two bends and using a momentum slit in the quadrupole channel. By using forward decay pions, a muon beam up to almost 180 MeV/c is possible. However, the beamline is normally used for "surface" muons and was not used in the forward decay mode for 20 years before MuScat!

INITIAL RESULTS FROM 2003

In 2003, MuScat had about 16 days of data-taking in the M20 beamline and recorded a total of 57M triggers, as shown in table 1. The analysis of these data so far has focussed

TABLE 1. Number of triggers recorded in 2003 for the different targets used.

Thick liquid hydrogen full	3.1M
Thick liquid hydrogen empty	3.9M
Thin liquid hydrogen full	7.9M
Thin liquid hydrogen empty	8.5M
Thick lithium	6M
Thin lithium	6M
Thick beryllium	3M
Thin beryllium	3M
Aluminium	3M
Carbon	2M
Polythene	2M
Thick iron	2M
Thin iron	2M
Empty	5M
Total	57M

on understanding the beam composition, the beam momentum and the performance of the tracking chambers.

Beam Composition

The time-of-flight distribution measured with the two bending magnets in M20 set to the same momentum and the momentum slit wide open is shown in figure 3(a) and under normal running conditions, with the bends set for foward decays and a narrow momentum slit, in figure 3(b). In the former it is possible to identify peaks due to electrons, muons and taus. There are a number of other peaks which correspond to earlier RF pulses of the cyclotron and these are believed to be due to the protons. There are more than one of these because the protons do not penetrate all the way through the detector to the second trigger scintillator plane. Thus a "proton" trigger actually results from a proton signal in the first trigger counter and an accidental coincidence with another particle, pion, muon or electron, hitting the second trigger counter.

In normal running conditions, using forward muon decays and a narrow momentum slit, it can be seen that the backgrounds from particles other than muons are essentially eliminated. The shoulder on the muon peak is currently being understood, but is believed to be due to pion decays after the momentum slit.

Additional particle identification is possible with TINA. Figure 4 shows energy depositions in the calorimeter and peaks due to electrons, muons and pions are clearly visible.

Beam Momentum

As M20 has not been used for forward muon decays for a long period, it is very important to have an independent measurement of the beam momentum. This is possible

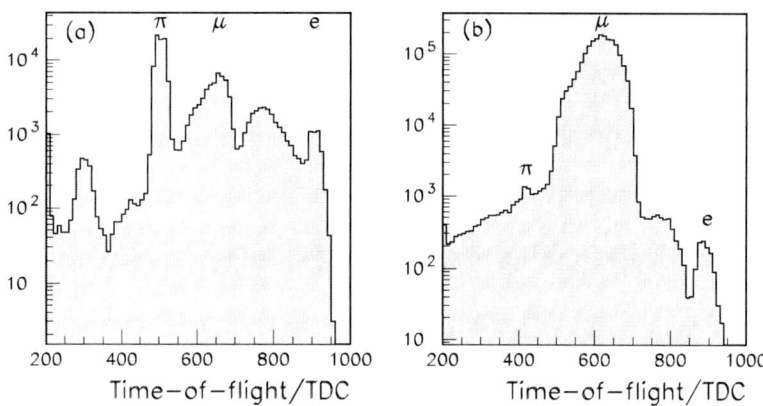

FIGURE 3. Time-of-flight plots: (a) with the bending magnets set to the same momentum and the momentum slit wide, and (b) with the beamline set for forward decays.

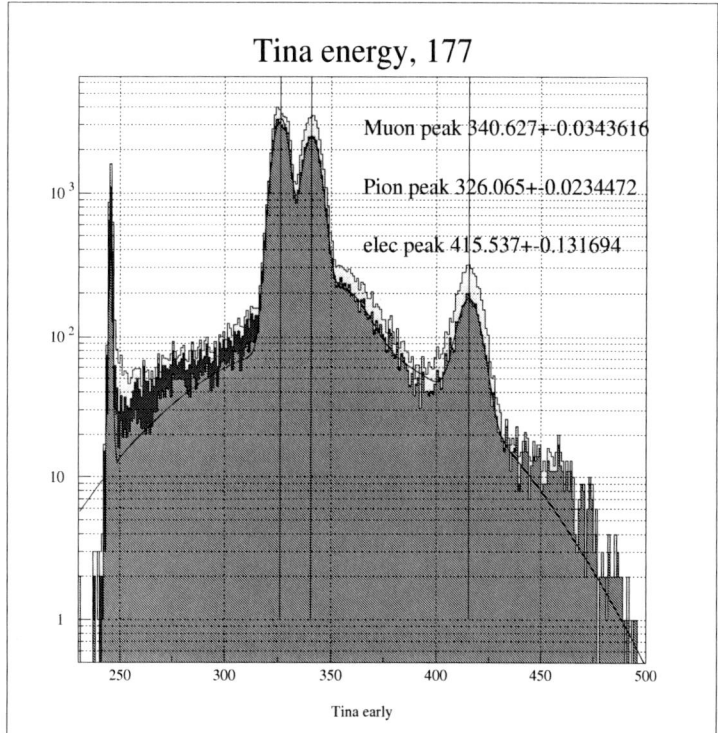

FIGURE 4. The energy measurement in TINA in counts under the same running conditions as figure 3(a).

119

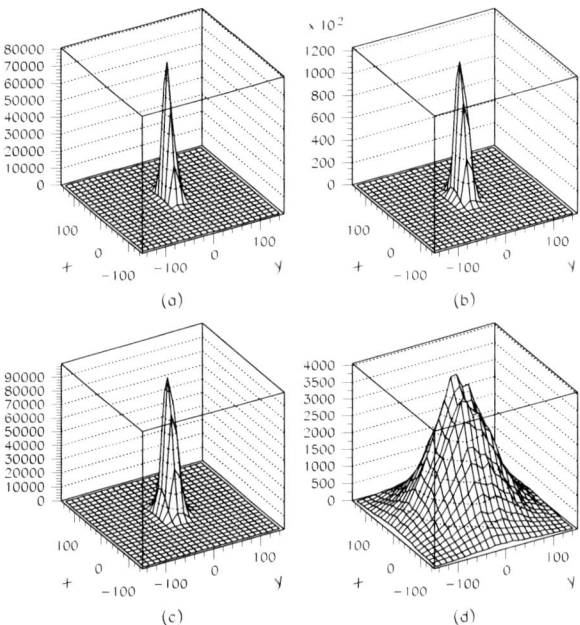

FIGURE 5. Raw hit distributions in the first scintillating fibre detector with (a) no target, (b) thin lithium, (c) thick lithium and (d) thick iron. The distributions are broader in 'x' as this is long dimension in the collimator slits. The scattering will be measured in 'y'.

in two ways: (1) using the time-of-flight of a number of particle types and (2) using the kinetic energy depositions in TINA for these particle types. Both of these are currently under investigation and give a preliminary measurement consistent with expectations from the M20 magnet settings.

Tracking

The tracking detectors are clearly the most important element of the experiment and a lot of work has been done to understand them. In particular, the pedestals and gains have been determined and the alignment, hit identification and error determination are almost final. Tests of the tracking algorithm are also almost complete. As an example, figure 5 shows very preliminary raw hit distributions, without tracking requirements, for running with solid targets. The algorithm for deconvoluting the real muon scattering distribution from these raw distributions has been developed and is also under test.

In addition, a detailed Geant4 simulation of the whole experiment has been written and is in the final stages of tuning. Figure 6 shows a comparison between the total signal

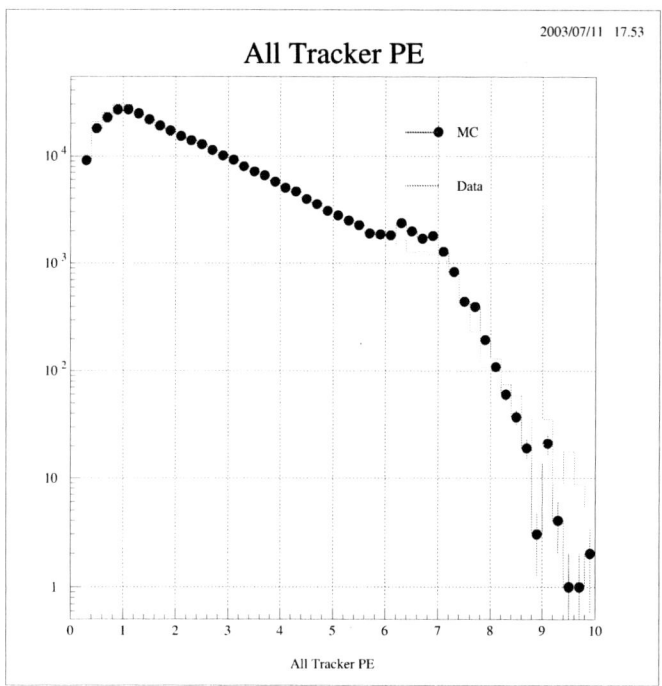

FIGURE 6. Comparison of total signals from clusters found in the tracking detector, measured in photo-electrons, between data (solid line) and Monte Carlo (points). Note the differences above 6pe is because a single saturation value for electronics is used for the Monte Carlo, while there are variations for the real data.

from "clusters" of anodes found in a single plane of the scintillating fibre detectors in data compared to the simulation. A method of using a number of scattering algorithms as input to Geant4 for comparison with the raw scattering distributions is also under development.

REFERENCES

1. MUCOOL Collaboration, Fermilab Proposal P904 (1998)
2. R.Fernow, MUCOOL Note 123 (2000)
3. L.Kulchitsky and G.Latyshev, Phys. Rev. 61 (1942) 254;
   A.I.Andrievsky et al, J. Phys. (USSR) 6 (1942) 279
4. V.G.Moliere, Z. Naturforschg. 3a (1948) 78;
   W.T.Scott, Rev. Mod. Phys. 35 (1963) 231
5. see http://cat1.hpk.co.jp/Eng/catalog/ETC/R5900U-L16_TPMH1146E06.pdf
6. C.E.Waltham et al, Nucl. Inst. Meth. A256 (1987) 91

Impact of Neutrino Oscillation Measurements on Theory

Hitoshi Murayama[†]

School of Natural Sciences, Institute for Advanced Study, Princeton, NJ 08540
Department of Physics, University of California, Berkeley, CA 94720

Abstract. Neutrino oscillation data had been a big surprise to theorists, and indeed they have ongoing impact on theory. I review what the impact has been, and what measurements will have critical impact on theory in the future.

INTRODUCTION

I was asked to comment on the impact of neutrino oscillation measurements on theory. It is completely clear that recent neutrino oscillation data had big impact on theory, and it will continue to do so. I will remind you about the ongoing impact. Then I will list measurements that will have critical impact on theory in the future.

Let me organize my discussion as "Past," "Present," and "Future."

PAST

It is useful to recall why theorists had always been interested in the small neutrino masses and their consequences on neutrino oscillation. It is because we are always interested in probing physics at as high energies as possible. One way to probe it is of course to go to the high-energy collider experiments and study physics at the energy scale directly. Another way is to look for rare and/or tiny effects coming from the high-energy physics. The neutrino mass belongs to the second category.

To study rare and/or tiny effects from physics at high energies, we can always parameterize them in terms of the power series expansion,

$$\mathcal{L} = \mathcal{L}_4 + \frac{1}{\Lambda}\mathcal{L}_5 + \frac{1}{\Lambda^2}\mathcal{L}_6 + \cdots. \tag{1}$$

The zeroth order term $\mathcal{L}_4$ is renormalizable and describes the Standard Model. On the other hand, the higher order terms are suppressed by the energy scale of new physics Λ. Possible operators can be classified systematically, which I believe was done first by Weinberg (but I couldn't find the appropriate reference). With two powers of suppression, there are many terms one can study:

$$\mathcal{L}_6 \supset QQQL, \bar{L}\sigma^{\mu\nu}W_{\mu\nu}He, W^\mu_\nu W^\nu_\lambda B^\lambda_\mu, \bar{s}d\bar{s}d, (H^\dagger D_\mu H)(H^\dagger D^\mu H), \cdots \tag{2}$$

The examples here contribute to proton decay, $g-2$, anomalous triple gauge boson vertex, K^0–$\overline{K}^0$ mixing, and the ρ-parameter, respectively. It is interesting that there is only one operator suppressed by a single power:

$$\mathscr{L}_5 = (LH)(LH). \tag{3}$$

After substituting the expectation value of the Higgs, the Lagrangian becomes

$$\mathscr{L} = \frac{1}{\Lambda}(LH)(LH) \to \frac{1}{\Lambda}(L\langle H\rangle)(L\langle H\rangle) = m_\nu \nu\nu, \tag{4}$$

nothing but the neutrino mass.

Therefore the neutrino mass plays a very unique role. It is the lowest-order effect of physics at short distances. This is a very tiny effect. Any kinematical effects of the neutrino mass are suppressed by $(m_\nu/E_\nu)^2$, and for $m_\nu \sim 1$ eV which we now know is already too large and $E_\nu \sim 1$ GeV for typical accelerator-based neutrino experiments, it is as small as $(m_\nu/E_\nu)^2 \sim 10^{-18}$. At the first sight, there is no hope to probe such a small number. However, any physicist knows that interferometry is a sensitive method to probe extremely tiny effects. For interferometry to work, we need a coherent source. Fortunately there are many coherent sources of neutrinos in Nature, the Sun, cosmic rays, reactors (not quite Nature), etc. We also need interference for an interferometer to work. Because we can't build half-mirrors for neutrinos, this could have been a show stopper. Fortunately, there are large mixing angles that make the interference possible. We also need long baselines to enhance the tiny effects. Again fortunately there are many long baselines available, such as the size of the Sun, the size of the Earth, etc. Nature was very kind to provide all necessary conditions for interferometry to us! Neutrino interferometry, a.k.a. neutrino oscillation, is therefore a unique tool to study physics at very high energy scales.

Indeed, the recently established neutrino oscillation results [1, 2]

$$\Delta m^2_{atm} \sim 0.002\,\text{eV}^2, \tag{5}$$
$$\Delta m^2_{solar} \sim 0.00007\,\text{eV}^2, \tag{6}$$

interpreted naively in a "hierarchical" mass scheme

$$m_3 \sim \sqrt{\Delta m^2_{atm}} \sim 0.04\,\text{eV}, \tag{7}$$
$$m_2 \sim \sqrt{\Delta m^2_{solar}} \sim 0.008\,\text{eV}, \tag{8}$$

suggests

$$\Lambda \sim \frac{\langle H\rangle^2}{m_3} \sim 8 \times 10^{14}\,\text{GeV}. \tag{9}$$

It is tantalizingly close to the energy scale of apparent gauge coupling unification in the Minimal Supersymmetric Standard Model, 2×10^{16} GeV. (See, Fig. 1.)

This way, the neutrino oscillation appears to provide us a unique window to physics at very high energies as their "leading order" effects. Indeed, theoretical estimates based on the seesaw mechanism in the grand unified theories [3] are practically confirmed!

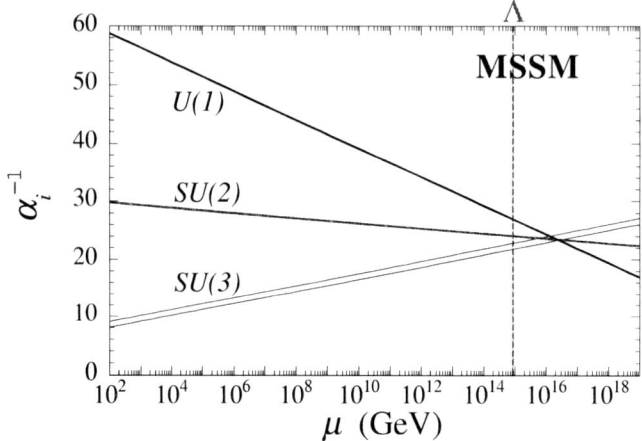

FIGURE 1. Apparent unification of gauge coupling unification in the MSSM at 2×10^{16} GeV, compared to the suggested scale of new physics from the neutrino oscillation data.

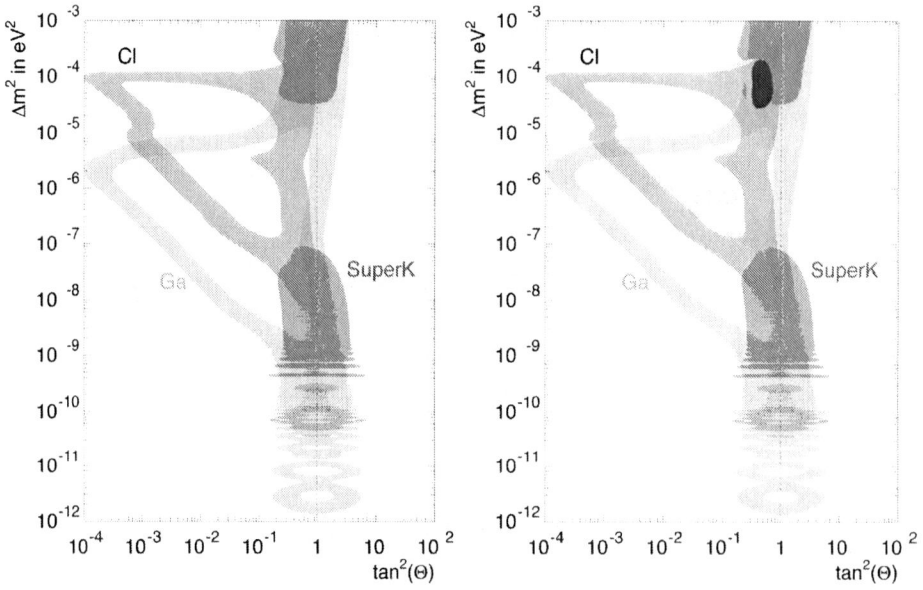

FIGURE 2. The progress in solar neutrino in the year 2002. Before March and after April [4].

PRESENT

The last year was an amazing year in neutrino physics. Before March, the situation of the solar neutrino data looked like the first plot in Fig. 2, and there had been overlaps

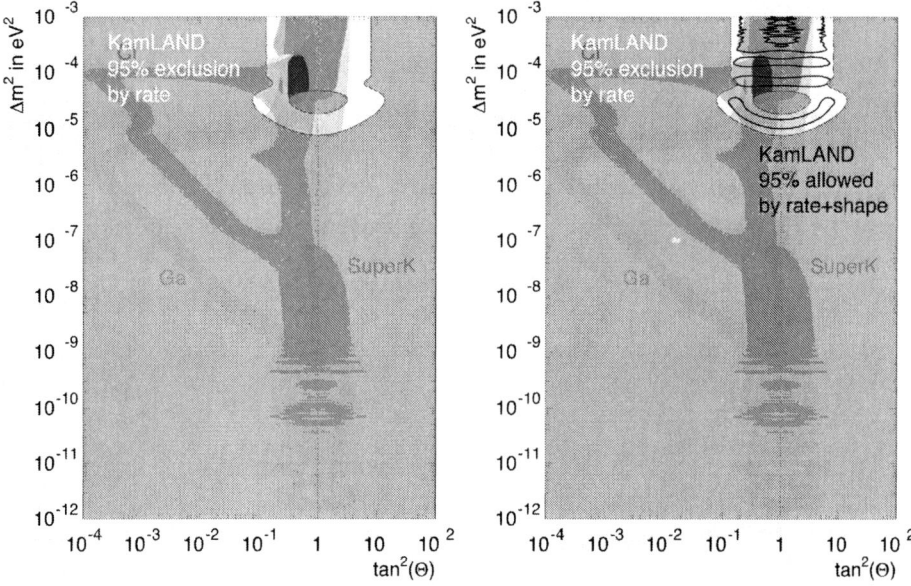

FIGURE 3. The comparison of the solar neutrino data and the reactor anti-neutrino data after December [4].

between SuperK, Homestake, and Gallium experiments in the LMA and LOW regions, some down in quasi-vaccum. After SNO neutral current result in April, the parameter space focused only on the LMA region shown in red in the second plot in Fig. 2. In December, KamLAND has excluded most of the parameter space as shown in the first plot in Fig. 3, while its preferred region (inside the blue contours in the second plot in Fig. 3) has consistent overlaps with the that preferred by the solar neutrino data. It was a tremendous convergence from the parameter space over many decades down to factors of a few.

It is useful to recall what a typical theorist used to say back around 1990.

- The solution to the solar neutrino problem must be the small mixing angle MSW solution because it is so beautiful.
- The natural scale for $v_\mu \to v_\tau$ oscillation is $\Delta m^2 \sim eV^2$ because it is the cosmologically interesting range.
- The angle θ_{23} must be of the same order of magnitude as V_{cb} because of the grand unification.
- The atmospheric neutrino anomaly must go away because it would require a large mixing angle to explain.

Needless to say, theorists have a very good track record in neutrino physics.

Indeed, the recent results from neutrino oscillation physics had surprised almost everybody. The prejudice has been that the mixing angles must be small because quark

TABLE 1. Prediction of different flavor symmetries on the neutrino mass-squared ratio and various mixing angles, taken from [6].

Model	parameters	d_{23}	$\Delta m_{12}^2/|\Delta m_{23}^2|$	U_{e3}	$\tan^2\theta_{12}$	$\tan^2\theta_{23}$
A	$\varepsilon = 1$	O(1)	O(1)	O(1)	O(1)	O(1)
SA	$\varepsilon = \lambda$	O(1)	$O(d_{23}^2)$	$O(\lambda)$	$O(\lambda^2/d_{23}^2)$	O(1)
H_{II}	$\varepsilon = \lambda^2$	$O(\lambda^2)$	$O(\lambda^4)$	$O(\lambda^2)$	O(1)	O(1)
H_I	$\varepsilon = \lambda^2$	0	$O(\lambda^6)$	$O(\lambda^2)$	O(1)	O(1)
IH (LA)	$\varepsilon = \eta = \lambda$	$O(\lambda^4)$	$O(\lambda^2)$	$O(\lambda^2)$	$1+O(\lambda^2)$	O(1)
IH (LOW)	$\varepsilon = \eta = \lambda^2$	$O(\lambda^8)$	$O(\lambda^4)$	$O(\lambda^4)$	$1+O(\lambda^4)$	O(1)

mixing angles are small, and the masses must be hierarchical because both quarks and charged lepton masses are hierarchical. Given that the LMA is now chosen, all mixing angles are large except for U_{e3} that must be small-ish (but the current limit is not very strong, $|U_{e3}| \lesssim 0.2$).

The natural question then is if this newly discovered surprising pattern of neutrino masses and mixings require a new symmetry or any special structure to explain.

In fact, the big question has always been *what distinguishes flavor?* Three generations share exactly the same quantum numbers. Yet, they have such different masses. The hierarchy with small mixings means that there is a need for some kind of ordered structure. The "common sense" in quantum mechanics is that states with the same quantum numbers should have similar energy levels (*i.e.* masses) and mix significantly under small perturbations. The observed patterns go against this "common sense." The hierarchical masses and small mixings among quarks and charged leptons had been a puzzle.

Therefore, there has been a strong suspicion that there is a new set of quantum numbers, *flavor quantum numbers*, that distinguish three generations of quarks and leptons. As Noether told us, a new quantum number requires a new symmetry, *flavor symmetry*. This new symmetry must allow the top quark Yukawa coupling because it is of the natural size, $y_t \simeq 1.0$. On the other hand, all the other Yukawa couplings are practically zero (as opposed to $O(1)$), and the flavor symmetry must forbid them. After the symmetry is broken by a small parameter, all the other Yukawa couplings become allowed, but suppressed [5]. The hope is to identify the underlying symmetry based on the data, similarly to what was done by Heisenberg (isospin) or Gell-Mann–Okubo (flavor $SU(3)$).

Indeed, the neutrino data had been already effective in narrowing down the possibilities of flavor symmetries. In Table 1, many proposed flavor symmetries are shown together with their predictions on the mass-squared ratio $\Delta m_{12}^2/\Delta m_{23}^2$, U_{e3}, θ_{12}, and θ_{23}, taken from [6] (October 2002).[1] Since then, models H_{II}, H_I, and IH (LOW) had been excluded by KamLAND.

---

[1] d_{23} in the model SA refers to a degree of accidental cancellation in the 23 sector, that is used to enhance θ_{12}.

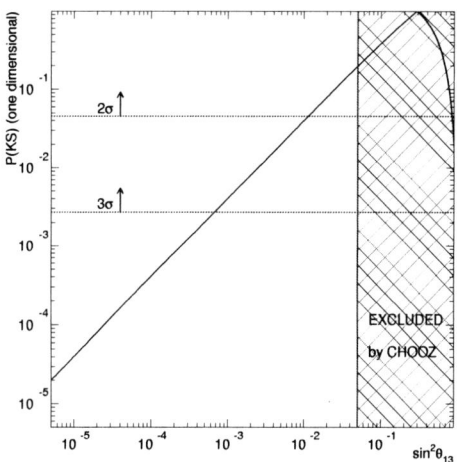

FIGURE 4. The one-dimensional KS probability on $\sin^2\theta_{13}$ based on anarchy, *i.e.*, no fundamental distinction among three neutrinos. Taken from [9].

Among them, I liked the model A the best, because it is *mine* [7]. It is called *anarchy*, based on the idea that neutrinos are actually normal, while quarks and charged leptons aren't. As I mentioned already, the hierarchical masses and small mixing angles are against the "common sense," while the neutrinos do not seem to have a large hierarchy and mix a lot. Maybe the *lack of flavor symmetry* can explain the data. Indeed, if there is no fundamental distinction among three neutrinos, or in other words if their flavor quantum numbers are all equal, the group theory of three-by-three unitary matrices uniquely determine the probability distribution of mixing angles [8]. Then all three angles, θ_{12}, θ_{23}, and θ_{13} are three random draws from the distribution $dP/dx \propto (1-x)^{-1/2}$ for $x = \sin^2 2\theta$. Because it is peaked towards the maximal angle $x = 1$, it is very plausible that two draws come out large, while one of them comes down the tail (but not expected *way* down the tail). Indeed, the Kolgomorv–Smirnov test suggests that the probability that three random draws come out worse than the actual data is 64%, and hence the observed pattern is completely natural if there is *no* fundamental distinction among three generations [9]. On the other hand, θ_{13} is expected to be not too far below the current limit. The one-dimensional KS probability is $P(KS) = 4(\sin^2\theta_{13} - \frac{1}{2}\sin^4\theta_{13})$, and hence we expect $\sin^2\theta_{13} > 0.013$ at "95% CL." The size of the CP-violation $\sin\delta$ is distributed as $1/|\cos\theta|$, and hence is expected to be large.

FUTURE

Having discussed the impact of neutrino data on theory so far, it is clear what will be the critical measurements in the future.

- $\sin^2 2\theta_{23} = 1.00 \pm 0.01$? If it comes out that precisely maximal, it surely will require a new symmetry.
- $\sin^2 \theta_{13} < 0.01$? If so, electron-neutrino must have a different flavor quantum number from muon and tau neutrinos.
- Normal or inverted hierarchy? Most flavor symmetries predict the normal hierarchy, but theorists had been wrong!
- CP Violation? Even though the CP violation in neutrino oscillation may not prove the relevant CP violation for leptogenesis, it will at least make it very plausible.

After going through the critical measurements, we hope to determine the underlying flavor symmetries behind neutrinos (and flavor in general). Then comes an even bigger question: can we understand the dynamics behind the flavor symmetry? In the case of the strong interaction, isospin and flavor $SU(3)$ are the flavor symmetries, while the QCD is the dynamics. Can we get to the same level? This question will depend crucially on what we will find at the TeV-scale. If it is supersymmetry, the answer may be anomalous U(1) gauge symmetry with the Green–Schwarz mechanism from the string theory [10]. If it is extra dimensions, the answer may be physical dislocation of different particles within a thick brane [11]. If it is technicolor, the answer may be new broken gauge symmetries at 100 TeV scale [12].

Of course one shouldn't forget LSND [13] because no theory fit the data very well. It is true that most theorists do not take the LSND evidence seriously at this moment, only data will decide. Currently all explanations have difficulties: sterile neutrino(s) [14], CPT violation [15], lepton-number violating muon decay [16]. But if any of them will turn out to be true, it will have a huge impact on theory.

CONCLUSION

Neutrino oscillation physics has had big impact on theory already. Yet, there is a lot more to learn. The (precise) measurements of θ_{23}, θ_{13}, the type of hierarchy, and the CP violation, will have critical impact. Through these measurements, we hope to determine the symmetries behind the neutrino masses and mixings or flavor in general. In conjunction with data from the energy frontier, we may even have access to understand dynamics behind the flavor. Depending on how things will turn out, there may well be even more surprises.

ACKNOWLEDGMENTS

This work was supported by the Institute for Advanced Study, funds for Natural Sciences, as well as in part by the DOE under contracts DE-FG02-90ER40542 and DE-

AC03-76SF00098 and in part by NSF grant PHY-0098840.

REFERENCES

1. M. Shiozawa, in this proceedings.
2. K. Nakamura, in this proceedings.
3. T. Yanagida, in *Proceedings of the Workshop on Unified Theory and Baryon Number of the Universe*, eds. O. Sawada and A. Sugamoto (KEK, 1979) p.95; M. Gell-Mann, P. Ramond and R. Slansky, in *Supergravity*, eds. P. van Niewwenhuizen and D. Freedman (North Holland, Amsterdam, 1979).
4. http://hitoshi.berkeley.edu/neutrino
5. C. D. Froggatt and H. B. Nielsen, Nucl. Phys. B **147**, 277 (1979).
6. G. Altarelli, F. Feruglio and I. Masina, JHEP **0301**, 035 (2003) [arXiv:hep-ph/0210342].
7. L. J. Hall, H. Murayama and N. Weiner, Phys. Rev. Lett. **84**, 2572 (2000) [arXiv:hep-ph/9911341].
8. N. Haba and H. Murayama, Phys. Rev. D **63**, 053010 (2001) [arXiv:hep-ph/0009174].
9. A. de Gouvêa and H. Murayama, Phys. Lett. B **573**, 94 (2003) [arXiv:hep-ph/0301050].
10. L. E. Ibañez and G. G. Ross, Phys. Lett. B **332**, 100 (1994) [arXiv:hep-ph/9403338]. P. Binétruy and P. Ramond, Phys. Lett. B **350**, 49 (1995) [arXiv:hep-ph/9412385].
11. N. Arkani-Hamed and M. Schmaltz, Phys. Rev. D **61**, 033005 (2000) [arXiv:hep-ph/9903417].
12. E. Eichten and K. D. Lane, Phys. Lett. B **90**, 125 (1980).
13. B. Louis, in this proceedings.
14. M. Maltoni, T. Schwetz, M. A. Tortola and J. W. F. Valle, Nucl. Phys. B **643**, 321 (2002) [arXiv:hep-ph/0207157].
15. M. C. Gonzalez-Garcia, M. Maltoni and T. Schwetz, Phys. Rev. D **68**, 053007 (2003) [arXiv:hep-ph/0306226].
16. B. Armbruster *et al.*, Phys. Rev. Lett. **90**, 181804 (2003) [arXiv:hep-ex/0302017].

Neutrinos in Cosmology

Georg G. Raffelt

Max-Planck-Institut für Physik (Werner-Heisenberg-Institut)
Föhringer Ring 6, 80805 München, Germany

Abstract. In the framework of the standard cosmological model, large-scale structure observations imply neutrino mass limits in the range $\sum m_\nu < 0.7$–2.1 eV (95% CL), depending on the included data sets and assumed priors on cosmological parameters. Future observations will further improve these sensitivities. The measured neutrino oscillation parameters in conjunction with the big-bang nucleosynthesis constraint on the primordial ν_e chemical potential exclude a significant deviation from the standard cosmological neutrino density. Therefore, large-scale structure limits on the hot dark matter fraction indeed translate directly into neutrino mass limits. While neutrinos with sub-eV masses are almost negligible for the cosmic dark matter inventory, they fit nicely with leptogenesis scenarios for creating the baryon asymmetry of the universe.

COSMOLOGICAL NEUTRINO MASS LIMIT

In a classic paper, Gershtein and Zeldovich (1966) for the first time used cosmological data to constrain neutrino masses, notably the mass of the muon neutrino that had just been discovered [1]. The cosmic number density of neutrinos plus anti-neutrinos per flavor is $n_\nu = \frac{3}{11} n_\gamma$ with n_γ the number density of cosmic microwave photons, and assuming that there is no neutrino chemical potential. With $T_\gamma = 2.728$ K this translates to $n_\nu = 112$ cm^{-3}. If neutrinos have masses one finds a cosmic mass fraction

$$\Omega_\nu h^2 = \sum_{\text{flavors}} \frac{m_\nu}{92.5 \text{ eV}}, \qquad (1)$$

where as usual h is the Hubble constant in units of 100 km s^{-1} Mpc^{-1}. The requirement that neutrinos do not "overclose" the universe then leads to the limit $\sum m_\nu \lesssim 40$ eV.

Later Cowsik and McClelland (1973) speculated that massive neutrinos could actually be the dark matter of the universe [2]. However, neutrinos are not good for this purpose. First, the phase space for neutrinos gravitationally bound to a galaxy is limited [3]. As a consequence, if massive neutrinos are supposed to be the dark matter in galaxies, they must obey a *lower* mass limit of some 30 eV for typical spirals, and even 100–200 eV for dwarf galaxies. The tritium limit on the "ν_e mass" of about 2.2 eV (95% CL) [4] and the very small neutrino mass differences implied by the oscillation experiments show that all neutrino masses are far too small to play a dominant role in galaxies.

The second argument against neutrino dark matter relies on cosmic large-scale structure. The observed structure in the distribution of cosmic matter, and notably in the distribution of galaxies, is thought to arise from the gravitational instability of primordial density fluctuations. The small masses of neutrinos imply that they stay relativistic for a long time after their decoupling ("hot dark matter"), allowing them to stream freely, thereby erasing the primordial density fluctuations on small scales [5]. While this effect does not preclude neutrino dark matter, it implies a top-down scenario for structure formation where large structures form first, later fragmenting into smaller ones. It was soon realized that the predicted properties of the large-scale matter distribution did not agree with observations and that neutrino dark matter was ruled out [6].

Today it is widely accepted that the universe has critical density and that its matter inventory sports several nontrivial components. Besides some 5% baryonic matter (most of it dark) there are some 25% cold dark matter in an unidentified physical form and some 70% of a negative-pressure component ("dark energy"). And because neutrinos do have mass, they contribute at least 0.1% of the critical density. This fraction is based on a hierarchical mass scenario with $m_3 = 50$ meV, the smallest value consistent with atmospheric neutrino oscillations.

An upper limit on the neutrino dark matter fraction can be derived from the measured power spectrum $P_M(k)$ of the cosmic matter distribution. Neutrino free streaming suppresses the small-scale structure by an approximate amount [7]

$$\frac{\Delta P_M}{P_M} \approx -8 \frac{\Omega_\nu}{\Omega_M}, \qquad (2)$$

where Ω_M is the cosmic mass fraction in matter, i.e. excluding the dark energy.

The exact limit on the neutrino masses from the absence of any noticable suppression of small-scale structure depends on the used data for the matter power spectrum and on additional information on global cosmological parameters. Perhaps the most conservative approach is to use only the 2dF galaxy redshift survey for the matter distribution and the recent WMAP measurements of the temperature fluctuation of the cosmic microwave background radiation (CMBR). In this case one finds [8, 9]

$$\sum_{\text{flavors}} m_\nu < 2.1 \text{ eV} \quad (95\% \text{ CL}). \qquad (3)$$

Including smaller-scale CMBR data improves this limit by about a factor of 2. Including further data/priors as described in the paper by the WMAP collaboration leads to $\sum m_\nu < 0.7$ eV (95% CL) [10]. This restrictive constraint is often quoted as the "WMAP limit on the neutrino mass"—a misnomer because the limit, of course, results primarily from the small-scale power spectrum of the matter distribution. The large-scale WMAP data play an auxiliary role by reducing the uncertainty of the global cosmological parameters.

In a recent paper, cosmological evidence for a positive indication of a nonvanishing neutrino mass $\sum m_\nu = 0.56^{+0.30}_{-0.26}$ eV was found when including the baryon mass fraction and X-ray luminosity in galaxy clusters as further observables [11].

The effect of such small neutrino masses on the power spectrum is quite subtle. Therefore, generally one may wish to be cautious of results that combine many heterogeneous and potentially inconsistent sets of cosmological data to extract a limit or a detection.

Within the framework of the standard theory of structure formation, one systematic uncertainty comes from the unknown biasing parameter b which relates the power spectrum of the galaxy distribution to that of the true underlying matter distribution, $P_{\text{Gal}}(k) = b^2 P_M(k)$. The biasing parameter is one of the quantities which must be taken into account when fitting all large-scale structure data to observations of the galaxy distribution and of the temperature fluctuations of the cosmic microwave background radiation. In future the Sloan Digital Sky Survey will have greater sensitivity to the overall shape of $P_{\text{Gal}}(k)$ on the relevant scales, allowing one to disentangle more reliably the impact of b and Ω_ν on $P_M(k)$. It is foreseen that one can then reach a sensitivity of $\sum m_\nu \sim 0.65$ eV [7]. In the more distant future, CMBR measurements with the PLANCK satellite may help to improve the $\sum m_\nu$ sensitivity of the SLOAN data to 0.12 eV [12]. A similar sensitivity may be achieved more directly by future weak-lensing studies that are directly sensitive to the matter power spectrum and thus unaffected by biasing [13].

Any neutrino mass limit or sensitivity forecast depends on the assumption that the standard cosmological model with its limited number of parameters correctly accounts for our universe. On the other hand, it is possible that the primordial spectrum of density fluctuations is not accounted for by a pure power law ("running spectral index") or that other subtle deviations from the standard assumptions exist that may compensate for the effect of a neutrino mass or mimic one. The observed universe as a neutrino-mass experiment may suffer from systematics that have not yet been fully appreciated. Surely, however, cosmology has begun to probe the sub-eV range of neutrino masses in earnest.

HOW MANY NEUTRINOS IN THE UNIVERSE?

One possible systematic uncertainty of the cosmological mass limits consists of the very number density n_ν of neutrinos in our universe. Even though there are exactly three neutrino flavors as indicated by the Z^0 decay width, the assumption of thermal equilibrium in the early universe alone does not fix the neutrino abundance. Each flavor is characterized by an unknown chemical potential μ_ν, or equivalently a degeneracy parameter $\xi_\nu = \mu_\nu/T$, the latter being invariant under cosmic expansion. While the very small baryon-to-photon ratio suggests $\xi \ll 1$ for all fermions, for neutrinos this is an assumption and not an established fact.

In the presence of a degeneracy parameter ξ_ν the thermal number and energy densities of one species of relativistic neutrinos plus anti-neutrinos are

$$n_\nu = T_\nu^3 \frac{3\zeta_3}{2\pi^2} \left[1 + \frac{2\ln(2)\,\xi_\nu^2}{3\zeta_3} + \frac{\xi_\nu^4}{72\zeta_3} + O(\xi_\nu^6) \right], \tag{4}$$

$$\rho_\nu = T_\nu^4 \frac{7\pi^2}{120} \left[1 + \frac{30}{7} \left(\frac{\xi_\nu}{\pi} \right)^2 + \frac{15}{7} \left(\frac{\xi_\nu}{\pi} \right)^4 \right]. \tag{5}$$

Therefore, n_ν can only be larger than the standard value. One may think that the cosmological mass limits are thus conservative in that a limit on the hot dark matter fraction would translate into a smaller limit on the neutrino mass. However, the increased radiation content affects the structure-formation limits in a counter-intuitive way. If the number density of neutrinos is larger than standard, the limit on $\sum m_\nu$ weakens [8]. Therefore, the allowed range of ξ_ν for the different flavors affects the cosmological mass limits.

Big-bang nucleosynthesis (BBN) is affected by ρ_ν in that a larger neutrino density increases the primordial expansion rate, thereby increasing the neutron-to-proton freeze-out ratio n/p and thus the cosmic helium abundance. Therefore, the observed helium abundance provides a limit on ρ_ν which corresponds to some fraction of an effective extra neutrino species. In addition, however, an electron neutrino chemical potential modifies $n/p \propto \exp(-\xi_{\nu_e})$. Depending on the sign of ξ_{ν_e} this effect can increase or decrease the helium abundance and can compensate for the ρ_ν effect of other flavors [14]. If ξ_{ν_e} is the only chemical potential, BBN provides the limit

$$-0.01 < \xi_{\nu_e} < 0.07. \tag{6}$$

Including the compensation effect, the only upper limit on the radiation density comes from precision measurements of the power spectrum of the temperature fluctuations of the cosmic microwave background radiation and from large-scale structure measurements so that the regions $-0.01 < \xi_{\nu_e} < 0.22$ and $|\xi_{\nu_{\mu,\tau}}| < 2.6$ are allowed [15].

However, the observed neutrino oscillations imply that the individual flavor lepton numbers are not conserved and that in true thermal equilibrium all neutrinos are characterized by one single chemical potential ξ_ν. If flavor equilibrium is achieved before n/p freeze-out the restrictive BBN limit on ξ_{ν_e} applies to all flavors, i.e. $|\xi_\nu| < 0.07$, implying that the cosmic number density of neutrinos is fixed to within about 1%. In that case the relation between Ω_ν and m_ν is uniquely given by the standard expression Eq. (1).

The approach to flavor equilibrium in the early universe by neutrino oscillations and collisions was recently studied by several groups [16, 17, 18, 19]. The detailed treatment is rather complicated and involves a number of subtleties related to the large weak potential caused by the neutrinos themselves as they oscillate. The intriguing phenomenon of synchronized flavor oscillations [20, 21] plays an important and subtle role. The practical bottom line, however, is rather simple. Effective flavor equilibrium before n/p freeze-out is reliably achieved if the solar oscillation parameters are in the favored LMA region. In the LOW region, the result depends sensitively on the value of the small but unknown third mixing angle Θ_{13}. In the SMA and VAC regions equilibrium is not achieved.

After the KamLAND measurements the solar LMA solution is now singled out as the final answer to the solar neutrino problem [22]. Therefore, if one accepts the above chain of arguments, neutrino chemical potentials no longer constitute a systematic uncertainty for the cosmological mass limit.

The situation is different if one speculates about the existence of additional neutrino states in the form of low-mass right-handed states (sterile neutrinos). They can be thermally excited in the early universe by oscillations and collisions if they mix with the ordinary active neutrinos—for a recent detailed study see Ref. [23]. A small ν_e chemical potential, perhaps equally shared among all active and sterile neutrinos, can compensate the BBN expansion-rate effect of the additional energy density in sterile neutrinos. Therefore, in this scenario it is not excluded that the radiation density of the early universe could be larger than is usually assumed and that the cosmological neutrino mass bounds are somewhat relaxed.

NEUTRINOS AND THE BARYONIC MATTER

Neutrino masses in the sub-eV range can play an interesting albeit indirect role for creating the baryon asymmetry of the universe (BAU) in the framework of leptogenesis scenarios [24]. The main ingredients are those of the usual see-saw scenario for small neutrino masses. Restricting ourselves to a single family, the relevant parameters are the heavy Majorana mass M of the ordinary neutrino's right-handed partner and a Yukawa coupling g_V between the neutrinos and the Higgs field Φ. The observed neutrino then has a Majorana mass

$$m_V = \frac{g_V^2 \langle \Phi \rangle^2}{M} \tag{7}$$

that can be very small if M is large, even if the Yukawa coupling g_V is comparable to that for other fermions. Here, $\langle \Phi \rangle$ is the vacuum expectation value of the Higgs field which also gives masses to the other fermions.

The heavy Majorana neutrinos will be in thermal equilibrium in the early universe. When the temperature falls below their mass, their density gets exponentially Boltzmann suppressed. However, if at that time they are no longer in thermal equilibrium, their abundance will exceed the equilibrium distribution. The subsequent out-of-equilibrium decays can lead to the net generation of lepton number. CP-violating decays are possible by the usual interference of tree-level with one-loop diagrams with suitably adjusted phases of the various couplings. The generated lepton number excess will be re-processed by standard-model sphaleron effects which respect $B-L$ but violate $B+L$. It is straightforward to generate the observed BAU by this mechanism.

The requirement that the heavy Majorana neutrinos freeze out before they are Boltzmann suppressed implies an upper limit on the combination of parameters g_V^2/M that also appears in the see-saw formula for m_V. The out-of-equilibrium condition thus implies an upper limit on m_V. Detailed scenarios for generic neutrino mass and mixing schemes have been worked out, see Ref. [25] for a recent review and citations of the large body of pertinent literature. For a broad range of models, one finds that thermal leptogenesis works only if all light neutrino masses are below 0.1 eV [26].

In summary, neutrino mass and mixing schemes suggested by the atmospheric and solar oscillation data are nicely consistent with plausible leptogenesis scenarios. Of course, it is an open question of how one would go about to verify or falsify leptogenesis as the correct baryogenesis scenario. Still, it is intriguing that massive neutrinos may have a lot more to do with the baryons than with the dark matter of the universe!

ACKNOWLEDGMENTS

This work was supported, in part, by the Deutsche Forschungsgemeinschaft under grant No. SFB-375 and by the European Science Foundation (ESF) under the Network Grant No. 86 Neutrino Astrophysics.

REFERENCES

1. Gershtein, S.S., and Zeldovich, Y.B., "Rest mass of muonic neutrino and cosmology," Pisma Zh. Eksp. Teor. Fiz. 4, 174 (1966) [JETP Lett. 4, 120 (1966)].
2. Cowsik, R., and McClelland, J., "Gravity of neutrinos of nonzero mass in astrophysics," Astrophys. J. 180, 7 (1973).
3. Tremaine, S., and Gunn, J.E., "Dynamical role of light neutral leptons in cosmology," Phys. Rev. Lett. 42, 407 (1979).
4. Weinheimer, C., "The neutrino mass direct measurements," hep-ex/0306057.
5. Doroshkevich, A.G., Zeldovich, Y.B., Syunyaev, R.A., and Khlopov, M.Y., "Astrophysical implications of the neutrino rest mass," Pisma Astron. Zh. 6, 457 (1980) [Sov. Astron. Lett. 6, 252 (1980)].
6. White, S.D.M., Frenk, C.S., and Davis, M., "Clustering in a neutrino-dominated universe," Astrophys. J. Lett. 274, L1 (1983).
7. Hu, W., Eisenstein, D.J., and Tegmark, M., "Weighing neutrinos with galaxy surveys," Phys. Rev. Lett. 80, 5255 (1998) [astro-ph/9712057].
8. Hannestad, S., "Neutrino masses and the number of neutrino species from WMAP and 2dFGRS," JCAP 0305, 004 (2003) [astro-ph/0303076].
9. Elgaroy, O., and Lahav, O., "The role of priors in deriving upper limits on neutrino masses from the 2dFGRS and WMAP," JCAP 0304, 004 (2003) [astro-ph/0303089].
10. Spergel, D.N., et al., "First Year Wilkinson Microwave Anisotropy Probe (WMAP) observations: Determination of cosmological parameters," astro-ph/0302209.
11. Allen, S.W., Schmidt, R.W., and Bridle, S.L., "A preference for a non-zero neutrino mass from cosmological data," astro-ph/0306386.
12. Hannestad, S., "Can cosmology detect hierarchical neutrino masses?," Phys. Rev. D 67, 085017 (2003) [astro-ph/0211106].
13. Abazajian, K.N., and Dodelson, S., "Neutrino mass and dark energy from weak lensing," Phys. Rev. Lett. 91, 041301 (2003) [astro-ph/0212216].
14. Kang, H.S., and Steigman, G., "Cosmological constraints on neutrino degeneracy," Nucl. Phys. B 372, 494 (1992).
15. Hansen, S.H., Mangano, G., Melchiorri, A., Miele, G., and Pisanti, O., "Constraining neutrino physics with BBN and CMBR," Phys. Rev. D 65, 023511 (2002) [astro-ph/0105385].
16. Lunardini, C., and Smirnov, A.Y., "High-energy neutrino conversion and the lepton asymmetry in the universe," Phys. Rev. D 64, 073006 (2001) [hep-ph/0012056].
17. Dolgov, A.D., Hansen, S.H., Pastor, S., Petcov, S.T., Raffelt, G.G., and Semikoz, D.V., "Cosmological bounds on neutrino degeneracy improved by flavor oscillations," Nucl. Phys. B 632, 363 (2002) [hep-ph/0201287].
18. Wong, Y.Y., "Analytical treatment of neutrino asymmetry equilibration from flavour oscillations in the early universe," Phys. Rev. D, 66, 025015 (2002) [hep-ph/0203180].
19. Abazajian, K.N., Beacom, J.F., and Bell, N.F., "Stringent constraints on cosmological neutrino antineutrino asymmetries from synchronized flavor transformation," Phys. Rev. D 66, 013008 (2002) [astro-ph/0203442].
20. Samuel, S., "Neutrino oscillations in dense neutrino gases," Phys. Rev. D 48, 1462 (1993).
21. Pastor, S., Raffelt, G.G., Semikoz, D.V., "Physics of synchronized neutrino oscillations caused by self-interactions," Phys. Rev. D 65, 053011 (2002) [hep-ph/0109035].
22. Eguchi, K., et al. [KamLAND Collaboration], "First results from KamLAND: Evidence for reactor anti-neutrino disappearance," Phys. Rev. Lett. 90, 021802 (2003) [hep-ex/0212021].
23. Dolgov, A.D., and Villante, F.L., "BBN bounds on active-sterile neutrino mixing," hep-ph/0308083.
24. Fukugita, M., and Yanagida, T., "Baryogenesis without grand unification," Phys. Lett. B 174, 45 (1986).
25. Buchmüller, W., and Plümacher, M., "Neutrino masses and the baryon asymmetry," Int. J. Mod. Phys. A 15, 5047 (2000) [hep-ph/0007176].
26. Buchmüller, W., Di Bari, P., and Plümacher, M., "The neutrino mass window for baryogenesis," Nucl. Phys. B 665, 445 (2003) [hep-ph/0302092].

Summary of Working Group 2
—Muon Physics—

Masaharu Aoki[1]* and Alessandro Baldini[†]

School of Science, Osaka University, Osaka 560-0043, Japan
[†]*INFN Pisa, Via F. Buonarroti 2, 56127 Pisa*

Abstract. This summary covers a part of topics discussed in the working group 2; the muon physics and relevant R&D works. It reviews current status of muon programs, and shows the importance of the muon program in the future of particle physics. Several technological developments commonly interested in between the muon programs and the neutrino-factory programs were also discussed. The idea of "Staging" towards the neutrino factories will be shown.

INTRODUCTION

Since muon was firstly discovered in cosmic rays at 1937, it has been playing an important role in the development of particle physics. For examples, the $V - A$ structure of the weak interaction was firmly established by measuring muon decay parameters precisely. The idea of two neutrinos was raised at early 60's because of the lack of $\mu \to e\gamma$ decay.

In the recent years, the importance of muon is increasing in three independent fields of science with three different roles:

- as a neutrino source with nicely controlled neutrino flavor and energy,
- as a direct object for the particle physics research,
- as a probe in the material/biological/chemical sciences.

Neutrino factories are good examples of the first role. The objective of this working group is to review later two roles, and find the common aspects between these three roles.

Muon Trio Experiments

Lepton flavor violation (LFV) is forbidden *a priori* in the Standard Model (SM). Even in the "new SM" which includes neutrino oscillation phenomena in the old SM, LFV process like $\mu \to e\gamma$ is greatly suppressed by GIM like mechanism to a level of 10^{-60}[1]. Thus, the experimental observation of LFV process directly indicates the

---

[1] Presentation at the conference was given by M. Aoki alone.

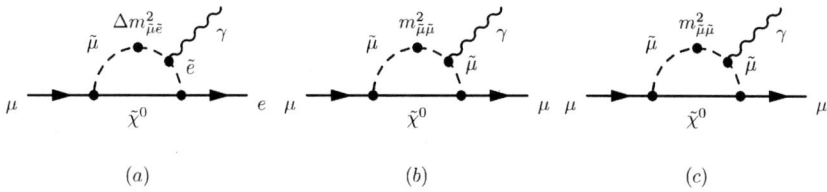

FIGURE 1. Typical Feynman diagrams for the "Muon Trio" processes; (a) photonic μ-LFV, (b) anomalous magnetic dipole moment ($a_\mu = (g-2)/2$), and (c) muon electric dipole moment (d_μ). The magnetic dipole moment comes from a real part of the diagonal muon interaction while the electric dipole moment is raised from the imaginary part.

physics beyond the SM. One of speculative models which predicts the LFV signal is based on Supersymmetry (SUSY) Grand Unified Theory (GUT). The calculations generally show branching ratios of only a few orders of magnitude below the current experimental limit. In this model, relatively large top mass results in substantial scale of the radiative corrections to the slepton mass matrix, and the off-diagonal element of the matrix might be large. As a result, smuon-selectron mixing may occur, and the μ-LFV process as shown in Figure 1(a) could be observed. In short words, a study of μ-LFV process will provide us an information of off-diagonal element of the slepton mass matrix.

On the contrary, muon $g-2$ provides information of on-diagonal element of the slepton mass matrix. Figure 1(b) shows a Supersymmetry process which could yield anomalous contribution to $g-2$. If we take the current experimental observation, in which the observed value is inconsistent with SM prediction, a generic slepton mass might be within a range of 150 – 670 GeV for $\tan\beta = 10 - 50$[2].

One can write both electric dipole moment (EDM) operator and magnetic dipole moment operator in a similar manner by using a generic dipole moment. The real part of the generic dipole moment corresponds to the magnetic dipole moment, thus contributes to $g-2$. The imaginary part of the generic dipole moment corresponds to CP violating dipole moment, thus EDM[3]. The diagram which yields the muon EDM (d_μ) in Supersymmetry is shown in Figure 1(c). This seems to be almost the same to that of $g-2$. In deed, they are about the same but the difference exists in the phases of the diagrams, and that leads us to the complex number of the generic dipole moment. While $g-2$ is related to the real part of the on-diagonal element of the slepton mass matrix, EDM should correspond to the imaginary part of the on-diagonal element. According to the current $g-2$ measurement, the muon EDM might be naturally in a range of $a_\mu = 10^{-22}$ ecm.

μ-LFV, muon $g-2$ and μ-EDM will provide us comprehensive information of the slepton mass matrix, thus these are called the golden "Muon Trio". The experimental study of these processes are thus quite important.

MEG experiment[4] is one of μ-LFV search programs recently under construction, and will start the data taking by the end of 2005. Its goal is to search for $\mu^+ \to e^+\gamma$ decay down to 4×10^{-14} at a single event sensitivity. MEG utilizes surface muons from proton cyclotron at PSI. The event signature of $\mu^+ \to e^+\gamma$ decay is a coincidence of both e^+ and γ going back-to-back at the monochromatic energy about $m_\mu/2$. The main background source may be an accidental coincidence of two individual decays of muons. Because of this, the maximum muon beam rate that MEG can handle strongly depends on the detector performance, and they developed a state-of-the-art technology of liquid Xe scintillation detector for this purpose.

MECO[5] is another effort to search for μ-LFV process, by looking at $\mu^- N \to e^- N$ conversion process. It will shortly start construction, and is planning to start the data taking around 2008. The event signature is a single electron with energy at $E_e = m_\mu - B_\mu$ where B_μ is a binding energy of μ^- atom. The background is high energy electron coming from μ^- decay in orbit, whose spectrum drops sharply as $(E - E_e)^5$. Thus, by limiting the detector acceptance close to E_e, the detector rate could be reduced down to the manageable level. This contrasts the case of MEG experiment, in which the background is limited by the accidental coincidence. Another advantage of measuring $\mu^- N \to e^- N$ conversion process over $\mu^+ \to e^+ \gamma$ is the sensitivity to non-photonic process. It is not only sensitive to the non-photonic process, but also is possible to identify the operator type of the process. Koike[6] reported that the atomic number dependence of the branching ratios of $\mu^- N \to e^- N$ conversion process are different for the various types of the operators. Thus, measuring the branching ratios for the different atoms will provide us the useful information to the type of the operator.

PRIME[7] is the next-generation experiment for $\mu^- N \to e^- N$ conversion search. PRIME utilizes high-intensity and high-brightness muon beam specially designed for this experiment, that is called Phase-Rotated-Intense-Slow-Muon (PRISM) beam. It will provide 10^{12}/sec of muons at 70 MeV/c with only 2–3% momentum spread. PRISM consists of 1) high power proton driver, 2) pion capture solenoid, 3) pion decay channel and 4) FFAG-based phase rotator. This is very similar to the front-end of the neutrino factories. The only difference is the muon beam energy; less than 100 MeV at PRISM and about 300 MeV at neutrino factories.

The current value of muon $g-2$ was measured at BNL[8] to be $a_\mu^{\text{exp}} = 11659203(8) \times 10^{-10}$, where $a_\mu = (g-2)/2$. Standard model prediction to a_μ is dominated by hadronic contribution, and that hadronic contribution is evaluated by using either $e^+ e^-$ cross section data or τ decay data. The former calculation gives $36.1(\pm 10.9) \times 10^{-10}$ discrepancy between the theory and experiment, while the later calculation is almost consistent with the data. Thus there should be further theoretical studies to sort out the discrepancy between $e^+ e^-$-based and τ-based calculations. This will naturally include the updates of $e^+ e^-$ scattering data from experiments. Improvement of the $g-2$ experiment will be also very important, and a plan of continuing the $g-2$ experiment at J-PARC with almost one order of magnitude better precision was reported[8]. Although the current experiment was limited by systematic errors, a part of these systematic errors could be reduced by using high power proton driver. That is, the error coming from the magnetic field non-uniformity will become smaller as the size of the aperture of the ring becomes smaller. The cost of using smaller-sized aperture is a drop of statistics, but that can be compensated by the increase of the density of muons in the beam; thank to the high-power proton driver.

The experiment measuring d_μ down to 10^{-24} is also planned at J-PARC[9]. The measurement requires excellent uniformity of both magnetic and electrostatic field for the muon storage ring. Thus, the emittance of the muon beam in the ring should be as small as possible. On the other hand, the figure of merit of the polarized muon beam should be $NP^2 \geq 10^{16}$ in order to achieve the goal, where N is the number of muons and P is the polarization of the muons. In order to realize such a muon beam, the current proposal utilizes a combination of pion decay solenoid and FFAG phase rotator ring called PRISM-II. The polarized muon will be collected by backward decay of pion in the pion decay solenoid channel. The momentum selection will be made by using the curved solenoid spectrometer. Finally, the momentum spread of the muon beam will be narrowed by the FFAG phase rotator. Note that the major difference between the PRISM ring and PRISM-II ring from the view point of the FFAG phase rotator design is their momentum; PRISM is for 100 MeV/c while PRISM-II is for 500 MeV/c.

μ-LFV and Neutrino Physics

There are models based on See-Saw mechanism with low energy SUSY[1], which provide quite sizable effect of the μ-LFV process. In these models, the slepton mass matrix carry information of neutrino yukawa couplings, thus we can draw close relations between μ-LFV process and neutrino oscillation phenomena. From this model, the branching ratio of $\mu \rightarrow e\gamma$ could be as large as 10^{-12} for the LMA solution.

There is other model based on SUSY with trilinear R-parity violation couplings which also draws relation between μ-LFV process and neutrino oscillation[1]. In general, LFV might be intimately related to the neutrino mixing, and in that case, LFV will be very important to understand the physics behind the neutrino masses.

Muon Precise Measurements

The precise measurements of muon decay parameters are reviewed in this working group. The topics covered were muon life time measurement, Michel parameter measurement and positron polarization measurement.

There are three fundamental inputs to the standard model; fine structure constant, Fermi constant and Z boson mass. The current precision of these parameters are 0.045 ppm, 9 ppm and 22 ppm, and the error on M_Z will be improved in near future. Thus, improving the Fermi constant is very important. Since the Fermi constant is determined by muon life time, the muon life time measurement is crucial. There are three on-going experiments for the precise measurement of the muon life time; these are RIKEN-RAL experiment, μLan, and FAST. RIKEN-RAL experiment already finished the data taking and its analysis is almost completed[10]. The error is dominated by the pile-up of background and the inefficiencies of gas chamber tracking system, and is expected to be a level of 10 ppm in total. μLan experiment only uses plastic scintillator and all the signals will be read by waveform digitizer[11]. This will reduce the systematics suffering the RIKEN-RAL experiment, and the goal of the μLAN is the measurement of muon life time at 1 ppm level. μLAN will start data taking from 2004.

TWIST is an experiment measuring Michel parameters, ρ, δ and $P_\mu \xi$ simultaneously[12]. The data taking was finished and the analysis is on going. The current data set should provide measurements of ρ and δ to 10^{-3}. The error of the measurement is dominated by systematics coming from the estimation of positron interactions in a detector.

A unique experiment at PSI measured positron polarization from muon decay[13]. It measured all the three components of the polarization, P_{T_1}, P_{T_2} and P_L. None-zero P_{T_2} would indicate time reversal violation at pure leptonic process, thus something beyond the standard model. The goal of the experiment is the measurements of P_{T_1} and P_{T_2} to 0.008, which is about one order of magnitude improvement over the last experiment.

Other Muon Programs

Muon is also interesting as a probe to other objects. Indeed, the applications of muon in the material/biological/chemical sciences are very unique and indispensable. This uniqueness is from a fact that positive-charged muon could be regarded as "light proton" while negative-charged muon could be taken as "heavy electron". In both cases, the spin motion of those "light proton" and "heavy electron" can be measured by using the Michel decay of them. Thus, muon is very useful to study the structure of atoms, molecules, proteins, polymer, metal and so on.

During the working group session, P. Strasser reported the status of his effort to form muonic atoms of unstable nuclei[14]. This will be realized by injecting negative muons and unstable nuclei simultaneously in the thin solid-Hydrogen sheet. In general, muons that captured by light atoms are effectively transfered to heavier atoms. Thus, muons captured by Hydrogen atom will quickly transfer to the unstable nuclei, and that improves the efficiency of forming the muonic unstable-nuclei. He could be able to provide unique measurement of the charge radii of the unstable nuclei with muonic X rays from such muonic atoms. This technique will be also useful to measure μ^--μ^+ conversion process[15]. Since the Q value of this process only depends on the difference in the muon binding energies between mother- and daughter-nuclei, the possible target is only three and all of them are unstable nuclei. Strasser's technique will be very crucial to increase the yield of the muonic unstable-nuclei.

TARGETRY AND PION CAPTURE DEVELOPMENTS

There are several key technologies that has to be developed for the future of muon program. These are 1) high power targetry, 2) large acceptance beam line, 3) muon beam manipulation. The muon beam manipulation includes the phase rotation and ionization cooling. Because that the same technologies are also indispensable to the neutrino factories, the collaboration in R&D between muon programs and neutrino programs might be very important to boost the developments. Based on this idea, talks related to the targetry and pion capture developments were reviewed in the joint session of working group 2 and 3.

Targetry

Targetry is a major issue that should be solved for any applications at 1-MW-or-more proton driver facilities. R. Bennett[16] reported a status of targetry study in Europe. H.G. Kirk reviewed an activity in USA[17]. Several different target design have been investigated, and will be kept considered. They are mercury jet target, contained flowing mercury target, granulated target, solid target and solid rotation ring target. High power beam tests of those designs were also performed for some of them. Both mentioned that the possibility of the collaborative R&D between spallation-neutrino-source facilities such as ESS and SNS. Such collaborations will be certainly very important to boost the R&D activity in this field.

H. Ohnishi and J. Popp told about the targetry R&D for their muon physics programs. In Ohnishi's report, he showed the result of heat load measurement performed at 12-GeV PS KEK[18]. He confirmed that the MARS calculation code is consistent with his measurement within 20% of accuracy. J. Popp showed his effort of building a pencil-shaped water-cooled target for MECO experiment[19].

In order to persuade the targetry design for the neutrino factories, high-power beam test facility is essential. Some of speakers mentioned that J-PARC will be one of such facilities.

Pion Capture and Decay

The design and optimization of the capture section requires iterations to many design parameters. In each iteration, Monte Carlo calculation has being used for the estimation of the muon yield, thus it is very time consuming work. It is also not so easy to understand the relations between several design parameters. K. Paul showed his effort of decay channel optimization based on such Monte Carlo methodology[20]. He found that the muon yield at the exit of the decay channel strongly depends on the magnetic field strength of the intermediate channel. F. Meót showed another effort of the decay channel design. He introduced a Mathematica based tool for the π/μ beam calculation[21]. His intention is to overcome the difficulties of Monte Carlo study by using analytical formula of the beam transport. The result of his analytical calculation showed a good agreement with the Monte Carlo calculation, and that seems very encouraging. It is very important to combine several different tools in order to boost the total design work.

One interesting proposal that raised in the working group session was a funneling of capture sections in order to reduce the engineering difficulties of the high power targetry[22].

Two reports on steady developments of the high intensity slow muon beam lines at KEK[23, 24] were shown. From the talks, it became much more clearer that the pion capture design is a common issue for both the muon- and neutrino-factory.

There was another report about the muon beam development for the muon program, which aims to produce very slow muon beam by laser-ionization of muonium atoms[25]. The test showed promising result and the further improvement of the total efficiency is awaited.

DISCUSSION AND CONCLUSIONS

The review of the recent muon physics program showed us that the muon is still very interesting particle by itself since the property of muon decay provide us many fundamental information of the standard model and its extensions. In particular, the importance of the "Muon Trio" experiments are exceptional. Muon is also very important in the fields of the material/biological/chemical sciences because of its unique ability of probing the magnetic property of the target material. These muon oriented programs require drastic improvements of the muon beam quality, and the individual efforts to develop such advanced muon beams were already started. These are DAI-OMEGA, Super-OMEGA, laser-ionization slow muon beam, PRISM, PRISM-II and so on. These facilities will provide us unique muon beams for various muon related scientific fields from particle physics to the material science. We call such a next-generation muon beam facilities as "Muon Factories".

The technology required to build such an advanced muon beam is sometimes very similar to that needed for the neutrino factories. PRISM and PRISM-II gives a very good example; it utilizes high power proton beam, pion capture solenoid, pion decay channel and phase rotator. This is like a front-end of the neutrino factories. Indeed, the phase rotator ring of the PRISM-II could be directly used as the first ring of the FFAG based neutrino factory.

Thus, it might be a good idea that the both muon-programs and neutrino-programs work together for the realization of the Muon Factories at first. Then, gradually upgrade it towards the Neutrino Factories. In each stage, there will be varieties of physics outcomes. This "Staging" idea was emerged in the workshop, and seems to be quite attractive. Note that the "Staging" in here should not necessary to mean the staging of facility construction. It rather aims the staging of technology developments. Thus the neutrino factories can be built at the different site than the muon factories. The serious discussion about such "Staging" should be kept continued.

REFERENCES

1. Gouvéa, A. D., Theoretical Aspects of (Charged-) Lepton Flavor Violation [CLFV], these proceedings.
2. Nomura, D., Theory of Muon $g-2$ and Muon EDM, these proceedings.
3. Feng, J. L., Matchev, K. T., and Shadmi, Y., *Nucl. Phys. B*, **613**, 366–381 (2001).
4. Baldini, A. M., Status of the MEG experiment, these proceedings.
5. Yamin, P., The MECO Experiment, Coherent $\mu \to e$ Conversion in the Field of a Nucleus, these proceedings.
6. Koike, M., Theoretical Study on the Lepton Flavor Violating $\mu - e$ Conversion in Nuclei, these proceedings.
7. Sato, A., PRISM/PRIME, these proceedings.
8. Roberts, L., $g-2$, these proceedings.
9. Morse, W., Measuring the Electric Dipole Moment of the Muon and Deuteron in Storage Rings, these proceedings.
10. Tomono, D., Precise Measurement of the Positive Muon Lifetime at RIKEN-RAL Muon Facility, these proceedings.
11. Onderwater, G., Towards an improved determination of the Fermi coupling constant from the μLan experiment, these proceedings.

12. Kitching, P., TWIST, A Precision Measurement of Muon Decay at TRIUMF, these proceedings.
13. Fetscher, W., Measurement of the polarization vector of the e^+ from the decay of polarized μ^+, these proceedings.
14. Strasser, P., Radioactive Muonic Atom Studies with Intense Muon Beams, these proceedings.
15. Aoki, M., *Nucl. Instr. and Meth. A*, **503**, 258–261 (2003).
16. Bennett, R., Review of Target Developments in Europe, these proceedings.
17. Kirk, H. G., Targetry Program in the US, these proceedings.
18. Ohnishi, H., Study of Pion Capture Solenoids for PRISM, these proceedings.
19. Popp, J. L., MECO Production Target Developments, these proceedings.
20. Paul, K., Decay Channel Optimization using MARS, these proceedings.
21. Autin, B., Méot, F., and Verdier, A., Calculated life of π/μ beams in a decay channel, these proceedings.
22. Autin, B., Sievers, P., and Verdier, A., Funneling π's and μ's, these proceedings.
23. Miyadera, H., Nagamine, K., Shimomura, K., Tanaka, H., Ikedo, K. N. Y., and Ishida, K., Development of high intense surface muon beam using a large solid angle axial focusing channel, these proceedings.
24. Shimomura, K., Muon Channel at Muon Science Facility and feasibility of super intense surface muon beam, these proceedings.
25. Matsuda, Y., Development of a point positive source by laser excitation of muonium atoms, these proceedings.

NuFact'03 Machine Working Group Summary

T.R.Edgecock[1], S.Machida[2] and R.A.Rimmer[3]

[1]*Rutherford Appleton Laboratory, Chilton, Didcot, Oxon, UK*
[2]*KEK, 1-1 Oho, Tsukuba-shi, Japan*
[3]*Jefferson Laboratory, Newport News, VA, USA*

Abstract. The machine working group sessions at NuFact workshops have always been characterized by the presentation and discussion of both new ideas and the developments in existing concepts and by lively debate. The machine sessions at NuFact'03 were no exception to this. In this article, we will try and summarize the work presented and the discussion that took place.

1. INTRODUCTION

The machine working group session at NuFact'03 demonstrated the continuing development in the ideas for the accelerator components of future high intensity neutrino facilities. In comparison to NuFact'02, there was more emphasis this year on the Neutrino Factory, with focus on the most challenging or expensive items of the accelerator complex, as determined by recent studies in the US [1] and Japan [2]. In particular, there were many presentations related to a Neutrino Factory in Japan based on FFAGs and centered around the J-PARC accelerator complex [3]. A staged approach to building the complete machine is envisaged, with physics at each stage. In the US, design work has been focused on cheaper alternatives for the muon frontend of a Neutrino Factory and on the muon acceleration. Cooling rings have received a lot of attention and the parallel program of R&D on cooling has made considerable progress through the MICE experiment and the MUCOOL collaboration. In addition, the development of a target and collector able to withstand the 4MW proton beam has continued both experimentally and theoretically, particularly in Europe and the US. Finally, the first ideas for a World Neutrino Factory Design Study, that would tie together much of this work, were considered at this workshop.

In this article, we will attempt to summarize this work and discussion that took place both during and between the working group sessions.

2. PROTON DRIVER

The aim of the proton driver is to deliver a 1 MW beam in the first phase and 4 or 5 MW with a future upgrade. The output energy should be several GeV to 50 GeV depending on the acceleration schemes. One way to realize the facility is to upgrade existing machines. Two examples were shown during the working group session. One is an ISIS upgrade with a new synchrotron [4] and the other is an AGS upgrade with a

1.2 GeV superconducting linac [5]. Of course, to make a new facility from scratch is another option. Although it is not solely for neutrino production, J-PARC in Japan will be a brand new proton driver aiming at 1 MW beam from the beginning [6]. Also as a brand new proton driver, a fixed field alternating gradient (FFAG) based scheme is considered although no detailed parameters are determined [7]. The energy and beam power of these candidates are listed in table 1.

Table 1: Proton drivers – energy and beam power

	Design goal	Future upgrade
ISIS with new synchrotron	1 MW with 8 GeV	5 MW
AGS with 1.2 GeV SCL	1 MW with 28 GeV	unknown
J-PARC	1 MW with 50 GeV	4 MW
FFAG	?	?

The ISIS upgrade plan has three phases. First, they will make a 3 GeV synchrotron with 50 Hz repetition rate as a spallation neutron source, which will also be able to accelerate a beam to 8 GeV at 16.7 Hz to give a test facility for a Neutrino Factory. Secondly, the present ISIS synchrotron will be decommissioned and two new booster synchrotrons and linac will be constructed. Finally, two main synchrotrons operated at 25 Hz each will give 5 MW beam power. In parallel with the machine design, beam studies such as bunch compression and simulation will be carried out in ISIS.

The AGS upgrade plan relies on an increase of the repetition rate. This can be done both by shortening the injection time and by higher ramping rate of the bending field. The former is only possible with direct injection, instead of stacking a few batches from the booster. Therefore, an idea of a 1.2 GeV linac as an injector comes up. There are several reasons why they have picked a superconducting linac instead of warm one. One is simply because of the on-going SNS project, which uses a cold linac structure. Therefore, the same technology can be applied. The other may be more critical, that is the limited real estate that can accommodate the new injector. In any event, the average current is low with a 0.179% duty factor and the choice of superconducting RF sounds reasonable.

Construction of J-PARC started in 2001 and it will deliver a first beam around 2008. J-PARC is multi-purpose facility, not only for particle physics, and consists of a 3 GeV rapid cycling synchrotron and a 50 GeV slow cycling one. The 3 GeV machine will have spallation neutron users and the 50 GeV will deliver a beam either to Kaon factory users by slow extraction or to neutrino users by fast extraction. The upgrade path to a few MW beam power is similar to the AGS upgrade, that is to increase the repetition rate.

Every one agrees that we know how to construct a 1 MW facility even though it is a big challenge. However, the future upgrade to a few MW seems to be still a big jump and, at the moment, no one has the confidence to make it. FFAGs, on the other hand, can be a breakthrough to solve all the problems of conventional high intensity synchrotrons. Table 2 illustrates the uniqueness of an FFAG.

As planned for the AGS and J-PARC, one way, and probably the only way, to increase beam power is to increase repetition rate. Because of space charge effects and beam instabilities, the number of particles per cycle will be limited. Unfortunately, the repetition rate will be also saturated because of the ramping speed of lattice magnets. An FFAG, on the other hand, uses a static magnetic field and the repetition rate can be very high such as 1000 Hz. Although it is still premature to say that an FFAG is able to deliver high power protons as a proton driver, we should keep eyes on the development because of its high potential.

Table 2: Proton drivers – energy and repetition rate

	Design goal	Future upgrade
ISIS with new synchrotron	16.7 Hz (8 GeV)	25 Hz (6 GeV)
AGS with 1.2 GeV SCL	2.5 Hz (28 GeV)	unknown
J-PARC	0.3 Hz (50 GeV)	1 Hz (50 GeV)
FFAG	1,000 Hz	10,000 Hz

3. TARGETRY

There was a lively discussion as always on the subject of targetry. The issues are centered on the extremely high peak and average power dissipation in any target suitable for a neutrino factory. These include shock from peak power heating, cooling of average power, radiation damage, material lifetime and shielding. There was no shortage of ideas and a wide range of schemes and experimental tests were presented and discussed. Scarcity of funding and limited availability of suitable test beams were common themes.

Bennett and Kirk gave summaries of the European and US targetry programs respectively [8,9], and there were also presentations on the PRISM target in Japan [11] and the MECO target at BNL [10]. A wide range of target options have been proposed including free and confined flowing mercury, figures 1-3, liquid or gas cooled granular targets, figure 4, solid targets of low expansion or shock resistant materials, figure 5, and various rotating band, levitated or projectile targets, figures 6-7. These schemes attempt to address the thermal shock and heat dissipation problems in various ways. The liquid jets are disrupted by the beam pulse but on a timescale much slower than the particle production time, and they are regenerated before the next beam pulse arrives. The constrained mercury attempts to solve the problem of material degradation by constantly circulating the target material, although the entrance window or containment may still be problematic. The solid targets use materials with low thermal expansion rates such as carbon composites or Invar to minimize the thermal shock. These are radiatively or water cooled but target lifetime is still a concern, however they look feasible at ~1 MW power levels. Rotating band or other moving solid targets attempt to address the heating and radiation damage problem by spreading successive beam pulses over a larger surface area by continuously moving the solid material. This may work as long as the single-shot damage threshold is not reached. Cooling and servicing the target material and mechanism would be challenging.

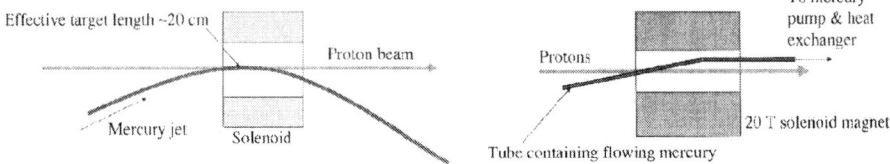

Figure 1: Schematic of a free mercury jet   Figure 2: A contained mercury target.

Good progress has been made experimentally to test several of these concepts. Thermal shock measurements have been made on metal and carbon solid targets at BNL, and material properties after high dosage exposure have also been measured, figure 8 [9]. Mercury jets have been tested at BNL with beam from the AGS (without solenoid field) [9], figures 9-10, and in a 20T solenoid at CERN/Grenoble (without beam), figure 11. The E951 experiment at BNL showed disruption of the mercury jet by the beam pulse but it was rapidly reformed and no damage was sustained by the nozzle. The CERN/Grenoble result shows that the strong magnetic field is helpful in damping surface instabilities in the jet. Both of these results are very encouraging. The logical next step is to perform a test with beam in a strong magnetic field. This will require a new magnet, figure 12, and beam time at a suitable facility such as AGS, PS or JHF.

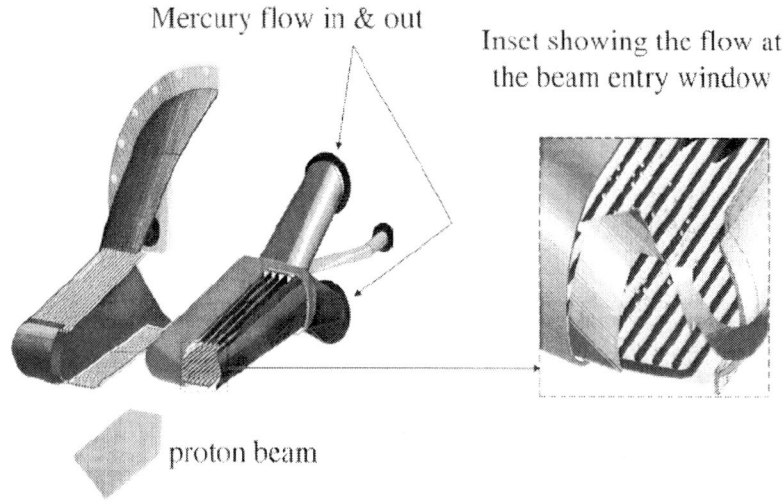

Figure 3: Proposed spallation neutron source target.

The MECO project, BNL E940, plans to use a water-cooled solid target [10]. They have already performed cooling tests on a sample using induction heating rather than a beam. The PRISM project in Japan has just started testing components of a current-carrying mercury loop target [11]. They have also directly measured the radiation heat load into a superconducting magnet with a beam test at KEK.

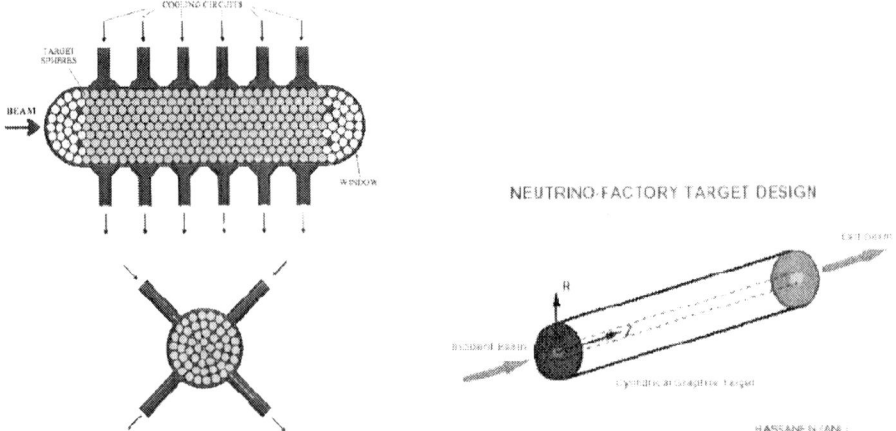

Figure 4: Granular target cooled by liquid   Figure 5: Solid carbon target.

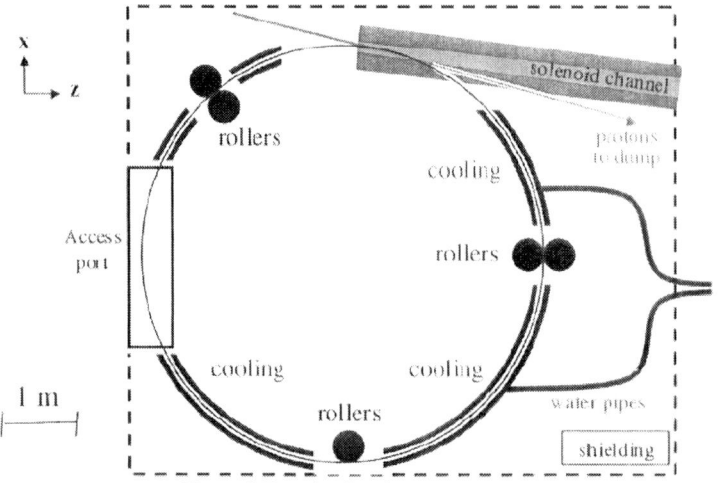

Figure 6: Rotating band target concept.

4. COOLING

4.1 Simulations

Raja et. al [12] presented results of simulations showing that Geant 3 is now ready and being used for cooling channel simulations. They showed examples of cooling rings with realistic field models, see figure 13, and that it is possible to model complex geometries with this package. ICOOL has been a stalwart of cooling channel development and is now being used for ring coolers. Kahn et. al. showed results for rings with realistic fields [13]. There is no longer any reason to be struggling with simulations with non-physical field approximations. Makino showed results from a first attempt to model a cooling ring using COSY [14]. The results showed a high sensitivity to magnet end effects. This work is continuing.

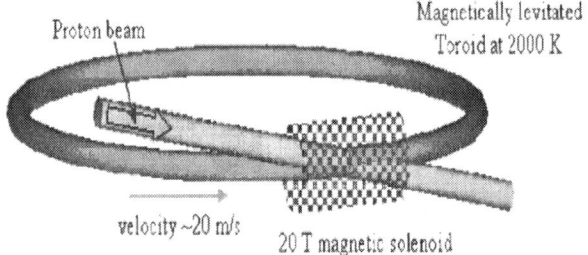

Figure 7: Magnetically levitated rotating toroid, radiatively cooled.

Figure 8: BLIP irradiation test target holder after 24 Rads.

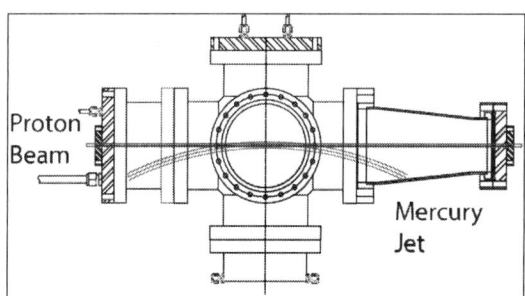

Figure 9: E951 Hg jet chamber (no magnetic field).

149

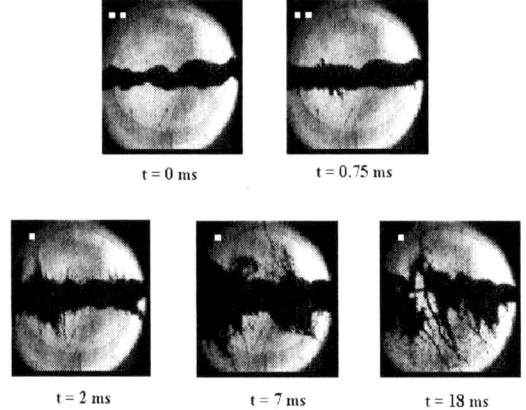

Figure 10: E951 Hg jet tests, 1 CM diameter jet, 16 GeV, 4 TP protons

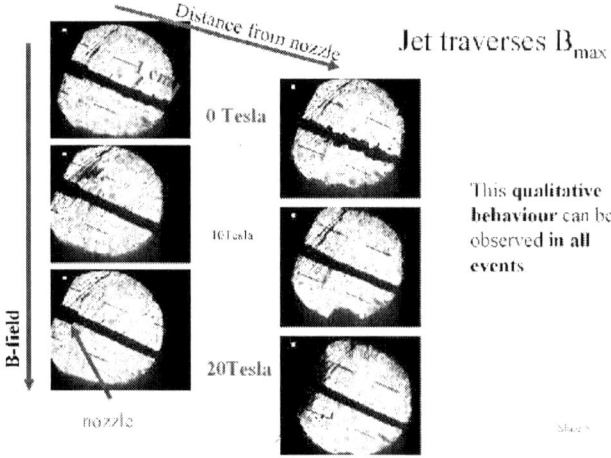

Figure 11: CERN/Grenoble test, 4mm jet, 12 m/s, 0, 10, 20 T, no beam.

4.2 Ring coolers

There was a lot of progress on ring coolers presented at this meeting. The Balbekov ring cooler, figure 14, has now been simulated with ICOOL and Geant with realistic fields [13] and there is good agreement between the codes. Significant cooling in 6D is being achieved in the simulations. Work is in progress to replicate this using COSY [14]. Work has continued apace on the RFOFO ring, figure 15. Fernow et. al. presented results from ICOOL showing at least a four fold increase in muon density [15]. The ring has also been simulated by Balbekov using his own code, and work is underway to model it in Geant. Palmer et. al. have looked at perturbations of the ideal ring including missing cells to allow for injection and extraction, various absorber materials and window thicknesses. The results are very encouraging. Kicker R&D is clearly needed for injection and extraction of such large beams but the parameters of these devices are not beyond the bounds of possibility [20]. Kirk et. al. discussed options for dipole and quadrupole rings, figure 16 [16]. These have the advantage that the RF is outside of the magnets but like the other rings a very demanding kicker is required for injection and extraction. With wedge absorbers good 6D cooling is seen in simulations. Berg pointed out that magnet edge effects are very important in this type of ring. A possible scheme for further transverse cooling is the lithium rod concept. Fukui and Garren showed how this could be used in a solenoid channel or a ring, figure 17 [16]. This method could be used downstream of the other cooling rings to achieve lower transverse emittances but presently does not provide any longitudinal cooling, in fact there is emittance growth in the longitudinal plane. The ring concept does include space for injection and extraction however.

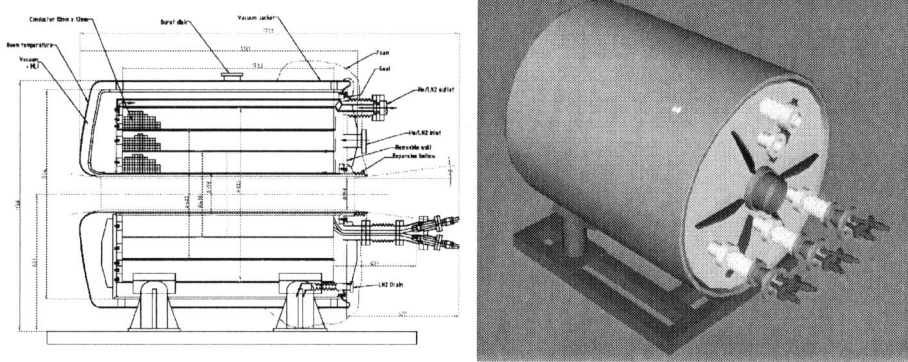

Figure 12: High field pulsed solenoid. 70° K Operation, 15 T with 4.5 MW Pulsed Power, 15 cm warm bore, 1 m long beam pipe.

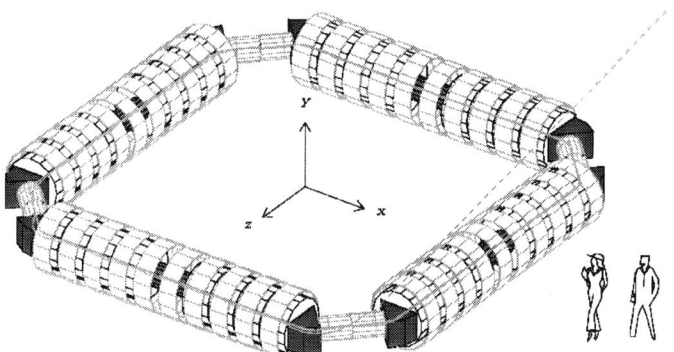

Figure 13: Cooling ring modeled in Geant 3.

Rees summarized the Muon frontend studies at RAL [17]. They have been looking at a racetrack cooling ring, figure 18, using solenoids and dipoles. It uses low-frequency RF, like the other ring coolers, and has room for injection and extraction in the straight sections of the racetrack. Simulations are under way using OPERA, Parmela and another new code at RAL.

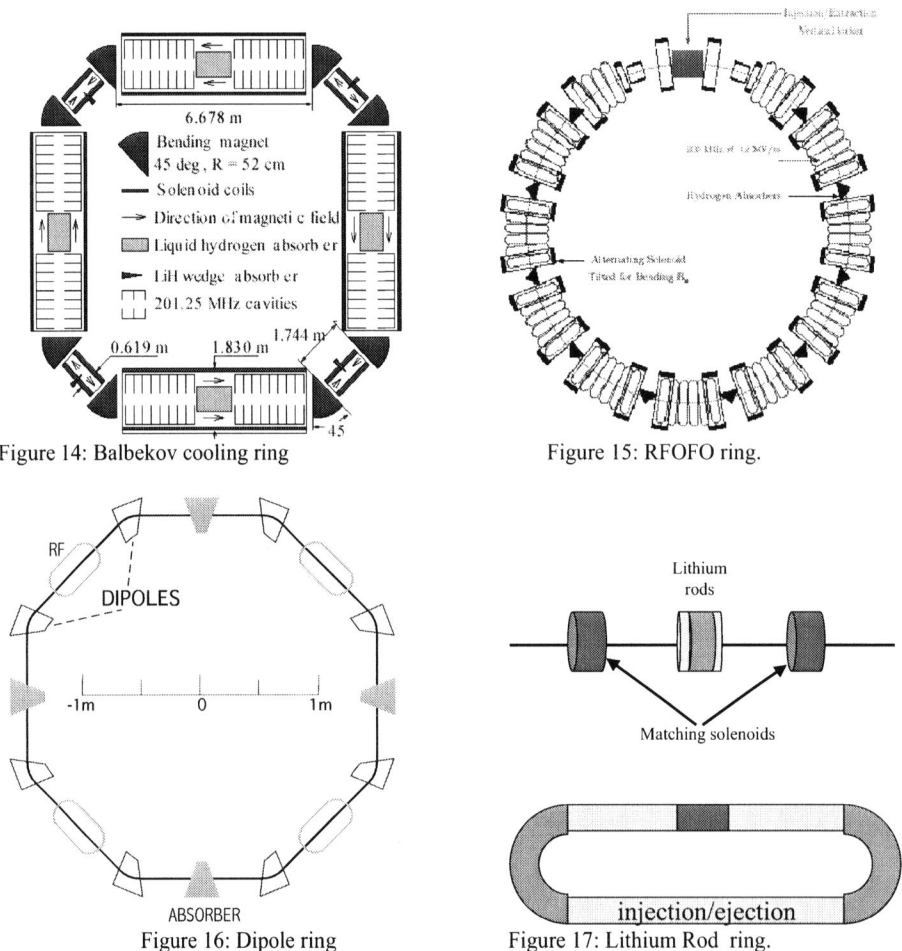

Figure 14: Balbekov cooling ring

Figure 15: RFOFO ring.

Figure 16: Dipole ring

Figure 17: Lithium Rod ring.

4.3 Alternative Cooling methods

Galea presented a method for frictional cooling in helium gas [18]. He showed simulation results that indicated that the method should work and attempted to verify this with an experiment to cool low energy protons. The experiment did not have a solenoid field and was not able to see the effect but provided some useful data to compare with simulations and some practical experience. Galea is planning a follow up experiment that will include the magnetic field.

Johnson et. al of Muons Inc. presented the latest results in the experimental demonstration of the feasibility of gaseous ionization cooling [21]. Perhaps their greatest success was in developing a new funding source (Muons Inc. has been awarded four small business grants to develop technology for gaseous ionization cooling). This scheme uses high-density gaseous hydrogen as the cooling medium. The hydrogen cools the muon beam but also allows for potentially higher gradient in the RF cavities because it suppresses breakdown and cools the cavity windows. Because the cooling medium is distributed within the RF cavities the packing factor is also greatly improved. Muons Inc. has built an 805 MHz test cell, figure 19, to measure the Paschen curve for gasses, particularly hydrogen, up to high gradients. The data agrees well with DC breakdown data in the literature and extends the range of measurements up to 50 MV/m, see figure 20. At this point the surface finish of the test cell electrodes appears to become significant. Future tests will attempt to improve this or condition the surface to allow higher gradients. Muons Inc. has also been awarded grants to study high-power pulse compression and 6D cooling schemes based on similar technology.

S = solenoid, A = absorber, 36 cavities in blocks of 3

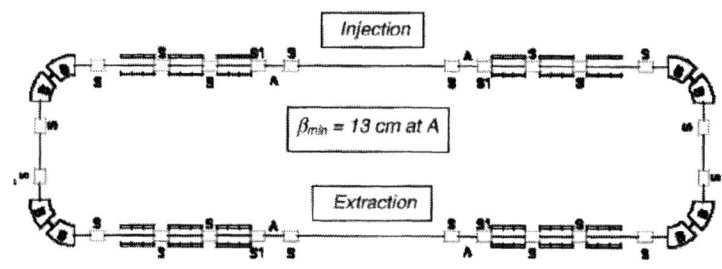

Figure 18: RAL racetrack cooling ring.

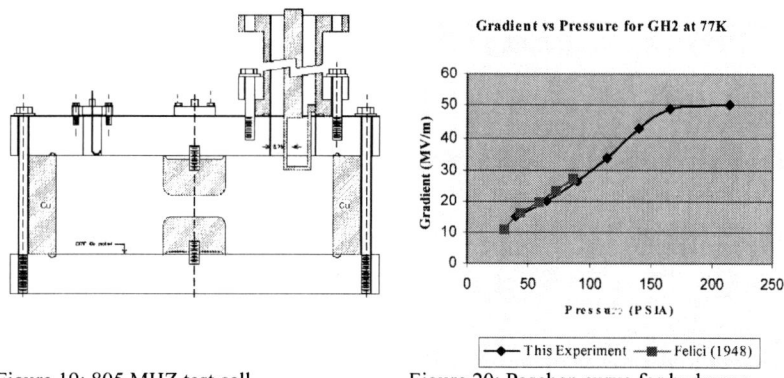

Figure 19: 805 MHZ test cell.   Figure 20: Paschen curve for hydrogen.

4.4 RF bunching and phase rotation

Some of the most expensive components of the study II Neutrino Factory were the induction linacs for phase rotation. Study II established feasibility but since then work has been ongoing to establish a more cost optimal solution. One approach to this is to use RF phase rotation and bunching, as proposed by Neuffer et. al. [19], see figure 21. Simulations show good performance for a system that would be substantially less expensive than that in study II. To get the best performance from this scheme the downstream cooling channel should be re-optimized. This method also captures both signs of muons. Work is continuing to look for the best combination of RF frequencies and to compare simulations with different codes. Iwashita showed an alternative RF bunching scheme using modulated RF [22]. This could be achieved using a few frequency components. Initial studies are promising and this work is also continuing.

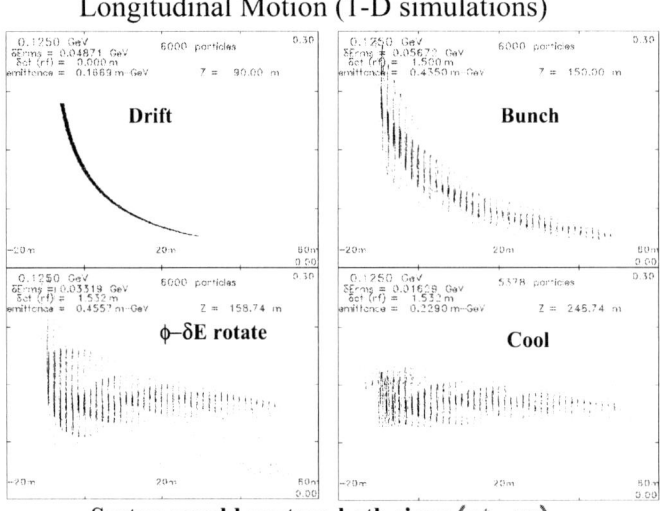

Figure 21: RF bunching and phase rotation [19].

4.5 Bunch compressor for a muon collider

Balbekov showed a complete end to end layout for a possible muon collider [23]. This consisted of a target station, phase rotation, bunch compressor ring, matching section, RFOFO cooling ring and lithium lens channel. The scheme is still missing kickers but the other components can now be simulated. This kind of end-to-end simulation will be vital for overall system optimization in the next design study.

4.6 MICE and MuCool

A cooling channel forms one of the technically most changing aspects of a Neutrino Factory accelerator complex. It has long been released that an R&D program is essential to demonstrate that such a channel can be built and will cool and to learn more about the cooling process. The MUCOOL collaboration [24] has existed since 1998 and is focusing on the construction of the elements of a cooling channel, the demonstration that they can work together as expected and the operation of a cooling cell in an intense proton beam. The MICE collaboration [25] will take the cooling cells developed in collaboration with MUCOOL and use a muon beam to show that they will cool and to learn more about the cooling process.

4.6.1 MuCool

The MUCOOL collaboration has already begun the development and construction of all elements of a cooling channel, in close collaboration with MICE. In particular, a 805MHz RF-cavity has been constructed and tested in Lab G in Fermilab partially inside a super-conducting coil very similar to a focus coil. This has produced results of great importance to MICE. It is well-known that RF-cavities produce electrons by field emission and the tests in Lab G have shown that very large numbers of particles can result, see figure 22. This is a problem for MICE because although these electrons will be stopped in the cooling cell absorber, they will produce x-rays which could form a significant background in the detectors used to track the muons (see section 4.6.2). In addition, the electrons are focused by the field from the super-conducting magnet and became intense enough to make a small hole in the window used to seal the vacuum in the cavity in the initial tests in Lab G. A program of measurements has been undertaken to better understand the electron production mechanism and to find ways of reducing it [26].

As well as tests with a 805MHz cavity, the design and construction of a 200MHz cavity, the correct frequency for a cooling cell, has also started [27].

Another area of particular interest has been the construction of liquid hydrogen absorbers [28]. These have two special problems as far as a Neutrino Factory is concerned: safety and removing the heat deposited by the muon beam. The most important safety problem is the windows of the absorber [29]. These need to be as thin as possible to minimize their heating effect on the beam, but they must also satisfy the hydrogen safety requirements. A number of window designs have been considered, all of which make the window thinner in the center than at the edges. Five torispherical windows, with a thickness of less than 400μm at the center, have been built and tested to destruction at Fermilab to satisfy the local safety requirements. Finite element calculations suggest that an alternative design with a thickness of less than 140μm will also satisfy these requirements. As far as the heat removal is concerned, FEA calculations of heat deposition are also being made and the proposed schemes for its removal tested [29].

To aid in the MUCOOL aims, a new, dedicated test area is being built in Fermilab at the end of linac [30]. This will allow testing of all elements of a cooling cell together and, in particular, the use of liquid hydrogen, which is not possible in Lab G.

At a later date, it will also be possible to address the final aim of MUCOOL, passing a high intensity proton beam through the cooling cell to test its operation under these conditions.

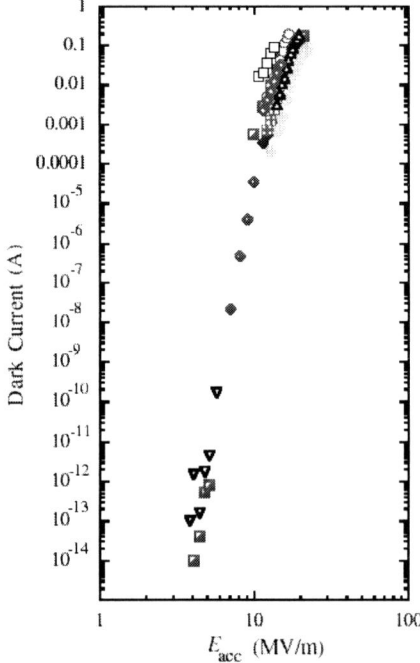

FIGURE 22. Electron current measurements made with the 805MHz cavity in Lab G at Fermilab under a variety of conditions (see [26] for more details).

4.6.2 MICE

As already mentioned, the aim of mice is to demonstrate that ionization cooling works and to learn more about the cooling process. It will use two cooling cells based on the SFOFO cells from US Study II [1] and place instrumentation around them to identify muons and measure their parameters entering and leaving these cells, see figure 23.

The two cooling cells will bring a reduction of about 10% in the 6D normalized muon emittance, which is not measurable by collective techniques. Therefore, MICE will use a single muon beam, with individual muons passing through the detector in a given time period [31]. The "beam" emittance will then be reconstructed offline from the measured parameters of the muons.

The collaboration submitted a proposal to run the experiment at the Rutherford Appleton Laboratory to an international review panel in January 2003 and defended it in February. The review panel recommended approval of the experiment in May and the collaboration is currently seeking funding to build it. Design and prototyping work continue in parallel.

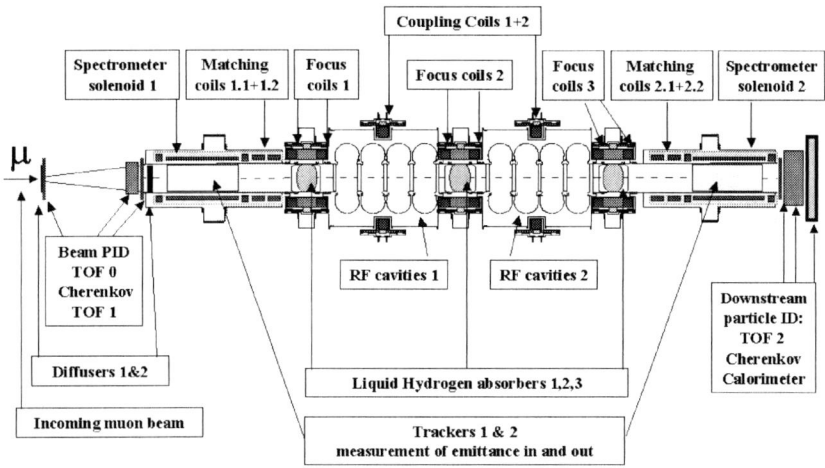

FIGURE 23. Schematic layout of the MICE experiment.

5. MUON ACCELERATION

As in the previous workshops, three schemes were presented to accelerate muons. One is a re-circulating linear accelerator (RLA), another is a very rapid cycling synchrotron (VRCS) and the last is a fixed field alternating gradient (FFAG) synchrotron.

5.1 Re-circulating linear accelerator (RLA)

There were three talks related to re-circulating linear accelerators (RLA). One was an overview talk on a European Neutrino Factory based on two RLAs by Meot [32]. Another was an optics update presented by Bogacz [33]. The third was on 200 MHz superconducting cavity development by Geng [34].

The so-called CERN design assumes three stages of acceleration of the muons. It starts with pre-acceleration by a linac from 0.3 to 3.0 GeV. The second stage utilizes an RLA that accelerates muons from 3 to 11 GeV in the same RF cavity four times. The final stage is again an RLA from 11 to 50 GeV followed by a 50 GeV triangular muon storage ring.

The initial beam emittance after the cooling channel and therefore before the RLA was updated assuming a ring cooler instead of a cooling channel which was proposed in the Study II. Now, both transverse and longitudinal emittance are reduced and optimization of a RLA based on those new parameters was done. In particular, the reduction of the longitudinal emittance, which is about a factor of 10, impacts on the RLA design a lot. Further optimization is in progress.

Superconducting RF cavity development is good news from a hardware aspect. The field gradient of 11 MV/m is achieved although some problems such as so-called Q-slope still remain.

5.2 Very rapid cycling synchrotron (VRCS)

A more detailed study of very rapid cycling synchrotron (VRCS) was presented by Summers [35]. Here "very rapid" means 4600 Hz, which is similar or even higher than the repetition rate of the FFAG proton driver discussed. One concern of such a high cycling operation is power consumption and voltage induced in a magnet. In fact, the energy stored in each magnet is around 100 kJ and the voltage is 120 kV. The power consumption is, however, still low due to low duty factor and a center tap of SCR stacks eases the voltage problem, namely +-60 kV. The other issue is an effect of eddy current on a beam. That needs more study.

5.3 Fixed field alternating gradient (FFAG)

This year, more than a half of the machine working group talks discussed a fixed field alternating gradient (FFAG) acceleration scheme. Palmer presented an overview of FFAG schemes in comparison to others [36]. Another talk was the issue of longitudinal dynamics of a non-scaling machine presented by Johnstone [37]. Four talks were on a scaling machine from several view points, such as updated lattice design and particle tracking study [38], superconducting and normal conducting magnet design [39], and one application as a phase rotator of muons which is called PRISM [40].

The first question was whether a scaling or a non-scaling FFAG is suitable for a muon acceleration ring. Besides the beam dynamics issues, Palmer claims that the superconducting magnets become cheaper in a non-scaling machine because of smaller aperture and simpler structure. The design of superconducting magnets for a scaling machine was presented and that shows there are not many technical differences. In any case, this needs further study for both scaling and non-scaling magnets when more detailed optics design is made.

The second question was whether a low or high frequency is adopted. At the moment, a low frequency RF system such as a few MHz is assumed in the scaling design and its gradient is around 1 MV/m. The non-scaling design takes high frequency such as 100 MHz or more and expects a higher gradient than the low frequency RF system. Higher gradient is preferable because of less beam loading and cost. Obviously, other combinations, for example a high frequency system with scaling machine, should be possible and needs to be examined. It is probably true that there is not much difference in cost.

An interesting observation was made by Palmer with respect to the third question, that is whether cooling is necessary or not. The question can be phrased in other words, namely which is smaller: the cost of a cooling system or the cost reduction due to smaller acceptance with a cooling system. Palmer's answer was that large acceptance magnets for non-cooled beams are cheaper. Furthermore, the cooling system can be a future option to obtain more muons if necessary. The audience seemed to agree.

The fourth issue is if a kicker is feasible for an FFAG design. The kicker is also an issue for a ring cooler. The general consensus was that the kicker for an FFAG is easier than one for a ring cooler for three reasons. The first is the smaller aperture in an FFAG. The second is the larger circumference and therefore longer rise time. The last is that in an FFAG the orbit kick is made not only by kickers but also by the main lattice magnets.

Finally, a staging scheme for an FFAG based design was discussed taking as an example a path starting from the PRISM project to a Neutrino Factory. It should be emphasized that physics outputs will be made in each stage of the path, see table 3.

6. DESIGN STUDY

So far, three Neutrino Factory feasibility studies have been undertaken, two in the US, in 2000 [41] and 2001 [1], and one in Japan in 2002 [2]. All had limited participation from outside the respective regions. It was recognized in the build up to NuFact'03 that the time is ripe to embark on a "third" study, but that this should be a World Study with participation from all three regions. As a result, a session of the machine working group was devoted to presentations of the form such a study should take as seen from each region and a discussion of how to proceed. In this section, we will give a brief summary of the presentations and outline the conclusions of the discussion.

6.1 View from Europe

The view of the World Design Study (WDS) from Europe is largely determined by a planned bid for funding from the European Union Framework 6 (FP6) Design Study program [42]. The Framework Programs are a method by which the European Commission (EC) provides funding for science in Europe. A number of activities are specified against which bids can be made and in FP6 this includes Design Studies. These will provide funding for both theoretical studies of new infrastructure, including detailed engineering, and R&D on new technologies involved in this infrastructure. About 10 MEUR per study is expected to be available over a time period of up to 4 years. Collaboration with partners outside Europe is strongly encouraged.

In Europe, it is planned to submit a proposal to this program. This is very important for two reasons: (1) to re-invigorate Neutrino Factory activities in Europe, which have suffered considerably following budget cuts at CERN, and (2) to provide funding for a number of planned R&D projects. As a result, Europe would envisage the WDS lasting 4 years, producing a feasibility study for a Neutrino Factory after 2 years and a report including more detailed engineering after a further 2. R&D would be an integral part of the study.

6.2 View from Japan

A first design study for a Neutrino Factory in Japan was completed in 2002. This is centered around the J-PARC facility already under construction at the Tokai campus of the JAERI and is based on FFAGs. A staged approach to building the Neutrino Factory is envisaged, with physics at each stage, as shown in table 3.

TABLE 3. Staged approach to the construction of a Neutrino Factory in Japan.

Stage	Facility	Main physics outcome
1	High power proton driver	Muon g-2
2	PRISM FFAG ring	Muon LFV decays
3	PRISM II ring	Muon EDM
4	1MW Neutrino Factory	Oscillation parameters
5	4.4MW Neutrino Factory	CP-violation
6	Muon collider	Higgs mass & width, etc.....

The plan in Japan for their contribution to the WDS, in the early stages at least, is to continue to investigate FFAGs, both theoretically and experimentally, to assess whether they are a realistic option for a Neutrino Factory. They would seek active participation from the other regions in this work.

6.3 View from the US

Two feasibility studies have taken place in the US, the first based at Fermilab [41] and the second at Brookhaven [1]. The aim of the first was simply to demonstrate the feasibility of the machine and cost was of secondary importance. This it achieved, though the performance was poor. The aim of the second study was to improve on the performance of the first, with the cost being an output of the study. This was again achieved, though the output cost was rather high. For the US, the aim of a third study would be to take the output of the second and reduce the cost, while maintaining the performance. This is seen as a theoretical design and engineering effort, with R&D being done in parallel, but not part of the study.

6.4 Conclusions from the Discussion

Although the discussion which followed the presentations was wide-ranging, considerable time was spent on reconciling the differences in the views summarized above. As a result, a number of important conclusions were drawn:
- There should be a world-wide design study
- A steering group should be created to start the study off
- The study should aim to produce a cost-optimized Neutrino Factory design
- It should include detectors as well as the accelerator

Following the discussion, a six person steering group was created with the membership shown in table 4 and tasked with starting the WDS.

ACKNOWLEDGMENTS

The conveners of Working Group 3 would particularly like to thank all the speakers for the work they put in to create excellent presentations and share their knowledge with us and for largely keeping to time! In addition, we would like to thank all the participants in the working group for their enthusiasm and important contributions to the discussion.

TABLE 4. WDS steering group

Region	
Europe	Alain Blondel
	Rob Edgecock (chair)
	Helmut Haseroth
Japan	Yoshi Kuno
	Yoshi Mori
	Kenzo Nakamura
US	Steve Geer
	Bob Palmer
	Mike Zisman

REFERENCES

1. Ozaki, S., Palmer, R., Zisman, M., and Gallardo, J., eds. *BNL-52623* (2001).
2. See http://www-prism.kek.jp/nufactj/nufactj.pdf
3. J-PARC Joint Project Team, *KEK Report 2002-13* (2002)
4. Prior, C., these proceedings.
5. Rugierro, A., these proceedings.
6. Mori, Y., these proceedings.
7. Nakano, J., these proceedings.
8. Bennett, R., these proceedings.
9. Kirk, H., these proceedings.
10. Popp, J., these proceedings.
11. Yoshimura, K., these proceedings.
12. Raja, R., these proceedings.
13. Kahn, S., these proceedings.
14. Makino, K., these proceedings.
15. Fernow, R., these proceedings.
16. Kirk, H., these proceedings.
17. Rees, G., these proceedings.
18. Galea, R., these proceedings.
19. Neuffer, D., these proceedings.
20. Palmer, R., these proceedings.
21. Johnson, R., these proceedings.
22. Iwashita, Y., these proceedings.
23. Balbekov, V., these proceedings
24. MuCool Collaboration, *Fermilab Proposal P904* (1998).
24. Edgecock, R., *J. Phys.* **G29** 1601-1611 (2003).
25. Norem, J, et al, *Phys. Rev. ST Accel. Beams* **6** 072001 (2003).
27. Li, D., these proceedings.
28. Cummings, M.A., these proceedings.
29. Lau, W., these proceedings.
30. Bross, A., these proceedings.
31. Drumm, P., these proceedings.
32. Meot, F., these proceedings.
33. Bogacz, A., these proceedings.
34. Geng, R., these proceedings.

35. Summers, D., these proceedings.
36. Palmer, R., these proceedings.
37. Johnstone, C., these proceedings.
38. Machida S., and Yokoi, T., these proceedings.
39. Yoshimoto, M., and Ogitsu, T., these proceedings.
40. Sato, A., these proceedings.
41. Finley, D., and Holtkamp, N., eds., see
http://www.fnal.gov/projects/muon_collider/nu-factory/fermi_study_after_april1st/
42. See http://fp6.cordis.lu/fp6/call_details.cfm?CALL_ID=76

WORKING GROUP 1

Global Fits to Neutrino Properties

M. Maltoni

Instituto de Física Corpuscular – CSIC/UVEG,
Edificio Institutos de Paterna, Apt. 22085, E-46071 Valencia, Spain

Abstract. In this talk we report on the present status of neutrino oscillations, in the context of both standard and non-standard neutrino physics. First, we consider the standard three–neutrino scenario, presenting a global fit of solar, atmospheric, reactor and accelerator neutrino experiments. Then, we show that neither four-neutrino models nor CPT-violating 3ν models can reconcile LSND with the results of all the other neutrino experiments. Finally, we concentrate on the analysis of solar and KamLAND data, showing that it is possible to extract from them meaningful bounds on the amplitude of matter-density fluctuations in the solar interior.

INTRODUCTION

In the last few years, a large amount of data on neutrino oscillations has been published. As a consequence, a rather clear picture of the neutrino sector is starting to emerge. In particular, the results of the KamLAND reactor experiment [1] have played an important role in selecting the large mixing angle MSW solution (LMA) as the preferred solution to the solar neutrino problem, and in confirming that the disappearance of solar electron neutrinos is mainly due to oscillations and not to other types of neutrino conversions. Presently, all the data collected by solar, atmospheric, reactor and accelerator neutrino experiments, with the only exception of LSND, can be very naturally accommodated within a three-neutrino framework. The level of accuracy reached by neutrino experiments is such that from now on neutrinos will no longer be regarded only as a subject of investigation, but also as a powerful research instrument.

In this work we report on the present status of neutrino phenomenology, in the context of both standard and non-standard neutrino physics. First, we present our results for a complete three–neutrino fit to the data collected so far by solar, atmospheric, reactor and accelerator neutrino experiments, leaving out only LSND. Then, we discuss two classes of non-standard scenarios, namely four-neutrino models and three-neutrino models with CPT-violation, which were originally proposed as an attempt to account also for the LSND result, and we prove that both of them are now ruled out. Finally, we focus on solar neutrino data, showing how they can provide new information about matter-density fluctuations in the solar core on much shorter scales than those which can be presently probed by existing constraints, such as helioseismology.

THREE-NEUTRINO OSCILLATIONS

In this section we present an updated analysis of solar, atmospheric, reactor and accelerator neutrino data (excluding LSND), in the framework of a three-neutrino scenario. The

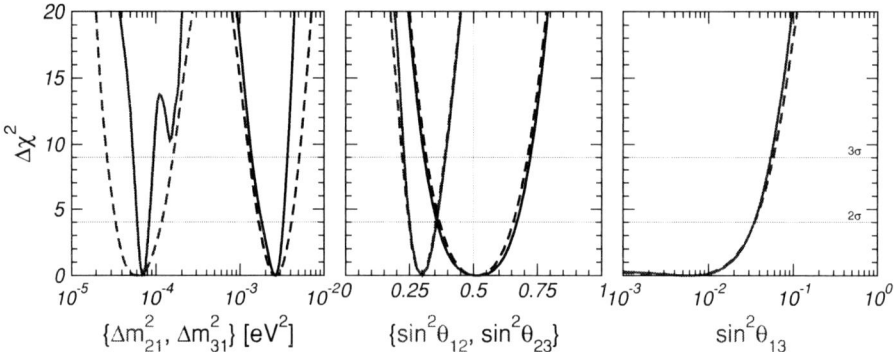

FIGURE 1. $\Delta\chi^2$ as a function of the five neutrino oscillation parameters Δm_{21}^2, Δm_{31}^2, s_{12}^2, s_{13}^2 and s_{23}^2. In each panel the undisplayed parameters are marginalized. Solid lines refer to the combination of solar, atmospheric, Chooz, KamLAND and K2K data. For dashed lines the KamLAND and K2K experiments are omitted.

details of our analysis are given in Ref. [2]. Our results are summarized in Fig. 1, where we show the projection of the $\Delta\chi^2$ function over the five neutrino oscillation parameters Δm_{21}^2, Δm_{31}^2, s_{12}^2, s_{23}^2, s_{13}^2. Solid lines refer to the combination of all data samples, while for dashed lines only solar, atmospheric and Chooz data are included.

From the first panel, we see that the inclusion of KamLAND and K2K data drastically improves the determination of the two mass-squared differences Δm_{21}^2 and Δm_{31}^2. In particular, K2K is responsible for the strong improvement of the upper bound on Δm_{31}^2, leaving the lower bound essentially unaffected. On the other hand, the recent KamLAND result single out LMA as the only viable oscillation solution to the solar neutrino problem. As noted in Ref. [3], the original LMA region is split by KamLAND into two separate islands, called LMA-I (characterized by $\Delta m_{21}^2 \approx 7 \times 10^{-5}$ eV^2) and LMA-II (with $\Delta m_{21}^2 \approx 1.5 \times 10^{-4}$ eV^2). The inclusion of the recent SNO-salt data practically rules out the LMA-II solution, which is now disfavored with a $\Delta\chi^2 = 10.3$ with respect to LMA-I. The best-fit point for the global analysis is located at $\Delta m_{21}^2 = 6.9 \times 10^{-5}$ eV^2 and $\Delta m_{21}^2 = 2.6 \times 10^{-3}$ eV^2, with allowed 3σ ranges 5.4×10^{-5} eV$^2 \le \Delta m_{21}^2 \le 9.5 \times 10^{-5}$ eV^2 and 1.4×10^{-3} eV$^2 \le \Delta m_{31}^2 \le 3.7 \times 10^{-3}$ eV^2. Note that we are *not* including in our calculations the changes in the detector simulation, data analysis and input atmospheric neutrino fluxes which the Super–Kamiokande collaboration has recently presented [4], since it is not possible to recover enough information to incorporate them into our codes. We note, however, that our value for Δm_{31}^2 is statistically compatible with the SK prediction $\Delta m_{31}^2 = 2 \times 10^{-3}$ eV^2.

The determination of the oscillation angles θ_{12} and θ_{23} (second panel) is dominated by solar and atmospheric data, respectively, while reactor and accelerator experiments only play a marginal role in this respect. The recent SNO-salt data drastically improves the upper bound on the solar angle θ_{12}, and maximal mixing $s_{12}^2 = 1/2$ is now ruled out at more than 5σ. Conversely, atmospheric data clearly prefer $\theta_{23} = 45°$. The best-fit point is characterized by $s_{12}^2 = 0.3$ and $s_{12}^2 = 0.52$, with allowed 3σ ranges $0.23 \le s_{12}^2 \le 0.39$

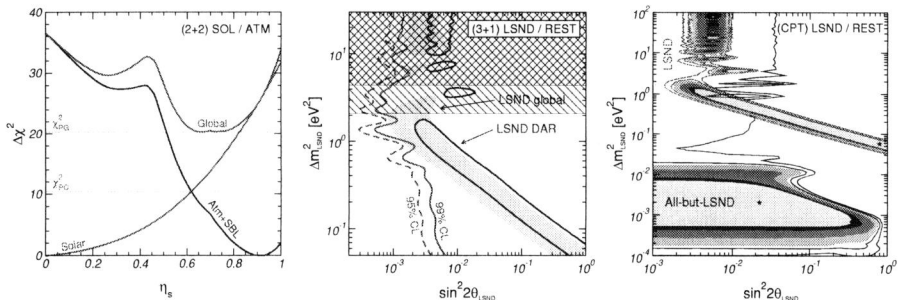

FIGURE 2. Left: $\Delta\chi^2_{\rm SOL}$ and $\Delta\chi^2_{\rm ATM}$ as a function of η_s, in (2+2) four-neutrino models. Middle: allowed regions from the analysis of LSND and of All-but-LSND results, in (3+1) four-neutrino models. The region above 2.1 (4.4) eV^2 is excluded at 2σ (3σ) by WMAP data. Right: allowed regions for LSND and All-but-LSND in CPT-violating three-neutrino models.

and $0.36 \leq s^2_{23} \leq 0.67$.

Finally, the bound on the reactor angle θ_{13} (third panel) is mainly dominated by the Chooz measurement, although solar and atmospheric data are essential to rule out those regions of the parameter space where the sensitivity of Chooz to θ_{13} is poor. Also, solar data play an important role in compensating to some extent the loosening of the Chooz bound on θ_{13} due to the downward shift of Δm^2_{31} reported by the Super–Kamiokande collaboration in their reanalysis of atmospheric data [4]. As for the best-fit point, we found $s^2_{13} = 0.006$; this deviation from zero is small and statistically irrelevant. The 3σ allowed range is $0 \leq s^2_{13} \leq 0.054$.

FOUR-NEUTRINO MODELS AND CPT VIOLATION

The LSND experiment observed an appearance signal in the channel $\bar{\nu}_\mu \to \bar{\nu}_e$. The interpretation of this evidence in terms of neutrino oscillations requires a mass-squared difference of the order of ~ 1 eV^2, which is incompatible with solar and atmospheric data in a three-neutrino scenario. A popular way to overcome this problem is to introduce of a fourth light neutrino, which must be sterile in order not to spoil the precise measurement of the Z decay width. Another possibility which does not require to enlarge the neutrino sector is to assume that CPT is violated. In this section, we show that none of these solutions is viable.

Our results are discussed in detail in Refs. [5] and [6], and summarized here in Fig. 2. Concerning four-neutrino models, we must distinguish between (2+2) schemes and (3+1) schemes. In the first panel of Fig. 2 we show the dependence of solar and atmospheric $\Delta\chi^2$ as a function of the fraction of sterile neutrino participating in *solar* oscillations, η_s, for (2+2) schemes. One feature of these schemes is that the fraction of sterile neutrino participating in *atmospheric* oscillations is bounded by unitarity to be $1 - \eta_s$. As a consequence, sterile neutrino always contributes significantly to either solar or atmospheric conversion, or eventually to both. Since both data samples strongly reject a sterile component, (2+2) schemes are essentially ruled out [5].

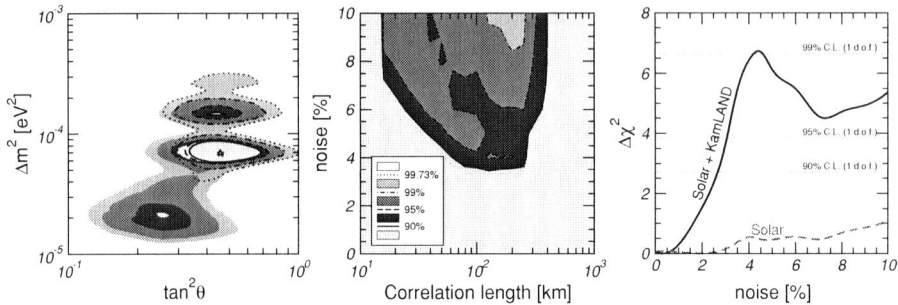

FIGURE 3. Left: allowed regions (2 d.o.f.) in the (Δm^2, $\tan^2\theta$) plane from the analysis of solar+KamLAND neutrino data, for the density profile predicted by the Standard Solar Model (lines) and in presence of matter-density fluctuations (shaded regions). Middle: allowed regions (2 d.o.f.) in the (L_0, ξ) plane from the analysis of solar+KamLAND neutrino data, in the presence of matter-density fluctuations. Right: $\Delta\chi^2$ as a function of the solar noise parameter ξ.

In (3+1) schemes there is no tension between solar and atmospheric data, however a tension appears between the LSND data and the results of short-baseline disappearance experiments. In the second panel of Fig. 2 we compare the LSND allowed region with the region allowed by the combination of all the other neutrino experiments (solar, atmospheric, reactor and accelerator). Values of $\Delta m^2_{\rm LSND} \geq 2.1$ (4.4) eV^2 are disfavored at more than 2σ (3σ) by the recent data of the WMAP satellite on Cosmic Microwave Background [7]. It is straightforward to see that there is no overlap of the two regions except at $\sim 99\%$ C.L. when the complete LSND data sample is used; when only Decay-At-Rest data are used, the disagreement is even stronger.

The situation is even worse for three-neutrino CPT violating models [6]. In this case, the problem arises since there are only two available anti-neutrino mass scales, whereas three would be required to accommodate simultaneously the KamLAND ($\Delta \bar{m}^2 \lesssim 10^{-3}$ eV^2), atmospheric ($\Delta \bar{m}^2 \gtrsim 10^{-3}$ eV^2) and LSND ($\Delta \bar{m}^2 \sim 1$ eV^2) data. As can be seen from the third panel of Fig. 2, the overlap between the LSND and All-but-LSND allowed regions occur only at $\Delta\chi^2 = 12.2$, slightly above the 3σ level. In the model considered here we have used the two anti-neutrino mass scales to account for KamLAND and LSND results, sacrificing atmospheric data. A variant of this model which tries to account for KamLAND and atmospheric data with a single mass scale is also discussed in Ref. [6], and leads to a similarly poor fit.

In summary, if the LSND anomaly is confirmed by the forthcoming experiment MiniBOONE, new and more sophisticated models will be needed to reconcile it with other experimantal evidences.

MATTER-DENSITY FLUCTUATIONS IN THE SUN

To conclude this brief review, in this section we will show how the determination of neutrino oscillation parameters from a combined fit of KamLAND and solar data strongly depends on the magnitude of solar matter-density fluctuations. Conversely, the fact that the KamLAND result largely support LMA neutrino oscillations can be used to provide new information about fluctuations in the solar core on much shorter scales than those which existing constraints (like helioseismology) can presently probe.

In Fig. 3 we show the results of our analysis of solar and KamLAND data in the presence of solar matter-density fluctuations. In addition to the usual oscillation parameters Δm^2 and θ, we denote by L_0 the fluctuation's correlation length and by ξ the fluctuation's amplitude. For details on our analysis and results, see Ref. [8].

From the first panel of Fig. 3, we see that the presence of density fluctuations in the solar core considerably affect the determination of neutrino parameters. Even after the inclusion of the KamLAND data a new region appears, with substantially lower Δm^2 value (around $\Delta m^2 = 2 \times 10^{-5}$ eV^2) and somewhat lower θ (around $\tan^2 \theta = 0.25$) than the standard solution. This new region is present even at 90% C.L. with 2 d.o.f.. This proves how a precise knowledge of the solar interior is required in order to sharpen the determination of neutrino oscillation parameters.

Conversely, solar neutrino data can be used to constrain the level of solar noise, as illustrated in the last two panels of Fig. 3. The inclusion of KamLAND is crucial to obtain a meaningful bound, and if the solar neutrino parameters were known with higher accuracy one could use them to probe the solar noise level with even better sensitivity. All considered, one finds that for a density fluctuation scale L_0 of 100 km or so the current limit to the noise amplitude is about 3% at 95% C.L.. Note that these bounds follow from the neutrino data themselves, and no helioseismological argument can at present rule out their existence.

ACKNOWLEDGMENTS

Talk based on the work performed in collaboration with C. Burgess, N.S. Dzhalilov, M.C. Gonzalez-Garcia, T.I. Rashba, T. Schwetz, V.B. Semikoz, M.A. Tórtola and J.W.F. Valle. This work was supported by the Spanish grant BFM2002-00345, by the European Commission RTN network HPRN-CT-2000-00148, by the European Science Foundation Neutrino Astrophysics Network No 86 and by the European Union Marie-Curie fellowship HPMF-CT-2000-01008.

REFERENCES

1. KamLAND collaboration, K. Eguchi *et al.*, Phys. Rev. Lett. **90**, 021802 (2003) [hep-ex/0212021].
2. M. Maltoni *et al.*, arXiv:hep-ph/0309130; see also references therein.
3. M. Maltoni *et al.*, Phys. Rev. D **67**, 093003 (2003) [arXiv:hep-ph/0212129].
4. Y. Hayato, Super–Kamiokande Coll., talk at the HEP2003 conference (Aachen, Germany, 2003).
5. M. Maltoni *et al.*, Phys. Rev. D **67**, 013011 (2003) [arXiv:hep-ph/0207227]; M. Maltoni *et al.*, Nucl. Phys. B **643**, 321 (2002) [arXiv:hep-ph/0207157]; M. Maltoni and T. Schwetz, Phys. Rev. D **68**, 033020 (2003) [arXiv:hep-ph/0304176].
6. M.C. Gonzalez-Garcia *et al.*, Phys. Rev. D **68**, 053007 (2003) [arXiv:hep-ph/0306226].
7. C.L. Bennett *et al.*, arXiv:astro-ph/0302207; D.N. Spergel *et al.*, arXiv:astro-ph/0302209; S. Hannestad, arXiv:astro-ph/0303076.
8. C. Burgess *et al.*, Astrophys. J. **588**, L65 (2003) [arXiv:hep-ph/0209094]; see also references therein.

Double Beta Decay and the Absolute Neutrino Mass Scale

Carlo Giunti

INFN, Sezione di Torino, and Dipartimento di Fisica Teorica,
Università di Torino, Via P. Giuria 1, I–10125 Torino, Italy

Abstract. After a short review of the current status of three-neutrino mixing, the implications for the values of neutrino masses are discussed. The bounds on the absolute scale of neutrino masses from Tritium β-decay and cosmological data are reviewed. Finally, we discuss the implications of three-neutrino mixing for neutrinoless double-β decay.

The marvelous recent results of neutrino oscillation experiments have given us important information on neutrino mixing. The data of solar neutrino experiments and of the KamLAND long-baseline $\bar{\nu}_e$ disappearance experiment show $\nu_e \to \nu_\mu, \nu_\tau$ transitions generated by the squared-mass difference Δm^2_{SUN} in one of the two ranges [1]

$$5.1 \times 10^{-5} < \Delta m^2_{\text{SUN}} < 9.7 \times 10^{-5} (\text{LMA-I}), \quad 1.2 \times 10^{-4} < \Delta m^2_{\text{SUN}} < 1.9 \times 10^{-4} (\text{LMA-II}), \tag{1}$$

at 99.73% C.L., with best-fit value $\Delta m^{2\text{bf}}_{\text{SUN}} \simeq 6.9 \times 10^{-5}$ (we measure squared-mass differences in units of eV^2). The effective solar mixing angle ϑ_{SUN} is constrained at 99.73% C.L. in the interval [1]

$$0.29 < \tan^2 \vartheta_{\text{SUN}} < 0.86, \tag{2}$$

with best-fit value $\tan^2 \vartheta^{\text{bf}}_{\text{SUN}} \simeq 0.46$. The results of atmospheric neutrino experiments and of the K2K long-baseline ν_μ disappearance experiment indicate $\nu_\mu \to \nu_\tau$ transitions generated by the squared-mass difference Δm^2_{ATM} in the 99.73% C.L. range [2]

$$1.4 \times 10^{-3} < \Delta m^2_{\text{ATM}} < 5.1 \times 10^{-3}, \tag{3}$$

with best-fit value $\Delta m^{2\text{bf}}_{\text{ATM}} \simeq 2.6 \times 10^{-3}$. The best-fit effective atmospheric mixing angle ϑ_{ATM} is maximal, $\sin^2 2\vartheta^{\text{bf}}_{\text{ATM}} \simeq 1$, with the 99.73% C.L. lower bound [2]

$$\sin^2 2\vartheta_{\text{ATM}} > 0.86. \tag{4}$$

These evidences of neutrino mixing are nicely accommodated in the minimal framework of three-neutrino mixing, in which the three flavor neutrinos ν_e, ν_μ, ν_τ are unitary linear combinations of three neutrinos ν_1, ν_2, ν_3 with masses m_1, m_2, m_3, respectively (see Ref. [3]). Figure 1 shows the two three-neutrino schemes allowed by the observed hierarchy $\Delta m^2_{\text{SUN}} \ll \Delta m^2_{\text{ATM}}$, with the massive neutrinos labeled in order to have

$$\Delta m^2_{\text{SUN}} = \Delta m^2_{21}, \quad \Delta m^2_{\text{ATM}} \simeq |\Delta m^2_{31}| \simeq |\Delta m^2_{32}|. \tag{5}$$

The two schemes in Fig. 1 are usually called "normal" and "inverted", because in the normal scheme the smallest squared-mass difference is generated by the two lightest neutrinos and a natural neutrino mass hierarchy can be realized if $m_1 \ll m_2$, whereas in the inverted scheme the smallest squared-mass difference is generated by the two heaviest neutrinos, which are almost degenerate for any value of the lightest neutrino mass m_3.

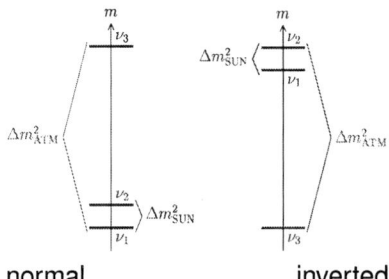

1: The two three-neutrino schemes allowed by the observed hierarchy $\Delta m^2_{\text{SUN}} \ll \Delta m^2_{\text{ATM}}$.

Solar neutrino oscillations depend only on the first row U_{e1}, U_{e2}, U_{e3} of the mixing matrix, and the hierarchy $\Delta m^2_{\text{SUN}} \ll \Delta m^2_{\text{ATM}}$ implies that neutrino oscillations generated by Δm^2_{ATM} depend only on the last column U_{e3}, $U_{\mu 3}$, $U_{\tau 3}$ of the mixing matrix. Hence, the only connection between solar and atmospheric neutrino oscillations is due to the element U_{e3}. The negative result of the CHOOZ long-baseline $\bar{\nu}_e$ disappearance experiment implies that electron neutrinos do not oscillate at the atmospheric scale and $|U_{e3}|$ is small: $|U_{e3}|^2 < 5 \times 10^{-2}$ at 99.73% C.L. [4]. Therefore, solar and atmospheric neutrino oscillations are practically decoupled [5] and the effective mixing angles in solar, atmospheric and CHOOZ experiments can be related to the elements of the three-neutrino mixing matrix by (see also Ref. [6])

$$\sin^2 \vartheta_{\text{SUN}} = \frac{|U_{e2}|^2}{1 - |U_{e3}|^2} \qquad \sin^2 \vartheta_{\text{ATM}} = |U_{\mu 3}|^2 \qquad \sin^2 \vartheta_{\text{CHOOZ}} = |U_{e3}|^2. \quad (6)$$

Taking into account all the above experimental constraints, we have reconstructed the

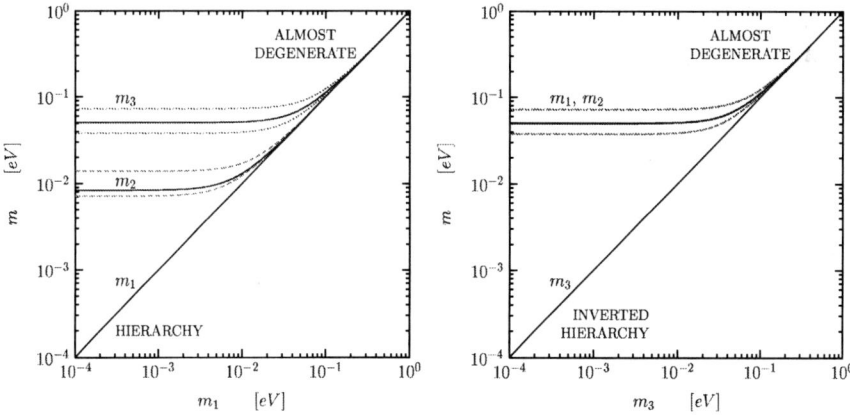

FIGURE 2. Allowed ranges for the neutrino masses as functions of the lightest mass m_1 and m_3 in the normal and inverted three-neutrino scheme, respectively.

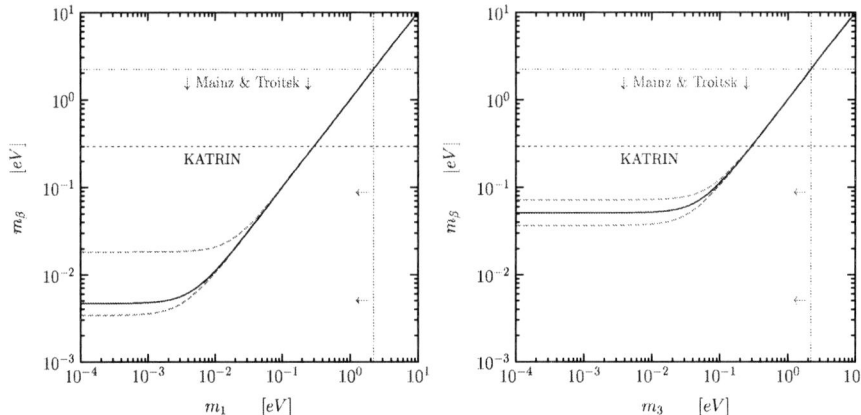

FIGURE 3. Effective neutrino mass m_β in Tritium β-decay experiments as a function of the lightest mass m_1 and m_3 in the normal and inverted three-neutrino scheme, respectively.

best-fit and allowed ranges for the elements of the mixing matrix (see Ref. [7] for a reconstruction taking into account the correlations among the mixing parameters):

$$U_{\text{bf}} \simeq \begin{pmatrix} -0.83 & 0.56 & 0.00 \\ 0.40 & 0.59 & 0.71 \\ 0.40 & 0.59 & -0.71 \end{pmatrix}, \quad |U| \simeq \begin{pmatrix} 0.71-0.88 & 0.46-0.68 & 0.00-0.22 \\ 0.08-0.66 & 0.26-0.79 & 0.55-0.85 \\ 0.10-0.66 & 0.28-0.80 & 0.51-0.83 \end{pmatrix}. \quad (7)$$

Such mixing matrix, with all elements large except U_{e3}, is called "bilarge". It is very different from the quark mixing matrix.

The absolute scale of neutrino masses is not determined by the observation of neutrino oscillations, which depend only on the differences of the squares of neutrino masses. Figure 2 shows the allowed ranges (between the dashed and dotted curves) for the neutrino masses obtained from the allowed values of the oscillation parameters in Eqs. (1)–(4), as functions of the lightest mass in the normal and inverted three-neutrino schemes. The solid lines correspond to the best fit values of the oscillation parameters. One can see that at least two neutrinos have masses larger than about 7×10^{-3} eV.

The most sensitive known ways to probe the absolute values of neutrino masses are the observation of the end-point part of the electron spectrum in Tritium β-decay, the observation of large-scale structures in the early universe and the search for neutrinoless double-β decay, if neutrinos are Majorana particles (see Ref. [8]; we do not consider here the interesting possibility to determine neutrino masses through the observation of supernova neutrinos).

Up to now, no indication of a neutrino mass has been found in Tritium β-decay experiments, leading to an upper limit on the effective mass

$$m_\beta = \sqrt{\sum_k |U_{ek}|^2 m_k^2} \quad (8)$$

of 2.2 eV at 95% C.L. [9], obtained in the Mainz and Troitsk experiments. After 2007, the KATRIN experiment [10] will explore m_β down to about $0.2 - 0.3$ eV. Figure 3

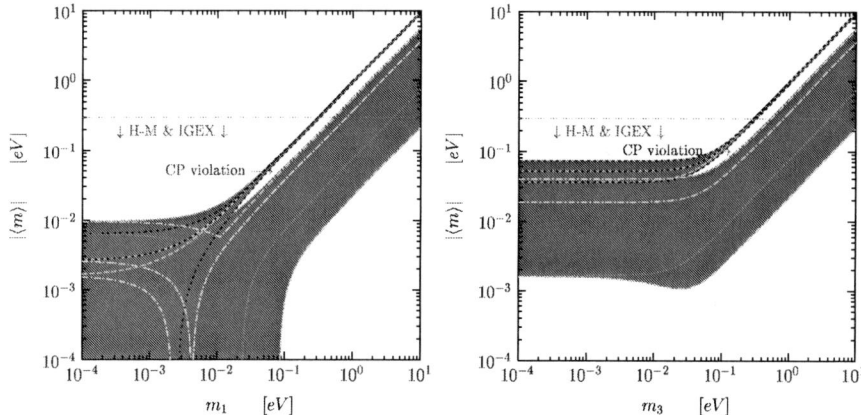

FIGURE 4. Effective Majorana mass $|\langle m \rangle|$ in neutrinoless double-β decay experiments as a function of the lightest mass m_1 and m_3 in the normal and inverted three-neutrino scheme, respectively.

shows the allowed range (between the dashed curves) for m_β obtained from the allowed values of the oscillation parameters in Eqs. (1)–(4), as a function of the lightest mass in the normal and inverted three-neutrino schemes. The solid line corresponds to the best fit values of the oscillation parameters. One can see that in the normal scheme with a mass hierarchy m_β has a value between about 3×10^{-3} eV and 2×10^{-2} eV, whereas in the inverted scheme m_β is larger than about 3×10^{-2} eV. Therefore, if in the future it will be possible to constraint m_β to be smaller than about 3×10^{-2} eV, a normal hierarchy of neutrino masses will be established.

The analysis of recent data on cosmic microwave background radiation and large scale structure in the universe in the framework of the standard cosmological model has allowed to establish an upper bound of about 1 eV for the sum of neutrino masses, which implies an upper limit of about 0.3 eV for the individual masses [11, 12]. This limit is already at the same level as the sensitivity of the future KATRIN experiment. Let us emphasize, however, that the KATRIN experiment is important in order to probe the neutrino masses in a model-independent way,

A very important open problem in neutrino physics is the Dirac or Majorana nature of neutrinos. From the theoretical point of view it is expected that neutrinos are Majorana particles, with masses generated by effective Lagrangian terms in which heavy degrees of freedom have been integrated out (see Ref. [13]). In this case the smallness of neutrino masses is naturally explained by the suppression due to the ratio of the electroweak symmetry breaking scale and a high energy scale associated with the violation of the total lepton number and new physics beyond the Standard Model.

The best known way to search for Majorana neutrino masses is neutrinoless double-β decay, whose amplitude is proportional to the effective Majorana mass

$$|\langle m \rangle| = \left| \sum_k U_{ek}^2 m_k \right|. \tag{9}$$

The present experimental upper limit on $|\langle m \rangle|$ between about 0.3 eV and 1.3 eV has been obtained in the Heidelberg-Moscow and IGEX experiments. The large uncertainty is due to the difficulty of calculating the nuclear matrix element in the decay. Figure 4 shows the allowed range for $|\langle m \rangle|$ obtained from the allowed values of the oscillation parameters in Eqs. (1)–(4), as a function of the lightest mass in the normal and inverted three-neutrino schemes (see also Ref. [14]). If CP is conserved, $|\langle m \rangle|$ is constrained to lie in a shadowed region. Finding $|\langle m \rangle|$ in an unshaded strip would signal CP violation. One can see that in the normal scheme large cancellations between the three mass contributions are possible and $|\langle m \rangle|$ can be arbitrarily small. On the other hand, the cancellations in the inverted scheme are limited, because v_1 and v_2, with which the electron neutrino has large mixing, are almost degenerate and much heavier than v_3. Since the solar mixing angle is less than maximal, a complete cancellation between the contributions of v_1 and v_2 is excluded, leading to a lower bound of about 1×10^{-3} eV for $|\langle m \rangle|$ in the inverted scheme. If in the future $|\langle m \rangle|$ will be found to be smaller than about 1×10^{-3} eV, it will be established that either neutrinos have a mass hierarchy or they are Dirac particles. Many neutrinoless double-β decay experiments are planned for the future, but they will unfortunately not be able to probe such small values of $|\langle m \rangle|$, extending their sensitivity at most in the 10^{-2} eV range (see Ref. [8]).

In conclusion, we would like to emphasize that, although the recent years have been extraordinarily fruitful for neutrino physics, yielding important information on the neutrino mixing parameters, still several fundamental characteristics of neutrinos are unknown. Among them, the Dirac or Majorana nature of neutrinos, the absolute scale of neutrino masses, the distinction between the normal and inverted schemes and the existence of CP violation in the lepton sector are very important for our understanding of the new physics beyond the Standard Model.

ACKNOWLEDGMENTS

I would like to thank Carlos Peña-Garay for enlightening discussions.

REFERENCES

1. Maltoni, M., Schwetz, T., and Valle, J., *Phys. Rev.*, **D67**, 093003 (2003), hep-ph/0212129.
2. Fogli, G. L., et al., *Phys. Rev.*, **D67**, 093006 (2003), hep-ph/0303064.
3. Bilenky, S. M., et al., *Prog. Part. Nucl. Phys.* **43**, 1 (1999), hep-ph/9812360.
4. Fogli, G. L., et al., *Phys. Rev.*, **D66**, 093008 (2002), hep-ph/0208026.
5. Bilenky, S. M., and Giunti, C., *Phys. Lett.*, **B444**, 379 (1998), hep-ph/9802201.
6. W.L. Guo and Z.Z. Xing, *Phys. Rev.* **D67**, 053002 (2003), hep-ph/0212142.
7. Gonzalez-Garcia, M. C., and Peña-Garay, C., hep-ph/0306001.
8. Bilenky, S. M., Giunti, C., Grifols, J. A., and Masso, E., *Phys. Rept.*, **379**, 69 (2003), hep-ph/0211462.
9. Weinheimer, C., hep-ex/0210050.
10. Osipowicz, A., et al., hep-ex/0109033.
11. Spergel, D. N., et al., astro-ph/0302209.
12. Hannestad, S., astro-ph/0303076.
13. Altarelli, G., and Feruglio, F., hep-ph/0306265.
14. Pascoli, S., Petcov, S. T., and Rodejohann, W., *Phys. Lett.*, **B549**, 177 (2002), hep-ph/0209059.

Natural Expectations for the Value of $|U_{e3}|$?

André de Gouvêa

*Department of Physics & Astronomy, Northwestern University
2145 Sheridan Road, Evanston, IL, 60208-3112, USA, and
Theoretical Physics Department, Fermilab, P.O. Box 500, Batavia, IL, 60510-0500, USA*

Abstract.
The discovery of neutrino masses has presented the theoretical physics community two interesting puzzles. One is the origin of the (very tiny) neutrino masses themselves, and the other is related to understanding why leptonic mixing is very "different" from quark mixing. Several different attempts which make use of several different grand-unified and/or family symmetries are well-documented and all explain the current data. They do differ, however, on what is predicted for the still-to-be-measured $|U_{e3}|$ element of the leptonic mixing matrix.

More importantly (in my opinion), I'll advocate that measuring $|U_{e3}|$ will help indicate whether there is indeed a hint for "symmetry" in the leptonic mixing matrix, or whether it is consistent with a random unitary matrix.

In the Standard Model of electroweak interactions (SM), neutrinos are strictly massless. This being the case, it is always possible to choose a basis where the weak eigenstates, *i.e.*, the fermions which are related via the weak charged-current interaction ($e \leftrightarrow \nu_e$, $\mu \leftrightarrow \nu_\mu$, $\tau \leftrightarrow \nu_\tau$,) coincide with the mass-eigenstates. The exciting development of particle physics in the past decade is the discovery that neutrinos do not behave as prescribed by the SM, and do indeed have mass. One of the key current theoretical issues is to understand the new physics behind neutrino masses and mixing, and to understand why neutrino masses are qualitatively much smaller than all charged fermion masses. The situation is depicted in the figure below.

Neutrino oscillation experiments are not only capable of measuring the neutrino mass-squared differences, but also indicate that neutrino mass-eigenstates differ from neutrino weak-eigenstates. The two "bases" are related by a unitary mixing matrix U:

$$\nu_\alpha = U_{\alpha i} \nu_i, \qquad \alpha = e, \nu, \tau \tag{1}$$

and $i = 1, 2, 3$. $\nu_{1,2,3}$ are the mass eigenstates, with corresponding mass-squared $m^2_{1,2,3}$. Choosing the PDG parameterization for U [1] and defining the mass eigenstates in the "usual way" [2], oscillation experiments have constrained $\sin^2\theta_{12} \sim 0.3$ (the "solar" angle) and $\sin^2\theta_{23} \sim 0.5$ (the "atmospheric" angle) [3]. The value of the one CP-odd phase to which oscillation experiments are sensitive is completely unknown, while the third mixing angle is only loosely bound [4]:

$$|U_{e3}|^2 \equiv \sin^2\theta_{13} < 0.07 \quad \text{(at } 3\sigma \text{ confidence level)}. \tag{2}$$

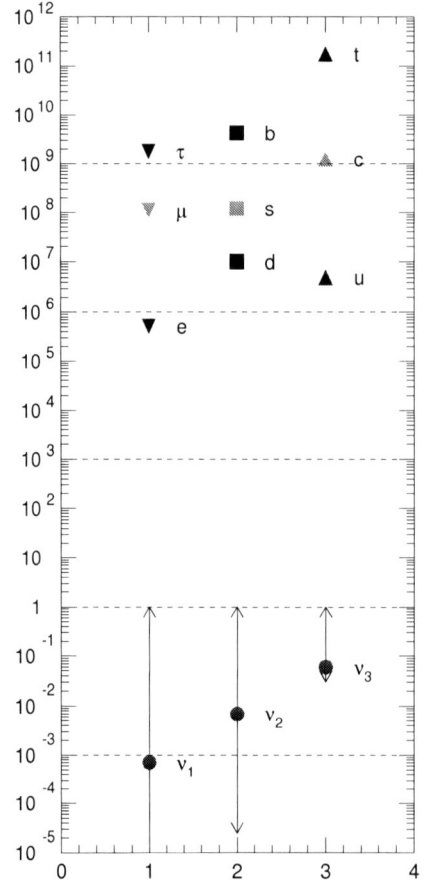

Masses of all standard model fermions, in eV. Since oscillation experiments are only sensitive to neutrino mass-squared differences, neutrino masses are only poorly constrained. The (conservative?) upper bound of 1 eV is provided by non-oscillation experiments [1, 5].

Hence, the leptonic mixing matrix U seems to have, qualitatively, the following "form"

$$|U| \simeq \begin{pmatrix} 1 & 1 & |U_{e3}| \\ 1 & 1 & 1 \\ 1 & 1 & 1 \end{pmatrix}. \qquad (3)$$

In the quark sector, quark weak-eigenstates are also different from quark mass-eigenstates. The unitary matrix that connects the two distinct bases is the CKM quark mixing matrix, which has been experimentally studied over the past several decades, and is known to have the following "suspiciously inviting" form:

$$|V_{CKM}| \simeq \begin{pmatrix} 1 & \lambda & \lambda^3 \\ \lambda & 1 & \lambda^2 \\ \lambda^3 & \lambda^2 & 1 \end{pmatrix}, \qquad (4)$$

where $\lambda \simeq 0.22$ is the sine of the Cabibbo angle.

The other key theoretical issue is related to understanding why Eq. (4) and Eq. (3) are qualitatively different. It was originally believed that the zeroth order prediction $U \simeq V_{CKM}$ should be correct. This naive guess is driven by grand unified theories (GUTs), which describe quarks and leptons as different components of more "fundamental" matter fields. This speculation helped drive the community to quickly dismiss the strong atmospheric neutrino anomaly observed by Kamiokande and IMB (which turned out to be confirmed by SuperKamiokande) and strongly prefer the small mixing angle solution to the solar neutrino puzzle (which, as we learned from more precise solar data and the KamLAND result, turned out to be incorrect). In all fairness, however, while GUTs did not successfully predict the leptonic mixing angles, they can accommodate ("post-dict") the CKM mixing matrix together with the two large leptonic mixing angles (see, for example, [6]) in a variety of ways.

Why is the issue of understanding mixing angles and masses important? One reason is the following: we believe that the curious pattern of V_{CKM} (see Eq. (4)) and the hierarchy of the quark masses is a consequence of some new organizing principle (symmetry) of Nature, which is perhaps broken at inaccessibly high energy scales. The fact that quark mixing angles are small and their masses are very different is interpreted as a consequence of some "flavor" symmetry that differentiates the top quark from, say, the charm quark (*i.e.*, the charm quark and top quark have some hidden quantum number that fundamentally tells them apart). The "traditional" approach to understanding neutrino mixing has been to apply the same logic to the leptonic sector. A lot of research effort is concentrated in trying to understand what is the organizing principle that tells a muon from a tau, and whether this flavor symmetry can explain why the atmospheric mixing angle is (almost?) $\pi/4$, etc [7].

We currently live in a specially interesting time as far as flavor model building is concerned: since we know two of the three mixing angles and the two mass-squared differences rather well, these should be considered "inputs" to flavor models. The "outputs" are, therefore, the currently unknown parameters: the mass-hierarchy, the CP-odd phase, and $|U_{e3}|$. Hence, the determination of these three parameters (especially $|U_{e3}|$) will have a deep impact on this type of research. No matter what $|U_{e3}|$ turns out to be, it is fair to say that a significant fraction of flavor models will be eliminated once it is determined (or tightly bound). This should be contrasted to the situation of flavor model building in the quark sector, where we already know all the mixing parameters – there are no (standard model) predictions to be made!

Here, however, I would like to point out that, perhaps, we are not asking the correct question as far as leptonic mixing is concerned. It is indeed interesting to first ask whether the data are pointing toward a structured mixing matrix, or whether the mixing matrix is "ordinary." Before attempting to discover the flavor symmetry that tells one lepton from the other, it is important to verify whether the data seem to indeed hint at such a distinction. It turns out, perhaps not surprisingly, that they currently do not [8]! As an aside, it is informative to note that when the Cabibbo angle was first introduced, the community was puzzled and surprised by the fact that it was very small – at that time, large mixing was expected![1] It is, perhaps, more appropriate to state that Eq. (3)

---

[1] I thank Stan Wojcicki for pointing this out.

is typical of what one should naively expect, while Eq. (4) is "unusual." The reason we today view large mixing as "strange" is related to the fact that we were exposed, for over thirty years, only to the small quark mixing angles.

For the purpose of establishing whether there is a hint of structure in the leptonic mixing matrix, it is imperative to determine whether $|U_{e3}|^2$ is "large" or "small." Recently, Hitoshi Murayama and I addressed this issue in a quantitative fashion [9], by performing a Kolmogorov–Smirnov test of the "anarchical hypothesis" (see [9] for details). We conclude that if $|U_{e3}|^2 < 0.011$ there is a very good chance (two sigma) that there is indeed structure in the leptonic sector. On the other hand, if $|U_{e3}|^2 > 0.011$ there is no obvious need for a fundamental distinction of the different neutrino fields. Hence, experiments are yet to decide whether we need new fundamental symmetries in the leptonic sector or not.

I conclude by stating that with the advent of neutrino masses and mixing, there are new "fundamental" Lagrangian parameters of the extended SM that remain unknown – most noticeably $|U_{e3}|$. Theorist do not "know" what to expect for the value of $|U_{e3}|$. Different models make wildly different predictions, and the only 100% unbiased statement one can make is $|U_{e3}|_{\text{theory}} = [0, \text{CHOOZ bound}]$. Another way of restating this is to point out that neutrino physics is a *data driven field*, meaning that theorists right now have little say regarding what $|U_{e3}|$ will turn out to be, while a measurement of $|U_{e3}|$ will qualitatively affect our understanding of the physics behind neutrino masses and leptonic mixing.

ACKNOWLEDGMENTS

I thank Manfred Lindner for the invitation to present this talk at the 2003 Neutrino Factory Conference. I also thank the organizers for putting together a successful and inspiring conference. I am indebted to Hitoshi Murayama for collaboration and enlightening discussions regarding some of the results presented here. This work was supported by the US Department of Energy Contract DE-AC02-76CHO3000.

REFERENCES

1. K. Hagiwara *et al.* [Particle Data Group Collaboration], Phys. Rev. D **66**, 010001 (2002).
2. See, for example, A. de Gouvêa, Nucl. Instrum. Meth. A **503**, 4 (2001).
3. See, for example, V. Barger, these proceedings.
4. G. L. Fogli *et al.*, hep-ph/0308055.
5. For the most recent bound provided by astrophysical observations see M. Tegmark *et al.* [SDSS Collaboration], astro-ph/0310723.
6. M. C. Chen and K. T. Mahanthappa, these proceedings [hep-ph/0311034].
7. For recent reviews of the status of flavor model building see, for example, S. F. King, hep-ph/0310204; V. Barger, D. Marfatia and K. Whisnant, Int. J. Mod. Phys. E **12**, 569 (2003); G. Altarelli and F. Feruglio, hep-ph/0206077.
8. L. J. Hall, H. Murayama and N. Weiner, Phys. Rev. Lett. **84**, 2572 (2000). For details see N. Haba and H. Murayama, Phys. Rev. D **63**, 053010 (2001). For an example with very hierarchical masses see A. de Gouvêa and J. W. F. Valle, Phys. Lett. B **501**, 115 (2001).
9. A. de Gouvêa and H. Murayama, Phys. Lett. B **573**, 94 (2003).

MINOS Status and Physics Goals

George S. Tzanakos

University of Athens, Department of Physics, Division of Nuclear and Particle Physics, Panepistimioupoli, Zografou, 15771 Athens, Greece

(For the MINOS Collaboration) [†]

Abstract. The MINOS experiment was designed to make a precise study of the "atmospheric" neutrino oscillations observed recently by underground experiments. It will use the NuMI neutrino beam and two neutrino detectors of 980 and 5400 tons, located at distances of 1 km and 735 km from the neutrino source, respectively. The physics goals and the status of the NuMI beam and MINOS detectors will be described.

PHYSICS GOALS OF THE MINOS EXPERIMENT

The MINOS experiment [1] was designed to make a precise study of the "atmospheric" neutrino oscillations observed in recent data from underground experiments [2]. MINOS will use the NuMI neutrino beam and two neutrino detectors with a mass of 980 ton and 5400 ton, located at distances of 1 km and 735 km from the neutrino source, respectively. One of the primary goals of the MINOS experiment is to demonstrate the oscillatory behavior of the neutrino flux. The experiment will measure and compare the CC neutrino energy spectrum between the near and far detectors, with a 2-4% of systematic uncertainty per 2 GeV bin of neutrino energy. MINOS shall make a precise determination of the oscillation parameters. The best-fit values of the oscillation parameters from the Super-Kamiokande data [3] are ($\sin^2(2\theta)$, Δm^2) = (1.0, 2.5 ×10^{-3} eV^2), while results from K2K [4] give best-fit at ($\sin^2(2\theta)$, Δm^2) = (1.0, 2.8 ×10^{-3} eV^2). MINOS expects to measure Δm^2 to about 10%. MINOS will make a precise determination of flavor participation in oscillations. Comparison of the number of CC ν_μ events in the near and far detectors is expected to be done to about 2%, from which the overall probability of $\nu_\mu \rightarrow \nu_x$ will be measured. Identification of ν_e events in both detectors will help determine the probability of $\nu_\mu \rightarrow \nu_e$ oscillation down to about 2%. This can provide discovery of the $\nu_\mu \rightarrow \nu_e$ oscillation and a first measurement of $|U_{e3}|$. Measurement and comparison of the total number of NC events in both detectors in combination with the CC ν_μ disappearance rates will measure the probability of $\nu_\mu \rightarrow \nu_{sterile}$ down to about 10%. By combining the information from all channels mentioned above we can indirectly determine the $\nu_\mu \rightarrow \nu_\tau$ oscillation probability with a precision of 5-10%. Finally, the considerable size and magnetic field of the far detector will make possible a comparison of survival probabilities between neutrinos and antineutrinos.

STATUS OF NEUTRINO BEAM AND DETECTORS

The NuMI beam and the MINOS detectors are described in some detail elsewhere [5]. The MINOS experiment will use neutrinos produced in the NuMI beamline by 120 GeV (1.9 s cycle time, 4×10^{13} ppp, 0.4 MW beam power) protons from the New Main Injector at Fermilab in a fast extraction mode (10 μs). The proton beam will be aimed at Soudan before it impacts on an external target to produce pions and kaons.

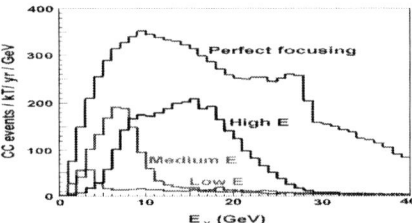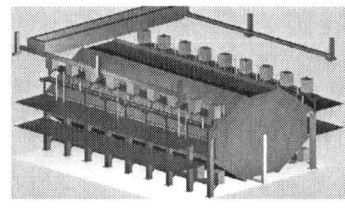

FIGURE 1. (Left) Low, medium, and high energy neutrino spectra using different tunes of the NuMI beam line. (Right) Pictorial representation of one supermodule of he Far MINOS detector.

A two-magnetic horn system followed by a 700 m long decay pipe and muon absorber, produces the ν_μ beam. The E_ν spectrum can be chosen by appropriate tuning of the distance of the target and second horn to the first horn. Fig.1 (left) shows the interacted ν_μ CC energy spectra at the Far detector for three different tunes of the NuMI beam, with corresponding numbers of 470, 1270, 2740 CC events/kt/year. The status of the NuMI beam in June 2003 is: The excavation of the NuMI beamline halls and tunnels is completed and the decay pipe is being installed along with the concrete shielding. The outfitting of the tunnels and halls are in progress expecting to finish by November 2003. Also the surface buildings are under construction. The NuMI beamline is expected to deliver the first protons on target in December 2004.

The MINOS experiment involves two detectors with the same basic structure of a segmented iron-scintillator calorimeter and magnetized muon spectrometer. The use of two detectors, one near the neutrino source (1 km) and another one far enough (735 km) is dictated by the need to measure neutrino disappearance and to control systematics. One of the two supermodules of the Far MINOS Detector is shown pictorially in Fig. 1 (right). The current status of the detectors is as follows: The near detector planes have been constructed and stored in the New Muon Laboratory at Fermilab, awaiting for the beneficiary occupancy of the MINOS Hall in early 2004. The Far Detector assembly finished just at the time of this conference. The magnetic field for the newly finished second supermodule was turned on in August 2003. Thus, the Far MINOS detector is now fully operational.

In order to ensure a 5% absolute energy calibration in each detector and also a 2% relative near/far calibration, a 8 ton calibration detector (CALDET) was constructed and exposed to positive and negative π, μ, e, and proton test beams at CERN, at energies 0.5-3.5 GeV (T11 beam) and 1-10 GeV (T7 beam). We have measured the muon, EM, and hadron energy response and topology. The CALDET response to pions and electrons (line shapes) is shown in Fig.2a. Also, the energy response is linear, with a resolution $\Delta E/E$ of $22\%/\sqrt{E}$ for electrons, and $55\%/\sqrt{E}$ for hadrons.

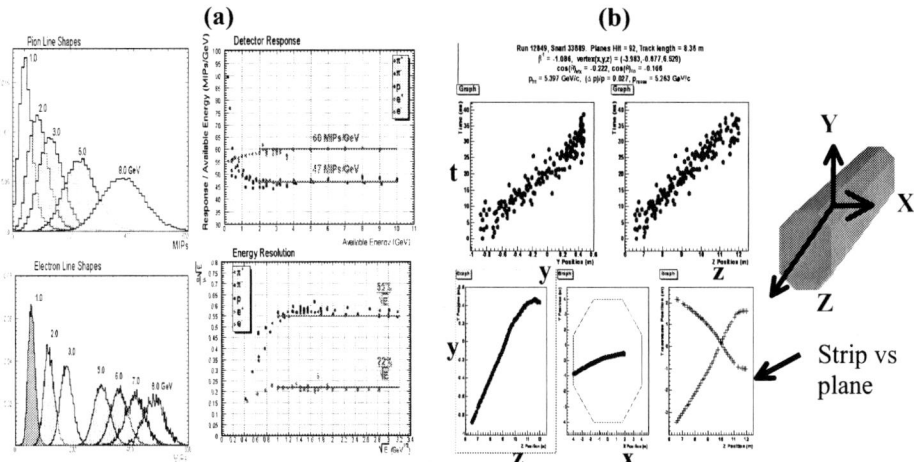

FIGURE 2. (a) CALDET response (Preliminary): pion and electron line shapes (left), linearity and energy resolution (right). (b)) MINOS Far Detector response: A 5.4 GeV/c up-going μ+ .

Cosmic ray muon data are extremely useful in measuring the light output of the scintillator system in all three detectors (Near, Far, CALDET) and are needed in order to intercalibrate the three detectors. In the Far detector we have measured a light output of 8.5 pe/cm, with a very good relative uniformity (11% (rms) variation).

The measured timing resolution of 2.6 ns/plane in the MINOS scintillator system and the magnetic field are crucial in the muon direction (upward/downward) as well as in the charge and momentum determination (good up to 70 GeV/c) in the far detector. Fig. 2b shows plots of an upgoing μ+ of 5.4 GeV/c. The Far MINOS detector has also a scintillator veto shield on the top (double) and on the sides (single), which will help to reduce the background from cosmics and enable the study of atmospheric neutrinos.

PHYSICS CAPABILITIES OF MINOS

The MINOS Far detector is the first underground detector which has a magnetic field. This, along with event reconstruction and timing allows complete measurement of the ratio L/E for atmospheric neutrinos. In addition, separation of ν_μ ($\bar{\nu}_\mu$) CC from NC and ν_e CC over a wide energy range above ~ 1 GeV will allow the measurement and direct comparison of atmospheric ν_μ and $\bar{\nu}_\mu$ oscillation. We estimate the number of reconstructed contained events with a muon to be 620 (400) ν_μ ($\bar{\nu}_\mu$) , and the number of events with upgoing muons to be 280 (120) ν_μ ($\bar{\nu}_\mu$), for 24 kton-years.

Original MINOS sensitivity curves [1,5] were based on 7.4×10^{20} POT. MINOS has provided updated physics sensitivity curves based on 7.4, 16, and 25×10^{20} protons on target, using Monte Carlo generated events, reconstructed using the MINOS event reconstruction. The study was done assuming standard oscillations with $\sin^2 2\theta = 1.0$ and two possible values of Δm^2, 0.0025 and 0.0016 eV^2. The unoscillated and oscillated ν_μ CC spectra versus visible energy were calculated at the Far detector and their ratios are plotted in Fig. 3a for 7.4×10^{20} (upper row) and 25×10^{20} (lower row) protons on target. From the neutrino energy at minimum and the value of the ratio at minimum we extracted the values of oscillation parameters Δm^2,

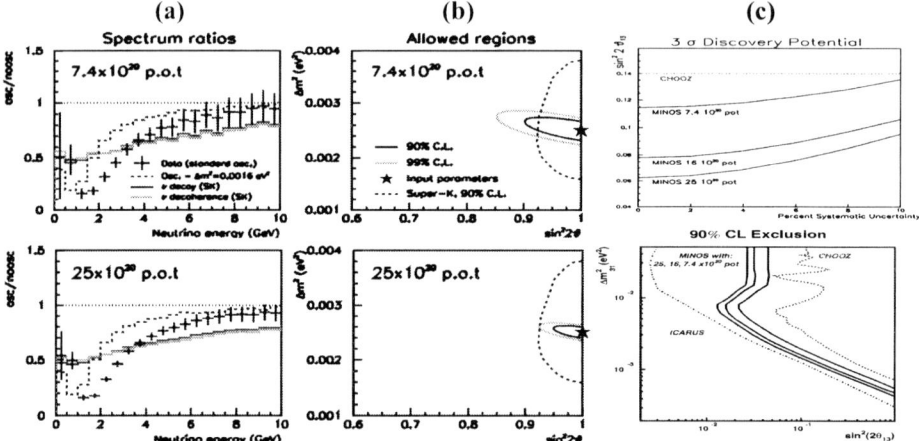

FIGURE 3 Results of MC calculations: a) The ratios of the oscillated/unoscillated ν_μ CC events for $(\sin^2 2\theta, \Delta m^2) = (1.0, 0.0025 \text{ eV}^2)$ (crosses, standard oscillations) for 7.4 and 25×10^{20} POT. The dashed line is for $\Delta m^2 = 0.0016 \text{ eV}^2$. The continouous lines (blue) refer to netrino decay and decoherence. b) Contours of oscillation parameters at 90% and 99% CL, along with the Super-K results. c) MINOS sensitivity to $\nu_\mu \to \nu_e$ oscillations: (Upper plot) The 3σ discovery potential of MINOS, for three different levels of POT, versus the percent systematic uncertainty on the background. (Lower plot) Corresponding 90% CL exlusion limits compared to the CHOOZ[6] results and ICARUS predictions.

and $\sin^2 2\theta$, respectively. Fig. 3b shows the 90% and 99% confidence level contours with which the MINOS experiment can determine these parameters. On the same figure the Super-Kamiokande 90% CL contour is also shown.

In addition, the sensitivity of MINOS in detecting θ_{13} via $\nu_\mu \to \nu_e$ is shown in Fig. 3. The 3σ discovery potential of the MINOS is shown in Fig. 3c (upper) for 7.4, 16, and 25×10^{20} protons on target. The corresponding 90%CL exclusion limits (lower) are shown with the CHOOZ [6] results and ICARUS expectations for 15 kt-years.

I would like to thank my colleagues of the MINOS collaboration. In particular D. Michael, J. Nelson, R. Plunkett, G. Bock, J. Thomas, K. Lang, D. Petyt, A. Weber, M. Kordosky, and P. Valle for providing material for this talk.

REFERENCES

(†) MINOS is an International Collaboration of about 215 physicists from 31 Institutions.
1. Fermilab Proposal P875, February 1995. The MINOS Technical Design Report, Fermilab, **NUMI-L-337**, October, 1998.
2. Y. Fukuda et al., PRL **81**, 1562 (1998); S. Fukuda et al., PRL **85**, 3999 (2000); M. Ambrosio et al., Phys. Lett. **B434**, 451 (1998), hep-ex/9807005; M. Ambrosio et al., Phys. Lett. **B478**, 5(2000), hep-ex/0001044; W.W.M.. Allison et al., Phys. Lett. **B449**, 137 (1999), hep-ex/9901024; D.A. Petyt et al., Nucl. Phys. Proc. Suppl. **110**, 349 (2002).
3. T. Kajita, J. Phys. G: Nucl. Part. Phys. **29**, 1471 (2003).
4. M.H. Ahn et al., (K2K) PRL **90**, 041801(2003).
5. D. Michael, (NEUTRINO 2002), Nucl. Phys. B (Proc. Suppl.) **118**, 189(2003); M. V. Diwan, (TAU 02) Nucl. Phys. B (Proc. Suppl.) **123**, 272(2003), hep-ex/0211026.
6. M. Apollonio et al., Phys. Lett. B420, 397(1998); M. Apollonio et al., Phys. Lett. B466, 415(1999).

SNO II: Salt Strikes Back
Update from the Sudbury Neutrino Observatory

J. A. Formaggio
for the SNO Collaboration

*Center for Experimental Nuclear Physics and Astrophysics, and Department of Physics
University of Washington, Seattle, WA 98195*

Abstract. The Sudbury Neutrino Observatory (SNO) is a multi-phase experiment designed to measure both the electron and total active neutrino flux coming from the sun. During the first phase of the experiment (the pure D_2O phase), it was definitively shown that neutrinos from the sun change flavors as they travel to the Earth. In this report, results from these measurements and the current status of the SNO detector are presented.

INTRODUCTION

Observation of solar neutrino data gathered over the past three decades has presented physicists with a difficult problem to solve. The so-called "solar neutrino puzzle" arose from the discrepancy between the measured electron neutrino flux from the sun and the flux measured by experiments. Results from Homestake [1], SAGE [2], GALLEX, [3], GNO [4], Kamiokande [5], and Super-Kamiokande [6] indicate that only between one-third to one-half of the neutrinos expected from solar models [7, 8] are actually detected (see Table 1). One explanation for all the observations is that v_e's from fusion reactions in the sun undergo matter enhanced neutrino oscillations.

TABLE 1. Summary of solar neutrino observations at different solar neutrino detectors. Measured fluxes are compared with theoretical expectations. Fluxes are presented either in Solar Neutrino Units (SNU), which is defined as 1 capture per second per 10^{36} target atoms, or in units of 10^6 $cm^{-2}s^{-1}$ (for Kamiokande and Super-Kamiokande)

Experiment	Measured Flux	SSM Flux
Homestake	2.56 ± 0.16 (stat.) ± 0.16 (sys.) SNU	$7.6^{+1.3}_{-1.1}$ SNU
SAGE	$69.1^{+4.3}_{-4.2}$ (stat.) $^{+3.8}_{-3.4}$ (sys.) SNU	128^{+9}_{-7} SNU
Gallex	77.5 ± 6.2 (stat.) $^{+4.3}_{-4.7}$ (sys.) SNU	128^{+9}_{-7} SNU
GNO	62.9 ± 5.4 (stat.) ± 2.5 (sys.) SNU	128^{+9}_{-7} SNU
Kamiokande	2.80 ± 0.19 (stat.) ± 3.3 (sys.)	$5.05(1^{+0.20}_{-0.16})$
SuperK	2.35 ± 0.03 (stat.) $^{+0.07}_{-0.06}$ (sys.)	$5.05(1^{+0.20}_{-0.16})$

SNO: AN EXPERIMENT IN THREE PHASES

The Sudbury Neutrino Observatory is located ~ 2 km (6020 m.w.e.) underground in INCO Ltd.'s Creighton Mine. The 1 kT of heavy water is contained in a 12 meter diameter transparent acrylic vessel. The Cerenkov light produced by neutrinos and radioactive backgrounds is detected by an array of about 9500 20-cm photomultiplier tubes (PMTs), supported by a stainless steel geodesic sphere measuring 17.8 m in diameter. Each PMT is surrounded by a light concentrator, which increases the photocathode coverage to nearly $\sim 55\%$. Further physical details regarding the detector can be found elsewhere[9].

The SNO experiment detects solar neutrinos through three distinct reactions:

$$v_e + d \to p + p + e^- \quad (\text{CC}) \tag{1}$$
$$v_l + e^- \to v_l + e^- \quad (\text{ES}) \tag{2}$$
$$v_l + d \to p + n + v_l \quad (\text{NC}) \tag{3}$$

The charged-current (CC) reaction on the deuteron is sensitive exclusively to electron neutrinos, while the neutral current (NC) reaction is sensitive to all active flavors equally. Elastic scattering (ES) on electrons is also sensitive to all active flavors, but with reduced sensitivity to v_μ and v_τ. SNO's simultaneous measurement of all three reactions allows it to monitor the oscillation of electron neutrinos to other active flavors. In addition, the neutral current reaction allows a direct measurement of the total ^{8}B solar neutrino flux above the 2.2 MeV reaction threshold.

The SNO experiment is divided into three distinct phases, each of which is intended to measure the neutral current component of the solar flux in an unique way. The following sections describe briefly the results of the first phase of SNO and gives an update on the current status of the detector.

THE FIRST PHASE: PURE D$_2$O

The first phase, known as the pure D$_2$O phase, consisted of filling the inside of the acrylic vessel with 99.92% isotopically pure heavy water. Under these conditions, the neutron released from the neutral current reaction is recaptured by the deuterium in the heavy water, producing a mono-energetic gamma ray at an energy of 6.25 MeV. The neutron detection efficiency for the pure D$_2$O phase is 14.4% (with volume and energy thresholds taken into account).

The data presented in this talk were recorded between November 2, 1999 and May 28, 2001, with over 450 million triggers recorded. The corresponding livetime was 306.4 days. Detector diagnostic triggers and instrumental background events were first removed in the online data processing. The latter reduction was accomplished by a set of algorithms that remove events that do not have the characteristics of Cherenkov light emission. For the events that passed the first online reduction, the calibrated times and positions of the hit photomultiplier tubes were used to reconstruct the vertex position and the direction of the particle. The energy estimator then assigned an effective kinetic energy, T_{eff}, to each event. Candidate events were required to have a kinetic energy above

5 MeV and be reconstructed within 550 cm of the center of the detector. A total of 2928 events passed all selection criteria.

In order to reduce the systematic uncertainties associated with event reconstruction, the SNO detector is calibrated using a variety of different energy sources. The sources that can be deployed include a light diffusing sphere that is connected to a multi-wavelength laser for measurements of optical parameters and PMT timing, a ^{16}N source which provides a triggered sample of 6.13 MeV γ's [11], a ^{8}Li source delivering β's with an endpoint near 14 MeV, a ^{3}H$(p,\gamma)^4$He ('pT') source [12] for high energy (19.8 MeV) γ's, ^{238}Cf sources for neutrons, and a variety of lower energy sources for background studies.

The primary calibration of SNO consists of deploying the diffusing laserball in approximately 20 positions at each of six optical wavelengths. Based on this information, we are able to use the measured occupancy to fit for the attenuations in the water and the photomultiplier tube angular response. This determines all of the free parameters in our optical model, with the exception of the absolute PMT quantum efficiency. We determine this through the comparison of the ^{16}N source at the center of the detector with a Monte Carlo model of the detector which is based upon the derived optical constants. The total uncertainty on the energy scale and resolution is $\pm 1.2\%$ and $\pm 4.5\%$, respectively.

TABLE 2. Measurements of the solar ^{8}B neutrino flux from SNO's pure D$_2$O phase. The neutrino flux is reported in units of 10^6 cm^{-2}s^{-1}

Experiment	Measured Flux
SNO ES	$2.39^{+0.24}_{-0.23}$ (stat.) ± 0.12 (sys.)
SNO CC	$1.76^{+0.06}_{-0.05}$ (stat.) ± 0.09 (sys.)
SNO NC	$5.09^{+0.44}_{-0.43}$ (stat.) $^{+0.46}_{-0.43}$ (sys.)

Radioactive decays of the daughters in the natural ^{232}Th and ^{238}U chains are the dominant backgrounds in the neutrino signal window. The levels of ^{232}Th and ^{238}U daughters in the detector were determined by radiochemical assays of the detector target[13] and by analyzing the Cherenkov signals from their decays in the neutrino data set. The radioactive backgrounds were found to be better than those specified in the design criteria of the detector.

To extract the contribution from the charged current, neutral current, and elastic scattering reactions, a maximum likelihood analysis is performed. The data are resolved into contributions from CC, ES, and NC events above threshold using probability distribution functions (pdf's) in T_{eff}, $\cos\theta_\odot$, and $(R/R_{\text{AV}})^3$ derived from Monte Carlo calculations generated assuming no flavor transformation and the standard ^{8}B spectral shape [14]. Background event pdf's are included in the analysis with fixed amplitudes determined by the background calibration. The extended maximum likelihood method used in the signal decomposition yields $1967.7^{+61.9}_{-60.9}$ CC events, $263.6^{+26.4}_{-25.6}$ ES events, and $576.5^{+49.5}_{-48.9}$ NC events. The extracted events are then compared to the expected neutrino flux rates from the ^{8}B spectrum, as shown in Table 2 and Figure 1. The charged current to neutral current ratio is 0.35, illustrating the change of electron neutrinos to other active flavors. The good agreement between the expected and measured total ^{8}B flux shows the flavor

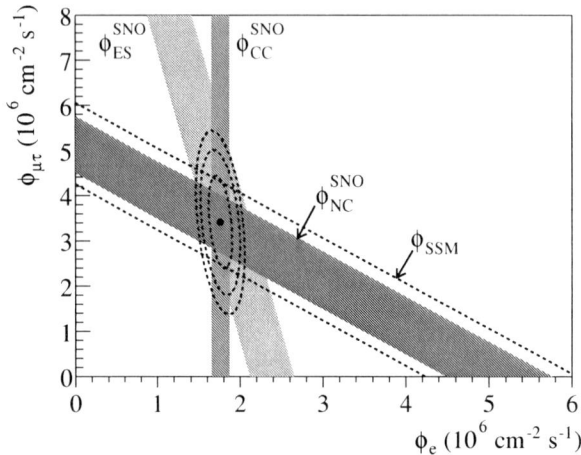

FIGURE 1. Flux of ^{8}B solar neutrinos which are μ or τ flavor vs flux of electron neutrinos. The diagonal bands show the total ^{8}B flux as predicted by the SSM [15] (dashed lines) and that measured with the NC reaction in SNO (solid band). The intercepts of these bands with the axes represent the $\pm 1\sigma$ errors. The bands intersect at the fit values for ϕ_e and $\phi_{\mu\tau}$, indicating that the combined flux results are consistent with neutrino flavor transformation assuming no distortion in the ^{8}B neutrino energy spectrum (D$_2$O data).

change takes place primarily to active neutrinos.

THE SECOND PHASE: UPDATE ON SALT

During the second phase of the experiment, named the salt phase, two metric tonnes of ultra pure sodium chloride have been added to the heavy water. The addition of salt presents clear advantages over the pure D$_2$O phase of the experiment. The neutron capture on ^{35}Cl has a three-fold increase in capture for thermal neutrons. The subsequent de-excitation of ^{36}Cl to the ground state releases a cascade of gamma rays with a total energy of 8.6 MeV, moving the energy profile of the captured neutron further above threshold. Finally, the gamma cascade topologically distinguishes neutral current events from the single-ring charged current Cerenkov ring, allowing SNO to use the event isotropy to separate classes of events.

The addition of salt, however, also presents a number of new challenges to the experiment. After the addition of NaCl into the D$_2$O target, the amount of detected Cherenkov light has dropped significantly over a short period of time. Following this sharp drop, there has been a gradual decline in the detector gain of about 2%/year (Figure 2). There are a number of contributors to the drop in detector gain. These include the degradation of the light concentrator surrounding each PMT, and the increase in the contamination of manganese and organic compounds in the heavy water. As a result, the calibration of the detectorŠs optical and energy response has been updated to include time variation of the water transparency measurements made at various

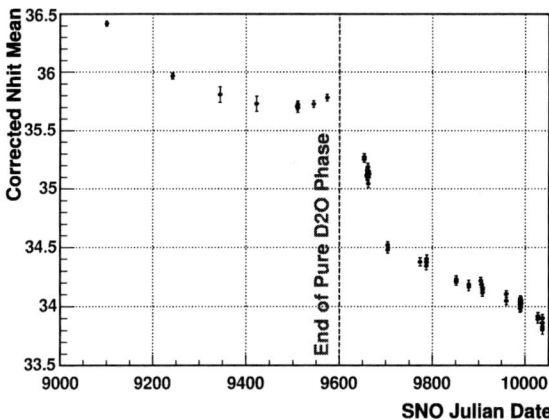

FIGURE 2. Time evolution of the detector energy response to ^{16}N calibration runs taken at the center of the detector. The number of fired PMTs, corrected for the variation of the number of online PMTs, is plotted against time. The end of the D$_2$O phase is on May 28, 2001.

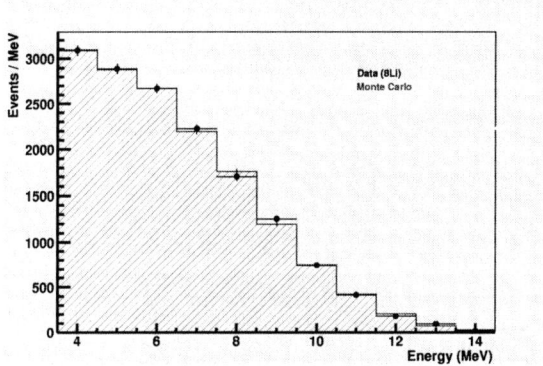

FIGURE 3. Data (black) and Monte Carlo (red) energy distribution for ^{8}Li calibrations above 4 MeV taken at the center of the acrylic vessel. The Monte Carlo is normalized relative to the data above 4 MeV.

wavelengths throughout the running period. Again, ^{16}N is used to test the stability of the reconstructed energy of the event. Other calibration sources, such as ^{8}Li and ^{252}Cf are used to further verify the energy response (see Figure 3). The energy scale uncertainty has been constrained to $\pm 1.1\%$, and the error on the energy resolution to $\pm 3.4\%$.

The degree of the Cerenkov light isotropy is reflected in the pattern of photomultiplier-tube hits. Event isotropy was characterized by parameters β_l, the average value of the Legendre polynomial P_l of the cosine of the angle between PMT hits. The combination $\beta_1 + \beta_4 = \beta_{14}$ was selected as the measure of event isotropy to optimize the separation of NC and CC events. Systematic uncertainty on β_{14} distributions generated by Monte Carlo for signal events was evaluated by comparing ^{16}N calibration data to Monte Carlo

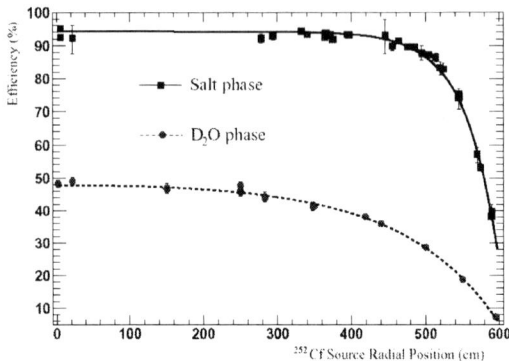

FIGURE 4. Neutron capture efficiency versus source radial position for the pure D2O phase (capture on D) and salt phase (capture on Cl or D) deduced from a ^{252}Cf source, with fits to an analytic function (salt) and to a neutron diffusion model.

calculations. The uncertainty on the mean value of β_{14} is $\pm 0.87\%$.

The neutron capture efficiency on ^{35}Cl is calibrated using essentially the same technique employed during the pure D$_2$O phase. A calibrated ^{252}Cf neutron source is used to map the neutron effiency at different locations in the detector. A total of 3.7676 ± 0.0047 neutrons/fission are emitted by ^{252}Cf, which serves as a handle on extracting the neutron capture profile. At the analysis threshold of $T_{eff} \geq 5.5$ MeV and reconstructed radius inside 550 cm, the neutron efficiency is $39.9 \pm 1.3\%$ (see Figure 4).

The inclusion on an additional parameter in the statistical separation of charged current and neutral current events greatly reduces the correlations between the errors of the two signatures. As event isotropy has a weak dependence on the energy of an event, it is possible to do both an energy-constrained and unconstrained analysis on the data. A constrained analysis requires that the energy spectrum conforms to the ^{8}B predicted shape, while the unconstrained analysis makes no explicit requirement on the energy spectrum. The ability of performing a shape-unconstrained analysis allows SNO to measure the phenomena of neutrino oscillations in a nearly model-independent way.

After this talk was given, the SNO collaboration presented the results of the salt phase of the experiment. Results on constrained and unconstrained analyses of the salt data can be found in [16].

THE THIRD PHASE: NEUTRAL CURRENT DETECTORS

In the third and final phase of the experiment, an array of 40 ^{3}He proportional counters –Neutral Current Detectors (NCDs)– will be added inside the acrylic vessel. These detectors are sensitive to the neutrons from the neutral current reaction via the reaction:

$$^3\text{He} + n \rightarrow {}^3\text{H} + p \qquad (4)$$

As such a reaction allows the detection of the neutron distinct from the Cerenkov signature of charged current events, the system enables the event-by-event separation of charged and neutral current neutrino interactions. The NCD array is scheduled to be deployed this winter.

CONCLUSION

The SNO experiment has demonstrated conclusively that solar ν_e's transform their flavor while in transit to the Earth. A high precision measurement of the total active solar neutrino flux without any assumption of the energy dependence of the neutrino flavor transformation probability is near completion at the time of this talk. In the next few months, the SNO physics program will enter its NCD phase. It is anticipated that the data from this NCD phase, along with data from the previous phases, will significantly enhance our understanding of the properties of neutrinos.

ACKNOWLEDGMENTS

This research was supported by: Canada: NSERC, Industry Canada, NRC, Northern Ontario Heritage Fund, Inco, AECL, Ontario Power Generation, HPCVL, CFI; US: Dept. of Energy; UK: PPARC. We thank the SNO technical staff for their strong contributions.

REFERENCES

1. B.T. Cleveland *et al.*, Astrophys. J. **496**, 505 (1998).
2. SAGE Collaboration, J. Exp. Theor. Phys. **95**, 181 (2002); J. N. Abdurashitov *et al.*, Phys. Rev. C. **60**, 055801 (1999).
3. W. Hampel *et al.*, Phys. Lett. B **447**, 127 (1999).
4. M. Altmann *et al.*, Phys. Lett. B **490**, 16 (2000); T. Kirsten, *Progress in GNO*, XXth Int. Conf. on Neutrino Physics and Astrophysics, Munich, May 25–30, 2002, to be published in Nucl. Phys. B Proc. Supp.
5. K.S. Hirata *et al.*, Phys. Rev. Lett. **65**, 1297 (1990); K.S. Hirata *et al.*, Phys. Rev. D **44**, 2241 (1991), **45** 2170E (1992); Y. Fukuda *et al.*, Phys. Rev. Lett. **77**, 1683 (1996).
6. S. Fukuda *et al.*, Phys. Rev. Lett. **86**, 5651 (2001) and Phys. Lett. B **539**, 179 (2002).
7. J.N. Bahcall, M. H. Pinsonneault, and S. Basu, astro-ph/0010346 v2. The reference [8] B neutrino flux is 5.05×10^6 cm^{-2}s^{-1}.
8. A.S. Brun, S. Turck-Chièze, and J.P. Zahn, Astrophys. J. **525**, 1032 (1999); S. Turck-Chièze *et al.*, Ap. J. Lett., v. **555** July 1, 2001.
9. The SNO Collaboration, Nucl. Instr. and Meth. A449, 172 (2000).
10. Q.R. Ahmad *et al.*, Phys. Rev. Lett. **89**, 011301 (2002) and Phys. Rev. Lett. **89**, 011302 (2002).
11. M. R. Dragowsky *et al.*, Nucl. Instr. Meth. **A481**, 284 (2002).
12. A. W. P. Poon *et al.*, Nucl. Instr. and Meth. **A449**, 172 (2000).
13. T. C. Andersenet *et al.*, Nucl. Instr. and Meth. **A501**, 386 (2003); T. C. Andersen *et al.*, Nucl. Instr. and Meth. **A501**, 399 (2003); I. Blevis *et al.*, arXiv:nucl-ex/0305022, submitted to Nucl. Instr. and Meth.
14. C. E. Ortiz *et al.*, Phys. Rev. Lett. **85**, 2909 (2000).
15. John N. Bahcall, M.H. Pinsonneault, and Sarbani Basu, Astrophys. J. **555**, 990(2001).
16. S. N. Ahmed *et al.* arXiv:nucl-ex/0309004 (2003).

Measurement of θ_{13} by reactor experiments

Osamu Yasuda[1]

Department of Physics, Tokyo Metropolitan University 1-1 Minami-Osawa, Hachioji, Tokyo 192-0397, Japan

Abstract. I describe how reactor measurements of $\sin^2 2\theta_{13}$ can be improved by a near-far detector complex. I show that in the Kashiwazaki plan it is potentially possible to measure $\sin^2 2\theta_{13}$ down to 0.02.

Introduction. After the successful experiments on atmospheric and solar neutrinos and KamLAND, the next step in neutrino oscillation physics is to determine θ_{13}. It has been known that the oscillation parameters θ_{jk}, Δm_{jk}^2, δ cannot be determined uniquely even if the appearance probabilities $P(\nu_\mu \to \nu_e)$ and $P(\bar{\nu}_\mu \to \bar{\nu}_e)$ are measured precisely from a long baseline accelerator experiment due to so-called parameter degeneracies, and several ideas have been proposed to solve the problem. Among others, combination of a reactor measurement and a long baseline experiment offers a promising possibility [2, 3, 1, 4]. In this talk I briefly explain how measurements of $\sin^2 2\theta_{13}$ in reactor experiments can be improved by a near-far detector complex. I also show that in the Kashiwazaki plan [5] the sensitivity to $\sin^2 2\theta_{13}$ is approximately 0.02.

Reactor measurements of θ_{13}. In the three flavor framework the disappearance probability of the reactor neutrinos is given by

$$P(\bar{\nu}_e \to \bar{\nu}_e) = 1 - \sin^2 2\theta_{13} \sin^2 \left(\frac{\Delta m_{13}^2 L}{4E}\right),$$

if the contribution from Δm_{21}^2 is negligible. So the analysis is reduced to that of the conventional two flavor framework in vacuum.

Let me start with the derivation of χ^2 used in [1] (See also [6]). For simplicity I assume one reactor and two detectors at near and far distances. χ^2 is defined as

$$\chi^2 \equiv \min_{\alpha, \alpha^n, \alpha^f} \left\{ \left[\frac{M^n - T^n(1 + \alpha + \alpha^n)}{T^n \sigma_{stat}^n}\right]^2 + \left[\frac{M^f - T^f(1 + \alpha + \alpha^f)}{T^f \sigma_{stat}^f}\right]^2 + \left(\frac{\alpha}{\sigma_c}\right)^2 + \left(\frac{\alpha^n}{\sigma_u}\right)^2 + \left(\frac{\alpha^f}{\sigma_u}\right)^2 \right\}, \quad (1)$$

---

[1] Based on the work [1] in collaboration with H. Minakata, H. Sugiyama, K. Inoue and F. Suekane. Talk presented at the 5th International Workshop on Neutrino Factories & Superbeams (NuFact'03), Columbia University, New York, USA, June 5-11, 2003.

where the superscripts n and f stand for the quantities at the near and far detectors, M and T stand for the measured and theoretical total numbers of events, $\sigma_{stat}^n = (T^n)^{-1/2}$ and $\sigma_{stat}^f = (T^f)^{-1/2}$ stand for the statistical errors, and α, α^n, α^f are the variables to introduce the correlated systematic error σ_c and the uncorrelated systematic error σ_u (I am assuming that the uncorrelated errors for the two detectors are the same). It is understood that the right hand side of Eq. (1) is minimized with respect to the three variables. After eliminating them I get

$$\chi^2 = \begin{pmatrix} y^n, & y^f \end{pmatrix} \begin{pmatrix} \sigma_c^2 + \sigma_u^2 + (\sigma_{stat}^n)^2 & \sigma_c^2 \\ \sigma_c^2 & \sigma_c^2 + \sigma_u^2 + (\sigma_{stat}^f)^2 \end{pmatrix}^{-1} \begin{pmatrix} y^n \\ y^f \end{pmatrix}, \quad (2)$$

where I have defined $y^n \equiv (M^n - T^n)/T^n$, $y^f \equiv (M^f - T^f)/T^f$. When the statistical errors are negligible, the error matrix in Eq. (2) is easily diagonalized and is expressed as

$$\chi^2 = \frac{1}{2\sigma_u^2}(y^f - y^n)^2 + \frac{1}{2\sigma_u^2 + 4\sigma_c^2}(y^f + y^n)^2. \quad (3)$$

The uncorrelated error $2\sigma_u^2$ in Eq. (3) is the sum of the contributions from the two detectors and $\sigma_{rel} \equiv \sqrt{2}\sigma_u$ is referred to as the relative normalization error in [7]. Assuming that one can extrapolate the reference values for the systematic errors of the Bugey experiment [7] to the CHOOZ detectors [8], the systematic errors are estimated to be $\sigma_u = 0.8\%/\sqrt{2} = 0.6\%$ and $\sigma_c = 2.6\%$. Eq. (3) indicates that the contribution from the sum $y^f + y^n$ is much smaller than that from the difference $y^f - y^n$. This was the reason why the $(y^f + y^n)^2$ term in χ^2 was ignored in [1]. It should be emphasized that Eq. (3) shows the advantage of a near-far detector complex, since the correlated systematic error σ_c is canceled in the denominator of the $(y^f - y^n)^2$ term [2].

From the expression (3) of χ^2 for the rate, let me define the following χ^2 for the spectrum analysis:

$$\chi^2 = \sum_j \frac{1}{\sigma_j^2} \left(\frac{M_j^f - T_j^f}{T_j^f} - \frac{M_j^n - T_j^n}{T_j^n} \right)^2,$$

where $M_j^{n,f}$ and $T_j^{n,f}$ stand for the measured and expected number of events at the near and far detectors for the j-th bin, and σ_j is the statistical error plus the uncorrelated systematic error for each bin: $\sigma_j^2 = 1/T_j^n + 1/T_j^f + 2(\sigma_{uj}^{bin})^2$. Here I assume that the uncorrelated systematic error is the same for all bins: $\sigma_{uj}^{bin} = \sigma_u^{bin}$, so σ_u^{bin} is estimated from the uncorrelated systematic error σ_u for the total number of events by

$$(\sigma_u^{bin})^2 = \sigma_u^2 \frac{(T_{tot}^f)^2}{\sum_j (T_j^f)^2}, \quad T_{tot}^f \equiv \sum_j T_j^f,$$

---

[2] The Krasnoyarsk proposal [9] also takes advantage of a near-far detector complex.

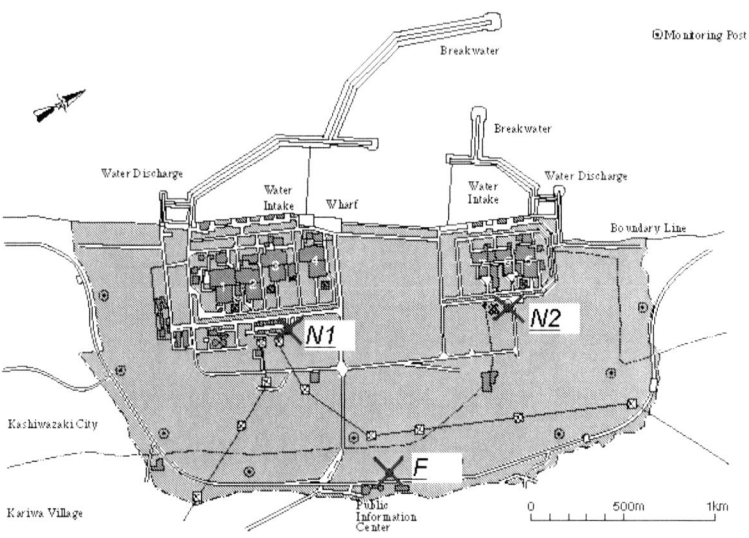

FIGURE 1. The location of the three detectors in the Kashiwazaki plan.

since the uncertainty squared of the total number of events is obtained by adding up the bin-by-bin systematic errors $(\sigma_u^{bin})^2(T_j^f)^2$. The ratio σ_u^{bin}/σ_u is approximately 3 in our analysis. Although the sensitivity to $\sin^2 2\theta_{13}$ is optimized at $L \simeq 1.7$km for $|\Delta m_{13}^2| = 2.5 \times 10^{-3}$eV^2 [1], the longest baseline for the far detector inside the campus of the Kashiwazaki-Kariwa nuclear power plant turns out to be 1.3km [5] (See Fig.1). In Fig. 2 the 90 % CL exclusion limits, which corresponds to $\chi^2 = 2.7$ for one degree of freedom, are presented for two sets of parameters (data size, σ_{rel})=(10 ton-year, 1 %), (40 ton-year, 0.5 %) and for two baselines $L = 1.3$km, 1.7km, where $\sigma_{rel} \equiv \sqrt{2}\sigma_u$ was introduced earlier [3]. Fig. 2 shows that the sensitivity to $\sin^2 2\theta_{13}$ does not decrease very much for $L = 1.3$km and that it is possible to measure $\sin^2 2\theta_{13}$ down to 0.02, provided the quoted values of the uncorrelated systematic error are realized.

Summary & Conclusions. In this talk I emphasized the advantage of a near-far detector complex in reactor measurements of $\sin^2 2\theta_{13}$. I showed that it is potentially possible to measure $\sin^2 2\theta_{13}$ down to 0.02 in the Kashiwazaki plan.

Acknowledgments. I would like to thank Hiroaki Sugiyama for many discussions on the derivation of χ^2. This work was supported by Grants-in-Aid for Scientific Research in Priority Areas No. 12047222 and No. 13640295, Japan Ministry of Education, Culture, Sports, Science, and Technology.

---

[3] For simplicity I have assumed in the calculation that there is only one reactor and two detectors, but the sensitivity with this simplification turns out to be almost the same as that with the exact calculation with seven reactors and three detectors [6].

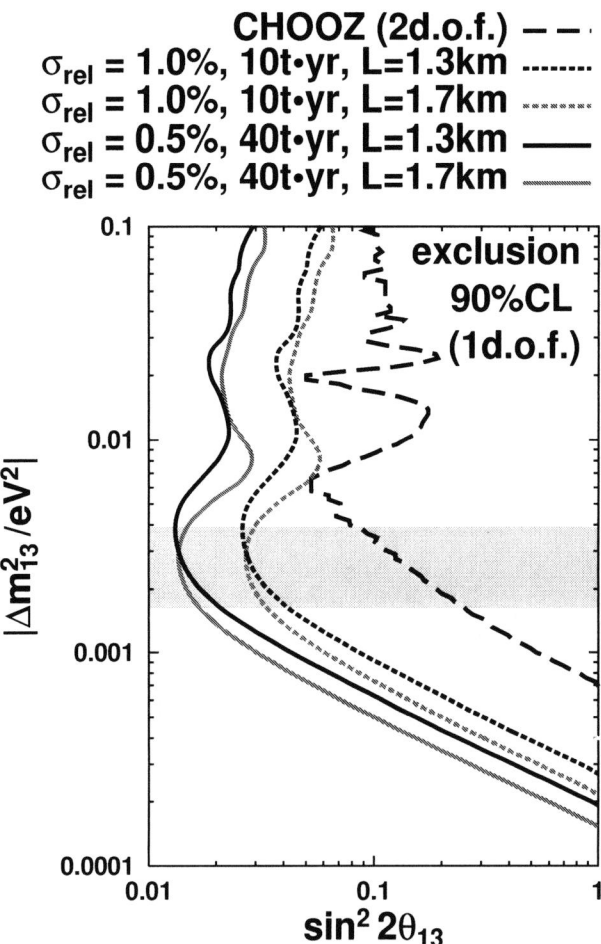

FIGURE 2. The 90% CL exclusion limits on $\sin^2 2\theta_{13}$ in the Kashiwazaki plan. The light shadowed band is the allowed region at 90%CL for $|\Delta m_{13}^2|$ from the atmospheric neutrino data.

REFERENCES

1. H. Minakata, H. Sugiyama, O. Yasuda, K. Inoue and F. Suekane, Phys. Rev. D **68**, 033017 (2003).
2. G. L. Fogli and E. Lisi, Phys. Rev. D **54**, 3667 (1996).
3. G. Barenboim and A. de Gouvea, arXiv:hep-ph/0209117.
4. P. Huber, M. Lindner, T. Schwetz and W. Winter, Nucl. Phys. B **665**, 487 (2003).
5. F. Suekane, K. Inoue, T. Araki and K. Jongok, arXiv:hep-ex/0306029.
6. H. Sugiyama, O. Yasuda, F. Suekane and G.A. Horton-Smith, to appear.
7. Y. Declais et al., Nucl. Phys. B **434**, 503 (1995).
8. M. Apollonio et al., Phys. Lett. B **420**, 397 (1998); Phys. Lett. B **466**, 415 (1999).
9. Y. Kozlov, L. Mikaelyan and V. Sinev, Phys. Atom. Nucl. **66**, 469 (2003) [Yad. Fiz. **66**, 497 (2003)]; V. Martemyanov, L. Mikaelyan, V. Sinev, V. Kopeikin and Y. Kozlov, arXiv:hep-ex/0211070.

The race for θ_{13} – Superbeams versus Reactors?

Patrick Huber[†]

Institut für Theoretische Physik, Physik-Department, James-Franck-Straße, D-85748 Garching
Max-Planck-Institut für Physik, Föhringer Ring 6, D-80805 München

Abstract. Recently a new reactor neutrino neutrino experiment has been proposed to improve the CHOOZ bound on θ_{13} by one order of magnitude. Here I discuss the potential of this kind of experiments and compare it to the one of superbeams. Furthermore I will emphasize the complementarity of the two approaches. Specifically I will discuss the possible synergies for the discovery of leptonic CP violation.

INTRODUCTION

The small angle θ_{13} is the missing link between atmospheric and solar neutrino oscillations. It is furthermore a prerequisite for the existence of sizeable CP effects and thus for the possibility to discover leptonic CP violation. The CHOOZ experiment [1] provides currently the most stringent bound on θ_{13} by using $\bar{\nu}_e$ disappearance. For a disappearance experiment which looks for small effects it is essential to have a very low level of systematical uncertainty and an excellent understanding of the various error sources. Recently it has been proposed to use an improved reactor experiment to get a better determination of θ_{13} [2, 3, 4, 5]. The key ingredient for a better control of systematical errors is to use a two detector setup, i.e. a near and far detector. In this way a large number of systematical errors cancel in the comparison of near and far detector event rates, e.g. the uncertainty on the total neutrino flux from the reactor station.

The big advantage of reactor experiments is that they observe the $\bar{\nu}_e \rightarrow \bar{\nu}_e$ transition which is free of any degeneracies. This allows a straightforward interpretation of the outcome of such an experiment in terms of $\sin^2 2\theta_{13}$. In the case of superbeam experiments which observe the $\nu_\mu \rightarrow \nu_e$ transition the problem arises that the interpretation of the results is plagued with ambiguities due to the lack of knowledge on the CP phase (see e.g. [6]).

In assessing the scientific value of a new reactor experiment basically two questions appear – How does the θ_{13} sensitivity compare to other approaches like superbeams? How can the reactor data improve the physics potential of superbeam facilities? Before dealing with those questions, the reactor setup and the corresponding assumptions on the level of systematics are introduced. The following discussion and results are based on [7].

REACTOR SETUP

The size of the event sample of a reactor experiment is determined by three factors – the fiducial mass of the detector, the thermal power of the reactor and the running time. Thus the statistical accuracy is given by the product of those three factors and we use the unit $t \cdot GW \cdot y$. Following [7] we consider two setups with different values for the integrated luminosity: **Reactor-I** with 400 t GW y and **Reactor-II** with 8000 t GW y. **Reactor-I** corresponds to a detector which is $2-4$ times larger than the CHOOZ detector whereas **Reactor-II** would be more of KamLAND-size. We do not consider any backgrounds at this point, mainly because the background level and spectral distribution is strongly dependent on the actual site and overburden. The two main contributions to the error budget are the normalization error σ_{norm} and the error in the energy calibration between the two detectors, called σ_{cal}. For σ_{norm} a value of 0.8% is assumed and for σ_{cal} a value of 0.5% is assumed. The size of the latter does however not affect the results notably as long as the oscillation dip is clearly within the spectral range of the experiment. The energy threshold of the detector for the visible energy is 1.0 MeV and the analysis extends up to 7.2 MeV. The energy resolution is taken to be $5\%/\sqrt{E}$, which is slightly better than the one of KamLAND [8]. More details concerning the statistical treatment of the two detectors and on various other systematical errors can be found in [7].

$SIN^2 2\theta_{13}$

The quest for θ_{13} is now the most pressing task in neutrino oscillations and therefore any next generation experiment has to address this issue. The sensitivity for θ_{13} at a new reactor experiment is determined by two factors – the statistical accuracy and the level of systematical uncertainty. For the **Reactor-I** setup the number of events is not sufficient to exploit the spectral information and it is therefore basically a counting experiment. For this reason the size of the normalization uncertainty σ_{norm} governs the final sensitivity. In case of the **Reactor-II** setup the number of events is large enough to fully exploit the spectral information and thus the normalization can be determined by the experiment itself. For a high statistics experiment like **Reactor-II** however the size and uncertainty of the backgrounds will be an important issue and has to be investigated in more detail than presented here.

The $\sin^2 2\theta_{13}$ sensitivity at 90% CL is shown in figure 1. In the left hand panel it is plotted as function of the atmospheric Δm^2 whereas in the right panel the dependence on the solar Δm^2 is depicted. The sensitivity for the **Reactor-I** setup is drawn as grey dashed line in both panels. For comparison also the sensitivity of JHF-SK is shown as thin solid black line. The reactor experiment exhibits a much weaker dependence on the atmospheric mass splitting than the superbeam experiment because of the larger range in L/E sampled by the reactor experiment. The sensitivity of a reactor experiment is independent of the solar mass splitting due the form of $P_{\bar{\nu}_e \to \bar{\nu}_e}$. Around the central values of both mass splittings the sensitivity of the reactor experiment is very similar to the one of superbeams and is roughly $1.6 \cdot 10^{-2}$.

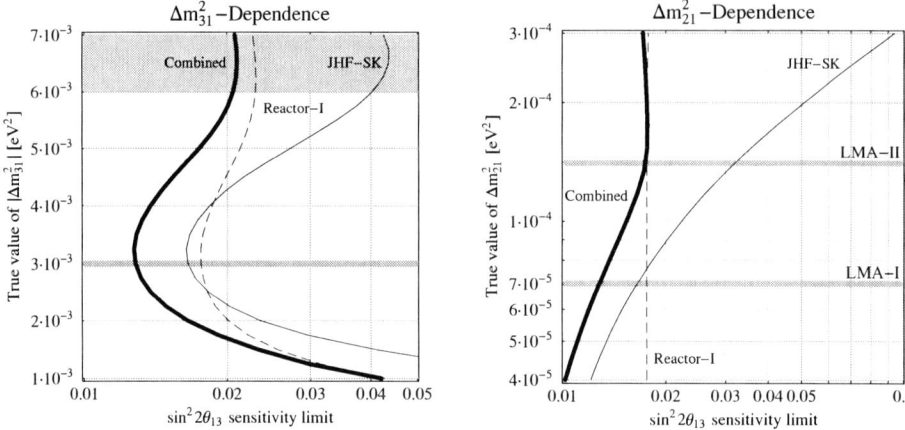

FIGURE 1. The sensitivity limits to $\sin^2 2\theta_{13}$ as functions of the true values of Δm_{31}^2 (left plot) or Δm_{21}^2 (right plot), respectively. They are shown for the JHF-SK and Reactor-I setups as well as their combination at the 90% confidence level. In the left figure, the atmospheric excluded region is shaded in gray, and in the right figure, only the KamLAND-allowed region is shown. This figure is taken from [7].

SYNERGY

The complementarity of reactor experiments can nicely be illustrated in the case of leptonic CP violation. In figure 2 the sensitivity to maximal CP violation at 90% CL for the case $\delta_{CP} = +\pi/2$ is shown as function of the true values of $\sin^2 2\theta_{13}$ and Δm_{21}^2. The thin dashed line indicates the region where JHF-SK is sensitive with neutrino running only. The thin solid line shows the region of sensitivity for the combined data of running with neutrinos and anti-neutrinos. The total running time is the same as in the case with neutrino running only and is divided such that the event rate for both channels is roughly the same. Clearly the combined mode enhances the sensitivity considerably. The thick solid line shows the obtainable sensitivity if the data of JHF-SK running exclusively with neutrinos and of Reactor-II are combined. In this case the sensitivity is much better than for the combined mode of JHF-SK although the reactor data does not carry any information on the CP phase. The reason is that the reactor data fixes θ_{13} and therefore makes the anti-neutrino running at JHF-SK superfluous. Thus JHF-SK can avoid the anti-neutrino run which enhances the statistics since the neutrino cross section is approximately three times larger than the anti-neutrino cross section. Now the LMA-II best fit value is clearly within the region of sensitivity.

CONCLUSION

A new reactor experiment offers the possibility to measure θ_{13} with excellent sensitivity, competitive to superbeams. It can, however, not replace a superbeam experiment since it has no sensitivity to the CP phase, the mass hierarchy and the atmospheric mixing

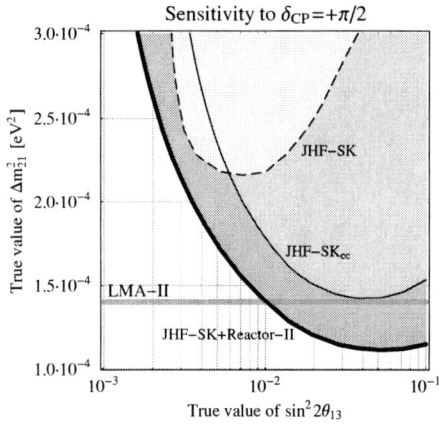

FIGURE 2. The sensitivity to maximal CP violation $\delta_{CP} = +\pi/2$ as function of the true values of $\sin^2 2\theta_{13}$ and Δm_{21}^2 within the KamLAND-allowed region. The sensitivity is at the 90% confidence level given on the upper sides of the curves. It is shown for JHF-SK (neutrino running only), JHF-SK$_{cc}$ (same overall running time split into neutrino and antineutrino running with about equal numbers of events), and JHF-SK (neutrino running only) combined with Reactor-II. The LMA-II best-fit value is marked by the horizontal gray line. This figure is taken from [7].

angle on its own. The argument works the other way round as well, since an initial stage superbeam experiment on its own cannot discover leptonic CP violation reliably. Moreover also for the measurement of the mass hierarchy the combined data of reactor and superbeam increase the sensitivity considerably, which was not discussed here due to limited space (see [7] for details). Thus the question raised in the title can be answered with "no" – it is not a matter of reactors versus superbeams but a matter of reactors *and* superbeams.

ACKNOWLEDGMENTS

I would like to thank the co-authors of reference [7] on which this work is based – M. Lindner, T. Schwetz and W. Winter.

REFERENCES

1. Apollonio, M., et al., *Eur. Phys. J.*, **C27**, 331–374 (2003).
2. Mikaelyan, L. A., and Sinev, V. V., *Phys. Atom. Nucl.*, **63**, 1002–1006 (2000).
3. Mikaelyan, L., *Nucl. Phys. Proc. Suppl.*, **91**, 120–124 (2001).
4. Martemyanov, V., Mikaelyan, L., Sinev, V., Kopeikin, V., and Kozlov, Y. (2002).
5. Minakata, H., Sugiyama, H., Yasuda, O., Inoue, K., and Suekane, F. (2002).
6. Barger, V., Marfatia, D., and Whisnant, K., *Phys. Rev.*, **D65**, 073023 (2002).
7. Huber, P., Lindner, M., Schwetz, T., and Winter, W., *Nucl. Phys.*, **B665**, 487–519 (2003).
8. Eguchi, K., et al., *Phys. Rev. Lett.*, **90**, 021802 (2003).

Precision measurement of oscillation parameters with reactors

Sandhya Choubey

INFN, Sezione di Trieste, Trieste, Italy
Scuola Internazionale Superiore di Studi Avanzati, I-34014, Trieste, Italy

Abstract. We review the potential of long and intermediate baseline reactor neutrino experiments in measuring the mass and mixing parameters. The KamLAND experiment can measure the solar mass squared difference very precisely. However it is not at the ideal baseline for measuring the solar neutrino mixing angle. If low-LMA is confirmed by the next results from KamLAND, a reactor experiment with a baseline of 70 km should be ideal to measure precisely the solar neutrino mixing angle. If on the contrary KamLAND re-establishes high-LMA as a viable solution, then a 20–30 km intermediate baseline reactor experiment could yield very rich phenomenology.

The first results from the KamLAND experiment in Japan [1] has showed that the electron antineutrinos undergo flavor oscillations on their way from their source to the detector. This result coupled with the assumption of CPT invariance has put to rest all speculations regarding the true solution of the long standing solar neutrino problem, where the electron neutrinos produced inside the Sun apparently *disappear* as they travel from the Sun to the Earth (see [2] for a recent review of the solar neutrino experiments). This disappearance of the solar neutrinos can now be attributed confidently to neutrino flavor mixing, with the mass squared difference same as that relevant for the KamLAND experiment. Earlier the spectacular evidence that these solar electron neutrinos do not really disappear, but rather appear disguised as a neutrino with a different active flavor, came from the first measurement of the total ^8B solar neutrino flux, through the neutral current (NC) reaction on deuterium, at the Sudbury Neutrino Observatory (SNO) [3]. The SNO NC data when combined with the data from all the other solar neutrino experiments picked the so called Large Mixing Angle (LMA) solution to the solar neutrino problem [3, 4]. This remarkable result has very recently been reinforced by the salt phase data from the SNO experiment [5]. Prior to the SNO salt phase results, KamLAND data when combined with the other solar neutrino results, allowed two sub-regions within the LMA allowed region at the 99% C.L.– which we choose to call low-LMA (with best-fit at $\Delta m_{21}^2 = 7.2 \times 10^{-5}$ eV^2 and $\sin^2\theta_{12} = 0.3$) and high-LMA (with best-fit at $\Delta m_{21}^2 = 1.5 \times 10^{-4}$ eV^2 and $\sin^2\theta_{12} = 0.3$) [6]. After the SNO salt phase results, the combined analysis using all available data now allows the high-LMA solution only at the 99.13% C.L. [7]. Thus high-LMA is now further disfavored compared to low-LMA, though still not ruled out comprehensively.

There exists also very strong evidence for $\nu_\mu \to \nu_\tau$ ($\bar{\nu}_\mu \to \bar{\nu}_\tau$) oscillations of the atmospheric ν_μ ($\bar{\nu}_\mu$) from the observed Zenith angle dependence of the μ–like events in the Super-Kamiokande (SK) experiment – with maximal mixing and 1.3×10^{-3}eV$^2 \lesssim |\Delta m_A^2| \lesssim 3.1 \times 10^{-3}$eV^2 (90% C.L.) [8].

Since the solar and the atmospheric neutrino "anomalies" involve two hierarchically different mass scales, simultaneous description of the two requires the existence of three-neutrino mixing. The solar neutrino data constrain the parameters $\Delta m_\odot^2 \sim \Delta m_{21}^2$ and $\theta_\odot \sim \theta_{12}$ while the atmospheric neutrino data constrain the parameters $\Delta m_{atm}^2 \sim \Delta m_{31}^2$ and $\theta_{atm} \sim \theta_{23}$. The two sectors get connected by the mixing angle θ_{13} which is at present constrained by the reactor data [9]. After these magnificent results the stage is set for the era of precision measurement of the oscillation parameters. We will discuss here what more can be achieved in this respect in reactor experiments sensitive to the Δm^2 driven distortions in the $\bar{\nu}_e$ spectrum due to oscillations.

The full expression for the $\bar{\nu}_e$ survival probability in the case of 3 flavor neutrino mixing and neutrino mass spectrum with normal hierarchy (NH) is given by [10]

$$P_{NH} = 1 - 2\sin^2\theta_{13}\cos^2\theta_{13}\sin^2\left(\frac{\Delta m_{31}^2 L}{4E_v}\right) - \frac{1}{2}\cos^4\theta_{13}\sin^2 2\theta_{12}\sin^2\left(\frac{\Delta m_{21}^2 L}{4E_v}\right)$$
$$+ 2\sin^2\theta_{13}\cos^2\theta_{13}\sin^2\theta_{12}\left(\cos\left(\frac{\Delta m_{31}^2 L}{2E_v} - \frac{\Delta m_{21}^2 L}{2E_v}\right) - \cos\frac{\Delta m_{31}^2 L}{2E_v}\right), \quad (1)$$

where E_v is the $\bar{\nu}_e$ energy. If the neutrino mass spectrum is with inverted hierarchy (IH), the $\bar{\nu}_e$ survival probability can be written in the form [10]

$$P_{NH} = 1 - 2\sin^2\theta_{13}\cos^2\theta_{13}\sin^2\left(\frac{\Delta m_{31}^2 L}{4E_v}\right) - \frac{1}{2}\cos^4\theta_{13}\sin^2 2\theta_{12}\sin^2\left(\frac{\Delta m_{21}^2 L}{4E_v}\right)$$
$$+ 2\sin^2\theta_{13}\cos^2\theta_{13}\cos^2\theta_{12}\left(\cos\left(\frac{\Delta m_{31}^2 L}{2E_v} - \frac{\Delta m_{21}^2 L}{2E_v}\right) - \cos\frac{\Delta m_{31}^2 L}{2E_v}\right), \quad (2)$$

Thus the $\bar{\nu}_e$ survival probability depends on the four continuous parameters Δm_{21}^2, $\sin^2\theta_{12}$, Δm_{31}^2, $\sin^2\theta_{13}$, and on a single "discrete" parameter — the type of the neutrino mass spectrum — NH or IH. Which of these parameters could be measured in a given reactor experiment depends on which of the terms in Eqs. (1) and Eqs. (2) dominate, which in turn depends crucially on the baseline. For a given Δm^2, if the baseline is such that $\sin^2(\Delta m^2 L/4E) \approx 1$, the neutrinos undergo maximum flavor oscillations, we have a trough in the resultant $\bar{\nu}_e$ spectrum and we call this a case of SPMIN (Survival Probability MINimum). If on the other hand the baseline corresponds to $\sin^2(\Delta m^2 L/4E) \approx 0$, we have a peak in the $\bar{\nu}_e$ spectrum and denote this case as SPMAX (Survival Probability MAXimum). Since the shape of the spectrum depends very crucially on the mass squared difference, the value of the relevant Δm^2 can be expected to be determined very accurately as long as there is an observable distortion, irrespective of whether the distortion corresponds to a SPMIN or SPMAX. However since for SPMAX, the survival probability can be written approximately as $P_{ee} \approx 1$, there is little sensitivity to θ if the SPMAX appears in the statistically most relevant part of the observed spectrum. On the other hand since for SPMIN the probability is $P_{ee} \approx 1 - \sin^2 2\theta$, one can expect maximum sensitivity to the mixing angle if the SPMIN is produced in the spectrum.

The KamLAND experiment in Japan is the world's first very long baseline reactor experiment which looks for disappearance of $\bar{\nu}_e$ from nuclear reactors all over Japan. The most powerful reactors are at a distance of about 160 km. Thus for Δm_{21}^2 in the

TABLE 1. The range of parameter values allowed at 99% C.L. and the corresponding spread.

Data set used	99% CL range of $\Delta m^2_{21} \times 10^{-5} eV^2$	99% CL spread of Δm^2	99% CL range of $\sin^2 \theta_{12}$	99% CL spread in $\sin^2 \theta_{12}$
only sol	3.2 - 17.0	68%	0.22 – 0.40	29%
sol+162 Ty	5.3 - 9.8	30%	0.22 – 0.40	29%
sol+1 kTy	6.7 - 8.0	9%	0.23 – 0.40	27%
sol+3 kTy	6.8 - 7.6	6%	0.24 – 0.37	21%

LMA range, this experiment is expected to put very stringent bounds on the allowed values of the solar neutrino oscillation parameters. In Table 1 we present the 99% C.L. allowed range for the parameters Δm^2_{21} and $\sin^2 \theta_{12}$, along with the corresponding % uncertainty ("spread"), obtained by taking various combination of data sets into account. We note that the uncertainty in Δm^2_{21} reduces from 68% from only the solar data to 30% by including first KamLAND data. This uncertainty would further go down to 9% (6%) after 1 kTy (3 kTy) data from KamLAND. However there seems to be little improvement in the uncertainty on the value of $\sin^2 \theta_{12}$, with increase in the KamLAND statistics.

The reason for this failure of KamLAND to measure $\sin^2 \theta_{12}$ accurately enough can be traced back to the fact that the KamLAND spectrum shows a peak in its survival probability (SPMAX) at around 3.6 MeV. Thus, as discussed before, this sensitivity to the spectral shape gives KamLAND the ability to accurately pin down Δm^2_{21}. However since the oscillatory term $\sin^2(\Delta m^2 L_i/4E)$, is close to zero, it smothers any $\sin^2 \theta_{12}$ dependence along with it. Therefore, we conclude that KamLAND probably is not at the ideal baseline for determining the solar neutrino mixing angle.

What is the baseline most suited for measuring $\theta_\odot$? For Δm^2_{21} in the low-LMA region, we expect to find a minimum in the survival probability (SPMIN) in the statistically most relevant part of the energy spectrum when the baseline $L \sim 70$ km. It was shown in [11] that for a new experimental set-up with a powerful reactor source, a la Kashiwazaki nuclear reactor in Japan with a maximum power generation of about 24.6 GW, producing a SPMIN in the detected spectrum in a KamLAND like detector at a distance of 70 km, $\sin^2 \theta_{12}$ can be measured to within 10% uncertainty, after 3 kTy of data.

On the other hand, if contrary to the trend emerging in the solar neutrino experiments, the next KamLAND spectral data conforms to a point in the high-LMA region, then we would need an intermediate baseline reactor experiment with $L \sim 20 - 30$ km to get a SPMIN in the resultant spectrum. We have shown in [12] that an experimental set-up with an intermediate baseline of 20–30 km, a reactor with power of 5 GW and with 3 kTy of statistics, we can measure both Δm^2_{21} and $\sin^2 \theta_{12}$ down to the few percent level. The impact of systematic uncertainties, baseline, statistics and energy threshold of the detector was studied [12]. It was concluded that as long as the baseline and the energy threshold allowed the experiment to observe the SPMIN, θ_{12} could be measured very accurately irrespective of the other conditions.

If in addition, the energy resolution of the detector is good enough to collect data in bins of 0.1 MeV width, then the intermediate baseline reactor experiment can observe the Δm^2_{31} driven subdominant oscillations – given by the second and the last terms in Eqs. (1) and (2). Thus, this experiment can also be used very effectively to extract information on Δm^2_{31}, $\sin^2 \theta_{13}$ and even the neutrino mass hierarchy. We have checked that an energy resolution of $\sigma(E)/E = 5\%/\sqrt{E}$, with E in MeV, should be good enough for this purpose. For $L = 20$ km, bin width 0.1 MeV, systematic uncertainty of 2% and 15 GWkTy statistic, one can put an upper bound of $\sin^2 \theta_{13} < 0.021(0.012)$ at $3\sigma(90\%)$ C.L.[12]. If on the other hand the true value of $\sin^2 \theta_{13}$ turns out to be large to produce observable effects in this experiment, then Δm^2_{31} can be measured to the percent level. For the measurement of $\sin^2 \theta_{13}$ and Δm^2_{31} we do not necessarily need the Δm^2_{21} to be in the high-LMA region. But if the condition for Δm^2_{21} to be in high-LMA is satisfied, then the difference between the solar and atmospheric neutrino mass scales is not too severe and the last terms in Eqs. (1) and (2) are non-zero. Since they are different for normal (NH) and inverted (IH) hierarchies for $\sin^2 \theta_{12}$ not maximal, we can gain some information on the neutrino mass spectrum. If the next KamLAND spectral data does bring back high-LMA as a viable solution, then if statistics are very high and the real value of $\sin^2 \theta_{13}$ (Δm^2_{31}) are high (low) enough, one can get some information on the neutrino mass hierarchy. For $\Delta m^2_{21} = 1.5 \times 10^{-4}$ eV^2, and statistics of 75 (125) GWkTy, one could distinguish the NH from the IH spectrum at 99.73% C.L. in the region of $\Delta m^2_{31} \lesssim 2.5 \times 10^{-3}$ eV^2 if $\sin^2 \theta \gtrsim 0.038$ (0.03) [12].

In conclusion, reactor neutrino experiments have great potential for precision measurement of the oscillation parameters. The KamLAND experiment can measure the solar mass squared difference very precisely. However it is not at the ideal baseline for measuring the solar neutrino mixing angle. If low-LMA is confirmed by the next results from KamLAND, a reactor experiment with a baseline of 70 km should be ideal to measure the solar neutrino mixing angle. If on the contrary KamLAND re-establishes high-LMA as a viable solution, a 20–30 km, intermediate baseline experiment could yield very rich phenomenology.

I acknowledge my collaborators, A. Bandyopadhyay, S. Goswami, S.T. Petcov and M. Piai and thank the organisers of the NuFact '03 workshop in New York for hospitality. Thanks are also due to S. Goswami for careful reading of the draft.

REFERENCES

1. K. Eguchi et al. [KamLAND Collaboration], Phys. Rev. Lett. **90**, 021802 (2003).
2. S. Goswami, arXiv:hep-ph/0303075.
3. Q. R. Ahmad et al. [SNO Collaboration], Phys. Rev. Lett. **87**, 071301 (2001).
4. A. Bandyopadhyay, S. Choubey, S. Goswami and D. P. Roy, Phys. Lett. B **540**, 14 (2002).
5. S. N. Ahmed et al. [SNO Collaboration], arXiv:nucl-ex/0309004.
6. A. Bandyopadhyay, S. Choubey, R. Gandhi,S. Goswami,D. P. Roy, Phys. Lett. B **559**, 121 (2003).
7. A. Bandyopadhyay, S. Choubey, S. Goswami, S. T. Petcov and D. P. Roy, arXiv:hep-ph/0309174.
8. Super-Kamiokande Coll., Y. Hayato et al., Talk given at the Int. EPS Conference on High Energy Physics, July 17 - 23, 2003, Aachen, Germany.
9. M. Apollonio et al., Eur. Phys. J. C **27**, 331 (2003); F. Boehm et al., Phys. Rev. D **64**, 112001 (2001).
10. A. Bandyopadhyay et al, Phys. Rev. D **65**, 073031 (2002); S. T. Petcov, M. Piai, Phys. Lett. B **533**, 94 (2002); S. M. Bilenky et al, Phys. Lett. B **538**, 77 (2002)
11. A. Bandyopadhyay, S. Choubey and S. Goswami, Phys. Rev. D **67**, 113011 (2003).
12. S. Choubey, S. T. Petcov and M. Piai, arXiv:hep-ph/0306017.

Scenarios for an entry-level neutrino Factory

Mario Campanelli

24, Q. Ernst Ansermet, Geneve Switzerland

Abstract. This paper presents cases where important physics can be extracted from an experimental setup using a reduced version of a neutrino factory. The main case for neutrino oscillations would be if new experimental findings lead to the need of non-standard scenarios; for instance, if a MiniBOONE confirmation of the LSND claim can be explained by sterile neutrinos. In that case, very interesting physics, and even sensitivity to CP violation, could be reached with much smaller machines than in the standard scenario. Other applications of muon storage rings are possible in the search of lepton-flavor violation and in the measurement of cross section in low-energy neutrino interactions.

Introduction

The concept of a neutrino beam produced by decay of muons stored in a ring can not only be applied to big projects aiming for long-baseline studies of neutrino oscillations or CP violation with high luminosity. From the accelerator point of view, smaller prototype-class machines could be built before a large-scale project is launched. We know that in the standard three-family oscillation scenario, with the mixing angles given by atmospheric and solar neutrino measurements, a neutrino factory with about 10^{20} integrated muon decays and detectors with masses of the order of 10 KTons are needed to push forward the present knowledge about the angle θ_{13}; or one order of magnitude more to start having sensitivity to the CP-violating phase δ.

The situation changes in a different scenario. If the LSND [1] claim is confirmed by MiniBOONE [2], the standard interpretation of three-family neutrino oscillations cannot be valid, and other models have to be taken into account. The most popular ones require the presence of additional neutrinos that, due to the LEP constraints to the couplings with the Z, have to be sterile, i.e. do not exhibit neutral weak interactions. Since the LSND effect can be explained with values of Δm^2 of the order of $1eV^2$, they can be studied at shorter baselines (so, with smaller neutrino fluxes) than in the case where the leading oscillation is governed by the atmospheric oscillation parameters.

In this paper we discuss possible interpretations of the LSND result and the requirements for testing them using a neutrino factory, and other applications, like measurement of the very badly-known low-energy neutrino cross section with an even smaller machine, and the possibility of observing (or having observed already) neutrinos from muon decays in the present Fermilab booster.

LSND result and its interpretation

The LSND experiment took data from 1993 to 1997 at LAMPF, searching for electron appearance in a v_μ beam. The baseline was 30 m, and the neutrino energy 36 Mev $< E_v <$ 52.8 MeV. An excess of electrons was reported in both neutrino and antineutrino channels. Antineutrinos are observed in the "classic" channel $\bar{v}_e p \to e^+ n$ followed by the neutron capture $np \to d\gamma$, while neutrinos in charged current reactions $v_e C \to e^- X$ with $60 MeV < E_e < 200 MeV$. In the two cases, the significant excess of electrons can be interpreted as a $v_\mu \to v_e$ oscillation probability of $0.31 \pm 0.12 \pm 0.05\%$ and $0.26 \pm 0.10 \pm 0.05\%$, respectively [1]. The Karmen experiment, covering a similar parameter space using a pulsed beam [3], did not confirm LSND findings. However, due to the slightly different L/E of the two experiments, a corner of the parameter space is still allowed by the combination of the two [4].

The final answer to this puzzle will come from the MiniBOONE experiment, currently running at Fermilab, where neutrinos with average energy of 1 GeV are produced from protons from the booster ($< E_p > = 8 GeV$). With respect to LSND, this experiment has 30 times more energy and 20 times longer baseline; due to the higher energy, Cerenkov light is four times larger than scintillation, and neutron capture is not used. The philosophy of the experiment is to have large electron excess(1000 events) in case LSND is correct, above a comparable background, controlled changing the length of the decay tunnel. In case of positive result, the MiniBOONE collaboration plans to build another detector (BOONE) to have a precision measurement of the oscillation parameters. A negative result would rule out the oscillation interpretation of the LSND result. A MiniBOONE run with antineutrinos is also foreseen; no signal in both modes would only leaving open very exotic effects or an experimental error.

If on the other hand MiniBOONE observes a signal, its interpretation needs a new machine to be fully understood. If the signal is seen only in antineutrino mode, its most natural interpretation would be violation of lepton flavor and/or CPT. If seen in both modes, the neutrino oscillation hypothesis would require introduction of sterile states.

Sterile neutrinos

It is likely that if sterile neutrinos exist at all, there is more than one flavor. We consider here the simplest case, with only one state. Depending on mass hierarchies, we can talk of 3+1 models, where the big mass difference gap is below or above the two of solar and atmospheric neutrinos, or 2+2 models, where the large LSND mass difference is between them. Pure 3+1 models are disfavored by short-baseline measurements, pure 2+2 models by the combination of atmospherics and solars; recently, 3+2 models have been proposed[5].

Neutrino Factory and sterile neutrinos

In order to simplify the number of parameters, we take the one mass dominance approximation. If CP is conserved, neglecting the atmospheric and solar mass difference squared (much smaller than that suggested by LSND), only the angles $\theta_{13}, \theta_{14}, \theta_{23}, \theta_{24}$ are relevant. According to [6], these angles can be studied in 2+2 and 3+1 schemes, using a Neutrino Factory with muon energy of 20 GeV, 10^{21} muons and a detector as small as 1 ton placed at 1 km of distance. Background was assumed to be at the 10^{-5} level. Using μ^+ appearance, sensitivities as low as 10^{-6} for $sin^2\theta_{23}$ can be reached for values of $\Delta m^2 \geq 10 eV^2$. In the μ^- disappearance channel, sensitivities to $sin^2\theta_{13}$ are a factor 100 worse.

Also the search for CP violation would be much simpler than in the standard 3-family scenario. In fact, while in the latter the magnitude of the effect depends on Δm_{12}^2, if a larger mass difference exists, the effect will depend on Δm_{23}^2. It means that it will be visible at smaller baselines, with smaller neutrino fluxes and, even more important, far less ambiguities due to matter effects. Defining the integrated CP asymmetries as $A^{CP} = (R^- - R^+)/(R^- + R^+)$, where $R^{\pm} = N(l_{\mu^+}^{\pm})/N(l_{\mu^-}^{\pm})$, we see that maximal CP violation could produce observable asymmetries between 5 and 10 for baselines smaller than 100 km, in both $\nu_e \to \nu_\mu$ and $\nu_e \to \nu_\tau$ appearance channels. Sensitivities in a 2+2 scheme are about a factor 2 worse. The τ channel is important for all studies mentioned above, and can discriminate between scenarios, so a detector with explicit τ identification would be a clear benefit.

Machine issues

If we consider a 100 ton detector, interesting parameter measurement can already be performed with 5 years runs of a machine with 2×10^{18} muons/year. Aiming at such low intensity, we can assume: no need for a new proton driver (JHF I or even the FNAL booster could be sufficient); no special radiation-hard targeting; no need for cooling. Also the energy could be further reduced, but not by large amounts since at the proposed baselines the beam width is already as small as 1 meter). Assuming as a baseline the US Study II cost table, items involved in these changes amount to about $900M out of a total of 1700. Overall, a total saving of about $600M can be envisaged, probably more due to the lower cost of all the other components.

Other uses of a small Neutrino Factory

If MiniBOONE finds nothing, LSND could still be right, and have seen instead of oscillations lepton-flavor violation, like $\mu^+ \to e^+ \nu_l \bar{\nu}_m u$ or $\mu^- \to e^- \bar{\nu}_e \nu_l$. It was shown [7] that a low-energy entry-level neutrino-factory and a 10 ton liquid Argon detector located 100 m from the machine improves limits on these decays, reaching sensitivities relevant to explain the LSND effect.

Even smaller machines, with decays of $10^{15} \mu$/year (i.e. using muons trapped in the CERN AD or in the Fermilab debuncher) have no interest for neutrino physics. However, such a machine could be used to measure neutrino cross section at low energy with a precision much better than the present data. For instance, it is reported in [8] that a 100 ton detector, located 10 meters from the end of the straight section of a machine that accelerates 10^{15} muons of 2 GeV, can measure the cross section with an error of about 10%.

It recently came out [9] that neutrinos from muons trapped in the Fermilab debuncher can be seen in MiniBOONE. This machine is used to debunch antiprotons at 8 GeV energy, but also many pions are trapped in the emittance. Pions decay within the first turn, while muons take 200 μs to decay. The debuncher captures 10^{11} antiprotons/hour; applying efficiencies for muon capture and detector acceptance, about 10^{12} neutrinos hit the MiniBOONE detector. Folding in the ν_e cross section at 2 GeV, 16 events/year can be observed from quasi-elastic and single π interactions. Such a small number of neutrinos can be used to calibrate detector response and crosscheck flux calculations.

Conclusions

After many years of data analysis and checks, the LSND effect is still alive and still represents a mystery in experimental neutrino physics. In case of positive MiniBOONE result, a new experimental program is needed to understand it better. The Neutrino Factory, coupled with a detector with τ identification, is the best machine to explore it. Due to the larger mass differences involved, physics reach would come in this scenario with a considerably smaller effort than in the 3-family case. In no signal is observed at MiniBOONE, a small machine could still explore the lepton flavor violation case. For even less-demanding tasks, like measuring low-energy neutrino cross section, very low-energy muon storage rings can work in parasitic mode from existing machines. However, it is clear that the main application of a small-scale neutrino factory would have sense in the case of a positive result from MiniBOONE.

REFERENCES

1. C. Athanassopoulos *et al*, (LSND Collaboration), *Phys. Rev.* C **54**, 2685 (1996);
2. *http://www.neutrino.lanl.gov/BooNE/*.
3. B.Armbruster *et al* (KARMEN coll.) Phys. Rev. **C57** 3414-3424 (1998)
4. FINAL LSND AND KARMEN-2 NEUTRINO OSCILLATION RESULTS Published in *Budapest 2001, High energy physics* hep2001/301
5. M.Sorel *et al* hep-ph/0305255
6. A.Donini, D.Meloni Eur.Phys.J. C **22** 179-186 (2001) hep-ph/0105863
7. A.Bueno *et al* JHEP 0106:032 (2001) hep-ph/0010308
8. M.Campanelli *et al* Nucl.Instrum.Meth.**A503** 151-153 (2001) hep-ph/0107221
9. B.Fleming, private communication

Overview of Degeneracies[1]

Hisakazu Minakata

Department of Physics, Tokyo Metropolitan University, Hachioji, Tokyo 192-0397, Japan

Abstract. A brief overview is given of the problem of parameter degeneracy in neutrino oscillation measurement of θ_{13}, θ_{23}, and the CP phase δ.

INTRODUCTION

After the KamLAND result [1] which uniquely selected out the LMA region and with the most recent solar neutrino analyses [2] we have pinned down the solution to the solar neutrino problem, the LMA-I. Together with the pioneering atmospheric neutrino observation by Super-Kamiokande [3], we now know the structure of lepton flavor mixing in the (1-2) and (2-3) sectors of the MNS matrix. Thus, we are left with the determination of the structure of its (1-3) sector, θ_{13} and the CP violating phase δ.

It has been recognized quite recently that their determination suffers from intrinsic ambiguity dubbed the "parameter degeneracy", and in fact it couples with the determination of θ_{23} if it is non-maximal. To explain the nature of the parameter degeneracy and to give an overview of this disease is the purpose of this talk. In its written form, this manuscript is meant for a "tour guide" for experimentalists who never worried about such a complicated question as the parameter degeneracy.

Let us start by describing some basic principle of determining neutrino mixing parameters θ_{23}, θ_{13}, and δ in long-baseline accelerator neutrino experiments. $\sin 2\theta_{23}$ can be determined by measuring disappearance probability $P(\nu_\mu \to \nu_\mu)$. The best sensitive place for the measurement is the dip at the oscillation maximum, whose depth will tell us $\sin 2\theta_{23}$ while its location (in energy) tells us Δm^2_{13}. The sensitivity to be reached in the JPARC-SK experiment [4] is expected to be $\sim 1\ \%$ (10 %) for $\sin^2 2\theta_{23}$ (Δm^2_{13}) at maximal mixing. Then, θ_{13}, and δ can be determined by measuring the electron and positron appearance oscillation probabilities $P(\nu_\mu \to \nu_e)$ and $P(\bar{\nu}_\mu \to \bar{\nu}_e)$.

WHAT IS THE PARAMETER DEGENERACY?

The parameter degeneracy is the phenomenon that measurement of oscillation probabilities $P(\nu_\mu \to \nu_\mu)$, $P(\nu_\mu \to \nu_e)$, and $P(\bar{\nu}_\mu \to \bar{\nu}_e)$ at a single baseline, no matter how accurate, does not allow us to determine uniquely θ_{23}, θ_{13}, and δ.

---

[1] Talk presented at the 5th International Workshop on Neutrino Factories & Superbeams (NuFact'03), Columbia University, New York, USA, June 5-11, 2003.

Let me try to explain the nature of the degeneracy. It can be characterized as follows; there is a fundamental degeneracy which can be called as (a) the intrinsic parameter degeneracy [5], which is duplicated by (b) the unknown sign of Δm_{13}^2 [6], and (c) two solution of θ_{23} for a given $\sin^2 2\theta_{23}$ for a non-maximal θ_{23} [7], the first and the second octant ambiguity. Notice that the degeneracy enhanced by the latter two factors is the "parity" degeneracy in nature, each of which multiply a factor 2 to the degeneracy. While it is first studied systematically in the context of neutrino factory in [5], the problem is even more relevant in conventional superbeam experiments [8]. In the rest of my talk I will try to explain the three elements of the parameter degeneracy one by one.

The intrinsic parameter degeneracy

The intrinsic parameter degeneracy arises because of the "elliptic nature" of the appearance oscillation probabilities

$$P(v) = A\cos\delta + B\sin\delta + C \qquad (1)$$

which was shown to hold in a quite general ground by Kimura, Takamura and Yokomakura [9]. The intrinsic parameter degeneracy is discovered and first analyzed by Burguet-Castell et al. [5]. The elliptic nature of the oscillation probabilities are best explained by the bi-probability plot introduced in [6]. As indicated in Fig. 1 (taken from [10]), a point in the bi-probability space determined by simultaneous measurement of oscillation probabilities $P(v_\mu \to v_e)$ and $P(\bar{v}_\mu \to \bar{v}_e)$ allows the interpretation by different two ellipses, each of which corresponds to a set of solution (θ_{13}, δ). The intrinsic degeneracy is the two-fold degeneracy which stems from the elliptic nature of the oscillation probabilities. This feature is evident from Fig. 1.

Duplication due to the unknown sign of Δm_{13}^2

It is also obvious from Fig. 1 that if we allow different sign of Δm_{13}^2 the degeneracy of the solution is duplicated. The existence of this type of degeneracy is first pointed out in [6]. The degeneracy disappears if the baseline of the experiment is long enough, typically $L > 1000 - 2000$ km. On the other hand, such long-baseline experiments are hard to carry out because huge detectors are required. Except possibly for neutrino factory, it seems to be difficult to design such an experiment that can determine sign of Δm_{13}^2 and at the same time can detect leptonic CP violation at more than 3 σ CL.

Duplication due to octant ambiguity of θ_{23}

Assuming that $\sin 2\theta_{23}$ is accurately determined by disappearance measurement of $P(v_\mu \to v_\mu)$, there is two solutions of θ_{23} for non-maximal value of θ_{23}, giving the degeneracy a two-fold discrete nature. The existence of this type of ambiguity was recognized early in 1996 by Fogli and Lisi [7]. This type of degeneracy produces the largest uncertainty in determination of θ_{13} with presently allowed region of $\sin 2\theta_{23}$, $0.90 < \sin^2 2\theta_{23} < 1.0$ at 90 % CL by SK [11].

GLOBAL FEATURE OF THE PARAMETER DEGENERACY

By collecting the elements described in the previous section it is clear that there exists eight ($= 2 \times 2 \times 2$) fold degeneracy in determination of θ_{23}, θ_{13}, and δ. Let us depict the existence of eight solutions for given measurement of neutrino and antineutrino appearance probabilities, P and $\bar{P}$. In Fig. 2 (taken from [12]) plotted is the set of allowed values of $\sin^2 2\theta_{23}$ and s_{23}^2 for given value of P and $\bar{P}$. It forms a closed contour in $\sin^2 2\theta_{23} - s_{23}^2$ plane for each sign of Δm_{13}^2, a closed one because δ is a phase variable. You can clearly see that there are eight solutions for non-maximal value of θ_{23}.

Some of the features of the eight-fold parameter degeneracy were analyzed in [13]. Then, the full structure of the degeneracy was worked out in [10] in which analytic solutions of the eight-fold parameter degeneracy is given under suitable approximation keeping leading-order terms of $\Delta m_{12}^2/\Delta m_{13}^2 \simeq 0.03$.

The analytic solutions are powerful because it allows us to derive general relations between degenerate solutions. While we refer to [10] for details we describe a concrete example. The relation between two solutions of δ for the same-Δm_{13}^2-sign case (the intrinsic degeneracy) is given by

$$\delta_2 = \pi - \delta_1 + \arccos((z^2 - 1)/(z^2 + 1)) \qquad (2)$$

where $z \equiv (X_- Y_+ - X_+ Y_-)/(X_- Y_+ + X_+ Y_-)$ and $X_\pm$ and $Y_\pm$ are the well known coefficients of θ_{13}^2 and θ_{13}, respectively. in the oscillation probability [5, 10]. In view of (2) the intrinsic degeneracy can in principle obscure the CP violation. However, the possibility is slim because the size of the extra term which describes deviation from the vacuum relation, $\delta_2 = \pi - \delta_1$, is small, less than a few %.

To give you a feeling of how serious is the problem of parameter degeneracy, we plot in Fig. 3 the updated results of fractional difference between two degenerate solutions, $(\Delta\theta/\theta)_{ij} \equiv 2(\theta_i - \theta_j)/(\theta_i + \theta_j)$ for the LMA-I solar neutrino parameters, the ones used in [14]. It is evident from Fig. 3 that $\Delta\theta/\theta$ is too large to ignore, but too small to measure. See [10] for more about global overview of the parameter degeneracy.

I concluded with a remark, though somewhat contradictory to the spirit of this talk, that what is important for experimentalists in coming 5 years is NOT to worry about the parameter degeneracy but to discover nonzero θ_{13}.

ACKNOWLEDGMENTS

This work was supported by Grants-in-Aid for Scientific Research in Priority Areas No. 12047222, Japan Ministry of Education, Culture, Sports, Science, and Technology.

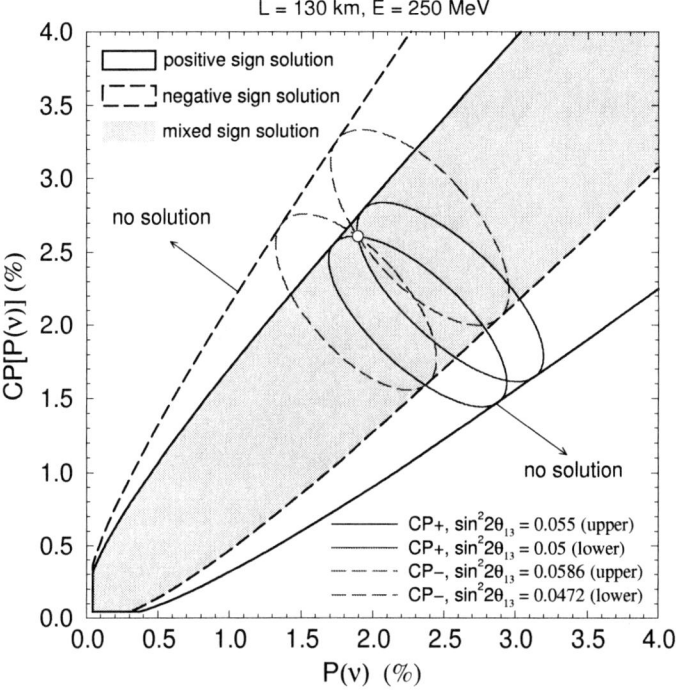

FIGURE 1. The intrinsic and the sign-of-Δm_{13}^2 degeneracies displayed on bi-probability diagram of [6].

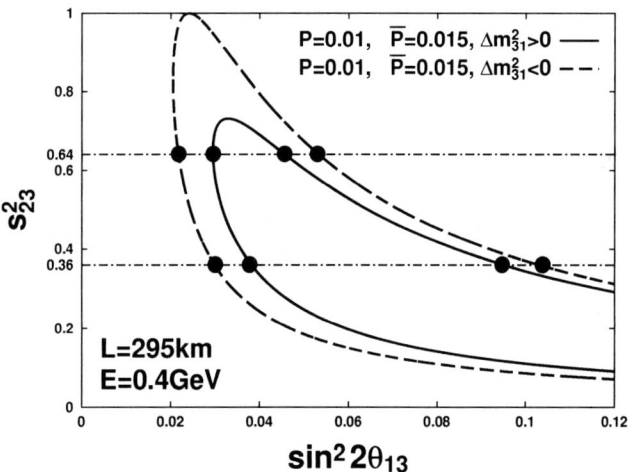

FIGURE 2. Depicted in the $\sin^2 2\theta_{13}$–s_{23}^2 plane are the contours determined by arbitrarily given values of the appearance probabilities $P \equiv P(\nu_\mu \to \nu_e) = 0.01$ and $\bar{P} \equiv P(\bar{\nu}_\mu \to \bar{\nu}_e) = 0.015$ with E/L off the oscillation maximum at the JPARC-SK experiment. The solid and the dashed lines correspond to positive and negative Δm_{31}^2, respectively.

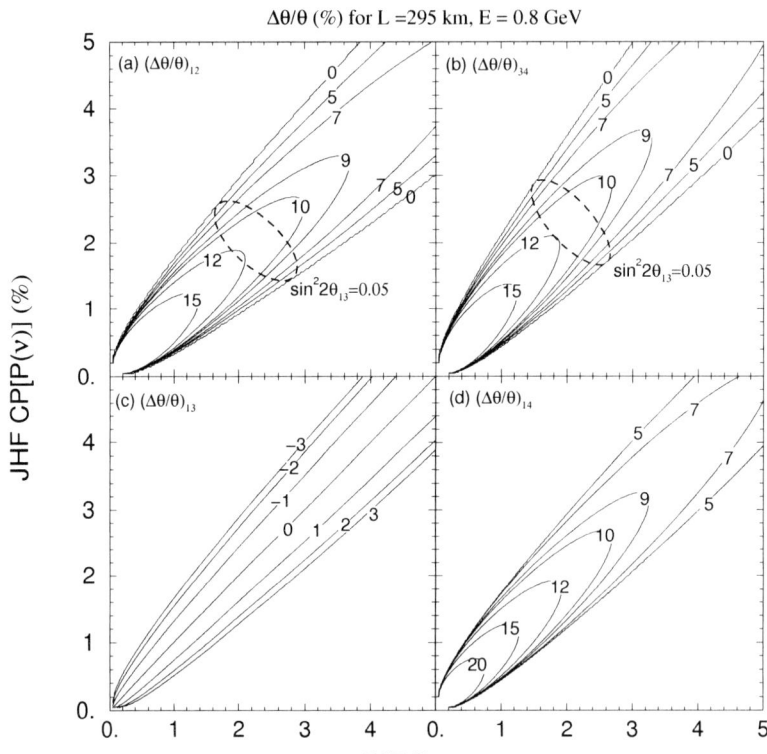

FIGURE 3. Isofractional difference between degenerate solutions of θ_{13} with the LMA-I parameters. $(i, j) = (1,2)$ and $(3,4)$ indicate the solutions with positive and negative Δm_{13}^2, respectively.

REFERENCES

1. K. Eguchi *et al.* [KamLAND Collaboration], Phys. Rev. Lett. **90**, 021802 (2003).
2. M. B. Smy *et al.* [Super-Kamiokande Collaboration], arXiv:hep-ex/0309011;
   S. N. Ahmed *et al.* [SNO Collaboration], arXiv:nucl-ex/0309004.
3. Y. Fukuda *et al.* [Kamiokande Collaboration], Phys. Lett. B **335**, 237 (1994);
   Y. Fukuda *et al.* [Super-Kamiokande Collaboration], Phys. Rev. Lett. **81**, 1562 (1998).
4. Y. Itow *et al.*, arXiv:hep-ex/0106019.
   For an updated version, see: http://neutrino.kek.jp/jhfnu/loi/loi.v2.030528.pdf
5. J. Burguet-Castell, M. B. Gavela, J. J. Gomez-Cadenas, P. Hernandez and O. Mena, Nucl. Phys. B **608**, 301 (2001).
6. H. Minakata and H. Nunokawa, JHEP **0110**, 001 (2001).
7. G. Fogli and E. Lisi, Phys. Rev. **D54**, 3667 (1996).
8. H. Minakata and H. Nunokawa, Phys. Lett. **B495** (2000) 369; J. Sato, Nucl. Instrum. Meth. **A472** (2001) 434; B. Richter, arXiv:hep-ph/0008222.
9. K. Kimura, A. Takamura and H. Yokomakura, Phys. Lett. **B544**, 286 (2002).
10. H. Minakata, H. Nunokawa, and S. J. Parke, Phys. Rev. D **66**, 093012 (2002).
11. Y. Hayato, Talk at International Europhysics Conference on High Energy Physics (EPS2003), July 17-23, 2003, Aachen, Germany.
12. H. Minakata, H. Sugiyama, O. Yasuda, K. Inoue and F. Suekane, Phys. Rev. D **68**, 033017 (2003).
13. V. Barger, D. Marfatia and K. Whisnant, Phys. Rev. D **65**, 073023 (2002).
14. H. Minakata, H. Nunokawa, and S. J. Parke, Phys. Rev. D **68**, 013010 (2003).

Resolving JHF degeneracy

Hiroaki Sugiyama[1]

Department of Physics, Tokyo Metropolitan University, 1-1 Minami-Osawa, Hachioji, Tokyo 192-0397, Japan

Abstract. I discuss the possibility to resolve the θ_{23} degeneracy by combining a reactor experiment with the JHF experiment. I show that the degeneracy can be resolved even if all of the errors in the experiment is taken into account.

1. INTRODUCTION

It is known that measurement of $P(\nu_\mu \to \nu_e)$ and $P(\bar{\nu}_\mu \to \bar{\nu}_e)$ can not determine uniquely θ_{13} and the CP phase δ because of the eight-fold parameter degeneracy problem[1]; The problem is caused by the unknown sign of $\pi/4 - \theta_{23}$ [2], the intrinsic degeneracy (θ_{13}-δ correlation)[3], and the uncertainty in the sign of Δm_{31}^2 [4]. By tuning the beam energy to that of the oscillation maximum, the intrinsic degeneracy can be reduced to $\delta \leftrightarrow \pi - \delta$ uncertainty which does not cause CP $\leftrightarrow$ CPV confusion. Concerning the JHF experiment[5], the sign of Δm_{31}^2 does not matter because of small matter effect in the experiment. Therefore, only the uncertainty in the sign of $\pi/4 - \theta_{23}$ may remain as the degeneracy problem in the experiment. Although a possibility to resolve the degeneracy was presented by combining a reactor experiment in Ref. [6], the treatment of the JHF experiment was not quantitative for simplicity. In this talk, I present the upgraded result with the quantitative analysis of the JHF experiment.(See also [7].)

2. ANALYSIS AND RESULTS

We assume 2 and 6.8 year measurement at the Hyper-Kamiokande with off-axis 2° 4MW ν_μ and $\bar{\nu}_\mu$ beam, respectively. Event cuts are performed for the signal and background events within each energy bin of 50MeV width. Once we have a Monte-Calro result of the event cuts for a set of parameter values, we can calculate results of the cuts for any set of parameter values by using the ratio of the numbers of events before cut as

$$N_{ac}(\delta = 0) = \frac{N_{bc}(\delta = 0)}{N_{bc}(\delta = \pi/4)} N_{ac}(\delta = \pi/4), \qquad (1)$$

where N_{bc} is the number of events before the cut and N_{ac} denotes that after the cut. The reactor experiment is assumed to be of 40t·yr measurement with a reactor of

---

[1] This work is collaboration with H. Minakata and O. Yasuda.

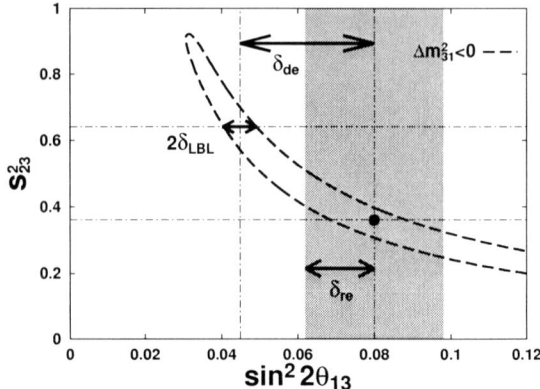

FIGURE 1. Surrounded by the dashed line is 90%CL allowed region of the JHF experiment for the true value shown by a blob. The fake solution is at $(s'_{23})^2 = 1 - s_{23}^2$ which is obtained by the constraint from the ν_μ dissapearance measurement. The shaded band is 90%CL allowed region of the reactor experiment. δ_{de} denotes a theoretical difference between the true and fake values of $\sin^2 2\theta_{13}$.

24.3GW$_{th}$ maximum power; The far detector is placed 1.7km away from the reactor, and the relative normalization error is assumed to be reduced to 0.8% by using an identical near (300m) detector.

When we calculate the values of $\Delta\chi^2$ to obtain the allowed region in the $\sin^2 2\theta_{13}$-s_{23}^2 space, the errors in the other parameters and the correlations of the errors are considered also. Note that those are unimportant for analyses of reactor experiments because the dissapearance probability does not include many parameters. That is a good point of reactor expeiments. We assume the following best fit values and 90%CL errors:

$$\Delta m_{21}^2 = (7.3 \pm 2.3) \times 10^{-5} \text{eV}^2, \quad |\Delta m_{31}^2| = (2.5 \pm 0.1) \times 10^{-3} \text{eV}^2,$$
$$\tan^2\theta_{12} = 0.36 \pm 0.17, \quad C = 1 \pm 0.05, \tag{2}$$

where C presents the coefficient of the matter effect ($\sqrt{2}CG_F N_e$). In order to deal with the parameter correlations, it is necessary for us to minimize $\Delta\chi^2$ with respect to the four parameters of 2 and unknown δ "at once".[2] The minimization is achieved numerically in reasonable calculation time (~ 1 hour) by virtue of the downhill simplex method[8] without which calculation would require infinite time by a brute force algorithm.

We obtain the allowed regions in the $\sin^2 2\theta_{13}$-s_{23}^2 space for given best fit values of $\sin^2 2\theta_{13}$, s_{23}^2, and the CP phase δ, namely given values of $P(\nu_\mu \to \nu_e)$ and $P(\bar{\nu}_\mu \to \bar{\nu}_e)$. Fig. 1 shows the allowed region, which is surrounded by the dashed line, with the best fit values $\sin^2 2\theta_{13} = 0.08$, $s_{23}^2 = 0.46$, and $\delta = \pi/2$ for $\Delta m_{31}^2 > 0$. The allowed region for the inverted hierarchy only is presented in the figure as the pessimistic case because

---

[2] One-by-one minimization corresponds to consideration of errors in parameter values only.

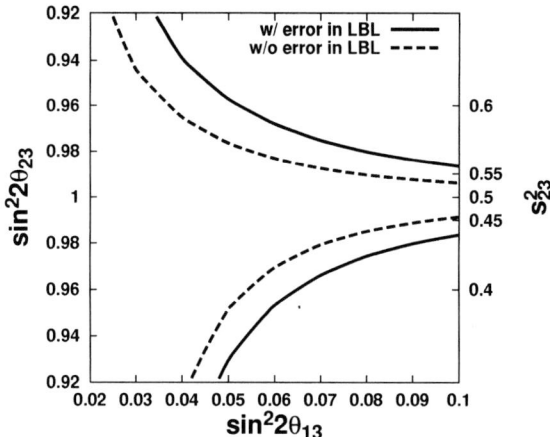

FIGURE 2. The fake θ_{23} can be excluded at 90%CL if true values of parameters live in the region (upper-right and bottom-right). The upper half of the figure corresponds to $\theta_{23} > \pi/4$. The dashed lines correspond to the old result in [6] without the errors in the JHF experiment, and solid lines show new result with the errors.

that for the normal hierarchy is inside of the region for the best fit values. When the measurement of $P(\nu_\mu \to \nu_\mu)$ determines the value of $\sin^2 2\theta_{23}$, the fake solution is at $s_{23}^2 = 0.64$ in the figure due to the degeneracy problem. We define the theoretical difference between true and fake $\sin^2 2\theta_{13}$ as

$$\delta_{de} \equiv |\sin^2 2\theta'_{13} - \sin^2 2\theta_{13}|. \tag{3}$$

The fake solution of θ_{23} can be excluded by the reactor experiment if the condition

$$\delta_{de} - \delta_{LBL} > \delta_{re} \tag{4}$$

is satisfied; δ_{re} denotes the 90%CL error in $\sin^2 2\theta_{13}$ of the reactor experiment, and δ_{de} does the 90%CL error, which was not considered in [6] for simplicity, around the fake solution of the JHF experiment. δ_{re} is almost independent of the true value of $\sin^2 \theta_{13}$; That is a feature of measurement of tiny dissapearance because the uncertainty of the number of events depends on the number of observed events. On the other hand, δ_{de} and δ_{LBL} largely depend on the true values of the paraeters.

Figure 2 presents the region where the θ_{23} degeneracy can be resolved in the space of true s_{23}^2 and $\sin^2 2\theta_{13}$; The upper half of the figure corresponds to $\theta_{23} > \pi/4$. The region is narrower than that of the old result in [6] because of the errors of the JHF experiment. There is an advantage of $\theta_{23} > \pi/4$ case even for the old result, and it is due to the θ_{23} dependance of δ_{de}. It can be seen in the figure that resolving the degeneracy is still possible for relatively large $\sin^2 2\theta_{13}$ and $\cos^2 2\theta_{23}$.

ACKNOWLEDGMENTS

I thank Kenji Kaneyuki and Yoshihisa Obayashi for helpful discussion and correspondence.

REFERENCES

1. V. Barger, D. Marfatia and K. Whisnant, Phys. Rev. D **65**, 073023 (2002) [arXiv:hep-ph/0112119].
2. G. L. Fogli and E. Lisi, Phys. Rev. D **54**, 3667 (1996) [arXiv:hep-ph/9604415].
3. J. Burguet-Castell, M. B. Gavela, J. J. Gomez-Cadenas, P. Hernandez and O. Mena, Nucl. Phys. B **608**, 301 (2001) [arXiv:hep-ph/0103258].
4. H. Minakata and H. Nunokawa, JHEP **0110**, 001 (2001) [arXiv:hep-ph/0108085].
5. Y. Itow *et al.*, arXiv:hep-ex/0106019.
6. H. Minakata, H. Sugiyama, O. Yasuda, K. Inoue and F. Suekane, Phys. Rev. D **68**, 033017 (2003) [arXiv:hep-ph/0211111].
7. P. Huber, M. Lindner, T. Schwetz and W. Winter, Nucl. Phys. B **665**, 487 (2003) [arXiv:hep-ph/0303232].
8. W.H. Press *et al.*, *Numerical Recipes in C, 2nd ed.* (Cambridge, 1992).

Combining Superbeams

K. Whisnant

Department of Physics and Astronomy, Iowa State University, Ames, IA 50010, USA

Abstract. The utility of combining data from more than one superbeam in order to resolve parameter degeneracies in long-baseline neutrino oscillation experiments is discussed.

INTRODUCTION

Data from atmospheric [1], solar and reactor [2] neutrino experiments give very strong indications for neutrino oscillations. Three neutrinos are sufficient to describe the results of these experiments [3], but not all of the three-neutrino oscillation parameters are determined by current data. Long-baseline neutrino oscillation experiments will be needed to further constrain the measured parameters and to measure or put stricter upper bounds on the currently unknown parameters. In this talk I will discuss the feasibility of combining results from the next generation of long-baseline experiments with conventional neutrino beams (superbeams) to unambiguously measure the neutrino oscillation parameters.

PARAMETER DEGENERACIES

The now-standard form of the three-neutrino mixing matrix is given in Ref. [4]. The mixing angle θ_{12} and the neutrino mass-squared difference δm_{21}^2, which describe the oscillations of solar and KamLAND neutrinos, and θ_{23} and δm_{31}^2, which describe the oscillations of atmospheric neutrinos, have been measured [1, 2]. The remaining parameters, the mixing angle θ_{13} and the CP phase δ, are as yet undetermined; there is an upper bound on $\sin^2 2\theta_{13}$ of approximately 0.1 from the CHOOZ reactor (the exact bound varies with δm_{31}^2).

The traditional method for measuring θ_{13} and δ is to measure the oscillation probabilities $P(\nu_\mu \to \nu_e)$ and $\bar{P}(\bar{\nu}_\mu \to \bar{\nu}_e)$ in a long-baseline experiment; which for $\delta m_{31}^2 > 0$ have the approximate form [5, 6]

$$P(\nu_\mu \to \nu_e) = B(\cos\delta \cos\Delta - \sin\delta \sin\Delta) + C, \qquad (1)$$

$$\bar{P}(\bar{\nu}_\mu \to \bar{\nu}_e) = \bar{B}(\cos\delta \cos\Delta + \sin\delta \sin\Delta) + \bar{C}, \qquad (2)$$

where $\Delta = |\delta m_{31}^2| L/(4E_\nu)$ and $B, C, \bar{B}$ and $\bar{C}$ are coefficients that depend on the neutrino mixing angles. Matter effects make $\bar{B}$ and $\bar{C}$ different from B and C, respectively. The probabilities for $\delta m_{31}^2 < 0$ are obtained by the exchanges $B \leftrightarrow \bar{B}, C \leftrightarrow \bar{C}$ and $\delta \to \pi - \delta$.

TABLE 1. Potential uncertainties due to parameter degeneracies for a simple measurement of P and $\bar{P}$. Uncertainties for the special case $\Delta = \frac{\pi}{2}$ are shown in parentheses.

ambiguity	θ_{13} variation	CPC/CPV confusion
(δ, θ_{13})	large (none)	large (none)
$\mathrm{sgn}(\delta m_{31}^2)$	small (small)	large (large)
$(\theta_{23}, \frac{\pi}{2} - \theta_{23})$	large (large)	large (small)

The two unknown parameters can then in principle be determined from these two measurements. However, there are three two-fold parameter degeneracies [7]:

- The (δ, θ_{13}) ambiguity, in which two different sets of the parameters δ and θ_{13} can give the same results for the two oscillation probabilities P and $\bar{P}$.
- The $\mathrm{sgn}(\delta m_{31}^2)$ ambiguity, which occurs since the sign of δm_{31}^2 is not determined in atmospheric neutrino experiments.
- The $(\theta_{23}, \frac{\pi}{2} - \theta_{23})$ ambiguity, which occurs since atmospheric neutrino experiments measure $\sin^2 2\theta_{23}$ and not $\sin \theta_{23}$. This ambiguity disappears for the experimentally favored value of $\theta_{23} (= \frac{\pi}{4})$.

Each ambiguity gives two different solutions for the parameters δ and θ_{13}. In most cases, the two solutions can have very different values of θ_{13}, and can mix CP conserving (CPC) solutions with CP violating (CPV) ones. Table 1 gives the amount of variation in θ_{13} and the amount of CPC/CPV confusion possible for each ambiguity. If all of the ambiguities occur simultaneously, there is a combined eight-fold ambiguity.

Strategies exist to reduce the effects of ambiguities for a single superbeam experiment [7]. If the leading oscillation argument Δ is set equal to $\frac{\pi}{2}$ (or, in general, any odd multiple of $\frac{\pi}{2}$, on a peak of the vacuum oscillation), then the $\cos \delta$ terms in Eqs. 1 and 2 drop out. For given values of $\mathrm{sgn}(\delta m_{31}^2)$ and θ_{23}, θ_{13} and $\sin \delta$ may then be unambiguously determined, thereby removing θ_{13} from the (δ, θ_{13}) ambiguity; a residual $(\delta, \pi - \delta)$ ambiguity remains since only $\sin \delta$ is measured. The other two ambiguities are not necessarily removed: the uncertainty in $\sin^2 2\theta_{13}$ due to the $\mathrm{sgn}(\delta m_{31}^2)$ ambiguity is still small, while for the $(\theta_{23}, \frac{\pi}{2} - \theta_{23})$ ambiguity it is approximately a factor of $\tan^2 \theta_{23}$. The effect of being on the peak for each of the three ambiguities is shown in Table 1.

Furthermore, if the baseline L is sufficiently long, then matter effects split apart the $\delta m_{31}^2 > 0$ and $\delta m_{31}^2 < 0$ solutions, thereby removing the $\mathrm{sgn}(\delta m_{31}^2)$ ambiguity. The required length increases as $\sin^2 2\theta_{13}$ decreases: $L = 1290$ km is sufficient for $\sin^2 2\theta_{13} = 0.01$ (before accounting for experimental uncertainties), while $L = 1770$ km is needed for $\sin^2 2\theta_{13} = 0.005$. Thus for long enough L, measuring P and $\bar{P}$ on the peak will only leave a $(\delta, \pi - \delta)$ ambiguity if $\theta_{23} \simeq \frac{\pi}{4}$; otherwise, there is also an uncertainty in θ_{13}, although the amount of CPC/CPV confusion will still be small.

Finally, another interesting scenario [7] is to choose $\sqrt{2} G_F N_e L = \pi$, where G_F is the usual weak coupling constant and N_e is the average electron number density, which corresponds to $L \simeq 7600$ km in a long-baseline experiment for the Earth's density profile. Then all dependence on δ and δm_{21}^2 drops out in Eqs. 1 and 2 and $P \simeq \bar{P} \simeq \sin^2 \theta_{23} \sin^2 2\theta_{13} \sin^2 \Delta$. There is no longer a (δ, θ_{13}) ambiguity, and θ_{13} can be directly measured, modulo the $(\theta_{23}, \frac{\pi}{2} - \theta_{23})$ ambiguity. Measurements at another baseline would be necessary to determine δ. Applications of this "magic baseline" scenario are discussed in Ref. [8] for a superbeam and in Ref. [9] for a neutrino factory.

COMBINING SUPERBEAM EXPERIMENTS

Existing or soon-to-be existing beamlines (JHF to Super-K, 295 km, and NuMI to Soudan, 730 km) are probably not long enough to have sufficient matter effects to remove the $\text{sgn}(\delta m_{31}^2)$ ambiguity. Therefore it is natural to ask if combining the results of superbeam experiments can improve this situation. In this section, several such possibilities that have been addressed in the literature are discussed.

In Ref. [10], the following scenario was studied: (i) a 4 MW, $2°$ off-axis beam from JHF to Super-K is run for 2 years with ν's and 6 years with $\bar{\nu}$'s ($L = 295$ km), and (ii) a 1.6 MW beam from NuMI to Soudan is run for 2 years with ν's and 5 years with $\bar{\nu}$'s. The off-axis angle (which determines the neutrino energy) and baseline for the NuMI beam were varied to obtain maximal sensitivity to CP violation and the sign of δm_{31}^2. It was found that for $\sin^2 2\theta_{13} \geq 0.03$ a good ability to determine $\text{sgn}(\delta m_{31}^2)$ *and* good *CPV* sensitivity was possible with an off-axis angle of $0.7° - 0.8°$ and $L \simeq 900$ km (see Figs. 2 and 4 of Ref. [10]). Furthermore, it was found that while the ability to determine $\text{sgn}(\delta m_{31}^2)$ with measurements at just one baseline depends strongly on the size of δm_{21}^2 (larger δm_{21}^2 making it more difficult), when measurements are made at two baselines the ability to determine $\text{sgn}(\delta m_{31}^2)$ is mostly insensitive to δm_{21}^2 (see Fig. 3 of Ref. [10]). The improved sensitivity to $\text{sgn}(\delta m_{31}^2)$ is not just due to improved statistics (more events), since the $\delta m_{31}^2 > 0$ and $\delta m_{31}^2 < 0$ regions strongly overlap for both JHF and NuMI, and could not always be distinguished by JHF or NuMI alone even with infinite statistics.

Scenarios where only one measurement each is made using the JHF and NuMI beams were studied in Ref. [11]. They showed that if only ν measurements, not $\bar{\nu}$, are made, then it may be possible to determine $\text{sgn}(\delta m_{31}^2)$ if the oscillation arguments are similar in the two experiments; however, sensitivity to the exact values of δ and θ_{13} are then lost. Similar results are possible if both beams are $\bar{\nu}$. If one beam, e.g., JHF, is ν and the other, NuMI, is $\bar{\nu}$, then the results are similar to those obtained with ν and $\bar{\nu}$ beams at a single baseline, with the advantage that the two experiments could be run simultaneously instead of sequentially. If the second baseline is increased substantially, e.g., by running ν's from JHF to Super-K and ν's from JHF to Beijing (with $L \simeq 2100$ km), the separation of the $\delta m_{31}^2 > 0$ and $\delta m_{31}^2 < 0$ solutions can be improved (see Fig. 10 of Ref. [12]).

A variety of two-measurement and four-measurement scenarios using the JHF and NuMI beams without luminosity upgrades were studied in Ref. [13]. They found that while sensitivity to $\text{sgn}(\delta m_{31}^2)$ and *CPV* are marginal to nonexistent with the JHF or NuMI beams alone, the combination of both could give good sensitivity to either $\text{sgn}(\delta m_{31}^2)$ or *CPV* (although not both). The best scenarios for determining $\text{sgn}(\delta m_{31}^2)$ used just ν beams from JHF and NuMI (with $L \sim 900$ km for the latter), while for measuring CP violation the most favorable scenarios used either ν's from JHF and $\bar{\nu}$'s from NuMI, or ν's and $\bar{\nu}$'s from both JHF and NuMI. When both ν's and $\bar{\nu}$'s are used from the same source, running time between the two was assumed split so as to give equal numbers of survival events in the absence of oscillations. The most versatile scenario, with good sensitivity to sgn(δm_{31}^2) and fair sensitivity to CP violation, used four measurements, ν and $\bar{\nu}$ from both JHF and NuMI, with $L = 950$ km for the latter.

A compromise between using two or four measurements would be to use three measurements. Theoretical studies show [14] that three measurements can eliminate all exact parameter degeneracies for $\sin^2 2\theta_{13} > 0.01$. A good determination of δ and θ_{13} is possible for $\sin^2 2\theta_{13} > 0.01$ in a JHF to Hyper-K (1000 kt) experiment using ν and $\bar{\nu}$ beams at one energy and a ν beam at a different energy [15]; however, the distance is too short to have sufficient matter effects to resolve sgn(δm_{31}^2) without a fourth measurement. Assuming a luminosity upgrade for JHF (to 4 MW), running ν's for 2 years and $\bar{\nu}$'s for 6 years from JHF to Hyper-K (450 kt fiducial volume) plus running ν's for 5 years from JHF to a 100 kt detector near Beijing can give a good determination of θ_{13} and $\sin\delta$, as well as resolve the sgn(δm_{31}^2) ambiguity, for $\sin^2 2\theta_{13} > 0.01$ [16]. A ($\delta, \pi - \delta$) ambiguity remains in the latter scenario, but by increasing the size of the Beijing detector ten-fold this final parameter degeneracy could be removed.

In summary, 1 + 1 is greater than 2 for superbeam experiments: combining the results of superbeam experiments may be able to determine the sign of δm_{31}^2 even if they are unable to do it separately. If one of the baselines is very long, with sufficient statistics a complete determination of the neutrino oscillation parameters may be possible.

ACKNOWLEDGMENTS

I thank V. Barger and D. Marfatia for collaboration.

REFERENCES

1. See plenary talk by M. Shiozawa, these proceedings.
2. See plenary talk by K. Nakamura, these proceedings.
3. See plenary talk by V. Barger, these proceedings.
4. K. Hagiwara et al. [Particle Data Group Collaboration], Phys. Rev. D **66**, 010001 (2002).
5. A. Cervera et al., Nucl. Phys. B **579**, 17 (2000) [Erratum-ibid. B **593**, 731 (2001)] [arXiv:hep-ph/0002108].
6. M. Freund, Phys. Rev. D **64**, 053003 (2001) [arXiv:hep-ph/0103300].
7. V. Barger, D. Marfatia and K. Whisnant, Phys. Rev. D **65**, 073023 (2002) [arXiv:hep-ph/0112119].
8. A. Asratyan et al., arXiv:hep-ex/0303023.
9. P. Huber and W. Winter, Phys. Rev. D **68**, 037301 (2003) [arXiv:hep-ph/0301257].
10. V. Barger, D. Marfatia and K. Whisnant, Phys. Lett. B **560**, 75 (2003) [arXiv:hep-ph/0210428].
11. H. Minakata, H. Nunokawa and S. Parke, Phys. Rev. D **68**, 013010 (2003) [arXiv:hep-ph/0301210].
12. Y. F. Wang et al., Phys. Rev. D **65**, 073021 (2002) [arXiv:hep-ph/0111317].
13. P. Huber, M. Lindner and W. Winter, Nucl. Phys. B **654**, 3 (2003) [arXiv:hep-ph/0211300].
14. V. Barger, D. Marfatia and K. Whisnant, Phys. Rev. D **66**, 053007 (2002) [arXiv:hep-ph/0206038].
15. M. Aoki, K. Hagiwara and N. Okamura, Phys. Lett. B **554**, 121 (2003) [arXiv:hep-ph/0208223].
16. K. Whisnant, J. M. Yang and B. L. Young, Phys. Rev. D **67**, 013004 (2003) [arXiv:hep-ph/0208193].

NUFACT'03: The Fate of the Clones

A. Donini[1]

IFT, Universidad Autonoma Madrid, Cantoblanco 28049 Madrid, Spain

Abstract. We present a Neutrino-Factory-based setup with three detectors of different kind in principle capable to solve the eightfold-degeneracy in the simultaneous measurement of θ_{13} and δ, for $\theta_{13} \geq 1°$ ($\sin^2(2\theta_{13}) \geq 10^{-3}$). Our setup includes a Superbeam-driven water Cherenkov (the Superbeam conceived as the first stage of the Neutrino Factory); two muon-storage-ring-driven detectors (namely, a large magnetized iron calorimeter and an emulsion cloud chamber) to take advantage of both the so-called "golden" ($\nu_e \to \nu_\mu$) and "silver" ($\nu_e \to \nu_\tau$) channels.

The planned long baseline experiments [1] will improve the measurement of Δm^2_{atm} and of θ_{23} and measure or increase the bound on θ_{13} [2, 3] (see also [4]). This new generation of experiments, however, is only the first step of a long-lasting experimental program including the development of some "superbeam" facilities (whose combination can strongly improve our knowledge on θ_{13}, see [5]) and, eventually, of a "Neutrino Factory" [6, 7]. One of the main goals of the Neutrino Factory program (see for example [8, 9] and refs. therein) would be the discovery of leptonic CP violation and, possibly, its study [10]-[11].

The transition probabilities $\nu_e \to \nu_\mu$ and $\nu_\mu \to \nu_e$ are extremely sensitive to θ_{13} and δ: this is what is called the *"golden measurement at the Neutrino Factory"* [11] and can be easily studied by searching for wrong-sign muons, provided the considered detector has a good muon charge identification capability. The determination of (θ_{13}, δ) from this channel is not at all free of ambiguities: in [12] it was shown that, for a given physical input parameter pair ($\bar\theta_{13}, \bar\delta$), measuring the oscillation probability for $\nu_e \to \nu_\mu$ and $\bar\nu_e \to \bar\nu_\mu$ will generally result in two allowed regions of the parameter space. The first one contains the physical input parameter pair and the second, the "intrinsic ambiguity", is located elsewhere. Worse than that, new degeneracies have later been noticed [13, 14], resulting from our ignorance of the sign of the Δm^2_{atm} squared mass difference and from the approximate $[\theta_{23}, \pi/2 - \theta_{23}]$ symmetry for the atmospheric angle. In general, for each physical input pair the measure of $P(\nu_e \to \nu_\mu)$ and $P(\bar\nu_e \to \bar\nu_\mu)$ will result in eight allowed regions of the parameter space, the *eightfold-degeneracy* [14].

From what learned in the previous papers [15]-[17] we conclude that the optimal combination to deal with the eightfold-degeneracy consists in taking advantage of all the neutrino beams produced in a Neutrino Factory Complex (i.e. factory plus detectors). The Neutrino Factory Complex that we consider consists of a SPL-like superbeam [18] and a 50 GeV muon storage ring [9], plus a network of three detectors of different

---

[1] E-mail andrea.donini@roma1.infn.it

technology:

1. a 40 Kton Magnetized Iron Detector (MID) at $L = 2810$ Km, [19];
2. a 4 Kton Emulsion Cloud Chamber (ECC) at $L = 732, 2810$ Km, [17];
3. a 400 Kton Water Cherenkov (WC) at $L = 130$ Km, [20].

This proposal, resulting from the combination of [12, 16] and [17], corresponds to the design of a possible CERN-based Neutrino Factory Complex, with detectors located at the Frejus (the WC), at Gran Sasso (the ECC) and at a third site to be defined (the MID and possibly the ECC). Each one of these detectors is especially optimized to look for a particular signal: $\nu_\mu \to \nu_e$ oscillations for the 400 Kton WC, $\nu_e \to \nu_\mu$ for the 40 Kton MID and $\nu_e \to \nu_\tau$ for the 4 Kton ECC.

The physical parameters to be measured at the Neutrino Factory Complex are, in the worst case (i.e. if the planned experiment are not able to measure some of them earlier), the two PMNS mixing matrix parameters $\bar{\theta}_{13}$ and $\bar{\delta}$, the sign of Δ_{atm}, $\bar{s}_{atm}$, and the θ_{23}-octant, $\bar{s}_{oct}$, where

$$\bar{s}_{atm} = sign[\Delta m^2_{atm}]; \qquad \bar{s}_{oct} = sign[\tan(2\theta_{23})]. \tag{1}$$

Both discrete variables can assume the values ± 1, depending on the physical assignments of the sign of Δm^2_{atm} and of the θ_{23}-octant ($s_{oct} = 1$ for $\theta_{23} < \pi/4$ and $s_{oct} = -1$ for $\theta_{23} > \pi/4$). The other parameters have been considered as fixed quantities, supposed to be known with good precision by the time when the Neutrino Factory will be operational. In particular: $\theta_{12} = 35°$ and $\Delta m^2_\odot = 7 \times 10^{-5}$ eV^2; $\theta_{23} = 40°, 50°$ (a generic value in the allowed region $\sin^2(2\theta_{23}) > 0.9$ with both possible octant choices) and $|\Delta m^2_{atm}| = 2.9 \times 10^{-3}$ eV^2; $A = 1.1 \times 10^{-4}$ eV^2/GeV.

The experimental information consists of the number of muons in the detector with charge opposite to that of the muons circulating in the storage ring. We group the events in bins of the final muon energy E_μ and call $N^{g,s}_i(\bar{\theta}_{13}, \bar{\delta})$ the number of "golden" or "silver" in the i-th energy bin for the input pair $(\bar{\theta}_{13}, \bar{\delta})$ [11]. In the case of the Superbeam, N represents the number of electrons in the water Cherenkov, grouped in one single bin. For a given energy bin and fixed input parameters $(\bar{\theta}_{13}, \bar{\delta})$, we can draw a set of curves of equal number of events [17] in the (θ_{13}, δ) plane,

$$N^i_{\mu\pm}(\bar{\theta}_{13}, \bar{\delta}; \bar{s}_{atm}, \bar{s}_{oct}) = N^i_{\mu\pm}(\theta_{13}, \delta; s_{atm} = \bar{s}_{atm}, s_{oct} = \bar{s}_{oct})), \tag{2}$$

$$N^i_{\mu\pm}(\bar{\theta}_{13}, \bar{\delta}; \bar{s}_{atm}, \bar{s}_{oct}) = N^i_{\mu\pm}(\theta_{13}, \delta; s_{atm} = -\bar{s}_{atm}, s_{oct} = \bar{s}_{oct})), \tag{3}$$

$$N^i_{\mu\pm}(\bar{\theta}_{13}, \bar{\delta}; \bar{s}_{atm}, \bar{s}_{oct}) = N^i_{\mu\pm}(\theta_{13}, \delta; s_{atm} = \bar{s}_{atm}, s_{oct} = -\bar{s}_{oct})), \tag{4}$$

$$N^i_{\mu\pm}(\bar{\theta}_{13}, \bar{\delta}; \bar{s}_{atm}, \bar{s}_{oct}) = N^i_{\mu\pm}(\theta_{13}, \delta; s_{atm} = -\bar{s}_{atm}, s_{oct} = -\bar{s}_{oct})), \tag{5}$$

Following the procedure outlined in [12, 17] we can numerically solve eqs. (2)-(5) and found the theoretical location of the clones in the (θ_{13}, δ) plane. We present in Fig.1 the outcome of this procedure for the different degeneracies with fixed $\bar{\delta} = 90°$ and changing $\bar{\theta}_{13} \in [0.1°, 10°]$. Apart from some exceptional abrupt change (remind 2π-periodicity in the δ axis), a small change in the input parameter $\bar{\theta}_{13}$ results in a small shift of the clone location. (Almost) continuous geometrical regions where degeneracies lie are defined for a given interval in $\bar{\theta}_{13}$, illustrating how the clones move due to a change in the input parameters: we will call this the "clone flow". In Fig. 1(left) we plotted the intrinsic clone flow for a set of different experiments and channels; in Fig. 1(right) the clone flows for the eightfold-degeneracy are presented, for the NF-golden channel only.

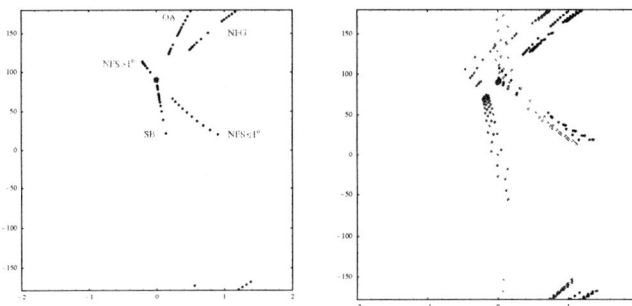

FIGURE 1. *(left) Intrinsic clone flows for the SPL and the NuMI Off-Axis superbeams and for the NF golden and silver channels; (right) The eightfold-degeneracy for the NF golden channel. The thick dot is the true solution, the thin dots the clones location for changing $\bar{\theta}_{13} \in [0.1°, 10°]$ and fixed $\bar{\delta} = 90°$.*

Fig. 1(left) shows that the combination of any two facilities solves the intrinsic degeneracy [16, 17]. More difficult is the case when all the degeneracies are treated on equal footing, Fig.1 (right), where the need of the combination of (at least) three facilities is manifest. This is exemplified in Fig. 2, where we present the outcome of combined χ^2 fits performed as in [11] for different combinations of the three detectors, for a fixed input pair $\bar{\theta}_{13} = 2°, \bar{\delta} = 90°$. In Fig. 2(a) four degeneracies can be seen when using the 40 Kton MID only; in Fig. 2(b) and Fig. 2(c) we notice how two of the degeneracies disappear when combining the 40 Kton MID with the 400 Kton WC or the 4 Kton ECC, respectively; eventually, in Fig. 2(d) the combination of the three detectors solve all the degeneracies reconstructing with a good precision the physical input values. CL contours up to 4 sigma are plotted.

Two comments are in order: first, the physical input pair $\bar{\theta}_{13} = 2°, \bar{\delta} = 90°$ is generic and similar results are obtained for different input parameters for $\bar{\theta}_{13} > 1°$; second, these results, although promising, are still preliminar and a new study where particular care is devoted to systematics in the three detectors is currently underway [21].

REFERENCES

1. F. Arneodo et al. [ICARUS Coll.], ICARUS-TM/2001-08 LNGS-EXP 13/89 add.2/01; M. Guler et al., OPERA Collaboration, CERN/SPSC 2000-028, SPSC/P318, LNGS P25/2000; E. Ables et al. [MINOS Coll.], FERMILAB-PROPOSAL-P-875; Y. Itow et al., KEK-REPORT-2001-4, arXiv:hep-ex/0106019.
2. M. Komatsu, P. Migliozzi and F. Terranova, J. Phys. G **29** (2003) 443 [arXiv:hep-ph/0210043].
3. M. Diwan et al., NuMI-NOTE-SIM-0714.
4. P. Migliozzi and F. Terranova, arXiv:hep-ph/0302274.
5. P. Huber et al., Nucl. Phys. B **654** (2003) 3; W. Winter, arXiv:hep-ph/0308227.
6. S. Geer, Phys. Rev. D **57** (1998) 6989 [Erratum-ibid. D **59** (1998) 039903].
7. A. De Rujula, M. B. Gavela and P. Hernandez, Nucl. Phys. B **547** (1999) 21.

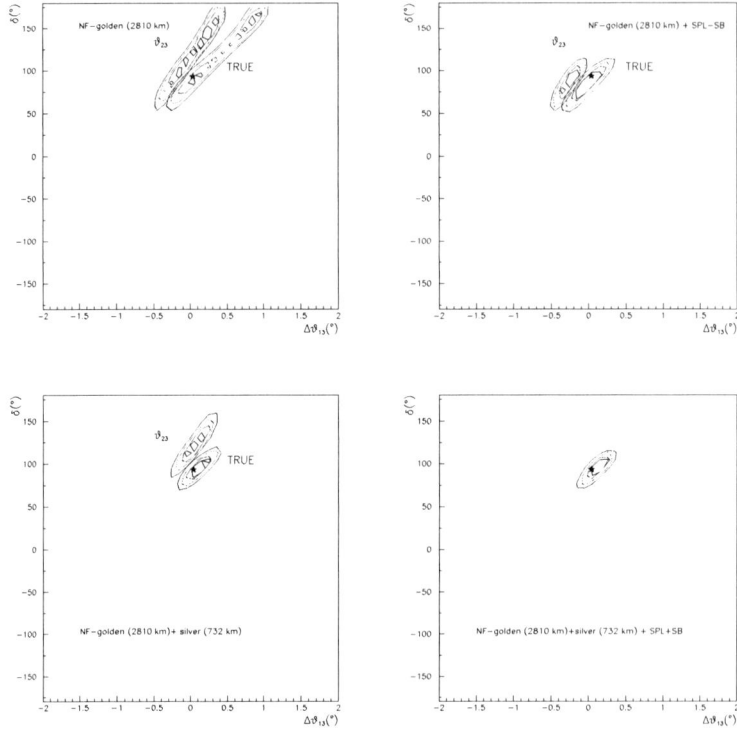

FIGURE 2. The results of a χ^2 fit for $\bar{\theta}_{13} = 2°; \bar{\delta} = 90°$. Four different combinations of experimental data are presented: a) MID; b) MID plus WC; c) MID plus ECC; d) The three detectors together.

8. A. Blondel et al., Nucl. Instrum. Meth. A **451** (2000) 102.
9. M. Apollonio et al., arXiv:hep-ph/0210192.
10. K. Dick et al., Nucl. Phys. B **562** (1999) 29; V. Barger, S. Geer and K. Whisnant, Phys. Rev. D **61** (2000) 053004; A. Bueno, M. Campanelli and A. Rubbia, Nucl. Phys. B **573** (2000) 27.
11. A. Cervera et al., Nucl. Phys. B **579** (2000) 17 [Erratum-ibid. B **593** (2001) 731].
12. J. Burguet-Castell et al., Nucl. Phys. B **608** (2001) 301.
13. H. Minakata and H. Nunokawa, JHEP **0110** (2001) 001.
14. V. Barger, D. Marfatia and K. Whisnant, Phys. Rev. D **65**, 073023 (2002).
15. H. Minakata et al., Phys. Rev. D **68** (2003) 013010; H. Minakata et al., Phys. Rev. D **68** (2003) 033017; M. Freund et al., Nucl. Phys. B **615** (2001) 331; A. Rubbia, arXiv:hep-ph/0106088; A. Bueno et al., Nucl. Phys. B **631** (2002) 239; T. Kajita et al., Phys. Lett. B **528** (2002) 245.
16. J. Burguet-Castell et al., Nucl. Phys. B **646** (2002) 301.
17. A. Donini, D. Meloni and P. Migliozzi, Nucl. Phys. B **646** (2002) 321; arXiv:hep-ph/0209240; D. Autiero et al., arXiv:hep-ph/0305185.
18. J. J. Gomez-Cadenas et al. arXiv:hep-ph/0105297.
19. A. Cervera, F. Dydak and J. Gomez Cadenas, Nucl. Instrum. Meth. A **451** (2000) 123.
20. A. Blondel et al., Nucl. Instrum. Meth. A **503** (2001) 173.
21. J. Burguet-Castell et al., in preparation.

The synergy of the golden and silver channels at the Neutrino Factory

Pasquale Migliozzi

I.N.F.N., Sezione di Napoli, Complesso Universitario di Monte Sant'Angelo, Via Cintia ed. G, I-80126 Naples, Italy

Abstract. In this paper we summarize the results of a detaled study of the so-called "silver channel" $\nu_e \to \nu_\tau$ [1, 2] and of the systematic errors associated both with the OPERA-like and the magnetized-iron detectors and their effects on the sensitivity.

1. INTRODUCTION

The present atmospheric, solar, accelerator and reactor[3, 4] neutrino data are strongly supporting the hypothesis of neutrino oscillations and can be easily accommodated, if LSND is discarded, in a three family mixing scenario. Finally, a comprehensive three-family analysis (including the negative CHOOZ results [5]) put a bound on θ_{13}, $\sin^2\theta_{13} \leq 0.02$.

The planned long baseline experiments [6]-[7] will improve the measurement of Δm^2_{atm} ($\Delta m^2_{atm} \simeq \Delta m^2_{23} \simeq \Delta m^2_{13}$) and of θ_{23}, and measure or increase the bound on θ_{13} [8, 9] (see also [10]). This new generation of experiments, however, is only the first step toward the ambitious goal measuring the leptonic CP-violating phase δ, which is the main goal of the Neutrino Factory program (see for example [11, 12] and refs. therein).

The most sensitive method to study this topic is the measure of the transition probability $\nu_e(\bar{\nu}_e) \to \nu_\mu(\bar{\nu}_\mu)$ (the *"golden measurement at the Neutrino Factory"* [13]). However, in [14] it was shown that, for a given physical input parameter pair (θ_{13}, δ), measuring the oscillation probability for $\nu_e \to \nu_\mu$ and $\bar{\nu}_e \to \bar{\nu}_\mu$ will generally result in two allowed regions of the parameter space. Worse than that, new degeneracies have later been noticed [15, 16], resulting from our ignorance of the sign of the Δm^2_{atm} squared mass difference and from the approximate $[\theta_{23}, \pi/2 - \theta_{23}]$ symmetry for the atmospheric angle.

In [1] it was noticed that muons proceeding from τ decay when τ's are produced via a $\nu_e \to \nu_\tau$ transition (the silver channel) show a different (θ_{13}, δ) correlation from those coming from $\nu_e \to \nu_\mu$. By using a lead-emulsion detector based on the Emulsion Cloud Chamber (ECC) technique, capable of the τ-decay vertex recognition, it is therefore possible to use the complementarity of the information from $\nu_e \to \nu_\tau$ and from $\nu_e \to \nu_\mu$ to solve the intrinsic (θ_{13}, δ) ambiguity. In the following, as in [1], we restrict ourselves to the (θ_{13}, δ) ambiguity, by fixing $\theta_{23} = 45°$ and by choosing a given sign for Δm^2_{atm}. How to solve all the ambiguities at the same time is discussed in [17].

2. SENSITIVITY TO (θ_{13}, δ)

The overall sensitivity achievable in the plane (θ_{13}, δ) is evaluated by combining golden muon events measured by a 40 Kton iron detector run with both beam polarities and silver muon events observed with a 5 Kton OPERA-like detector run only with the μ^+ polarity. For a detailed description of the OPERA-like detector and a thoroughly discussion on the foreseen scanning load we refer to [2].

2.1. Signal and background uncertainties

The total number of golden and silver muons, background events expected for the iron and the ECC detectors, and the details of the analyses are reported in [2, 18]. Here, we focus on the corresponding systematic uncertainties for both detectors.

The main sources of systematic uncertainties for the "silver" signal events in the OPERA-like detector are the knowledge of the emulsion scanning efficiency and the cross-section ratio σ_τ/σ_μ. In the following we assume an overall systematic uncertainty of 15% which is consistent with the one used in Ref. [19].

For an OPERA-like detector three main background contributions are present: the muonic decay of τ^+ events from $\bar{\nu}_\mu \to \bar{\nu}_\tau$ oscillations, the anti-neutrino induced charm-production and the decay in flight of h^- and punch-through h^-. Since a specific short-baseline program is foreseen at the Neutrino Factory [20]. The uncertainty on the background induced by τ^+ decays is the same as the one on the "silver" signal. The last background contribution is dominated by the poor knowledge of the hadronization and of the hadronic re-interaction processes. We conservatively assume a 50% uncertainty on this background source.

In a preliminary study of the performance of a magnetized iron detector for the $\nu_e \to \nu_\mu$ search at a Neutrino Factory [18] no systematic uncertainty was assigned to backgrounds and signal events in the calculations. In order to obtain realistic estimates of the experimental performance we try to include such uncertainties in the present work. In the following we assume a 10% uncertainty on the signal, taking into account the coarse granularity of the detector and the absence of an *in situ* calibration beam line. As far as the background is concerned, we assume a 50% systematic uncertainty, consistently with the treatment of the OPERA-like detector, although the overall sensitivity only mildly depends on this parameter.

2.2. Combining an OPERA-like detector at 732 km and an iron detector at 3000 km

The numbers of expected events, see Tables 5 and 6 of Ref. [2], show that the sensitivity of the OPERA-like detector degrades considerably at small values of both θ_{13} and δ. However, its contribution to the overall measurement is very relevant due to the complementary oscillation pattern [1]. Therefore, the effect of combining both "golden" and "silver" channels can be already seen for $\theta_{13} > 1°$.

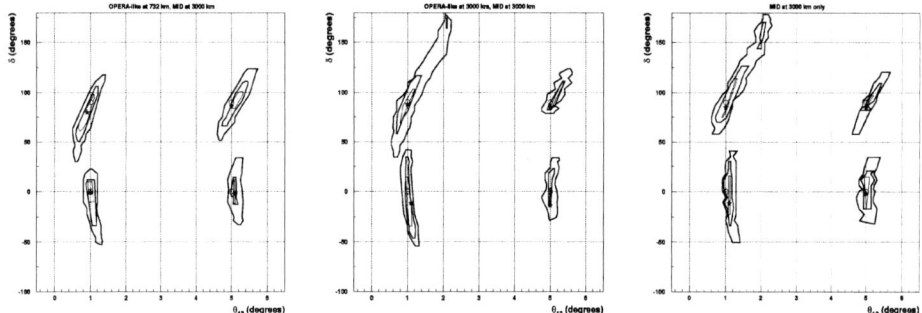

FIGURE 1. Expected 68.27%, 90% and 99% confidence regions in the (θ_{13}, δ) plane corresponding to four simulated points. Left panel: OPERA-like detector at 732 km and iron detector at 3000 km; Middle panel: OPERA-like detector at 3000 km and iron detector at 3000 km; Right panel: iron detector at 3000 km. The stars denote the best fit points, while the open circles are the true simulated points.

Fig. 1 shows the extracted 68.27%, 90% and 99% confidence regions in the (θ_{13}, δ) plane corresponding to four simulated points, if the OPERA-like detector is placed at $L = 732$ km from the neutrino source. It can be noticed that all the displayed curves are connected.

Since the expected background is small (< 10 events) in all bins, the systematic uncertainties on this component have a minor effect on the overall sensitivity. The same consideration applies to the "silver" signal in the OPERA-like detector. However, systematic uncertainties greater than 5% on the golden signal can reduce the sensitivity of the iron detector. This is particularly important at small values of δ. In our calculation we assumed a conservative 10% uncertainty, which could be further reduced by an accurate work on the detector. The dependence of the iron detector sensitivity on the systematic error is discussed in Ref.[2].

Given the small number of expected signal events in the silver channel for $\theta_{13} = 1°$, the ECC sensitivity is limited by the background. The possibility to further reduce the silver channel background is discussed in Ref.[2].

2.3. Combining an OPERA-like detector and an iron detector both at 3000 km

A location of the OPERA-like detector at 3000 km would reduce considerably most background contributions (proportional to $1/L^2$), with the exclusion of the muonic decay of τ^+ events from $\bar{\nu}_\mu \to \bar{\nu}_\tau$ oscillations.

In Fig. 1 we show the extracted 68.27%, 90% and 99% confidence regions in the (θ_{13}, δ) plane corresponding to the same four simulated points of Section 2.2. In this case, the curves extracted for the point $\bar{\theta}_{13} = 1°, \bar{\delta} = 90°$ are not fully connected, mainly due to the tiny signal contribution expected from the "silver" channel. However, values of $\theta_{13} \sim 1°$ are close to the intrinsic limit of the experimental sensitivity, producing wide fluctuations on the number of observed events. Actually, by increasing the number of simulated experiments for $\bar{\theta}_{13} = 1°, \bar{\delta} = 90°$, in few cases we observe the presence of disconnected curves also with the 732 km baseline for the OPERA-like detector, regardeless of the additional background reduction mentioned in Section 2.2. For comparison, in Fig. 1 we give the corresponding results when only the magnetized iron detector at $L = 3000$ km is used (still considering two beam polarities).

3. CONCLUSIONS

We show that the intrinsic ambiguity problem is solved for $\theta_{13} > 1°$ when using a 40 Kton magnetized iron detector, to deal with the "golden" muon signal, and a 5 Kton ECC OPERA-like detector, to measure the "silver" muon signal. We also include systematic effects in the treatment of the "golden" muon sample at the MID, previously not considered. We present (Section 2) a refined statistical analysis of the simulated data for a MID at $L = 3000$ km and an ECC either at $L = 732$ km or at $L = 3000$ km, taking advantage of the different energy distribution of "silver" signal and backgrounds. Below $\theta_{13} = 1°$, the "silver" muon sample at the ECC detector becomes statistically negligible. With both the MID and the ECC located at $L = 3000$ km we expect a significant decrease in the background of the "silver" channel, as previously mentioned. This translates, for $\theta_{13} > 1°$, in a reduction of the confidence intervals with respect to the configuration with the ECC at $L = 732$ km.

REFERENCES

1. Donini, A., Meloni, D., and Migliozzi, P., *Nucl. Phys.*, **B646**, 321–349 (2002), hep-ph/0206034.
2. Autiero, D., et al. (2003), hep-ph/0305185.
3. Shiozawa, M., "Atmospheric neutrino and K2K results and prospects," These Proceedings, 2003.
4. Nakamura, K., "Solar neutrino results, KamLAND results and prospects," These Proceedings, 2003.
5. Apollonio, M., et al., *Eur. Phys. J.*, **C27**, 331–374 (2003), hep-ex/0301017.
6. Aprili, P., et al., *CERN-SPSC-2002-027* (2002).
7. Itow, Y., et al., *hep-ex/0106019* (2001).
8. Komatsu, M., Migliozzi, P., and Terranova, F., *J. Phys.*, **G29**, 443 (2003), hep-ph/0210043.
9. Diwan, M., et al., *NuMI-NOTE-SIM-0714* (2001).
10. Migliozzi, P., and Terranova, F., *Phys. Lett.*, **B563**, 73–82 (2003), hep-ph/0302274.
11. Blondel, A., et al., *Nucl. Instrum. Meth.*, **A451**, 102–122 (2000).
12. Apollonio, M., et al., *hep-ph/0210192* (2002).
13. Cervera, A., et al., *Nucl. Phys.*, **B579**, 17–55 (2000), hep-ph/0002108.
14. Burguet-Castell, J., Gavela, M. B., Gomez-Cadenas, J. J., Hernandez, P., and Mena, O., *Nucl. Phys.*, **B608**, 301–318 (2001), hep-ph/0103258.
15. Minakata, H., and Nunokawa, H., *JHEP*, **10**, 001 (2001), hep-ph/0108085.
16. Barger, V., Marfatia, D., and Whisnant, K., *Phys. Rev.*, **D65**, 073023 (2002), hep-ph/0112119.
17. Donini, A., "Combining superbeams and Neutrino Factory," These Proceedings, 2003.
18. Cervera, A., Dydak, F., and Gomez Cadenas, J., *Nucl. Instrum. Meth.*, **A451**, 123–130 (2000).
19. Guler, M., et al., *CERN-SPSC-2000-028* (2000).
20. Mangano, M. L., et al., *hep-ph/0105155* (2001).

Resolving degeneracies for different values of θ_{13}

Walter Winter

*Institut für theoretische Physik, Physik–Department, Technische Universität München (TUM),
James–Franck–Strasse, 85748 Garching bei München, Germany*

Abstract. We discuss options to resolve correlations and degeneracies with combinations of future neutrino long-baseline experiments. We use a logarithmic scale of $\sin^2 2\theta_{13}$ as a representation for a systematical classification of the experiments.

The most interesting parameters to be determined by future long-baseline experiments are θ_{13}, δ_{CP}, and the mass hierarchy. All of these parameters are suppressed by the size of θ_{13} itself, which is constrained by the CHOOZ experiment [1] to below $\sin^2 2\theta_{13} \sim 10^{-1}$. From statistics and systematics only, each type of long-baseline experiment has (to a first crude approximation) a characteristic scale of θ_{13} which it can access with respect to these measurements, *i.e.*, the sensitivity reach in θ_{13}. Unfortunately, the measurements are spoilt by multi-parameter correlations and intrinsic degeneracies, which are the $(\delta_{CP}, \theta_{13})$ [2], $\text{sgn}(\Delta m_{31}^2)$ [3], and $(\theta_{23}, \pi/2 - \theta_{23})$ [4] degeneracies leading to an overall "eight-fold" degeneracy [5]. In this talk, we discuss options to resolve correlations and degeneracies from the point of view of the yet unknown true value of θ_{13}.

We define future long-baseline experiments as neutrino oscillation experiments which are using an artificial neutrino source (*e.g.*, accelerator or reactor) and are sensitive to atmospheric neutrino oscillations, *i.e.*, $\Delta m_{31}^2 L/E = \mathcal{O}(1)$. Compared to a natural neutrino source, such as the atmosphere or the sun, the artificial neutrino source produces a better known neutrino flux. In addition, a near detector is often proposed for a better control of the systematics. Beyond conventional beam experiments, such as K2K [6], MINOS [7], and CNGS [8], future superbeam experiments [9, 10, 11] are designed to find $\sin^2 2\theta_{13}$ down to $\sim 10^{-2}$. Superbeam upgrades[1], such as the JHF to Hyper-Kamiokande superbeam [9], could even access $\sin^2 2\theta_{13}$ down to $\sim 10^{-3}$. A useful experiment type below $\sin^2 2\theta_{13} \sim 10^{-3}$ (or $\theta_{13} \sim 1°$) is the neutrino factory (see, for example, Ref. [12]). A different type of long-baseline experiment, which also fits our definition, is a reactor experiment with a near and far detector [13, 14, 15, 16], which may find $\sin^2 2\theta_{13}$ down to 10^{-2} independent of correlations and degeneracies. However, such an experiment does not have sensitivities to δ_{CP} or the mass hierarchy at all.

There are essentially six different impact factors, which determine the performance of future long-baseline experiments (for a more detailed discussion, see Secs. 3 and 5 of

---

[1] In this talk, we refer to "superbeam upgrades" as superbeams with target powers in the megawatt, and fiducial detector masses in the megaton region, whereas "superbeams" refer to the size of the first-generation experiments.

Ref. [17]):

1. **Statistical errors** describe the experiment performance from statistics only.
2. **Systematics** makes the statistical errors somewhat larger and is determined by the experiment itself.
3. **Correlations** are connected degenerate solutions (at the chosen confidence level). The measurement error of the quantity of interest is usually obtained as the projection of the n-dimensional connected manifold onto the respective axis. It can even be orders of magnitude larger than the original error.
4. **Degeneracies** are disconnected degenerate solutions (at the chosen confidence level). Their treatment in the results depends on the definition of the quantity of interest.
5. **External input** can partially resolve correlations and degeneracies. For the long-baseline experiments, the external input usually includes the solar measurements and the knowledge about the matter density, whereas the atmospheric oscillation parameters are normally obtained by the disappearance channels.
6. **The true parameter values** are only known with a certain precision before the experiments are built. It can be shown that for future long-baseline experiments (depending on the measurement) especially the true values of Δm_{21}^2 (within the KamLAND-allowed range), Δm_{31}^2, δ_{CP}, and θ_{13} itself determine the performance.

These six different impact factors can be arranged in three different groups:

1. **(Statistics)** and 2. **(Systematics)** are determined by the R&D of the experiment.
3. **(Correlations)** and 4. **(Degeneracies)** are reducible by clever choices of baseline, energy, and combinations of experiments.
5. **(External input)** and 6. **(True parameter values)** are not controllable by the considered experiment at all.

Experimental results often include statistics and systematics, but not (or only partially) correlations and degeneracies. From the theoretical point of view, however, especially the correlations and degeneracies are relevant for the optimization of experiments. Thus, it is necessary that a quoted measurement error be clearly defined in order to be comparable to that of other analyses. In this talk, we only refer to the second group, *i.e.*, the reduction of correlations and degeneracies by the choices of baseline, energy, and combinations of experiments.

As far as correlations and degeneracies are concerned, there is an important difference between the analysis of existing and future experiments. An existing experiment would obtain one or more regions in parameter space fitting the data, such as the formerly allowed solar solutions. The purpose of the analysis of future experiments is, however, to minimize the extension (*i.e.*, correlations) and number (*i.e.*, degeneracies) of the disconnected solutions *before* the experiments are going to be built. In addition, the risk with respect to the yet unknown true parameter values within their allowed regions should be minimized. Thus, condensing the information as function of the most relevant parameters by a reasonable inclusion of correlations and degeneracies in the analysis is crucial for the optimization of future experiments.

TABLE 1. Logarithmic scale of the true value of $\sin^2 2\theta_{13}$, the corresponding values in degrees, a possible timescale, and the experiments, which are sensitive in the respective intervals. Note that the interval limits are only a crude approximation, and that in some cases the experiments do not entirely cover the whole intervals.

$\sin^2 2\theta_{13}$	$\sim 10^{-1} - 10^{-2}$	$\sim 10^{-2} - 10^{-3}$	$\sim 10^{-3} - 10^{-4}$	$< 10^{-4}$
θ_{13} [Degrees]	$\sim 9° - 3°$	$\sim 3° - 1°$	$\sim 1° - 0.3°$	$< 0.3°$
Timescale???	2000 – 2010	2010 – 2025	2025 – 2040	> 2040
Sensitive experiments:	- Conventional beams (partially) - Reactor experiments - First-generation superbeams	- Superbeam upgrades - Neutrino factories - Reactor upgrades?	- Neutrino factories	- Neutrino factories? - Theoretical reason for $\sin^2 2\theta_{13} \equiv 0$?

In Table 1, a logarithmic scale of $\sin^2 2\theta_{13}$ is shown together with the relevant future long-baseline experiments for the considered ranges in the respective columns. Note that the ranges are only crude approximations for the $\sin^2 2\theta_{13}$, CP violation, and mass hierarchy sensitivity reaches under optimal conditions, *i.e.*, statistics and systematics only. Table 1 implies that it only makes sense to combine experiments with similar capabilities, which are reflected by $\sin^2 2\theta_{13}$ as well as the timescale. Thus, by combining experiments from different columns in Table 1 the results are normally dominated by the statistics of the better experiment. Thus, one could also say that the true value of $\sin^2 2\theta_{13}$ "selects" the experiment types sensitive to it, which can be read off Table 1. Of course, this picture is quite simple and does not take into account continuously adjustable luminosities of the different experiment types. However, it is quite illustrative to classify possible experiment combinations to resolve correlations and degeneracies, which are so far:

$\sin^2 2\theta_{13} \sim 10^{-1} - 10^{-2}$: In this range, one can combine two first-generation superbeams [18, 19, 20, 21, 22] or one or two superbeams with a reactor experiment [14, 15] to resolve the degeneracies.

$\sin^2 2\theta_{13} \sim 10^{-2} - 10^{-3}$: One may combine two superbeam upgrades [23], a superbeam upgrade with a neutrino factory [24], the "golden" $v_e \leftrightarrow v_\mu$ appearance channel at a neutrino factory or superbeam upgrade with the "silver" channel (v_τ detection) at a neutrino factory [25, 26], or operate a superbeam upgrade at the "magic baseline" [27] (see also below).

$\sin^2 2\theta_{13} \sim 10^{-3} - 10^{-4}$: Since so far only neutrino factories have been demonstrated to operate efficiently in this range, one can only combine two neutrino factory baselines, which are naturally obtained from one neutrino factory. It has been realized in numerical (see, for example, Refs. [2, 28]) and analytical [5, 29] analyses that at a baseline $L_{\text{magic}} \sim 7300\,\text{km}$ all correlations and degeneracies involving the CP phase vanish independent of the energy and the oscillation parameters. It has therefore been called "magic baseline" [30, 31]. In Ref. [32], it has finally been demonstrated that the $\sin^2 2\theta_{13}$, (maximal) CP violation, and mass hierarchy sensitivities are all good in the considered range $\sin^2 2\theta_{13} \sim 10^{-3} - 10^{-4}$ for the combination of the two baselines $L_1 \sim 3000$ km and $L_2 = L_{\text{magic}}$.

In summary, we have discussed the combination of experiments to resolve degeneracies based upon the true value of $\sin^2 2\theta_{13}$, which "selects" the experiments sensitive to the quantities of interest. These sensitive experiments can then be combined to resolve correlations and degeneracies. The discussion in this talk is different from a strategical one, since the strategy depends on when $\sin^2 2\theta_{13}$ is found and the information and experiments available at that time. Right now, of course, the most interesting range is $\sin^2 2\theta_{13} \sim 10^{-1} - 10^{-2}$, which means that later decisions will depends on the results within this range and new technologies could still change the following options. Finally, we have discussed a logarithmic scale of $\sin^2 2\theta_{13}$, which is not necessarily an appropriate representation for all purposes. For example, a linear scale may be more plausible from linear mass models in flavor space.

I would like to thank P. Huber and M. Lindner for useful discussions and comments. This work was supported by the "Studienstiftung des deutschen Volkes" and the "SFB 375 für Astro-Teilchenphysik der DFG".

1. M. Apollonio *et al.* (CHOOZ Collaboration), Eur. Phys. J. **C27**, 331 (2003), hep-ex/0301017.
2. J. Burguet-Castell, M. B. Gavela, J. J. Gomez-Cadenas, P. Hernandez, and O. Mena, Nucl. Phys. **B608**, 301 (2001), hep-ph/0103258.
3. H. Minakata and H. Nunokawa, J. High Energy Phys. **10**, 001 (2001), hep-ph/0108085.
4. G. L. Fogli and E. Lisi, Phys. Rev. **D54**, 3667 (1996), hep-ph/9604415.
5. V. Barger, D. Marfatia, and K. Whisnant, Phys. Rev. **D65**, 073023 (2002), hep-ph/0112119.
6. K. Nakamura (K2K Collaboration), Nucl. Phys. Proc. Suppl. **91**, 203 (2001).
7. V. Paolone (MINOS Collaboration), Nucl. Phys. Proc. Suppl. **100**, 197 (2001).
8. A. Ereditato, Nucl. Phys. Proc. Suppl. **100**, 200 (2001).
9. Y. Itow *et al.*, Nucl. Phys. Proc. Suppl. **111**, 146 (2001), hep-ex/0106019.
10. D. Ayres *et al.* (2002), hep-ex/0210005.
11. D. Beavis *et al.* (2002), hep-ex/0205040.
12. M. Apollonio *et al.* (2002), and references therein, hep-ph/0210192.
13. V. Martemyanov, L. Mikaelyan, V. Sinev, V. Kopeikin, and Y. Kozlov (2002), hep-ex/0211070.
14. H. Minakata, H. Sugiyama, O. Yasuda, K. Inoue, and F. Suekane (2002), hep-ph/0211111.
15. P. Huber, M. Lindner, T. Schwetz, and W. Winter, Nucl. Phys. B (to be published), hep-ph/0303232.
16. M. H. Shaevitz and J. M. Link (2003), hep-ex/0306031.
17. P. Huber, M. Lindner, and W. Winter, Nucl. Phys. **B645**, 3 (2002), hep-ph/0204352.
18. Y. F. Wang, K. Whisnant, Z.-h. Xiong, J. M. Yang, and B.-L. Young (VLBL Study Group H2B-4), Phys. Rev. **D65**, 073021 (2002), hep-ph/0111317.
19. K. Whisnant, J. M. Yang, and B.-L. Young, Phys. Rev. **D67**, 013004 (2003), hep-ph/0208193.
20. V. Barger, D. Marfatia, and K. Whisnant, Phys. Rev. **D66**, 053007 (2002), hep-ph/0206038.
21. P. Huber, M. Lindner, and W. Winter, Nucl. Phys. **B654**, 3 (2003), hep-ph/0211300.
22. H. Minakata, H. Nunokawa, and S. Parke, Phys. Rev. **D68**, 013010 (2003), hep-ph/0301210.
23. V. Barger, D. Marfatia, and K. Whisnant, Phys. Lett. **B560**, 75 (2003), hep-ph/0210428.
24. J. Burguet-Castell, M. B. Gavela, J. J. Gomez-Cadenas, P. Hernandez, and O. Mena, Nucl. Phys. **B646**, 301 (2002), hep-ph/0207080.
25. A. Donini, D. Meloni, and P. Migliozzi, Nucl. Phys. **B646**, 321 (2002), hep-ph/0206034.
26. D. Autiero *et al.* (2003), hep-ph/0305185.
27. A. Asratyan *et al.* (2003), hep-ex/0303023.
28. S. T. Petcov Prepared for 9th Internat. Symp. on Neutrino Telescopes, Venice, Italy, 6-9 Mar 2001.
29. P. Lipari, Phys. Rev. **D61**, 113004 (2000), hep-ph/9903481.
30. P. Huber, J. Phys. **G29**, 1853 (2003), hep-ph/0210140.
31. M. Lindner, Nucl. Phys. (Proc. Suppl.) **B118**, 199 (2003), hep-ph/0210377.
32. P. Huber and W. Winter, Phys. Rev. **D68**, 037301 (2003), hep-ph/0301257.

CNGS, OPERA and ICARUS Status

Koichi Kodama[†]

OPERA Collaboration
Aichi University of Education

Abstract. A current status of OPERA experiment is reported. CNGS and ICARUS status are also described briefly. An emphasis is put on a progress in OPERA emulsion analysis.

INTRODUCTION

OPERA and ICARUS are long baseline ν_τ appearance experiments designed to be sensitive in the parameter region indicated by the deficit of atmospheric muon neutrinos and by its zenith angle dependence as observed by the Super-Kamiokande experiment [1]. They are placed in the underground lab at Gran-Sasso which is about 730km away from CERN. CNGS is a project to provide neutrino beam for these experiments using CERN SPS. CNGS first beam to Gran-Sasso is planned in May 2006 and both experiments are now in construction phase.

CNGS

Neutrino beam line to Gran-Sasso using 400GeV/c protons accelerated by the CERN SPS is under construction. CNGS [1] is a name of this project [2]. SPS provides 4.5×10^{19} p.o.t./year in shared mode operation. Major component of the neutrino beam is ν_μ with an average energy of 17GeV. Relative abandance of other neutrino species is $\nu_\mu : \bar{\nu}_\mu : \nu_e : \bar{\nu}_e = 2 \times 10^{-2} : 8 \times 10^{-3} : 5 \times 10^{-4}$. At Gran-Sasso laboratory, 2600 ν_μ CC events/kton/year is expected and, assuming oscillation parameters of $sin^2 2\theta = 1$ and $\delta m^2 = 3 \times 10^{-3} eV^2$, 22 ν_τ CC events/kton/year is expected.

OPERA [3]

OPERA is a Japan-Europe collaboration. New emulsion technique based on the Track Selector (i.e. automatic emulsion track recognition system) is used in OPERA to detect τ decay vertices in ν_τ charge current interactions. This technique have been developed and used in DONUT experiment at FNAL, by which the first direct observation of

---

[1] http://proj-cngs.web.cern.ch/proj-cngs/

v_τ charge current interactions have been done [4]. An example of observed v_τ charge current interaction is shown in fig.1.

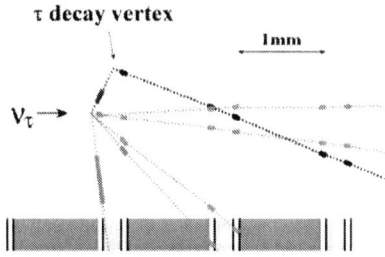

FIGURE 1. One of v_τ CC events observed in DONUT experiment. Red dashed line is a τ which decays into a charged particle (black dashed line). Flight length of τ is $280\mu m$, $\theta_{KINK} = 90 mrad$ and decay $P_T = 414^{+144}_{-81} MeV/c$.

Emulsion Preparation and Detector Construction

An emulsion brick is a basic component of OPERA detector. A brick is made up by stacking 56 lead plates of 1mm thick interleaved with thin emulsion sheets. It has a weight of about 8.5kg, a thickness of about $10X_0$ and a cross section of $102mm \times 127mm$. About 200k such bricks are used to build OPERA detector, which has a mass of about 1.8kton in total. A Target Tracker made of scintillator strips are inserted between each brick walls and used to identify a *fired* brick where a tagged neutrino interaction occured. Muon spectormeters are also placed in downstream of the target.

Soon after emulsion sheets are produced, compton electrons and cosmic rays entering them begin to be recorded. They act as noise hits and have to be avoided especially for electron energy measurement in OPERA. A new emulsion sheet has been developed in collaboration with Fuji Co.Ltd., in which such noise hits can be erased under high temperature and high humidity condition. This operation is called *refresh*. An underground facility for this *refresh* operation is prepared in Tono mine near Nagoya in Japan. Refreshed emulsion sheets are to be sent to Gran-Sasso and are made up to bricks in underground lab.

Detector construction in Gran-Sasso underground lab Hall C is in progress. A first piece of OPERA detector has been put in March 2003.

τ Decay Analysis

We can be rather relaxed in τ decay analysis in OPERA. It is based on the same emulsion technique as in DONUT experiment. And the number of neutrino interactions to be analyzed in OPERA, about 30k events in 5 years, is less than that in CHORUS, about 140k neutrino interactions to publish physics papers [5].

Although we are testing and developing techniques to improve our efficiency and to reduce backgrounds utilizing more information obtained from emulsion. A current status of 2 of such techniques are described below.

An energy loss (dE/dx) of a charged particle had been known to be a measureable quantity in emulsion. It is true also in our automated scanning scheme using the Track Selector. A quantity called *pulse height* obtained from the Track Selector has strong correlation to dE/dx. To test this correlation, 1.2GeV/c positively charged particles (i.e. mixture of pions and protons) have been exposed to an emulsion brick at KEK in November 2002. *Pulse height* averaged over 29 emulsion sheets along each track showed 2 clearly separated peaks as shown in fig.2. These peaks correspond to different dE/dx of pions and protons both at the same momentum. In fact, $p\beta$ measured by Multiple Coulomb Scattering in emulsion for tracks in these 2 peaks were consistent with those expected for pions and protons respectively. This method is to be used to reduce remaining charm background quoted in the proposal by identifing low momentum muons which are difficult to identify by simply measureing a track length.

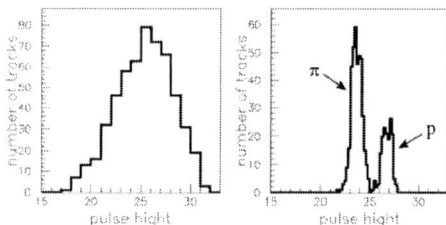

FIGURE 2. *Pulse height* distribution in a single emulsion layer (left) and averaged over 29 emulsion sheets along each track (right) are shown. In the averaged distribution, 2 clearly separated peaks each corresponding to pions and to protons both at 1.2GeV/c can be seen.

Electron identification is done by detecting shower tracks and by measureing energy loss due to bremsstrahlung. To test this method, an emulsion brick has been exposed to a mixture of electrons and pions at 2 and 4GeV/c using CERN PS in May 2002 [6]. Result was well described with Monte Carlo and gives about 90% efficiency and about 5% mis-identification probability when a window for shower track detection is chosen to be $\delta\theta < 75mrad$ and $\delta R < 45\mu m$. It should be noted that identification efficiency and mis-identification probability can be much better in real OPERA bricks as quoted in the proposal because, unlike as in the test, there will be no overlap among showers and thus a window size for shower track detection can be large enough in real OPERA bricks.

Event Location

Unlike τ decay analysis, event location in OPERA is a challenging job. A tagged neutrino interaction has to be located in an emulsion brick with rather poor assistance of tracking information available from the Target Tracker. Moreover it must begin from the day when the beam is on and when we started to extract *fired* bricks. About 60 bricks are expected to be extracted every day in shared mode operation of CERN SPS.

The Target Tracker is designed to identify a brick where a tagged neutrino interaction occurs. Thus a tagged neutrino interaction somewhere in an identified brick should be located by emulsion itself. Scanning power is crutial especially for this event location procedure and a faster scanning system called S-UTS is being developed in Japan. It is designed to have a scanning speed of $> 20 cm^2/h$, which is about 20 times faster or more than the current system. 30k frames/sec high speed CCD camera and syncronized z axis control to realize image data taking at 60 views/sec have been tested already. First working S-UTS is now being prepared. A similar scanning system is also being developed in Europe.

From our experience in CHORUS and DONUT experiments, 1 to 2 years of rehearsal must be organized for OPERA event location before the beam is on to be sure that it works properly from the beginning. In fact, it took about 1 to 2 years in CHORUS and in DONUT before event location procedure became stable. We plan to do this reheasal at KEK PS in Japan using a setup called Mini-OPERA to provide enough sample bricks to scan. The setup has already been completed and preliminary tests are in progress. Unfortunately CERN accelerators will be shutdown in 2005.

ICARUS

Liquid Ar TPC technique have been developed and are used in ICARUS experiment [2]. Neutrino interactions occured in ICARUS look like those in a bubble chamber but at different resolution. Their tracking resolution is about $\sigma_{X,Y} \sim 1mm$ and $\sigma_Z \sim 0.4mm$. ν_τ detection in ICARUS is done not by detecting τ decay vertices but by kinematical analysis. 600t module has been successfully operated and 5 such modules, 3000t in total, will be placed in Hall B in Gran-Sasso underground lab.

REFERENCES

1. The Super-Kamiokande Collaboration, Phys. Rev. Lett. 81 (1998)1562
2. G.Acquistapace et al., CERN 98-02, INFN/AE-98/05, 19 May 1998.
3. OPERA proposal, CERN/SPSC 2000-028 SPSC/P318, LNGS P25/2000, 10 July 2000
4. DONUT collaboration, Phys. Lett. B504(2001)218
5. CHORUS collaboration, Phys. Lett. B527(2002)173, B549(2002)48, B555(2002)156
6. K.Kodama et al., Review of Scientific Instruments vol.74(2003)53

---

[2] http://www.lngs.infn.it/site/exppro/icarus/icarus.html

JHF Sensitivity and the 2km Intermediate Detector

J. Burguet-Castell[*,†] and D. Casper[†]

[*]*Dept. de Física Atómica y Nuclear and IFIC, Universidad de Valencia, Spain*
[†]*Dept. of Physics and Astronomy, University of California, Irvine, CA 92697, USA*

Abstract. The current proposal for the J-PARC-Nu project includes an intermediate water Cerenkov detector at a distance of 2 km to reduce systematics, but only for the precision measurements on its 2nd stage.
  We show some preliminary results that suggest the convenience, and maybe the need, of the intermediate detector to achieve the objectives set for the 1st stage of the project.

PROJECT GOALS AND THE INTERMEDIATE DETECTOR

The J-PARC-Nu is a next-generation long-baseline neutrino oscillation experiment. It will use a beam from the JHF 50 GeV proton sincroton, and Super-Kamiokande as far detector at 295 km.

The expected sensitivities in the first stage, after a running period of 5 years, are:

1. Discovery of $\nu_\mu \to \nu_e$ at $\Delta m^2 \sim 3 \times 10^{-3} eV^2$ down to $\sin^2 2\theta_{13} \sim 0.006$, a factor of 20 improvement in sensitivity over past experiments.
2. Precision measurements of oscillation parameters in ν_μ disappearance down to $\delta(\Delta m^2_{23}) = 10^{-4} eV^2$ and $\delta(\sin^2 2\theta_{23}) = 0.01$.
3. Search for sterile component in ν_μ disappearance by detecting neutral current events.

After successful completion of the first stage, a second stage is envisaged, with an upgraded beam and detector. The aim would be to find CP violation through the appearance $\nu_\mu \to \nu_e$ search. The goal for it is a 2% subtraction error, and to achieve it an intermediate detector is planned. It would be a water Cerenkov detector situated at about 2 km from the source.

SYSTEMATIC UNCERTAINTIES

There are several sources of systematic uncertainties:

- Beam: Details in the beam modelization, distribution of pions in the decay volume, horn current and displacements.
- Detector: Interaction models (cross-sections, QE and π^0 production), particle identification (e/μ), pattern recognition (e/π^0), energy scale, fiducial volume.

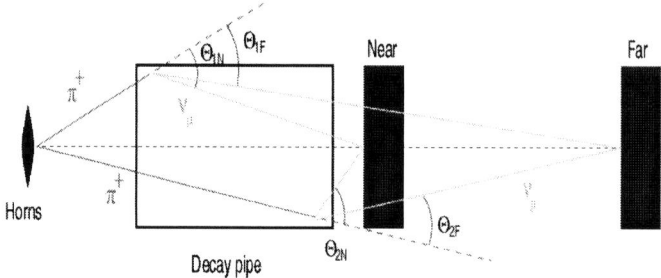

FIGURE 1. The closer location of the decay pipe introduces a large and complicated far-to-near spectrum ratio.

Controlling Systematics

The precision on the measurement of the oscillation parameters is mainly limited by high-energetic neutrinos with inelastic interactions. By means of an off-axis beam it is possible to narrow the energy spectrum and significantly lower the high-energy tail.

To further control the systematics, a near detector will be built at 280 m from the target. Its role is to predict the expected neutrino flux at the far detector. A fully-active fine-grained scintillator tracker identifies quasi-elastic charged currents, ν_μ and ν_e inelastic events and neutral currents, and measures the neutrino spectrum at the near location.

Caveats

Ideally all systematics cancel out using the measured spectra in the near detector. In reality, the near detectors are different from the far detector in terms of material, size (radiation length) and responses. The closer location to the decay pipe also introduces a large and complicated far-to-near spectrum ratio. As shown in Fig. 1, the neutrino source is point-like for the far detector, but the length of the decay pipe is not negligible for the near detector.

THE INTERMEDIATE DETECTOR

A water Cerenkov detector has the same cross-section than Super-Kamiokande. A location at 2 km and in the same off-axis angle as SK, makes the energy spectrum of the neutrino beam very similar to SK. Clearly this will help reduce considerably the systematics, but what is not clear is if there is a need for such a good control of systematics in the first stage of the project.

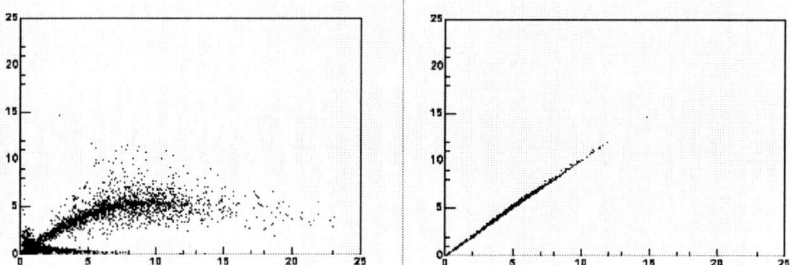

FIGURE 2. Correlation matrix that transforms the neutrino flux, at the near detector (left) and at the intermediate detector (right), to the flux at Super-Kamiokande. The simpler distribution for the intermediate detector suggests an easier reconstruction of the flux, which should also be less affected by systematics.

Analysis

To measure the effect of systematics, we have simulated independently the number of events per energy bin in the case of no systematics and the same values for systematics such as horn displacements. Instead of the double-ratio method, we have used the near-to-far correlation matrix. Figure 2 shows the values of the matrix elements for both the matrix that transforms from the near detector flux to Super-Kamiokande and the one that transforms directly from the intermediate detector. The non-zero elements for the intermediate detector lie very close to the diagonal, both making the analysis simpler and less sensitive to systematics.

IMPORTANCE OF SYSTEMATICS IN THE 1ST STAGE

We have studied the convenience of having the intermediate detector ready for the first stage of J-PARC-Nu. The fear is that even to get the precision expected for the first stage, the systematics can be problematic, in a way difficult to correct. Our preliminary results suggest that this may well be the case.

In Fig. 3 we can see that a 4 mm horn displacement would result in an appreciably distorted prediction of the neutrino spectrum at Super-Kamiokande, if we extrapolate, using the correlation matrix, the flux from the known at the near detector. For the same displacement, the prediction from the known flux at the intermediate detector is much close to the actual one at SK.

In a typical reconstruction, we have found, as expected, that the use of the *near* detector to control the systematics does improve significantly the results compared to what an experiment without near or intermediate detectors would be. On the other hand, it produces a bias in the reconstruction of θ_{23} and Δm^2_{23} that is not present if we perform the same analysis with the information available at the *intermediate* detector (Fig. 4).

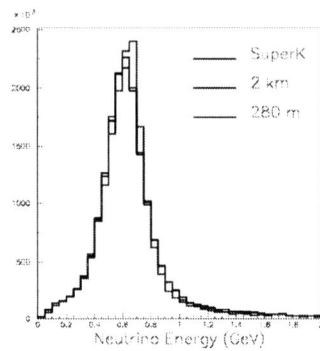

FIGURE 3. Predicted neutrino spectrum at SK compared to the real one, for a horn displacement of 4 mm. The prediction using the intermediate detector is much closer to the real one.

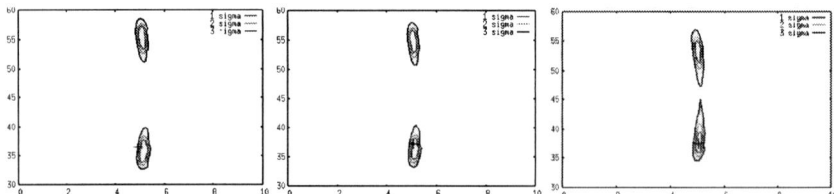

FIGURE 4. A typical reconstruction on the plane $(\theta_{23}, \Delta m_{23}^2)$, for no systematics control (left), control with only the near detector (center), and using the intermediate detector (right). The near detector data help improve the accuracy of the results, but it shows a bias that the intermediate detector corrects.

CONCLUSIONS

The beam extrapolation and background measurement are essential to the sensitivity of J-PARC-Nu.

The 2 km detector will dramatically simplify the systematic error analysis for both the appearance and disappearance searches. It may also be needed to achieve the precision expected for the first stage, a possibility that deserves more thorough studies.

BNL Very Long Baseline Experiment With a Super Neutrino Beam

Stephen Kahn

Brookhaven National Laboratory, Upton, NY, 11973

Abstract. An upgrade to the BNL AGS could produce a very intense proton source at a relatively low cost. This proton source could produce a conventional neutrino beam with a very significant flux at large distances from the laboratory. In this paper we examine the possibility of using this neutrino beam for a very long baseline oscillation experiment where a 500 kiloton water Cherenkov detector is situated at the Homestake mine in South Dakota. We study the physics potential of a high intensity neutrino oscillation experiment with a 2540 km baseline and a peak neutrino energy of ~1 GeV.

INTRODUCTION

Recent results have shown that neutrinos can oscillate between flavor states. This has created interest in understanding the fundamental aspects of neutrino oscillations. To explore the physics of the neutrino masses and mixing angles will require new facilities with intense proton beam sources. A working group at Brookhaven National Laboratory is studying the feasibility of upgrading the AGS to a 1.0 MW proton sources and using it to create an intense neutrino super-beam capable of producing a significant flux at a large distance from BNL. The AGS upgrade would increase the repetition rate of the accelerator from 0.5 to 2.5 Hz with 8.9×10^{13} protons per pulse. The expected integrated intensity for a typical year of operation (10^7 sec.) is expected to be 2.2×10^{21} 28 GeV protons on target. A description of the accelerator, target and horn system, and ν beamline of this super-beam facility is given in a report by the working group [1]. The physics potential of this proposed long baseline experiment is described in another report [2, 3]. The goal of this facility is to have the ability to measure all of the parameters of the SMN matrix in a single experiment. This includes:

- A precise determination of the oscillation disappearance parameters Δm_{32}^2 and $\sin^2 2\theta_{23}$.
- The detection of the appearance oscillation $\nu_\mu \to \nu_e$ and the measurement of $\sin^2 2\theta_{13}$.
- The measurement of $\Delta m_{21}^2 \sin^2 2\theta_{12}$ in a $\nu_\mu \to \nu_e$ appearance mode.
- The verification of the matter enhancement and the sign of Δm_{32}^2.
- The determination of the CP-violation parameter δ_{CP} in the neutrino sector.

The useful neutrino energy range is restricted at ~1 GeV on the low end by Fermi motion in the nucleus and by ~6 GeV on the high end by background from inelastic ν interactions. Placing the detector at the Homestake Mine, which is 2540 kilometers from BNL, allows the oscillation phase to vary from below $\pi/2$ to above $5\pi/2$ for $\Delta m_{32}^2 = 0.0025$ eV^2, which would clearly demonstrate oscillations in the energy distribution and provide a precise measurement of Δm_{32}^2. For the purposes of calculating expected event rates and backgrounds a 0.5 megaton water Cherenkov detector is used. A neutrino beam directed at the Homestake Mine must be inclined into the ground by 11.3° with respect to the surface. This incline restricts the location of the close-in detector to be just outside the beam dump at 275 m from the target.

FLUX CALCULATIONS AND EVENT ESTIMATES

Neutrino flux spectra have been calculated for the upgraded AGS with a horn design proposed for this beam [4]. Fig 1a shows the expected ν_μ and ν_e flux distributions for this beam in units of ν/GeV/m^2/Proton-on-target at 1 km from the target. The expected ν_e event contamination is ~1% in this wide-band beam. At distances greater than 1 km the spectrum is not strongly dependent on the on the ν source position and the consequently the flux at the detector scales with r^{-2}. The table in Fig 1b shows the number of events expected at 2540 km in a 0.5 megaton water Cherenkov detector for a 5×10^7 sec running period. We expect to see 52000 charged current and 17000 neutral current events during that run. The event analysis in a water Cherenkov detector will concentrate on the quasi-elastic samples since the neutrino energy is better known and the backgrounds in the ν_e appearance are better understood. A liquid argon detector would be a better detector to analyze the multiple particle channels.

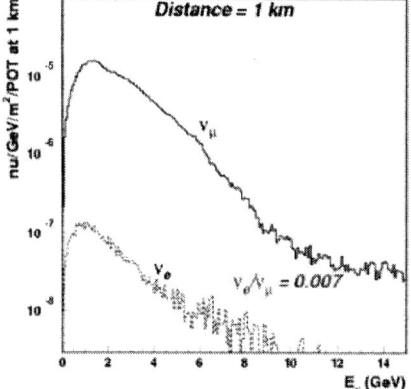

Channel	Number of Events
CC $\nu_\mu N \rightarrow \mu^- X$	51800
NC $\nu_\mu N \rightarrow \nu_\mu X$	16908
CC $\nu_e N \rightarrow e^- X$	380
QE $\nu_\mu n \rightarrow \mu^- p$	11767
QE $\nu_e n \rightarrow e^- p$	84
CC $\nu_\mu N \rightarrow \mu^- \pi^+ N$	14574
NC $\nu_\mu N \rightarrow \nu_\mu N \pi^0$	3178
NC $\nu_\mu O^{16} \rightarrow \nu_\mu O^{16} \pi^0$	574
CC $\nu_\mu N \rightarrow \pi X$	~110

Figure 1a: The ν_μ and ν_e flux spectra from 28 GeV protons on a graphite target seen at 1 km. **b:** Table of events seen in neutrino induced channels at 2540 km during a 5×10^7 sec run in a 0.5 megaton water Cherenkov Detector.

ν_μ DISAPPEARNCE

Figure 2a shows the energy spectrum of ν_μ quasi-elastic channel for a running period of 5×10^7 seconds with a 1 MW beam and a 0.5 megaton water Cherenkov detector at 2540 km. The top curve shows the event spectrum without oscillations. The middle curve with error bars shows the event spectrum with oscillations when $\Delta m_{32}^2=0.0025$ eV^2 and $\sin^2 2\theta_{23}=1$. In this figure we have assumed that the resolution in the reconstructed ν energy (E_ν) is 10%, which should be achievable with 10% photo multiplier tube coverage. This does not include a systematic error from the calibration of the overall detector energy scale. A great advantage of the very long baseline and the multiple oscillation pattern is that the systematic errors for flux normalization, background subtraction, and nuclear effects are small. The background shown in the lower curve of the figure comes from non quasi-elastic charged current events that also oscillate and will tend to smear the nodal pattern.

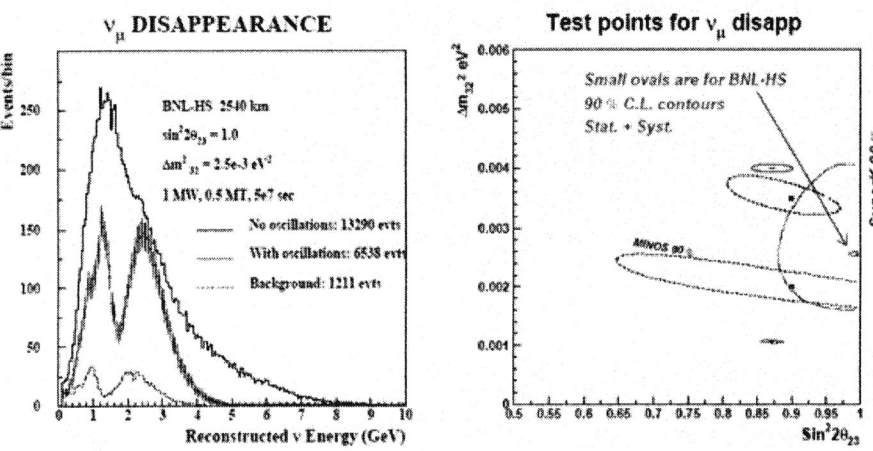

Figure 2a: ν_μ Quasi-elastic event spectrum. Curves show events without oscillations, with oscillations, and the background to oscillated QE events. The oscillated events assume that $\Delta m_{32}^2=0.0025$ eV^2. **b:** A comparison of the precision of the measurement of Δm_{32}^2 and $\sin^2 2\theta_{23}$ for this experiment and that expected by the MINOS experiment and seen by the Super-Kamiokande experiment.

Figure 2b shows a comparison of the experimental precision of Δm_{32}^2 and $\sin^2 2\theta_{23}$ expected in this experiment with those seen in the Super-Kamiokande experiment and expected in the MINOS experiment.

$\nu_\mu \rightarrow \nu_e$ APPEARANCE

The measurement of θ_{13}, δ_{CP} and Δm_{21}^2 and the sign of Δm_{32}^2 can be extracted from the wideband ν_e appearance spectrum. The probability for the appearance of $\nu_\mu \rightarrow \nu_e$ oscillations from 3-generation mixing including matter effects can be expressed analytically [3]. Figure 3a shows the expected ν_e event distribution for $\sin^2 2\theta_{13}=0.04$

and $\Delta m_{32}^2 = 0.0025$ eV^2 and the same running conditions previously mentioned. The matter enhancement (suppression) for the normal (reversed) Δm_{32}^2 sign ordering dominates the spectrum for $E_\nu > 3$ GeV. Sensitivity to the CP phase is greatest for E_ν range from 1 to 3 GeV. The effect of the solar oscillation from Δm_{21}^2 is largest at low E_ν. Figure 3b shows the event spectrum with $\sin^2 2\theta_{13} = 0$. The appearance mode is more sensitive to the presence of backgrounds. The main sources of background come from ν_e contamination in the beam and neutral current reactions that have a single π° that are misidentified as electrons. An estimate of these backgrounds is shown in figure 3. There is concern at this point that the detection efficiency of π° is overestimate in the current simulation. This is being investigated.

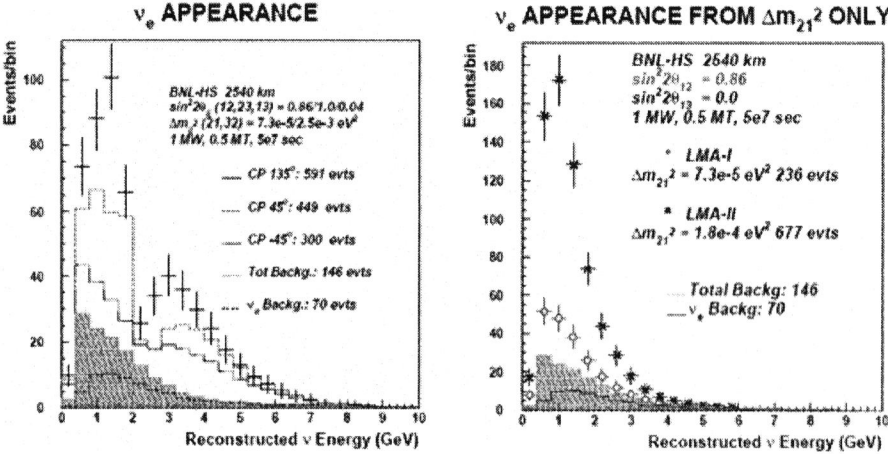

Figure 3a: The expected ν_e event spectrum for $\sin^2 2\theta_{13} = 0.04$ and for 3 values of δ_{CP}. The figure also shows the total background for ν_e. **b:** The expected spectrum for $\sin^2 2\theta_{13} = 0$. The resultant spectrum shows Δm_{21}^2 oscillations.

ACKNOWLEDGMENTS

This work was performed with the support of the US DOE under Contract No. DE-AC02-98CH10886.

REFERENCES

1. *The AGS-Based Super Neutrino Beam Facility, The BNL Neutrino Working Group Report II,* BNL-71228-2003-IR.
2. *Report of the BNL Neutrino Working Group: Very Long Baseline Neutrino Oscillation Experiment for Precise Determination of Oscillation Parameters and Search for $\nu_\mu \to \nu_e$ Appearance and CP Violation,* BNL Report 69395 (2002).
2. M.V. Diwan et al., Physical Review D **68**, 012002 (2002)
4. S.A. Kahn et al., *Focusing Horn System for the BNL...,* Proc of the 2003 Particle Accelerator Conference, BNL-71411-2003-CP.

India-based Neutrino Observatory

G.Rajasekaran
Institute of Mathematical Sciences
Madras-600 113,India

Abstract:An introduction to India-based Neutrino Observatory and a brief status report are presented.The two possible sites are described along with their special advantages.The proposed detector and its physics capabilities for atmospheric neutrinos and long-base-line experiments are discussed.

Introduction: Historically,the Indian initiative in Cosmic Rays and Neutrino Physics experiments goes back several decades.In fact,atmospheric neutrinos were first detected in the Kolar Gold Fields (KGF) experiments in India almost 4 decades ago. KGF were one of the deepest mines in the world.When the cosmic ray muon experiments were set up at deeper and deeper levels in the mines, the counters fell silent at a particular depth.It was realised that at those depths and beyond, atmospheric neutrinos could be detected. They went ahead and detected them.That was in 1965 and it was the beginning of atmospheric neutrino physics.However, the deeper levels of KGF are now closed.

Sometime ago it was decided to revive neutrino experiments in India and a major collaboration involving about 12 institutions has been formed.This is the India-based Neutrino Observatory (INO) project.More than 60 scientists have already joined and feasibility studies are in progress. Two possible sites for the underground laboratory have been located. A magnetised tracking iron calorimeter of 30-50 Kton with RPC detector elements is under design and prototyping.In the first stage the aim will be to study atmospheric neutrinos and in the next stage this detector is envisaged as the far detector for a long-base-line neutrino experiment.International collaboration is invited.

A Tale of Two Sites: We give more details about the two sites: PUSHEP (Lat 11.5 deg N, Long 76.6 deg E): Under the Nilgiri Mountains in South India; adjacent to a hydel project PUSHEP (Pykara Ultimate Stage HydroElectric Project); vertical overburden in the range 1.3- 1.4 Km and all-around cover of more than 1 Km; laboratory cavern to be dug at the end of a tunnel of length about 2 Km;located in the Southern Peninsular Shield; uniform rock medium of mean density 2.8 gms/cc; seismic zone 2;close to the Cosmic Ray Laboratory of TIFR and the Radio Astronomy Centre of TIFR,both in the hill station Ooty; close to big cities (with airports) like Coimbatore and Bangalore,with excellent industrial and academic infrastructure.Detailed survey of the region is complete.

RAMMAM (Lat 27 deg N, Long 88 deg E):Under the Himalayas, in the Darjeeling District of West Bengal; a tunnel of 3-5 Km can reach an overburden of 1.5-1.85 Km or even more; seismic zone 4; detailed survey is in progress.

Both sites are excellent for an underground neutrino laboratory. After a critical evaluation of the relative advantages and disadvantages of both sites with respect to the physics goals and practical aspects, one of them will be chosen.

Detector: To start with, it is proposed to build a magnetised tracking iron calorimeter, based on Monolith design. It will consist of 140 layers of 6 cm thick iron plates interleaved with 2.5 cm air gap containing the active detector elements. The dimensions will be 15 m X 32 m X 12 m (ht) and the weight about 30Ktons. The detector elements will be glass RPC's (Resistive Plate Chambers), with nanosecond timing, to provide up-down discrimination. The detector will be sensitive to muons and other charged particles.

A magnetic field of 1-1.3 Tesla will be an important feature of the detector. This will provide efficient energy-momentum resolution and also charge-identification which is an essential requirement for a far-end detector in a long-base-line experiment.

The detector will be constructed in a modular fashion, so that additional modules can be added in future, to augment its capability for the long-base-line experiment and other experiments. Emulsion sandwich is also being considered as a possibility with the aim of detecting the tau leptons.

Physics possibilities at INO: From the long term point of view, a neutrino detector located in India will have several advantages. A solar neutrino detector at a low latitude (PUSHEP at 11.5 deg) will detect solar neutrinos passing through the core of the Earth. A geoneutrino detector at RAMMAM will detect the geoneutrinos from the unusually thick continental crust below the Tibetan plateau. Very long base lines from neutrino factories become possible with detectors located in India. The baseline lengths from a neutrino factory at JHF(Japan) to PUSHEP or RAMMAM are 6,595 Km (31 deg) and 4,880 Km (22 deg) respectively, the brackets showing the required dip angle of the muon decay pipe. The corresponding numbers from a neutrino factory at CERN are 7,145 (34) and 6,890 (33) and from a factory at Fermilab are 11,300(62) and 10,500(55). The importance of multiple number of long base lines for neutrino physics is now well recognised. Some of the above distances are close to the "magic" baseline length of 7,200 Km. The very long base line of 11,300 Km passes through 3000 Km of Earth's core and hence is likely to play a major role in future neutrino tomography of the Earth.

In phase I, INO will be studying atmospheric neutrinos. The aim will be to establish neutrino oscillations by showing the rise in the neutrino flux after the minimum and to reduce the present uncertainty in the value of Δm^2_{32}. In phase II, INO can play the role of the far detector in a long-base-line experiment. The major tasks in this phase would be probing θ_{13}, determination of the sign of Δm^2_{32} and gaining a first glimpse of the CP violating phase δ.

We shall now show a few results of preliminary calculations by Raj Gandhi and Anindya Datta on the physics capabilities of INO both in phase I and phase II. The inputs used are muon detection threshold of 2 GeV and muon energy resolution of 5 percent. All measurements in phase II involve wrong sign muon detection and so backgrounds are low.

Fig 1 shows the up/down ratio of atmospheric nu as a function of L/E for 200 Kton-yr of operation of the INO detector. It is clear that oscillations can be established.

Fig 2 gives the reach of $\sin\theta_{13}$ for nu factory experiments as a function of the muon detection threshold energy. The reach is defined as the value that will yield 10 signal events (wrong sign muons) for a given kT-yr exposure. We have shown the plots for Japan-RAMMAM and Fermilab-PUSHEP baselines.

Fig 3 shows the number of wrong-sign muon events as a function of Δm^2_{23} for three base lines from a neutrino factory in Japan, the detector being at Beijing, Rammam or PUSHEP. The sign discriminating capability for either of the two sites Rammam or PUSHEP is clearly demonstrated.

Because of lack of space we shall not show the graphs for the CP violating phase, but calculations have been done for the ratio of wrong-sign muon events for a run with negatively charged muons in the storage ring to that for a run with positively charged muons. Although the CP violating effect is weak for the Japan-PUSHEP baseline, it is clearly present for the Japan-Rammam baseline.

Finally, it must be emphasised that we need the support and cooperation of neutrino enthusiasts all over the world. International collaboration for the INO project is invited. Naba Mondal (nkm@tifr.res.in) is the spokesperson for the project. More information on INO is available at the INO web-site ⟨⟨ www.imsc.res.in/~ino ⟩⟩.

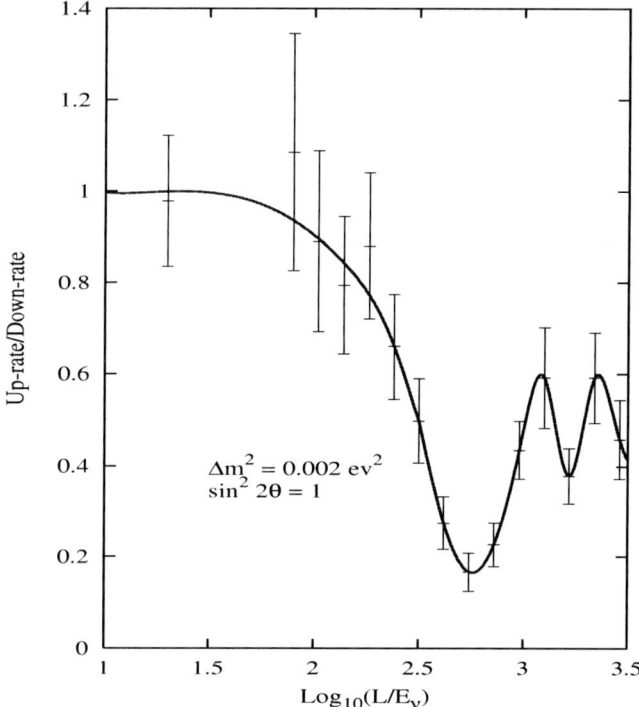

Figure 1: The up/down ratio of atmospheric ν vs L/E for 200 Kton-year.

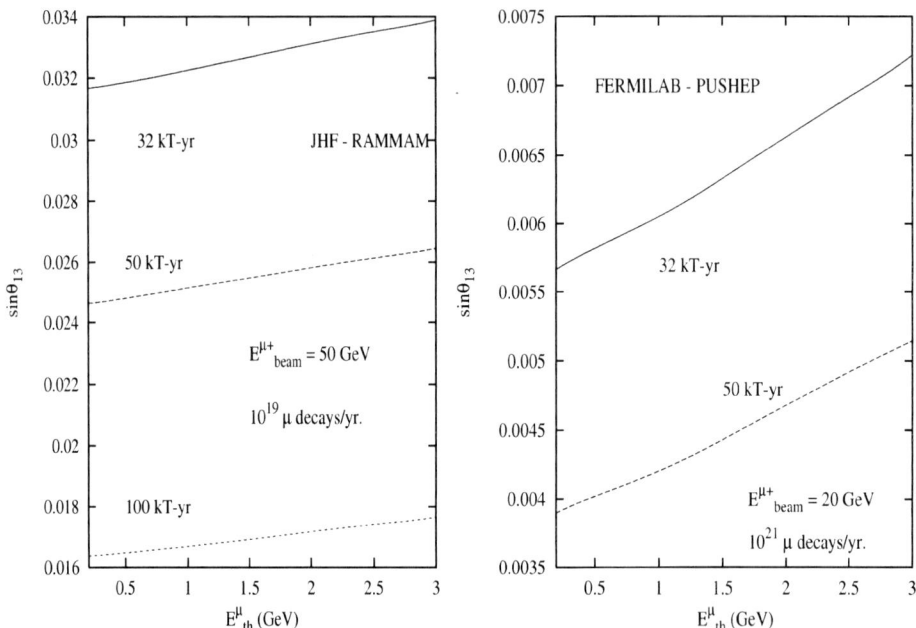

Figure 2: $\sin\theta_{13}$ reach as a function of the muon threshold energy. Left panel is for JHF to Rammam baseline. Right panel is for Fermilab to PUSHEP baseline.

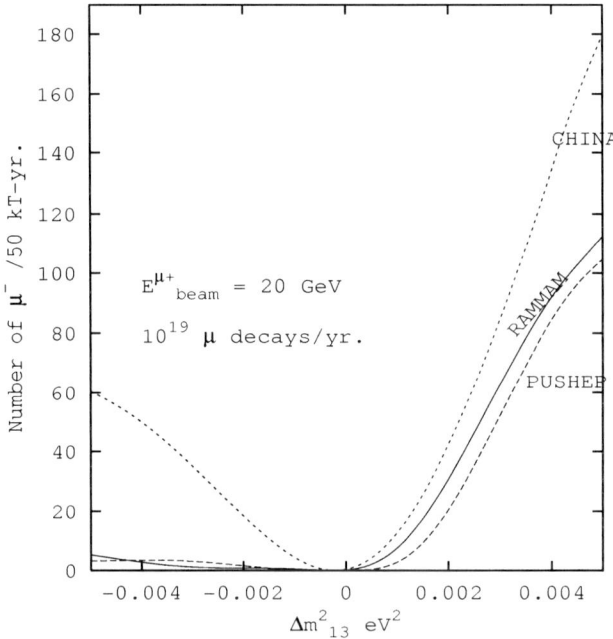

Figure 3: The number of wrong-sign muon events vs Δm_{23}^2 corresponding to baselines from JHF to Beijing, Rammam and PUSHEP.

Low Energy Neutrino-Nucleus Interactions

C.W. Walter[*], K. McConnel[†] and M. Sakuda[**]

[*]590 Commonwealth Ave. Boston MA 02215 (corresponding author)
[†]Department of Physics, Columbia University, New York, NY 10027
[**]KEK, IPNS, Tsukuba-shi, 305-0801, Japan

Abstract.
As the precision of neutrino oscillation experiments increases, correctly modeling neutrino-nucleus interactions is becoming ever more important. Uncertainties in nuclear form factors and nuclear structure result in uncertainties in determining oscillation parameters.

NEUTRINO-NUCLEUS INTERACTIONS

Water Cherenkov detectors like Super-Kamiokande [1] have been extremely successful as neutrino detectors. Water is an inexpensive target which also serves as Cherenkov radiator. However, because of the nature of the Cherenkov process not all particles produced in neutrino interactions are visible in a water Cherenkov detector. The momentum threshold for producing Cherenkov light in water is .6 MeV for electrons, 120 MeV for muons, and 1.1 GeV for protons.

So, while all electrons, and most muons are visible most protons are not. Fortunately, if the reaction is quasi-elastic(QE), the kinematics of the the event and the incoming neutrino energy can be reconstructed using *only the energy and angle with respect to the beam of the produced lepton*. Equation 1 shows the relationship between the incoming neutrino energy and the reconstructed momentum of the produced lepton.

$$E_v = \frac{m_N E_\mu - m_\mu^2/2}{m_N - E_\mu + p_\mu \cos(\theta_\mu)}, \qquad (1)$$

At neutrino energies less than 1 GeV the cross-section is dominated by QE interactions. This is shown in figure 1 which is taken from [2].

Equation 1 allows one to reconstruct the energy of the incoming neutrino even if all of the reaction particles aren't visible. However, it should be noted that, in a real experiment, one cannot know a priori which events come from QE interactions and which come from non-QE interactions. If one uses equation 1 for all events, those events which are QE will reconstruct with the proper energy while non-QE events wont. Those events which are not QE will have their energies systematically underestimated.

Figure 2 shows the effect on mis-reconstruction on an oscillation experiment. The top panel shows the energy of Super-K events from the K2K beam MC with oscillations applied using the true energy of the MC for the axis. As can be expected for $\sin^2 2\theta = 1$ the effect of oscillations is to maximally suppress the flux of v_μ neutrinos all the way to zero at about 700 MeV(which is the oscillation maximum at 3×10^{-3} and 250 km.).

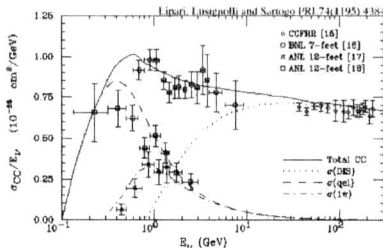

FIGURE 1. The neutrino nucleon cross section as a function of neutrino energy [2]. Below 1 GeV the cross section is dominated by QE interactions. For these interactions the neutrino energy can be reconstructed using only the outgoing lepton.

The bottom panel shows what happens if the reconstructed energy is used instead. The non-QE events have their energy mis-reconstructed to lower energy. Since almost all of the non-QE events have energies greater than then where the oscillations take place they are not suppressed and they "fill in" the region of the oscillation signal. This effect is why although the neutrino mixing can be maximal the reconstructed energy spectra can still have many reconstructed events in the region of the oscillation dip.

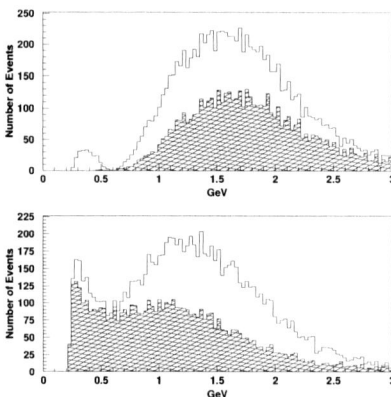

FIGURE 2. The top panel shows the Monte Carlo K2K spectrum at Super-K with oscillations applied. The oscillation dip at 700 MeV maximally suppresses the flux of ν_μ neutrinos. The non-QE interactions(hatched region) are un-effected by oscillations because their energy is too high. The bottom panel shows the same thing using reconstructed energy. The non-QE interactions "fill in" the oscillation dip.

For this reason, it is quite important to accurately model the fraction and shape of both the QE "signal" and this non-QE "background". The parameter $\sin^2 2\theta$ determines the overall normalization of the oscillation suppression, with $\sin^2 2\theta = 1$ resulting in a complete suppression of the flux. If the amount of non-QE interactions is not-properly modeled, then the overall suppression in the oscillation region will not be modeled properly either, and the less than maximal suppression will be incorrectly interpreted as a $\sin^2 2\theta$ less than unity [3]. The shape of the non-QE background is affected by, among other items, the relative proportions of the interaction modes, the axial mass of the nucleon [3], and the effect of Pauli blocking [4].

In order to determine parameters such as the axial mass it is useful not only to analyze the new data that is now available from K2K [5], and will soon be available from MiniBooNE [6], but also the still existing bubble chamber data from previous experiments[7]. While fitting the axial mass, the most effective technique is to fit to the shape of the Q^2 distribution, since a shape test alone is insensitive to errors in the absolute cross-section.

ELASTIC FORM FACTORS

The cross section for neutrino-nucleon quasi-elastic scattering[8] is expressed in terms of the form factors for the vector and axial currents. These form factors account for the fact that the interaction is taking place on nucleons rather than on a point like object. At zero Q^2, the vector electric and magnetic form factors are normalized by the electric charge and anomalous magnetic moments of the neutron and proton. Electron-nucleon scattering experiments in the 1960s showed that these form factors roughly followed a dipole form and the Q^2 dependence of the vector form factors was parameterized by Galster [9] with the form $D = 1/(1+Q^2/M_V^2)^2$, with the vector mass $M_V = .843\ (GeV/c^2)$. As early as the 1970's it became clear that this parameterization was only good to the 10-20% level [10].

Nonetheless, the axial form factor was parameterized in the same way with the both the normalization at zero Q^2, and the associated axial mass measured in neutrino scattering experiments. In addition, there is a pseudo-scaler form factor for neutrino scattering, but this term(along with the electric form factor of the neutron) was ignored in the calculations of neutrino cross section.

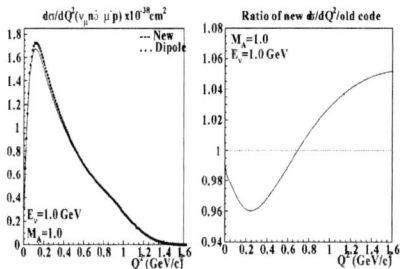

FIGURE 3. a) Differential cross section $d\sigma/dq^2(\nu_\mu + n \rightarrow \mu^- + p)$ at $E_\nu = 1\,GeV$ with both the Galster dipole form factors and new vector form factors and (b) the ratio.

After new electron scattering experiments were recently preformed with polarized electron beams new parameterizations were proposed [11, 12]. In addition, an accurate measurement of the electric charge distribution of the neutron($G_E^N(Q^2)$) has now been made [13]. At NuInt02, Budd [14] showed that the effect of the new parameterizations was to change the total neutrino-nucleon cross-section by a few percent. Figure 3 shows the differential cross section for quasi-elastic scattering at a neutrino energy of 1 GeV both with the old [9] and the new [11] vector form factors. Also shown are their ratio. The effect of the new form factors is to change the cross section by 4-5% in a Q^2 dependent manner. As expected, fitting the axial mass using neutrino data with the new vector form factors also results in approximately a 5% change in the fitted value.

SPECTRAL FUNCTIONS

When modeling neutrino interactions inside of the nucleus, both the binding energy and the initial Fermi momentum of the target nucleon must be considered. Many neutrino interaction Monte Carlos consider the nucleons to have flat Fermi momentum distribution up to the level of the Fermi sea, and a fixed nuclear potential. Modeling errors on the binding energy and initial momentum of the target nucleon will cause the outgoing lepton in the neutrino interaction to have the wrong energy. This is equivalent to introducing a error in the energy scale. The corresponds directly to an error on the reconstructed Δm^2 since the mass difference is determined by the location of the dip in the oscillation pattern, which is a function of L and E.

In order to more accurately model these effects a spectral function [14, 15] can be used. The spectral function is a density map of the probability of finding a nucleon at a particular momentum and binding energy. It is formed by using both electron scattering data and calculation. Figure 4 shows the Fermi momentum distribution for oxygen [14] as taken from a spectral function. The solid line is calculated using the local density approximation(LDA) [14, 15] while the points are from a MC calculation by Piper et. al. [16]. Also shown is a flat Fermi momentum distribution. While the average Fermi momentum is similar to that of the spectral function the long tail of high momentum is missing for the case of the flat distribution.

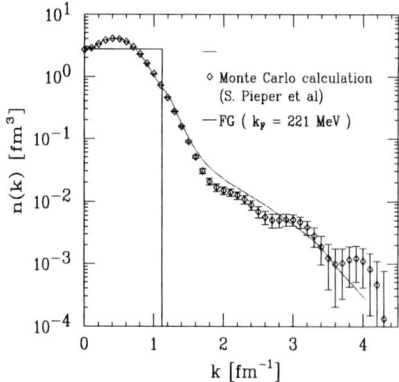

FIGURE 4. Fermi momentum distribution of oxygen [14]. The solid line (LDA) represents the spectral function calculation.

A comparison with experimental electron scattering data [17, 18, 19] on oxygen with a calculation [20] using the Smith-Moniz relativistic Fermi-gas model [21] and momentum distribution based on spectral functions shows good agreement between data and MC. Models based on a flat Fermi momentum distribution are missing the high energy tail seen in the data and spectral function calculations.

Summary

Uncertainties in our knowledge of neutrino-nucleus interactions are a important source of systematic errors when trying to make precision measurements of neutrino oscillation parameters. Properly modeling Pauli blocking, vector and axial form vectors, and the momentum and binding energy distribution of nucleons inside of a nucleus are just some of the effects which are now being studied to address this issue.

There are other issues which I did not have space to address in these proceedings such as the transition between the resonance and DIS region [22], the difference between parton distribution functions in nucleon and in the nucleus where there are many nucleons [23, 24, 25, 26], and new models of pion production [27].

ACKNOWLEDGMENTS

The author would like to thank M. Sakuda(KEK) for useful discussions and contributions to this proceeding, and K. McConnel for the calculations of figure 3.

REFERENCES

1. S. Fukuda et al. Nucl. Instrum. Meth. A **501**, 418 (2003).
2. P. Lipari, M. Lusignoli and F. Sartogo, Phys. Rev. Lett. **74**, 4384 (1995) [arXiv:hep-ph/9411341].
3. C. W. Walter, NuInt02 workshop, Irvine CA. December 12-15, 2002.
4. M. Sakuda, NuInt02 workshop, Irvine CA. December 12-15, 2002.
5. Y. Itow, NuInt02 workshop, Irvine CA. December 12-15, 2002.
6. H. Tanaka, NuInt02 workshop, Irvine CA. December 12-15, 2002.
7. K. Furuno, NuInt02 workshop, Irvine CA. December 12-15, 2002.
8. C.H.Llewellyn Smith, Phys. Rep. **C3**, 261 (1972).
9. S.Galster et al., Nucl. Phys. **B32**, 221 (1971).
10. M.Gourdin, Phys. Rep. **C11**, 29 (1974).
11. P.E. Bosted, Phys. Rev. C **51**, 409 (1995).
12. E.J.Brash et al., Phys. Rev. **C65**, 051001 (2002).
13. H.Gao, Int. J. Mod. Phys. **E12** ,1 (2003).
14. H. Budd, NuInt02 workshop, Irvine CA. December 12-15, 2002.
15. O. Benhar, NuInt02 workshop, Irvine CA. December 12-15, 2002.
16. S.C.Pieper,R.B.Wiringa and V.R.Pandharipande, Phys. Rev. **C46**, 1741 (1992).
17. M.Bernheim et al., Nucl. Phys. **A375**, 381 (1982).
18. J.S. O'Connell et al., Phys. Rev. **C35**,1063 (1987).
19. M.Anginolfi et al., Nucl. Phys. **A602**, 405 (1996).
20. H.Nakamura and R.Seki, Nucl. Phys. B (Proc.Suppl.) **112**, 197 (2002).
21. R.A.Smith and E.J.Moniz, Nucl. Phys. **B43**, 605 (1972).
22. A.Bodek and U-K.Yang, Nucl. Phys. B (Proc.Suppl.) **112**, 70 (2002).
23. E.Paschos, Nucl. Phys. B (Proc.Suppl.) **112**, 89 (2002).
24. S.Kumano, Nucl. Phys. B (Proc.Suppl.) **112** 42 (2002).
25. J.Eskola et al. Nucl. Phys. **B535**, 351 (1998).
26. K.McFarland, Nucl. Phys. B(Proc.Suppl.) **112**, 226 (2002).
27. E. A. Paschos, J. Y. Yu and M. Sakuda, [arXiv:hep-ph/0308130].

Lepton flavor violation in a long-baseline experiment

T. Ota[*] and J. Sato[†]

[*]Depertment of physics, Osaka University, Machikaneyama 1-1, Toyonaka, Osaka, 560-0043, Japan
[†]Depertment of physics, Faculty of science, Saitama University, Saitama, 338-8570, Japan

Abstract. The size of the signal of the lepton flavor violation with neutrinos induced by the model, MSSM+right-handed neutrinos, is estimated. The effects can be detactable in the next-generation oscillation experiments.

INTRODUCTION

The experiments made in the last decade established the neutrino oscillation. This fact means the existence of the neutrino masses and mixings and makes us expect that the sizable lepton flavor violation (LFV) effects. The LFV with charged lepton, such as $\mu \to e + \gamma$, $\mu \to ee\bar{e}$, μ-e conversion, is one of the most interesting subject to explore beyond the standard model. We, here, deal with the LFV processes but with neutrinos. Now, all we are waiting for the era of the precision measurement of the oscillation parameters. However, do the oscillation experiments only measure the oscillation parameters ? There are some articles which suggest that the sensitivity reach of such future experiment has the potential to search for not only the oscillation parameters but also the LFV effcet with neutrinos (n-LFV) [1–4]. We here discuss the question whether the effect of n-LFV induced by the model, MSSM+right-handed neutrinos, is detectable or not.

DETECTABLE SIGNAL: MODEL INDEPENDENT ANALYSIS

Following Ref.[1], we recupturate the basics of our method and model independent analyses, and show what kind of n-LFV effect can be observed in the oscillation experiment.

The long-baseline experiment has the advantage of observing the interference terms between amplitude induced by the standard interaction and that including n-LFV process in comparison with the direct LFV search. We can parametrize the exotic interactions by using the ratio of the coupling of the standard interaction and that of non-standard ones, which is denoted by ε following Refs.[2] and [3]. If such exotic interactions exsit, the neutrino is produced as the mixed flavor state in the beam production process:

$$|v_\alpha^s\rangle = |v_\alpha\rangle + \sum_{\beta \neq \alpha} \varepsilon_{\alpha\beta}^s |v_\beta\rangle, \qquad (1)$$

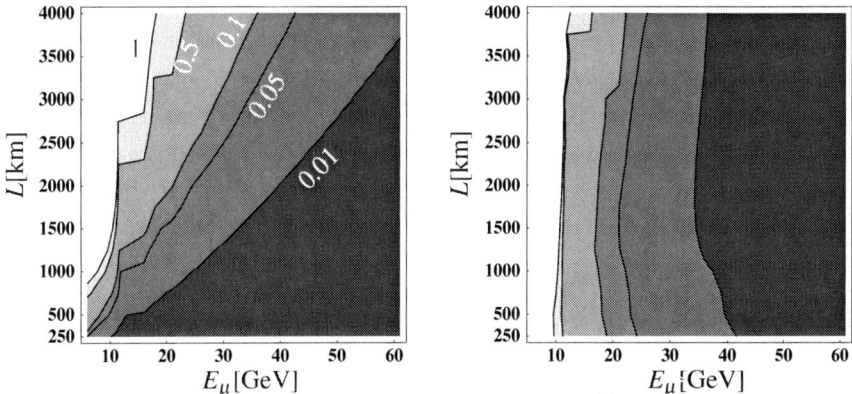

FIGURE 1. Contour plot of the required data size in unit of 10^{21} muon decays and 100 kt detector to observe the effect of $\varepsilon_{\mu\tau}^{s,m}$ at 90% confidence level in $\nu_\mu \to \nu_\tau$. The left plot for $(\varepsilon_{\mu\tau}^s, \varepsilon_{\mu\tau}^m) = (3.0 \times 10^{-3}i, 0)$, and the right for $(\varepsilon_{\mu\tau}^s, \varepsilon_{\mu\tau}^m) = (0, 3.0 \times 10^{-3})$ It is assumed that each oscillation parameter including the matter effect has 10% uncertainty.

and the neutrino feel the extra matter effect potential in the propagation process:

$$H = \frac{1}{2E}\left\{ \begin{pmatrix} 0 & & \\ & \Delta m_{21}^2 & \\ & & \Delta m_{31}^2 \end{pmatrix} + a \begin{pmatrix} 1+\varepsilon_{ee}^m & \varepsilon_{e\mu}^m & \varepsilon_{e\tau}^m \\ \varepsilon_{e\mu}^{m*} & \varepsilon_{\mu\mu}^m & \varepsilon_{\mu\tau}^m \\ \varepsilon_{e\tau}^{m*} & \varepsilon_{\mu\tau}^{m*} & \varepsilon_{\tau\tau}^m \end{pmatrix} \right\}. \quad (2)$$

We investigated the effect originated from these exotic interactions in all oscillation channels with both neutrino factory and super-conventional beam scheme in Ref.[1]. We arrived at the conclusions that only the signal induced by $\varepsilon_{\alpha\beta}^{s,m}$ can be observed by utilizing its charactaristic energy dependence in $\nu_\alpha \to \nu_\beta$ channel. Even if the uncertainty of the oscillation parameters is taken into account, the signal can not vanish. In $\nu_e \to \nu_\mu$, so called *golden channel*, the signals of $\varepsilon_{e\mu}^{s,m}$ take the distinctive energy dependence, however, the LFV processes with charged lepton put the strict bounds to their size. Therefore, we turn to $\varepsilon_{\mu\tau}^{s,m}$ in $\nu_\mu \to \nu_\tau$. The sensitivity reach are shown in Fig.1 and it tells the effects of $|\varepsilon_{\mu\tau}^{s,m}| \sim \mathcal{O}(10^{-4})$ are detectable enough.

MODEL: MSSM WITH RIGHT-HAND NEUTRINOS

In this section, we now turn to the subject whether the model can induce the detectable size of $\varepsilon_{\mu\tau}^{s,m}$. Firstly we roughly estimate the size using the mass insertion method and next we will see the numerical calculation supports our estimation.

Figure 2 is one of the diagrams which contribute to $\varepsilon_{\mu\tau}^{s,m}$. It shows that the origin of the n-LFV effect is the off-diagonal term of the s-lepton mass matrix, and the contribution

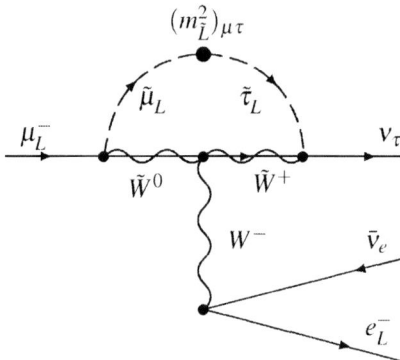

FIGURE 2. One of the diagram which contribute $\varepsilon^s_{\mu\tau}$ in the neutrino factory scheme.

from this diagram is approximated as

$$\varepsilon^s_{\mu\tau} \sim \frac{g^4}{16\pi^2} \frac{(m^2_{\tilde{L}})_{\mu\tau}}{m^4_S} \Big/ G_F, \qquad (3)$$

where m_S denotes the typical value of the super-particle masses. Some assumption[1] allows us to estimate the off-diagonal elements of the slepton mass as

$$(m^2_{\tilde{L}})_{\alpha\beta} \simeq -\frac{6+2a^2}{16\pi^2} \ln\left(\frac{M_X}{M_R}\right) m^2_0 (Y^\dagger_\nu Y_\nu)_{\alpha\beta}, \qquad (4)$$

where M_X is the scale of gravitation and m_0 is the universal soft-mass and a is the constant for the universal A parameter. Therefore, $\varepsilon^s_{\mu\tau}$ can become

$$\varepsilon^s_{\mu\tau} \sim \mathcal{O}(10^{-4}), \qquad (5)$$

which can make a detectable size signal as we exactly refered in the previous section.

The numerical calculation is shown in Fig.3. Here, we take account of all the diagrams which contribute to $\varepsilon^s_{\mu\tau}$. The behavier of this plot can be understand by above rough estimation.

CONCLUSION

The typical model, here we check MSSM with right-hand neutrino, can induce the detectable size of n-LFV effect. The next generation long-baseline experminet can work

---

[1] Concretely, we here assume mSUGRA senario and the SeeSaw mechanism with $(M_R)_{ij} = M_R \delta_{ij}$ and ignore the Majorana phases of U_{MNS} and phases for V_R which is defined as $Y_\nu = V^\dagger_R Y^{\text{diag}}_\nu U_L$.

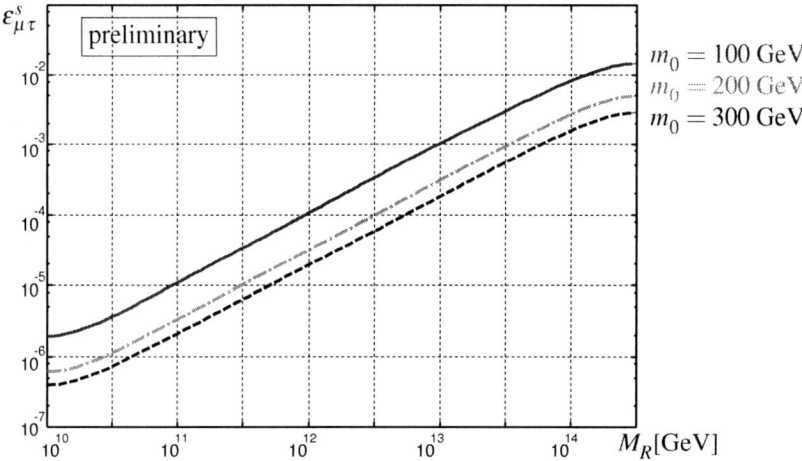

FIGURE 3. The M_R dependence of $\varepsilon^s_{\mu\tau}$. Here, we assume $M_{1/2} = 160$ GeV, $a = 0$, and $\tan\beta = 10$ in mSUGRA scenario.

not only as the precision measurement of the oscillation parameter but also as one of the method to explore beyond the standard model, the same as the search for the LFV with charged leptons.

ACKNOWLEDGMENTS

The one of the author TO is greatful to M. Koike and D. Nomura for useful comments which are given during this conference. The work of TO is supported by the JSPS Research Fellowships for Young Scientists. This work was supported by the Grant-in-Aid for Scientific Research, the Ministry of Education, Science and Culture, Japan (Grant Numbers 14039209, 14046217, 1474068 and 15540283).

REFERENCES

1. T. Ota, J. Sato, and N. Yamashita, Phys. Rev. **D65** (2002) 093015 [hep-ph/0112329], T. Ota, and J. Sato, Phys. Lett.**B545** (2002) 367 [hep-ph/0202145].
2. Y. Grossman, Phys. Lett. **B359** (1995) 141 [hep-ph/9507344].
3. A. M. Gago, M. M. Guzzo, H. Nunokawa, W. J. C. Teves, and R. Zukanovich Funchal, Phys. Rev. **D64** (2001) 073003 [hep-ph/0105196].
4. M. C. Gonzalez-Garcia, Y. Grossman, A. Gusso, and Y. Nir, Phys. Rev. **D64** 096006 [hep-ph/0105159].
   P. Huber, T. Schwetz, and J. W. F. Valle, Phys. Rev. **D66** (2002) 013006 [hep-ph/0202048].
   J. W. F. Valle and P. Huber, Phys. Lett. **B523** (2001) 151 [hep-ph/0108193].
   M. Campanelli and A. Romanino, Phys. Rev. **D66** (2002) 113001 [hep-ph/0207350].
   G. L. Fogli, E. Lisi, A. Mirizzi, and D. Montanino, Phys. Rev. **D66** (2002) 013009 [hep-ph/0202269].
   S. Davidson, C. Pena-Garay, N. Rius, and A. Santamaria, JHEP **0303** (2003) 011 [hep-ph/0302093].
   T. Hattori, T. Hasuike, and S. Wakaizumi, [hep-ph/0210138].

The Cosmological Energy Density of Neutrinos from Oscillation Measurements

Kevork Abazajian

Theoretical Division, MS B285, Los Alamos National Laboratory, Los Alamos, New Mexico 87545

Abstract. The emerging structure of the neutrino mass matrix, when combined with the primordial element abundances, places the most stringent constraint on the flavor asymmetries in the cosmological neutrino background and therefore its energy density. I review the mechanism of synchronized neutrino oscillations in the an early universe with degenerate (asymmetric) neutrino and antineutrino densities and the implications of refined measurements of neutrino parameters.

INTRODUCTION

The dawn of the era of precision cosmology has come with the observations of anisotropies in the cosmic microwave background (CMB) with the Wilkinson Microwave Anisotropy Probe (WMAP) over the whole sky to better than fundamental uncertainty over a wide range in anisotropy scale [1]. Combined with the three-dimensional galaxy distribution of the Sloan Digital Sky Survey [2], a consistent picture has emerged for the standard concordance cosmology: a universe dominated by dark matter and dark energy with structure growing from nearly scale-invariant adiabatic Gaussian density perturbations. In the simplest models, WMAP and SDSS measure the cosmological matter density to nearly 10% [3].

Given the success of the standard concordance cosmology, it is tempting to assume that the density of all cosmological matter and radiation components of the universe are known to great precision. However, the neutrino density, often simply assumed to be fixed to its standard model value, is actually only known to factors of its own magnitude when using the WMAP data alone [4].

One can hope to do better with primordial nucleosynthesis. During primordial nucleosynthesis, the nucleon beta-equilibrium weak interaction rates are sensitive to the electron neutrino and antineutrino densities. The cosmic expansion rate depends on the overall neutrino density, which sets when nuclear reactions freeze-out. These two effects can compensate each other and can produce primordial element abundances for deuterium, helium and lithium that are consistent with their observed abundances, as long as the nucleon density is increased to allow the nuclear rates to keep up with the required increased expansion rate [5]. The non-zero neutrino chemical potentials (or degeneracy parameters) of this model led to its description as degenerate big bang nucleosynthesis (DBBN). Since the nucleon (baryon) density is independently constrained by the CMB, the magnitude of deviations from non-zero neutrino chemical potentials was appreciably constrained from the original DBBN models, but still allowed neutrino densities over twice that of the standard value [6].

With the emergence of the mass and mixing spectrum of the active neutrino flavors, particularly in the large to maximal mixing angles of the solar and atmospheric neutrino oscillation solutions, it was proposed that the mixing could lead to the equilibration of neutrino asymmetries prior to nucleosynthesis in the studies of Refs. [7, 8].

The first attempt to solve the full evolution equations for the active neutrino system using was performed numerically by Dolgov, Hansen, Pastor, Petcov, Raffelt, and Semikoz [9], who found that the maximal mixing solution of the atmospheric results and large mixing angle solution of the solar neutrino problem invariably led to a near equalization of neutrino asymmetries between flavors. Therefore, DBBN, which required a large disparity between electron and muon or tau neutrino densities, would not be viable in a universe with the observed neutrino mass and mixing matrix. Analytic insight into the flavor asymmetries' equalization and a quantification of changes within the range of mixing parameters was studied by Wong [10], and Abazajian, Beacom and Bell [11]. The constraint imposed by the resulting equalizing transformations excludes DBBN and requires neutrino densities to be within $\sim 3\%$ of the standard value. Therefore, any nonstandard cosmic radiation energy density must come from a more exotic phenomenon than photons and neutrinos.

SYNCHRONIZED OSCILLATIONS

In an elegant paper, Pastor, Raffelt & Semikoz [13] showed that the synchronization mechanism, initially studied in Refs. [12] can be framed in the representation of synchronized dipoles precessing in a magnetic field, with the orientation of the dipole representing the flavor content.

The system of mixed neutrinos in a dense, scattering, self-refractive environment must be handled in a density matrix formalism. The two-flavor neutrino density matrix is

$$\rho(p) = \begin{pmatrix} \rho_{\alpha\alpha} & \rho_{\alpha\beta} \\ \rho_{\beta\alpha} & \rho_{\beta\beta} \end{pmatrix} = \frac{1}{2}\left[P_0(p) + \sigma \cdot \mathbf{P}(p)\right], \quad (1)$$

where $\mathbf{P}(p)$ as the neutrino's "polarization" vector, which can be represented as an individual "magnetic-dipole." The polarization vector describes asymmetries in flavor densities, such that $\mathbf{P}(p)^{initial} \propto \left[f_e(p,\xi_e) - f_\mu(p,\xi_\mu)\right]$, where $f_\alpha(p,\xi_\alpha)$ is the Fermi-Dirac distribution for a neutrino of flavor α with degeneracy parameter ξ.

The synchronized transformation can be described by the vector equations

$$\begin{aligned}\partial_t \mathbf{P}_p &= +\mathbf{A}_p \times \mathbf{P}_p + \alpha(\mathbf{J}-\bar{\mathbf{J}}) \times \mathbf{P}_p, \\ \partial_t \bar{\mathbf{P}}_p &= -\bar{\mathbf{A}}_p \times \bar{\mathbf{P}}_p + \alpha(\mathbf{J}-\bar{\mathbf{J}}) \times \bar{\mathbf{P}}_p,\end{aligned} \quad (2)$$

where neutrino scattering is negligible, $\mathbf{J}$ denotes the individual neutrino polarizations integrated over momentum, over-bars refer to antineutrino quantities, and α is the strength of the neutrino self-potential: $\alpha(\mathbf{J}-\bar{\mathbf{J}}) \times \mathbf{P}_p$.

The general "magnetic field" vector $\mathbf{A}_p$ includes terms incorporating vacuum mixing, a thermal potential from the charged-lepton background, and a potential due to asymmetries between the charged leptons, $\mathbf{A}_p = \vec{\Delta}_p + \left[V^T(p) + V^B\right]\hat{\mathbf{z}}$. Vacuum mixing is incorporated by

$$\vec{\Delta}_p = (\delta m_0^2/2p)(\sin 2\theta_0 \hat{\mathbf{x}} - \cos 2\theta_0 \hat{\mathbf{z}}), \tag{3}$$

where $\delta m_0^2 = m_2^2 - m_1^2$ and θ_0 are the vacuum oscillation parameters.

The thermal potential V^T arises from the finite-temperature modification of the neutrino mass due to the presence of thermally populated charged leptons, and V^B is the background potential arising due to asymmetries in charged leptons. V^B is the crucial term in the case of the sun, but is negligible in the early universe.

If one ignores the non-linear neutrino self-potential, the evolution of the system is trivial: the "magnetic-field" vector points in the direction of the charged lepton thermal potential, in the $\pm \hat{\mathbf{z}}$ direction, which is also the initial direction of the polarization vectors in a flavor-asymmetric system, as in DBBN. The thermal potential initially dominates but decreases as the universe cools, eventually becoming comparable to $\vec{\Delta}_p$, the vacuum term. $\vec{\Delta}_p$ points in a direction determined by the vacuum mixing angle (Eq. 3), which for large mixing is close to the $\hat{\mathbf{x}}$ direction. Each neutrino polarization (the flavor descriptor) then follows its respective "magnetic-field," whose final orientation is in the direction of $\vec{\Delta}_p$, and thus the cosmic flavor content, simply depends on the vacuum mixing angle.

When including the neutrino self-potential, the explicit solution can only be calculated numerically. Ref. [11] found that with the self-potential, the collective system behaves on average identically with the case when the self-potential is flatly ignored, even though the self-potential dominates all other terms by five or more orders of magnitude. Refs. [10, 11] showed that under certain approximations, the effect of the neutrino self-potential is to force all neutrino polarizations to follow a specific synchronization momentum's $\mathbf{A}_p$, whose value is $\frac{p_{\text{sync}}}{T} = \pi\sqrt{1+\xi^2/2\pi^2} \simeq \pi$, which is coincidentally very close to the average momentum of the Fermi-Dirac distribution $\langle p/T \rangle \simeq 3.15$. Of course, this is what the system average would follow without self-potential.

This remarkable coincidence allows for a dramatic simplification of the apparently initially intractable nonlinear evolution equations and allows a straightforward visualization of the general behavior of the neutrino gas for a variety of cases and mixing parameters. As described above, the transformation that leads to total or partial flavor equalization occurs at a temperature where the vacuum term and thermal potential are comparable. Since the vacuum term $\vec{\Delta}$ is proportional to δm^2, larger δm^2 leads to transformations at higher temperature. And, since the final orientation of the flavor polarization vectors is in the $\vec{\Delta}$ direction, the level of total or partial flavor equalization is determined by the vacuum mixing angle [11].

OSCILLATION PARAMETERS AND THE EARLY UNIVERSE

The consequences of the emerging neutrino mass matrix structure for a universe that contains neutrino degeneracies is quite rich. The implications for each of the mass scales in a three-neutrino mixing frame-work and their mixings is as follows:

Atmospheric Neutrinos, δm_{23}^2 and θ_{23}: for the range of δm^2 preferred by the oscillation solution to the atmospheric neutrino results by Super-Kamiokande [14], flavor equilibration occurs at a temperature $T \sim 12\,\text{MeV}$ due to the presence of equilibrating scatterings, and maximal mixing produces absolute equalization of flavor density asymmetries. If precision measurements of θ_{23} reveal a non-maximal angle, the equalization of neutrino density would be very close though not necessarily perfect, and an explicit calculation would be necessary since scattering is not negligible at $T \sim 12\,\text{MeV}$.

Solar Neutrinos, δm_{12}^2 and θ_{12}: δm^2 for the large mixing angle solution to the solar neutrino problem is much smaller than that of the atmospheric scale, so that the thermal potential dominates until a lower temperature. The transformation in this case occurs at $T \sim 2\,\text{MeV}$, sufficiently before the start of nucleosynthesis at $T \sim 1\,\text{MeV}$, disallowing DBBN. The level of equalization is dependent on the orientation of $\vec{\Delta}$, i.e., how "large" the large mixing angle is. Precise measurements of θ_{12} would determine the final vacuum vector orientation, what neutrino asymmetries can be accommodated by primordial nucleosynthesis [11], and therefore the maximum allowed cosmic neutrino density.

Neutrino Factories, Reactors and Long-Baseline Experiments, θ_{13}: a non-zero value θ_{13} close to the current upper limit can lead to equalization at higher temperatures than that from the solar scale [9]. Also, for an inverted neutrino mass hierarchy, a very small but non-zero θ_{13} can lead to a resonance at $T \sim 5\,\text{MeV}$ that would also enhance equalization [11]. An appreciable θ_{13} or inverted hierarchy would further tighten the limits on the maximum cosmic neutrino density.

In summary, the intertwining of cosmic neutrino scattering, decoupling, weak beta-equilibrium freeze-out, and primordial nucleosynthesis with the mass and mixing scales for neutrino transformations in degenerate cosmic neutrino scenarios is exciting, particularly since the mass scales could have placed the transformations much higher or lower than the primordial nucleosynthesis scale. Therefore, the exact nature of the neutrino mass and mixing matrix, especially if it contains further surprises, will illuminate exactly what cosmic neutrino scenarios are plausible.

REFERENCES

1. D. N. Spergel *et al.*, Astrophys. J. Suppl. **148**, 175 (2003).
2. M. Tegmark *et al.* [SDSS Collaboration], in press, Astrophys. J., arXiv:astro-ph/0310725.
3. M. Tegmark *et al.* [SDSS Collaboration], arXiv:astro-ph/0310723.
4. P. Crotty, J. Lesgourgues and S. Pastor, Phys. Rev. D **67**, 123005 (2003); S. Hannestad, JCAP **0305**, 004 (2003); E. Pierpaoli, Mon. Not. Roy. Astron. Soc. **342**, L63 (2003).
5. R. V. Wagoner, W. A. Fowler and F. Hoyle, Astrophys. J. **148**, 3 (1967).
6. J. P. Kneller, R. J. Scherrer, G. Steigman and T. P. Walker, Phys. Rev. D **64**, 123506 (2001); S. H. Hansen, G. Mangano, A. Melchiorri, G. Miele and O. Pisanti, *ibid.* **65**, 023511 (2002); M. Orito, T. Kajino, G. J. Mathews and Y. Wang, *ibid.* **65**, 123504 (2002).
7. M. J. Savage, R. A. Malaney and G. M. Fuller, Astrophys. J. **368**, 1 (1991).
8. C. Lunardini and A. Y. Smirnov, Phys. Rev. D **64**, 073006 (2001).
9. A. D. Dolgov, S. H. Hansen, S. Pastor, S. T. Petcov, G. G. Raffelt, and D. V. Semikoz, Nucl. Phys. **B632**, 363 (2002).
10. Y. Y. Y. Wong, Phys. Rev. D **66**, 025015 (2002).
11. K. N. Abazajian, J. F. Beacom and N. F. Bell, Phys. Rev. D **66**, 013008 (2002).
12. S. Samuel, Phys. Rev. D **48**, 1462 (1993); V. A. Kostelecky and S. Samuel, *ibid.* **52**, 621 (1995); S. Samuel, *ibid.* **53**, 5382 (1996); J. Pantaleone, *ibid.* **58**, 073002 (1998).

13. S. Pastor, G. G. Raffelt, and D. V. Semikoz, Phys. Rev. D **65**, 053011 (2002).
14. R. J. Wilkes [Super-Kamiokande and K2K Collaborations], eConf **C020805**, TTH02 (2002) [arXiv:hep-ex/0212035].

Neutrino Masses in Theories with Dynamical Symmetry Breaking

Thomas Appelquist* and Robert Shrock[†]

*Physics Department, Sloane Laboratory
Yale University
New Haven, CT 06520
[†]C. N. Yang Institute for Theoretical Physics
State University of New York
Stony Brook, N. Y. 11794

Abstract. We address the problem of accounting for light neutrino masses in theories with dynamical electroweak symmetry breaking. We discuss this in the context of a class of extended technicolor (ETC) models. As a possible solution, we propose a combination of suppressed Dirac masses and a seesaw involving dynamically generated condensates of standard-model singlet, ETC-nonsinglet fermions. We show how this can be realized in an explicit ETC model. An important feature of this proposal is that, because of the suppression of Dirac neutrino mass terms, a seesaw yielding realistic neutrino masses does not require superheavy Majorana masses. We then generalize this mechanism to theories with extended strong-electroweak gauge symmetries.[1]

INTRODUCTION

A basic question arising from the recent evidence for neutrino masses and lepton mixing from solar, atmospheric, accelerator, and reactor data concerns the underlying explanation for these masses. Recently we have proposed a possible explanation in the context of theories with dynamical electroweak symmetry breaking [1, 2, 3], specifically extended technicolor (ETC) theories [2] Accounting for light neutrino masses has been a longstanding challenge for models with dynamical electroweak symmetry breaking, since these have no very large mass scale analogous to the grand unification scale M_{GUT} that enters in the seesaw mechanism yielding a Majorana mass $m_\nu \sim m_D^2/m_R$, where m_D is a Dirac mass and $m_R \sim M_{GUT}$ is the mass characterizing electroweak-singlet neutrinos.

Suppose that the technicolor gauge group is taken to be $SU(N_{TC})$. The set of technifermions includes, as a subset, one family, viz., $Q_L = \binom{U}{D}_L$, $L_L = \binom{N}{E}_L$, U_R, D_R, N_R, E_R transforming according to the fundamental representation of $SU(N_{TC})$ and the usual representations of $G_{SM} = SU(3) \times SU(2)_L \times U(1)_Y$ (color and TC indices are suppressed here). To satisfy constraints from flavor-changing neutral-current processes, the ETC vector bosons, which can mediate generation-changing transitions, must have

---

[1] Talk by R. Shrock at the NuFact03 Workshop, Columbia University, June, 2003.
[2] Recent reviews of ETC theories include R. S. Chivukula, hep-ph/0011264, K. Lane, hep-ph/0202255, and C. Hill and E. Simmons, hep-ph/0203079.

large masses. These arise from self-breaking of the (strongly coupled) ETC gauge symmetry, which occurs in stages. The $SU(N_{ETC})$ group has the related property that $N_{ETC} = N_{TC} + N_{gen.}$, where $N_{gen.} = 3$ denotes the number of SM fermion generations.

A particularly attractive choice for the technicolor group is $SU(2)_{TC}$, which has the appeal that it minimizes the TC contributions to the electroweak S parameter and can yield slow gauge coupling evolution over a certain energy range ("walking"), allowing for realistically large quark and charged lepton masses. With $N_{gen.} = 3$, the choice $N_{TC} = 2$ corresponds to $N_{ETC} = 5$. For $N_f \simeq 8$ vectorially coupled technifermions in the fundamental representation, studies suggest that this $SU(2)_{TC}$ theory could have an (approximate) infrared fixed point, which yields the walking behavior. The scale Λ_{TC} at which the TC coupling gets large, producing technifermion condensation is $\Lambda_{TC} \simeq 2f_F\sqrt{3/N_{TC}}$ and is fixed by the relation $m_W^2 = (g^2/4)(N_c f_Q^2 + f_L^2) \simeq (g^2/4)(N_c+1)f_F^2$, where f_F is the technicolor pseudoscalar decay constant. This gives $\Lambda_{TC} \sim 300$ GeV.

ETC MODEL

Our model is based on the gauge group $G = SU(5)_{ETC} \times SU(2)_{HC} \times G_{SM}$. An additional gauge interaction, $SU(2)_{HC}$ (HC = hypercolor), is introduced along with $SU(5)_{ETC}$ and G_{SM}. Both the $SU(2)_{HC}$ and $SU(5)_{ETC}$ interactions become strong, triggering a sequential breaking pattern. The fermion content of this model is listed below, where the numbers indicate the representations under $SU(5)_{ETC} \times SU(2)_{HC} \times SU(3)_c \times SU(2)_L$ and the subscript gives the weak hypercharge:

$$(5,1,3,2)_{1/3,L}, \quad (5,1,3,1)_{4/3,R}, \quad (5,1,3,1)_{-2/3,R}$$

$$(5,1,1,2)_{-1,L}, \quad (5,1,1,1)_{-2,R}, \quad (\overline{10},1,1,1)_{0,R}, \quad (10,2,1,1)_{0,R}. \quad (1)$$

Thus the fermions include quarks and techniquarks in the representations $(5,1,3,2)_{1/3,L}$, $(5,1,3,1)_{4/3,R}$, and $(5,1,3,1)_{-2/3,R}$, left-handed charged leptons and neutrinos and technileptons in $(5,1,1,2)_{-1,L}$, and right-handed charged leptons and technileptons in $(5,1,1,1)_{-2,R}$, together with SM-singlet fermions $\psi_{ij,R}$ and $\zeta_R^{ij,\alpha}$ transforming as $(\overline{10},1,1,1)_{0,R}$ and $(10,2,1,1)_{0,R}$. We additionally envision including fermions $\omega_{p,R}^\alpha$, $p = 1,2$ transforming as $(1,2,1,1)_{0,R}$.[3] Here, to fix the convention for the lepton number assigned to $\psi_{ij,R}$, we take it to be $L = 1$ in order that Dirac terms $\bar{n}_{i,L}\psi_{jk,R}$ conserve lepton number. The full model is a chiral gauge theory, while the $SU(2)_{HC}$ and $SU(2)_{TC}$ subsectors are vectorial. This model has some features in common with that of Ref. [4], but has a different gauge group and different fermion content.

To analyze the stages of symmetry breaking, we identify plausible preferred condensation channels using a generalized most-attractive-channel (GMAC) approach. We envision that as the energy decreases from high values down to

---

[3] For further related work, see T. Appelquist, M. Piai, and R. Shrock, hep-ph/0308061, submitted after NuFact03. We thank M. Piai for valuable discussions.

$E \sim \Lambda_1 \sim 10^3$ TeV, the coupling α_{ETC} is sufficiently large to produce condensation in the attractive channel $(\overline{10},1,1,1)_{0,R} \times (\overline{10},1,1,1)_{0,R} \to (5,1,1,1)_0$, breaking $SU(5)_{ETC} \to SU(4)_{ETC}$. With respect to the unbroken $SU(4)_{ETC}$, we have $(\overline{10},1,1,1)_{0,R} = (\bar{4},1,1,1)_{0,R} + (\bar{6},1,1,1)_{0,R}$; we denote the $(\bar{4},1,1,1)_{0,R}$ as $\alpha_{1i,R} \equiv \psi_{1i,R}$ for $2 \leq i \leq 5$ and the $(\bar{6},1,1,1)_{0,R}$ as $\xi_{ij,R} \equiv \psi_{ij,R}$ for $2 \leq i,j \leq 5$. The associated condensate is then

$$\langle \varepsilon^{lijk\ell} \xi_{ij,R}^T C \xi_{k\ell,R} \rangle = 8 \langle \xi_{23,R}^T C \xi_{45,R} - \xi_{24,R}^T C \xi_{35,R} + \xi_{25,R}^T C \xi_{34,R} \rangle . \quad (2)$$

This condensate and the resultant dynamical Majorana mass terms for the six components of ξ in eq. (2) produce a violation of total lepton number by $|\Delta L| = 2$ units.

At lower scales, depending on relative strengths of couplings, different symmetry-breaking sequences can occur. We focus on one here, in which at a scale $\Lambda_{BHC} \lesssim \Lambda_1$ (BHC = broken HC), the $SU(4)_{ETC}$ interaction produces a condensation in the channel $(6,2,1,1)_{0,R} \times (6,2,1,1)_{0,R} \to (1,3,1,1)_0$ with condensate $\langle \varepsilon_{1ijk\ell} \zeta_R^{ij,1\,T} C \zeta_R^{k\ell,2} \rangle + (1 \leftrightarrow 2)$. This breaks $SU(2)_{HC} \to U(1)_{HC}$ and gives dynamical masses $\sim \Lambda_{BHC}$ to the twelve $\zeta_R^{ij,\alpha}$ fields involved. We let $\alpha = 1,2$ correspond to $Q_{HC} = \pm 1$ under the $U(1)_{HC}$. This gives dynamical masses $\sim \Lambda_{BHC}$ to the twelve $\zeta_R^{ij,\alpha}$ fields involved. At the lower scale Λ_{23} there is condensation in the $4 \times 4 \to 6$ channel with condensate $\langle \varepsilon_{\alpha\beta} \zeta_R^{12,\alpha\,T} C \zeta_R^{13,\beta} \rangle$, which then breaks $SU(4)_{ETC} \to SU(2)_{ETC}$ and is $U(1)_{HC}$-invariant. The $U(1)_{HC}$ interaction naturally also forms the condensates $\langle \zeta_R^{12,\alpha\,T} C \omega_{p,R}^\beta \rangle$ and $\langle \zeta_R^{13,\alpha\,T} C \omega_{p,R}^\beta \rangle$ with $p = 1,2$, $\alpha \neq \beta$. Finally, technifermion condensation occurs at Λ_{TC}.

CALCULATIONS AND RESULTS

We calculate the neutrino masses and mixing by diagonalizing the full mass matrix of neutrino-like states. The nonzero entries in this matrix arise in two different ways: (i) directly, as dynamical masses associated with various condensates, and (ii) via loop diagrams involving dynamical mass insertions on internal fermion lines. In the effective low energy theory at scales below Λ_{TC}, the light neutrinos include the three electroweak-doublet neutrinos $\nu_{\ell,L}$ with $\ell = e,\mu,\tau$, arising from a 5 of $SU(5)_{ETC}$, and two electroweak-singlet neutrinos, $\alpha_{1j,R}$, $j = 2,3$ arising from a $\overline{10}$ of $SU(5)_{ETC}$. The fact that these transform as different representations leads to a strong suppression of Dirac neutrino mass terms, as in Ref. [4], because the relevant entries require mixing of ETC gauge bosons on internal lines of these loop diagrams. The relevant Dirac neutrino mass matrix is denoted b_{ij}, with $i = 1,2,3$ and $j = 2,3$. The dynamically generated technineutrino mass terms rely on the fact that we use $SU(2)_{TC}$ so that 2 and $\bar{2}$ representations are equivalent. For the given symmetry breaking sequence we find

$$|b_{23}| = |b_{32}| \propto \frac{\Lambda_{TC}^4}{\Lambda_{23}^3} \quad (3)$$

with comparable values for b_{22} and b_{33}, and much smaller values for b_{1j}, $j = 2, 3$. The relevant Majorana mass matrix r_{ij} is defined by the operator product $\alpha_{1i,R}^T C r_{ij} \alpha_{1j,R}$, $i, j = 2, 3$. We find $r_{22} = r_{33} = 0$ and $r_{23} \propto \Lambda_{BHC}^n \Lambda_{23}^n / \Lambda_1^{2n-1}$ where n has the respective values 2 and 3 in a theory with walking all the way up to Λ_1 or just up to Λ_{23}. The Majorana mass r_{23} also requires ETC gauge boson mixing and is suppressed relative to the largest ETC scale.

We find that the electroweak-doublet neutrinos are, to a very good approximation, linear combinations of three mass eigenstates having a normal hierarchy, with

$$m(\nu_j) \propto \frac{(b_{23} \pm b_{22})^2}{r_{23}} \qquad (4)$$

where $j = 3, 2$ correspond to the $+, -$. Hence, for the respective cases of a theory with walking up to Λ_1 and only up to Λ_{23}, we have $m(\nu_3) \propto (1+y)^2 \Lambda_{TC}^8 \Lambda_1^3 / (\Lambda_{23}^8 \Lambda_{BHC}^2)$ and $m(\nu_3) \propto (1+y)^2 \Lambda_{TC}^8 \Lambda_1^5 / (\Lambda_{23}^9 \Lambda_{BHC}^3)$, where $y = b_{22}/b_{23}$. In both cases, for fixed Λ_1 and Λ_{TC}, one can envision models with scales Λ_{BHC} and Λ_{23} that yield $m(\nu_3) \simeq 0.005$ eV, in agreement with the value extracted from the experimental measurement of $|\Delta m_{atm}^2|$ assuming a hierarchical neutrino mass spectrum. The prediction $m(\nu_2)/m(\nu_3) = (b_{23} - b_{22})^2/(b_{23} + b_{22})^2$ can account for the ratio of $m(\nu_2)/m(\nu_3) \sim 0.2$ extracted from the data on $|\Delta m_{atm}^2|$ and Δm_{sol}^2. The model naturally yields a large ν_μ, ν_τ mixing angle θ_{23}.

We have also shown [2, 3] how our mechanism for explaining light neutrinos can be realized in theories with the extended strong-electroweak gauge groups $G_{LR} = SU(3)_c \times SU(2)_L \times SU(2)_R \times U(1)_{B-L}$, where B and L denote baryon and lepton number, and $G_{422} = SU(4)_{PS} \times SU(2)_L \times SU(2)_R$, where the Pati-Salam group $SU(4)_{PS}$ unifies $U(1)_{B-L}$ with $SU(3)_c$.

CONCLUSIONS

We have proposed a mechanism that explains light neutrino masses in theories with dynamical electroweak symmetry breaking and have shown how it is realized in an explicit class of ETC models. This mechanism involves a seesaw, but one with no GUT-type mass scales.

This research was partially supported by the grants DE-FG02-92ER-4074 (T.A.) and NSF-PHY-00-98527 (R.S.).

REFERENCES

1. Appelquist, T., and Shrock, R., *Phys. Lett. B*, **548**, 204 (2002).
2. Appelquist, T., and Shrock, R., "Neutrino Masses in Theories with Dynamical Breaking of Electroweak and Extended Gauge Symmetries," in *Proceedings of SCGT02, Int'l Workshop on Strongly Coupled Gauge Theories, Nagoya*, edited by K. Yamawaki, 2003.
3. Appelquist, T., and Shrock, R., *Phys. Rev. Lett.*, **90**, 201801 (2003).
4. Appelquist, T., and Terning, J., *Phys. Rev. D*, **50**, 2116 (1994).

Extrinsic CPT Violation in Neutrino Oscillations

Tommy Ohlsson[1]

*Division of Mathematical Physics, Department of Physics, Royal Institute of Technology (KTH) –
Stockholm Center for Physics, Astronomy, and Biotechnology (SCFAB), Roslagstullsbacken 11,
SE-106 91 Stockholm, Sweden*

Abstract. In this talk, we investigate extrinsic CPT violation in neutrino oscillations in matter with three flavors. Note that extrinsic CPT violation is different from intrinsic CPT violation. Extrinsic CPT violation is one way of quantifying matter effects, whereas intrinsic CPT violation would mean that the CPT invariance theorem is not valid. We present analytical formulas for the extrinsic CPT probability differences and discuss their implications for long-baseline experiments and neutrino factory setups.

Introduction. Recently, there have been several studies on CPT violation in order to incorporate the results of the LSND experiment [1], which require a third mass squared difference. However, this is not compatible with three neutrino flavors. Therefore, in most of the phenomenological studies on CPT violation, different mass squared differences and mixing parameters for neutrinos and antineutrinos are introduced by hand leading to four mass squared differences and eight mixing parameters. Thus, it is possible to include the results of the LSND experiment. Note that the results of the LSND experiment will be tested by the MiniBooNE experiment (September 2002 → ~ 2005) [2]. Furthermore, note that another possible description of the results of the LSND experiment are sterile neutrinos. However, sterile neutrinos have, in principle, been excluded by the SNO experiment [3]. Moreover, the first KamLAND data are consistent with the LMA solution [4], which means that there is no need for fundamental CPT violation.

Eccentric or extrinsic CPT violation? Let us denote the neutrino oscillation transition probabilities by $P_{\alpha\beta} \equiv P(\nu_\alpha \to \nu_\beta)$. Then, the CP, T, and CPT probability differences (pds) are defined as $\Delta P^{\rm CP}_{\alpha\beta} \equiv P_{\alpha\beta} - P_{\bar\alpha\bar\beta}$, $\Delta P^{\rm T}_{\alpha\beta} \equiv P_{\alpha\beta} - P_{\beta\alpha}$, and $\Delta P^{\rm CPT}_{\alpha\beta} \equiv P_{\alpha\beta} - P_{\bar\beta\bar\alpha}$.

Now, intrinsic (eccentric) CPT violation (or fundamental or genuine CPT violation) is due to violation of the CPT invariance theorem, whereas extrinsic CPT violation (or matter-induced or fake CPT violation) is due to presence of ordinary matter. Here, we will assume that the CPT invariance theorem is valid. This implies for the CP and T pds that the intrinsic and extrinsic effects are mixed, whereas for the CPT pds there are extrinsic effects only. Therefore, non-zero (extrinsic) CPT pds show matter effects, and thus, they are one way of quantifying such effects.

From conservation of probability, we obtain $\sum_{\alpha=e,\mu,\tau,...} \Delta P^{\rm CPT}_{\alpha\beta} = \sum_{\beta=e,\mu,\tau,...} \Delta P^{\rm CPT}_{\alpha\beta} = 0$.

---

[1] In collaboration with: Magnus Jacobson.

Note that not all of these equations are linearly independent. For three neutrino flavors, we have nine CPT pds for neutrinos. However, only four are linearly independent. Choosing, e.g., $\Delta P^{\text{CPT}}_{ee}$, $\Delta P^{\text{CPT}}_{e\mu}$, $\Delta P^{\text{CPT}}_{\mu e}$, and $\Delta P^{\text{CPT}}_{\mu\mu}$ as the known ones, the other five can be expressed in terms of these. Furthermore, we have $\Delta P^{\text{CPT}}_{\alpha\beta} = -\Delta P^{\text{CPT}}_{\bar\beta\bar\alpha}$, i.e., the CPT pds for antineutrinos do not give any further information.

The CPT probability differences. In vacuum, the CPT pds are $\Delta P^{\text{CPT}}_{\alpha\beta} = 0$, $\alpha,\beta = e,\mu,\tau$, whereas, in matter, they are given by $\Delta P^{\text{CPT}}_{\alpha\beta} = |[S_f(t,t_0)]_{\beta\alpha}|^2 - |[\bar S_f(t,t_0)]_{\alpha\beta}|^2$, where $S_f \equiv S_f(t,t_0)$ and $\bar S_f \equiv \bar S_f(t,t_0)$ are the evolution operators for neutrinos and antineutrinos, respectively. We have calculated S_f and $\bar S_f$ explicitly using first order perturbation theory in the small leptonic mixing angle θ_{13}. These explicit expressions for S_f and $\bar S_f$ can be found in Ref. [5].

Two of the CPT pds (with an arbitrary matter density profile) are: $\Delta P^{\text{CPT}}_{ee} \simeq |\bar\beta|^2 - |\beta|^2$ and $\Delta P^{\text{CPT}}_{e\mu} \simeq c_{23}^2 \left(|\beta|^2 - |\bar\beta|^2\right) - 2 c_{23} s_{23} \Im\left(\beta f C - \bar\beta \bar f^* \bar A^*\right)$, where β and $\bar\beta$ describe a part of the two flavor neutrino evolution in the (1,2)-subsector, f and $\bar f$ are some functions, and $\bar A$ and C are complicated functions that can be found in Ref. [5].

In matter of constant density in the low-energy region ($V \lesssim \delta \ll \Delta$), the CPT pds $\Delta P^{\text{CPT}}_{ee}$ and $\Delta P^{\text{CPT}}_{\mu e}$ are calculated to be [5]

$$\Delta P^{\text{CPT}}_{ee} \simeq 8 s_{12}^2 c_{12}^2 \cos 2\theta_{12} \left(\delta L \cos \frac{\delta L}{2} - 2 \sin \frac{\delta L}{2}\right) \sin \frac{\delta L}{2} \frac{V}{\delta} + \mathcal{O}\left((V/\delta)^3\right), \quad (1)$$

$$\begin{aligned}\Delta P^{\text{CPT}}_{\mu e} \simeq\ & -8 s_{12}^2 c_{12}^2 c_{23}^2 \cos 2\theta_{12} \left(\delta L \cos \frac{\delta L}{2} - 2 \sin \frac{\delta L}{2}\right) \sin \frac{\delta L}{2} \frac{V}{\delta} \\ & - 16 s_{12} c_{12}^3 s_{13} s_{23} c_{23} \cos\delta_{\text{CP}} \cos 2\theta_{12} \left(\delta L \cos \frac{\delta L}{2} - 2 \sin \frac{\delta L}{2}\right) \sin \frac{\delta L}{2} \frac{V}{\delta} \\ & + 16 s_{12} c_{12} s_{13} s_{23} c_{23} \sin\delta_{\text{CP}} \Bigg\{\cos 2\theta_{12} \left[\delta L \cos\delta L - \cos\Delta L\right. \\ & \times \left.\left(\delta L \cos \frac{\delta L}{2} - 2 \sin \frac{\delta L}{2}\right) - \sin\delta L\right] + \delta L \sin \frac{\delta L}{2} \sin\Delta L \Bigg\} \frac{V}{\delta} + \mathcal{O}\left((V/\delta)^3\right), \quad (2)\end{aligned}$$

where $\delta \equiv \frac{\Delta m^2_{21}}{2E_\nu}$, $\Delta \equiv \frac{\Delta m^2_{31}}{2E_\nu}$, E_ν is the neutrino energy, L is the baseline length, and V is the matter potential. Note that if one makes the replacement $\delta_{\text{CP}} \to -\delta_{\text{CP}}$, then $\Delta P^{\text{CPT}}_{e\mu} \to \Delta P^{\text{CPT}}_{\mu e}$ and $\Delta P^{\text{CPT}}_{e\tau} \to \Delta P^{\text{CPT}}_{\tau e}$ and, in the case that $\delta_{\text{CP}} = 0$, one has $\Delta P^{\text{CPT}}_{e\mu} = \Delta P^{\text{CPT}}_{\mu e}$ and $\Delta P^{\text{CPT}}_{e\tau} = \Delta P^{\text{CPT}}_{\tau e}$. We observe also that in $\Delta P^{\text{CPT}}_{ee}$ there are no δ_{CP} terms, whereas in $\Delta P^{\text{CPT}}_{\mu e}$ there are both $\sin\delta_{\text{CP}}$ and $\cos\delta_{\text{CP}}$ terms. Therefore, it would be possible to extract δ_{CP} from $\Delta P^{\text{CPT}}_{\mu e}$. Actually, for symmetric matter density profiles it can be shown that the $\Delta P^{\text{CPT}}_{\alpha\beta}$'s are always odd functions with respect to the matter potential V [6, 5].

In the case of a step-function matter density profile in the low-energy region ($V_{1,2} \lesssim \delta \ll \Delta$), $\Delta P^{\text{CPT}}_{ee}$ is found to be [5]

$$\Delta P^{\text{CPT}}_{ee} \simeq 8 s_{12}^2 c_{12}^2 \cos 2\theta_{12} \left[\delta\left(L_1 \frac{V_1}{\delta} + L_2 \frac{V_2}{\delta}\right) \cos \frac{\delta(L_1+L_2)}{2}\right.$$

TABLE 1. Extrinsic CPT pds for some past, present, and fututre long-baseline experiments.

Experiment	CPT probability differences		Experiment	CPT probability differences	
	Quantities	Numerical value		Quantities	Numerical value
BNL NWG	$\Delta P^{CPT}_{\mu e}$	0.010	KamLAND	ΔP^{CPT}_{ee}	-0.033
BNL NWG	$\Delta P^{CPT}_{\mu e}$	0.032	LSND	$\Delta P^{CPT}_{\mu e}$	$4.8 \cdot 10^{-15}$
BooNE	$\Delta P^{CPT}_{\mu e}$	$6.6 \cdot 10^{-13}$	MINOS	$\Delta P^{CPT}_{\mu e}$	$1.9 \cdot 10^{-4}$
MiniBooNE	$\Delta P^{CPT}_{\mu e}$	$4.1 \cdot 10^{-14}$		$\Delta P^{CPT}_{\mu\mu}$	$-1.1 \cdot 10^{-5}$
CHOOZ	ΔP^{CPT}_{ee}	$-3.6 \cdot 10^{-5}$	NuMI I	$\Delta P^{CPT}_{\mu e}$	0.026
ICARUS	$\Delta P^{CPT}_{\mu e}$	$4.0 \cdot 10^{-5}$	NuMI II	$\Delta P^{CPT}_{\mu e}$	$2.6 \cdot 10^{-3}$
	$\Delta P^{CPT}_{\mu\tau}$	$-3.8 \cdot 10^{-5}$	NuTeV	$\Delta P^{CPT}_{\mu e}$	$1.6 \cdot 10^{-18}$
JHF-Kamioka	$\Delta P^{CPT}_{\mu e}$	$3.8 \cdot 10^{-3}$	NuTeV	$\Delta P^{CPT}_{\mu e}$	$8.2 \cdot 10^{-20}$
	$\Delta P^{CPT}_{\mu\mu}$	$-1.3 \cdot 10^{-4}$	OPERA	$\Delta P^{CPT}_{\mu\tau}$	$-3.8 \cdot 10^{-5}$
K2K	$\Delta P^{CPT}_{\mu e}$	$1.0 \cdot 10^{-3}$	Palo Verde	ΔP^{CPT}_{ee}	$-1.2 \cdot 10^{-5}$
	$\Delta P^{CPT}_{\mu\mu}$	$-5.3 \cdot 10^{-5}$	Palo Verde	ΔP^{CPT}_{ee}	$-2.2 \cdot 10^{-5}$

$$- 2\left(\frac{V_1}{\delta}\sin\frac{\delta L_1}{2}\cos\frac{\delta L_2}{2} + \frac{V_2}{\delta}\sin\frac{\delta L_2}{2}\cos\frac{\delta L_1}{2}\right)\right]\sin\frac{\delta(L_1+L_2)}{2}. \quad (3)$$

Note that this formula is completely symmetric with respect to interchange of layers 1 and 2 and, in the limit $V_{1,2} \to V$ and $L_{1,2} \to L/2$, one has ΔP^{CPT}_{ee}(step-function) $\to \Delta P^{CPT}_{ee}$(constant).

Similarly, in the case of the T probability difference in the low-energy region ($\delta = \Delta m^2_{21}/(2E_\nu) \gtrsim V_{1,2}$), one finds [7]

$$\Delta P^T_{\alpha\beta} \simeq \cos\delta_{CP} \cdot \underbrace{8 s_{12} c_{12} s_{13} s_{23} c_{23} \frac{\sin(2\theta_1 - 2\theta_2)}{\sin 2\theta_{12}}}_{J_{\text{eff}}} \{s_1 s_2 [Y - \cos(\Delta_1 L_1 + \Delta_2 L_2)]\}$$
$$+ \sin\delta_{CP} \cdot 4 s_{13} s_{23} c_{23} X_1 [Y - \cos(\Delta_1 L_1 + \Delta_2 L_2)], \quad (4)$$

where J_{eff} is an effective Jarlskog invariant. (See Ref. [7] for the definitions of the different quantities in this formula.) Here the $\cos\delta_{CP}$ term is due to matter-induced T violation, whereas the usual $\sin\delta_{CP}$ term is due to fundamental T violation.

Numerical calculations and implications. Using the present best-fit values of the fundamental neutrino parameters, $\Delta m^2_{21} \simeq 7.1 \cdot 10^{-5}\,\text{eV}^2$, $|\Delta m^2_{31}| \simeq 2.5 \cdot 10^{-3}\,\text{eV}^2$, $\theta_{12} \simeq 34°$, $\theta_{23} \simeq 45°$, and, in addition, choosing a normal mass hierarchy spectrum [sgn(Δm^2_{31}) = 1], $\theta_{13} = 9.2°$, and $\delta_{CP} = 0$, we have calculated the CPT pds for some past, present, and future long-baseline experiments. The results are presented in Table 1. Note that ΔP^{CPT}_{ee} for the KamLAND experiment is $|\Delta P^{CPT}_{ee}| \sim 3$ %, which means that extrinsic CPT violation is non-negligible for this experiment. The problem is just how one should obtain P_{ee} for the same neutrino energy and baseline length. Furthermore, we have calculated $\Delta P^{CPT}_{\mu e}$ for two neutrino factory setups, using the following parameter values: $\rho = \rho_{\text{mantle}} \simeq 4.5\,\text{g/cm}^3$, $E_\nu = 50\,\text{GeV}$, $L \in \{3\,000, 7\,000\}\,\text{km}$. Then, $L = 3\,000\,\text{km}$ leads to $\Delta P^{CPT}_{\mu e} \simeq 3.0 \cdot 10^{-5}$, whereas $L = 7\,000\,\text{km}$ leads to $\Delta P^{CPT}_{\mu e} \simeq 1.8 \cdot 10^{-5}$. Therefore, extrinsic CPT violation is practically negligible for a future neutrino factory. In Fig. 1, ΔP^{CPT}_{ee} and $\Delta P^{CPT}_{\mu e}$ are shown plotted as functions of

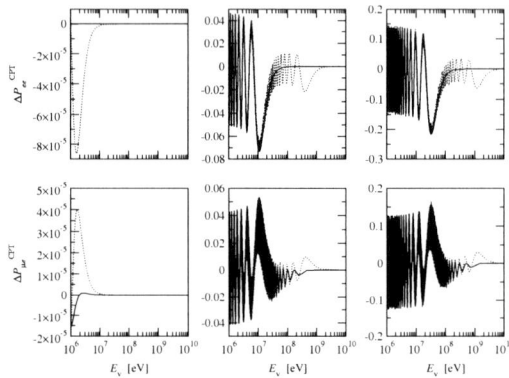

FIGURE 1. The CPT pds $\Delta P_{ee}^{\text{CPT}}$ and $\Delta P_{\mu e}^{\text{CPT}}$ as functions of the neutrino energy E_ν for baseline lengths $L \in \{1, 250, 750\}$ km. The solid curves show analytical results, whereas the dotted curves show numerical results. The fast oscillations present in the numerical results are averaged out in the analytical calculations.

E_ν for $L \in \{1, 250, 750\}$ km. We note that an increasing L implies an increasing values of the $\Delta P_{\alpha\beta}^{\text{CPT}}$'s and when $E_\nu \to \infty$ we observe that $\Delta P_{\alpha\beta}^{\text{CPT}} \to 0$.

Summary & Conclusions. In conclusion, we have studied extrinsic CPT violation in three flavor neutrino oscillations assuming the CPT invariance theorem. In general, the (extrinsic) CPT pds for an arbitrary matter density profile have been derived. In particular, first order perturbation theory formulas for constant and step-function matter density profiles have been calculated as well as low-energy approximations. Furthermore, implications for accelerator and reactor long-baseline experiments as well as neutrino factory setups have been presented. For certain experiments the CPT pds can be as large as $|\Delta P_{\alpha\beta}^{\text{CPT}}| \sim 5 \%$. In general, the CPT pds increase with increasing baseline length and decrease with increasing neutrino energy.

Acknowledgments. I would like to thank S.M. Bilenky, M. Jacobson, R. Johansson, M. Lindner, H. Minakata, G. Seidl, H. Snellman, and W. Winter for useful discussions and comments. This work was supported by the Swedish Research Council (Vetenskapsrådet), Contract No. 621-2001-1611, 621-2002-3577, the Magnus Bergvall Foundation (Magn. Bergvalls Stiftelse), and the Wenner-Gren Foundations.

REFERENCES

1. Athanassopoulos, C., et al., *Phys. Rev. Lett.*, **77**, 3082–3085 (1996); **81**, 1774–1777 (1998); Aguilar, A., et al., *Phys. Rev.*, **D64**, 112007 (2001).
2. http://www-boone.fnal.gov/.
3. Bahcall, J. N., Gonzalez-Garcia, M. C., and Peña-Garay, C., *J. High Energy Phys.*, **07**, 054 (2002).
4. Bahcall, J. N., Gonzalez-Garcia, M. C., and Peña-Garay, C., *J. High Energy Phys.*, **02**, 009 (2003).
5. Jacobson, M., and Ohlsson, T., hep-ph/0305064.
6. Minakata, H. (2003), private communication.
7. Akhmedov, E. K., Huber, P., Lindner, M., and Ohlsson, T., *Nucl. Phys.*, **B608**, 394–422 (2001).

An Overview of Neutrino Masses and Mixing in SO(10) Models

Mu-Chun Chen* and K.T. Mahanthappa†

*HET Group, Physics Department, Brookhaven National Laboratory, Upton, NY 11973
†Department of Physics, University of Colorado, Boulder, CO 80309

Abstract. We review in this talk various SUSY SO(10) models. Specifically, we discuss how small neutrino masses are generated in and generic predictions of different $SO(10)$ models. A comparison of the predictions of these models for $\sin^2 \theta_{13}$ is given.

The flavor problem with hierarchical fermion masses and mixing has attracted a great deal of attention especially since the advent of the atmospheric neutrino oscillation data from Super-Kamiokande indicating non-zero neutrino masses. The non-zero neutrino masses give support to the idea of grand unification based on $SO(10)$ in which all the 16 fermions (including the right-handed neutrinos) can be accommodated in one single spinor representation. Furthermore, it provides a framework in which seesaw mechanism arises naturally. Models based on $SO(10)$ combined with a continuous or discrete flavor symmetry group have been constructed to understand the flavor problem, especially the small neutrino masses and the large leptonic mixing angles. These models can be classified according to the family symmetry implemented in the model. We review in this talk how small masses and large mixing angles in the neutrino sector are generated in SO(10) models, and the unique predictions of each class of models. We also discuss other mechanisms that have been proposed to solve the problem of neutrino masses and mixing. For a more exhaustive list of references and detailed discussion, we refer the readers to our recent review[1] on which this talk is based.

Symmetric textures: This type of models have been considered, for example, in Ref.[2-4]. SO(10) breaks down through the left-right symmetry breaking chain, which ensures the mass matrices are symmetric. The Higgs content of this type of models contains fields in 10, 45, 54, 126 representations, with 10, 126 breaking EW symmetry and generating fermions masses, and 45, 54, 126 breaking SO(10). The mass hierarchy can arise if there is an $SU(2)_H$ symmetry acting non-trivially on the first two generations such that the first two generations transform as a doublet and the third generation transforms as a singlet under $SU(2)_H$, which breaks down at two steps, $SU(2) \overset{\varepsilon M}{\to} U(1) \overset{\varepsilon' M}{\to} nothing$ where $\varepsilon' \ll \varepsilon \ll 1$. The mass hierarchy is generated by the Froggatt-Nielsen mechanism which requires the flavon fields acquiring VEV's along the directions specified in Ref.[2-4]. The resulting mass matrices at the GUT scale are given by

$$M_{u,\nu_{LR}} = \begin{pmatrix} 0 & 0 & \langle 10_2^+ \rangle \varepsilon' \\ 0 & \langle 10_4^+ \rangle \varepsilon & \langle 10_3^+ \rangle \varepsilon \\ \langle 10_2^+ \rangle \varepsilon' & \langle 10_3^+ \rangle \varepsilon & \langle 10_1^+ \rangle \end{pmatrix} = \begin{pmatrix} 0 & 0 & r_2 \varepsilon' \\ 0 & r_4 \varepsilon & \varepsilon \\ r_2 \varepsilon' & \varepsilon & 1 \end{pmatrix} M_U \quad (1)$$

$$M_{d,e} = \begin{pmatrix} 0 & \langle 10_5^-\rangle \varepsilon' & 0 \\ \langle 10_5^-\rangle \varepsilon' & (1,-3)\langle \overline{126}^-\rangle \varepsilon & 0 \\ 0 & 0 & \langle 10_1^-\rangle \end{pmatrix} = \begin{pmatrix} 0 & \varepsilon' & 0 \\ \varepsilon' & (1,-3)p\varepsilon & 0 \\ 0 & 0 & 1 \end{pmatrix} M_D \qquad (2)$$

The right-handed neutrino mass matrix is of the same form as $M_{\nu_{LR}}$

$$M_{\nu_{RR}} = \begin{pmatrix} 0 & 0 & \langle \overline{126}_2'^0\rangle \delta_1 \\ 0 & \langle \overline{126}_2'^0\rangle \delta_2 & \langle \overline{126}_2'^0\rangle \delta_3 \\ \langle \overline{126}_2'^0\rangle \delta_1 & \langle \overline{126}_2'^0\rangle \delta_3 & \langle \overline{126}_1'^0\rangle \end{pmatrix} = \begin{pmatrix} 0 & 0 & \delta_1 \\ 0 & \delta_2 & \delta_3 \\ \delta_1 & \delta_3 & 1 \end{pmatrix} M_R \qquad (3)$$

Note that, since we use $\overline{126}$-dimensional representations of Higgses to generate the heavy Majorana neutrino mass terms, R-parity is preserved at all energies. The effective neutrino mass matrix is

$$M_{\nu_{LL}} = M_{\nu_{LR}}^T M_{\nu_{RR}}^{-1} M_{\nu_{LR}} = \begin{pmatrix} 0 & 0 & t \\ 0 & 1 & 1+t^n \\ t & 1+t^n & 1 \end{pmatrix} \frac{d^2 v_u^2}{M_R} \qquad (4)$$

giving rise to maximal mixing angle for the atmospheric neutrinos and LMA solution for the solar neutrinos. The form of the neutrino mass matrix in this model is invariant under the seesaws mechanism. The value of U_{e3} is predicted to be large, close to the sensitivity of current experiments. This is a consequence of the solar angle being large. The prediction for the rate of $\mu \to e\gamma$ is about two orders of magnitude below the current experimental bound.

Lopsided/Asymmetric textures: This type of models have been considered, for example, in Ref.[5-12]. In this case, $SO(10)$ breaks down to SM through the $SU(5)$ breaking chain. The Higgs sector of the model contains $10, 16, 45, 54$, with $<16_{H_1}>$ breaking $SO(10)$ down to $SU(5)$ and $<16_{H_2}>$ breaking the EW symmetry. The lopsided textures arise due to the operator $\lambda(16_i 16_{H_1})(16_j 16_{H_2})$ which gives rise to mass terms for the charged leptons and down quarks, satisfying the SU(5) relation $M_d = M_e^T$. When other operators are included, the lopsided structure of M_e results, provided the coupling λ is of order 1,

$$M_{u,\nu_{LR}} = \begin{pmatrix} \eta & 0 & 0 \\ 0 & 0 & (1/3,1)\varepsilon \\ 0 & -(1/3,1)\varepsilon & 1 \end{pmatrix} \cdot m_u \qquad (5)$$

$$M_d = \begin{pmatrix} \eta & \delta & \delta' e^{i\phi} \\ \delta & 0 & \lambda+\varepsilon/3 \\ \delta' e^{i\phi} & -\varepsilon/3 & 1 \end{pmatrix} \cdot m_d, \quad M_e = \begin{pmatrix} \eta & \delta & \delta' e^{i\phi} \\ \delta & 0 & -\varepsilon \\ \delta' e^{i\phi} & \lambda+\varepsilon & 1 \end{pmatrix} \cdot m_d. \qquad (6)$$

The large mixing in $U_{e,L}$ contributes to large leptonic mixing matrix, leading to the large atmospheric mixing angle. Meanwhile, because large mixing in $U_{e,L}$ corresponds to large mixing in $U_{d,R}$, the CKM mixing angles can be retained small. An unique prediction of the lop-sided models is the large branching ratio for LFV processes, e.g. $\mu \to e\gamma$. By considering neutrino RH Majorana mass term of the following form, large solar mixing angle can arise for some choice of the parameters in $M_{\nu_{RR}}$, leading to LMA solution for solar neutrinos,

$$M_{\nu_{RR}} = \begin{pmatrix} c^2\eta^2 & -b\varepsilon\eta & a\eta \\ -b\varepsilon\eta & \varepsilon^2 & -\varepsilon \\ a\eta & -\varepsilon & 1 \end{pmatrix} \cdot \Lambda_R, \quad M_\nu^{eff} = \begin{pmatrix} 0 & -\varepsilon & 0 \\ -\varepsilon & 0 & 2\varepsilon \\ 0 & 2\varepsilon & 1 \end{pmatrix} m_u^2/\lambda_R. \quad (7)$$

The value for $|U_{e3}|$ is predicted to be small.

Large mixing from RGE's: This class of models have been considered, for example, in [13]. The RG evolution of the effective neutrino mass matrix is given by, $dm_\nu/dt = -\{\kappa_u m_\nu + m_\nu P + P^T m_\nu\}$ where $P \simeq -\frac{1}{32\pi^2} \frac{h_t^2}{\cos^2\beta} diag(0,0,1)$, $\kappa_u \simeq \frac{1}{16\pi^2}[\frac{6}{5}g_1^2 + 6g_2^2 - 6\frac{h_t^2}{\sin^2\beta}]$. If one assumes nearly degenerate mass pattern and same Majorana CP phases at the GUT scale, the parameters s_{12} and s_{23} are driven to be large, while corrections to m_i are small. Assuming leptonic mixing matrix is identical to the CKM matrix at the GUT scale. Starting with $s_{12}^0 \simeq \lambda$, $s_{23}^0 \simeq \mathcal{O}(\lambda^2)$, and $s_{13}^0 \simeq \mathcal{O}(\lambda^3)$, one obtains $\sin^2 2\theta_{atm} = 0.99$, $\sin^2 2\theta_\odot = 0.87$, $\sin\theta_{13} = 0.08$ at the weak scale. These GUT scale conditions can be understood in Type II seesaw mechanism, with $M_{\nu_{LL}}$ term (due to the coupling to an SU(2) triplet Higgs) dominates: because $M_{\nu_{LL}} \sim I \cdot m_{\nu_{LL}}$ dominates, one obtains the nearly degenerate mass spectrum; as the flavor mixing is due to the usual seesaw term, the mixing angle of the resulting mass matrix is CKM-like. Thus the two conditions needed for enhancing the mixing angles are satisfied. The U_{e3} element is also amplified by the RG flow, with the low energy prediction close to the sensitivity of current experiments.

Large mixing from $b - \tau$ unification: This class of models have been considered, for example, in Ref.[14, 15]. It has a minimal Higgs sector which contains $\{10, 126, 45, 54\}$. The following mass relations arise due to SO(10) symmetry, $M_u = f<10> + h<\overline{126}>, M_d = f<10> + h<\overline{126}>, M_e = f<10> - 3h<\overline{126}>, M_{\nu_{LR}} = f<10> - 3h<\overline{126}>$. As there are only one 10 and one 126 Higgs representations, all mass terms are governed by two Yukawa matrices, f and h. The small neutrino masses are explained by the Type II see-saw mechanism with the assumption that the LH Majorana mass term dominates over the usual Type I see-saw term, $M_\nu^{eff} = M_{\nu_{LL}} - M_{\nu_{LR}} M_{\nu_{RR}}^{-1} M_{\nu_{LR}}^T$. The mass terms $M_{\nu_{LL}}$ and $M_{\nu_{RR}}$ are both due to the coupling to $\overline{126}$, leading to $M_{\nu_{LL}} \sim h v_{ew}^2/v_R$ and $M_{\nu_{RR}} \sim h v_R$. In this minimal scheme, we have the following sum-rule $M_\nu^{eff} = c(M_d - M_e)$. The down-type quark and charged lepton mass matrices can be parameterized in terms of Wolfenstein parameter as

$$M_{b,\tau} \sim \begin{pmatrix} \lambda^3 & \lambda^3 & \lambda^3 \\ \lambda^3 & \lambda^2 & \lambda^2 \\ \lambda^3 & \lambda^2 & 1 \end{pmatrix} m_{b,\tau} \quad (8)$$

For some value of $\tan\beta$ (small values are preferred), the deviation from $b - \tau$ unification at the GUT scale is $m_b(M_{GUT}) - m_\tau(M_{GUT}) \simeq \mathcal{O}(\lambda^2) m_\tau$ which leads to a bi-large mixing pattern in M_ν. Generic predictions of this model are $\sin^2 2\theta_{23} < 0.9$ and $\sin^2 2\theta_{12} > 0.9$, making the model testable. The prediction of this model for the value of $|U_{e3}|$ is large. This is a consequence of the atmospheric mixing angle being maximal.

Comparison of models: Predictions of selected models for $\sin\theta_{13}$ are summarized in Table 1. A general observation is the following: (i) large $U_{e3} \sim \mathcal{O}(0.1)$ (can be probed by conventional/superbeam) arises in models with symmetric texture, in models based on anarchy, in models with RG enhanced leptonic mixing, and in minimal models with

TABLE 1. Predictions for $\sin\theta_{13}$ of various models. The upper bound from CHOOZ experiment is $\sin\theta_{13} < 0.24$. First nine models use $SO(10)$. Last two models are not based on $SO(10)$.

Model	family symmetry	solar solution	$\sin\theta_{13}$
Albright-Barr[5]	$U(1)$	LMA	0.014
Babu-Pati-Wilczek[6]	$U(1)$	SMA	5.5×10^{-4}
Blazek-Raby-Tobe[7]	$U(2) \times U(1)^n$	LMA	0.049
Berezhiani-Rossi[8]	$SU(3)$	SMA	$\mathcal{O}(10^{-2})$
Chen-Mahanthappa[4]	$SU(2)$	LMA	0.149
Kitano-Mimura[9]	$SU(3) \times U(1)$	LMA	$\sim \lambda \sim 0.22$
Maekawa[10]	$U(1)$	LMA	$\sim \lambda \sim 0.22$
Raby[11]	3×2 seesaw with $SU(2)_F$	LMA	$\sim m_{\nu_2}/2m_{\nu_3} \sim \mathcal{O}(0.1)$
Ross-Velasco-Sevilla[12]	$SU(3)$	LMA	0.07
Frampton-Glashow[16]-Yanagida	3×2 seesaw	LMA	$\sim m_{\nu_2}/2m_{\nu_3} \sim \mathcal{O}(0.1)$
Mohapatra-Parida[13]-Rajasekeran	RG enhancement	LMA	$0.08 - 0.10$

approximate $b-\tau$ unification; (ii) intermediate U_{e3} value $\sim(0.05-0.07)$ (can be probed by superbeam) arises in models with asymmetric texture; (iii) small U_{e3} (neutrino factory may be needed) arises in models with lop-sided texture. The mass hierarchy can be probed by long baseline experiments with distance greater than $900Km$, which can then be used to distinguish different models. A general observation is (i) normal hierarchy arise in SO(10) models with $SU(3)_H$, $SU(2)_H$, $U(1)_H$, in minimal SO(10) models with $b-\tau$ unification, and in SO(10) models with 3×2 see-saw; (ii) inverted hierarchy arise in models with $L_e-L_\mu-L_\tau$ horizontal symmetry; (iii) nearly degenerate arises in SO(10) models with RG enhanced lepton mixing and in models with anarchy. This work was supported by US DOE under Grant No. DE-AC02-98CH10886 and DE-FG03-95ER40894.

REFERENCES

1. M. -C. Chen and K. T. Mahanthappa, arXiv:hep-ph/0305088.
2. M. -C. Chen and K. T. Mahanthappa, *Phys. Rev.* **D62**, 113007 (2000).
3. M. -C. Chen and K. T. Mahanthappa, *Phys. Rev.* **D65**, 053010 (2002).
4. M. -C. Chen and K. T. Mahanthappa, *Phys. Rev.* **D68**, 017301 (2003).
5. C. H. Albright and S. M. Barr, *Phys. Rev.* **D64**, 073010 (2001).
6. K. S. Babu, J. C. Pati and F. Wilczek, *Nucl. Phys.* **B566**, 33 (2000).
7. T. Blazek, S. Raby and K. Tobe, *Phys. Rev.* **D62**, 055001 (2000).
8. Z. Berezhiani and A. Rossi, *Nucl. Phys.* **B594**, 113 (2001).
9. R. Kitano and Y. Mimura, *Phys. Rev.* **D63**, 016008 (2001).
10. N. Maekawa, *Prog. Theor. Phys.* **106**, 401 (2001).
11. S. Raby, *Phys. Lett.* **B561**, 119 (2003).
12. G. G. Ross and L. Velasco-Sevilla, *Nucl. Phys.* **B653**, 3 (2003).
13. R. N. Mohapatra, M. K. Parida and G. Rajasekaran, arXiv:hep-ph/0301234.
14. B. Bajc, G. Senjanovic and F. Vissani, *Phys. Rev. Lett.* **90**, 051802 (2003).
15. H. S. Goh, R. N. Mohapatra and S. P. Ng, arXiv:hep-ph/0303055.
16. P. H. Frampton, S. L. Glashow and T. Yanagida, *Phys. Lett.* **B548**, 119 (2002).

WORKING GROUP 2

Theoretical Aspects of Charged-Lepton Flavor Violation[1]

André de Gouvêa

*Department of Physics & Astronomy, Northwestern University
2145 Sheridan Road, Evanston, IL, 60208-3112, USA, and
Theoretical Physics Department, Fermilab, P.O. Box 500, Batavia, IL, 60510-0500, USA*

Abstract.
With the discovery of neutrino masses, it is clear that lepton-flavor is no longer a conserved quantity. This means that charged-lepton flavor violating processes (CLFV) including rare muon and tau decays (e.g. $\mu \to e\gamma$, $\tau \to \mu e^+ e^-$) are, in principle, observable. I review some phenomenological aspects of CLFV (concentrating on rare muon processes) and briefly discuss a few Standard Model extensions that predict a "large" rate for CLFV. In several of these extensions there are deep relations between the physics that generates neutrino masses and mixing and CLFV.

In the Standard Model (SM), muon number is exactly conserved. When neutrino masses are added and neutrino oscillations take place, muon-number violating processes involving charged leptons become possible as well. However, because of the smallness of neutrino masses, the rates for these processes are unobservable [2, 3]; for instance

$$B(\mu \to e\gamma) = \frac{3\alpha}{32\pi} \sum_i \left| V_{\mu i}^* V_{ei} \frac{m_{\nu_i}^2}{M_W^2} \right|^2 \sim 10^{-60} \left| \frac{V_{\mu i}^* V_{ei}}{10^{-2}} \right|^2 \left(\frac{m_{\nu_i}}{10^{-2} \text{ eV}} \right)^4. \quad (1)$$

The observation of muon number violation in charged muon decay would, therefore, serve as an unambiguous sign of new physics and indeed, a number of SM extensions may be probed sensitively by the study of rare muon decay [4].

First, it is useful to discuss such processes in a model-independent way. Although a purely model-independent analysis based on effective operators cannot make any prediction for the absolute rate of CLFV processes, it can be very useful in determining the relative rates. I will concentrate on the rates for $\mu^+ \to e^+\gamma$, $\mu^+ \to e^+ e^- e^+$, and $\mu^- - e^-$ conversion. In a large class of models, the dominant source of individual lepton number violation comes from a flavor non-diagonal magnetic-moment transition. Hence, consider the effective operator

$$\mathcal{L} = \frac{m_\mu}{\Lambda^2} \bar{\mu}_R \sigma^{\mu\nu} e_L F_{\mu\nu} + \text{h.c.} \quad (2)$$

---

[1] Based on the Report of the Stopped Muons Working Group for the ECFA - CERN study on Neutrino Factory and Muon Storage Rings at CERN [1].

This interaction leads to the following results for the branching ratios of $\mu^+ \to e^+\gamma$ ($B(\mu \to e\gamma)$) and $\mu^+ \to e^+e^-e^+$ ($B(\mu \to 3e)$), and for the rate of $\mu^- -e^-$ conversion in nuclei normalized to the nuclear capture rate ($B(\mu N \to eN)$):

$$B(\mu \to e\gamma) = \frac{3(4\pi)^2}{G_F^2 \Lambda^4}, \tag{3}$$

$$\frac{B(\mu \to 3e)}{B(\mu \to e\gamma)} = \frac{\alpha}{3\pi}\left(\ln\frac{m_\mu^2}{m_e^2} - \frac{11}{4}\right) = 6 \times 10^{-3}, \tag{4}$$

$$\frac{B(\mu N \to eN)}{B(\mu \to e\gamma)} = 10^{12} B(A,Z) \frac{2G_F^2 m_\mu^4}{(4\pi)^3 \alpha} = 2 \times 10^{-3} B(A,Z). \tag{5}$$

Here $B(A,Z)$ is an effective nuclear coefficient which is of order 1 for elements heavier than aluminium [5]. The logarithm in Eq. (4) is an enhancement factor for $B(\mu \to 3e)$, which is a consequence of the collinear divergence of the electron-positron pair in the $m_e \to 0$ limit. Nevertheless, because of the smaller phase space and extra power of α, $B(\mu \to 3e)$ and $B(\mu N \to eN)$ turn out to be suppressed with respect to $B(\mu \to e\gamma)$ by factors of 6×10^{-3} and $(2-4) \times 10^{-3}$, respectively.

Next, one can include an effective four-fermion operator which violates individual lepton number

$$\mathscr{L} = \frac{1}{\Lambda_F^2} \bar{\mu}_L \gamma^\mu e_L \bar{f}_L \gamma_\mu f_L + \text{h.c.}, \tag{6}$$

where f is a generic quark or lepton. The choice of the operator in Eq. (6) is made for concreteness, and generic results do not depend significantly on the specific chiral structure of the operator. First, consider the case in which f is neither an electron nor a light quark, and therefore $\mu^+ \to e^+e^-e^+$ and $\mu^- -e^-$ conversion occur only at the loop level. Comparing the $\mu^+ \to e^+\gamma$ rate in Eq. (3) with the contributions from the four-fermion operator to $B(\mu \to 3e)$ and $B(\mu N \to eN)$, one finds

$$\frac{B(\mu \to 3e)}{B(\mu \to e\gamma)} = \frac{8\alpha^2 N_f^2}{9(4\pi)^4}\left(\frac{\Lambda}{\Lambda_F}\right)^4 \left[\ln\frac{\max(m_f^2, m_\mu^2)}{M_F^2}\right]^2, \tag{7}$$

$$\frac{B(\mu N \to eN)}{B(\mu \to e\gamma)} = 10^{12} B(A,Z) \frac{32 G_F^2 m_\mu^4 N_f^2}{9(4\pi)^6}\left(\frac{\Lambda}{\Lambda_F}\right)^4 \left[\ln\frac{\max(m_f^2, m_\mu^2)}{M_F^2}\right]^2. \tag{8}$$

Here N_f is the number of colors of the fermion f and M_F is the heavy-particle mass generating the effective operators (typically M_F is much smaller than Λ or Λ_F because of loop factors and mixing angles). The logarithms in Eqs. (7) and (8) correspond to the anomalous dimension mixing of the operator in Eq. (6) with the four-fermion operator generating the relevant rare muon process [6]. If $\Lambda \sim \Lambda_F$, then the contributions from the four-fermion operator are irrelevant, since the ratios in Eqs. (4) and (5) are larger than those in Eqs. (7) and (8). More interesting is the case in which the four-fermion operator in Eq. (6) is generated at tree level, while the magnetic-moment transition in Eq. (2) is generated only at one loop, as in models with R-parity violation [7]. In this case,

$$\left(\frac{\Lambda}{\Lambda_F}\right)^4 \simeq \frac{(4\pi)^3}{\alpha}. \tag{9}$$

If Eq. (9) holds and $M_F \simeq 1$ TeV, then the ratios in Eqs. (7) and (8) become of order unity, so the different rare muon processes have comparable rates.

Alternatively, if the fermion f in Eq. (6) is an electron (or a light quark), the effective operator can mediate $\mu \to 3e$ (or $\mu^- -e^-$ conversion) at tree-level, and the corresponding process can dominate over the others [8]. Figure 1 summarizes the behavior of the ratio of branching ratios as a function of the relative strength of the effective operators in Eq. (2) and Eq. (6), when the fermion f in Eq. (6) is an electron, as in the left part of Fig. 1, or a combination of first generation quarks, as in the right part of Fig. 1. It can easily be seen from the plots that when the magnetic moment operator dominates ($\Lambda^2 \ll \Lambda_F^2$) the ratio of branching ratios saturates at several times 10^{-3}, while it grows like $(\Lambda^2/\Lambda_F^2)^2$ when the four-fermion operators are dominant ($\Lambda^2 \gg \Lambda_F^2$).

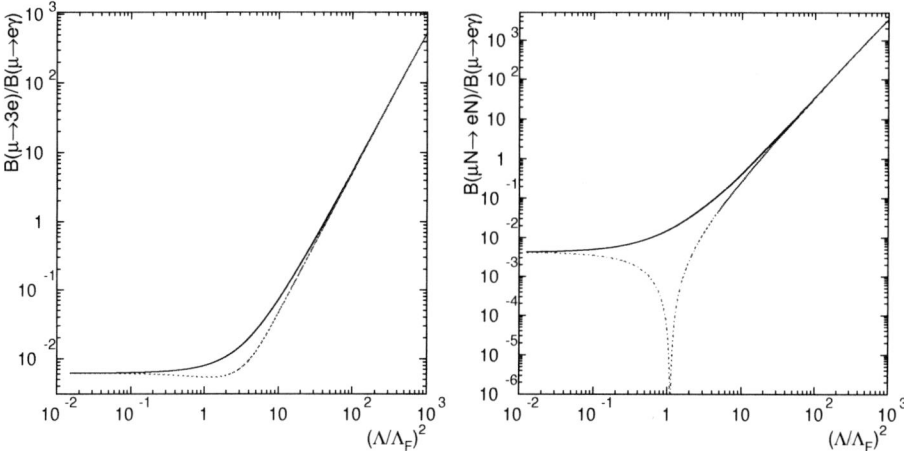

FIGURE 1. Branching ratios normalized to $B(\mu \to e\gamma)$ as a function of the ratio of the couplings of effective dimension-5 and dimension-6 operators, for $\mu^+ \to e^+e^-e^+$ and $\mu^- N \to e^- N$ conversion in ^{48}Ti. The solid (dashed) curves apply when the two operators interfere constructively (destructively).

Specif extensions of the SM will lead to different expectations for CLFV processes. Here I'll briefly address Supersymmetric theories and theories with extra-dimensions.

Supersymmetric extensions of the SM provide a very promising way of rendering the hierarchy of physical mass scales more natural. Moreover, low-energy supersymmetry often leads to large sources of individual lepton number violation. It provides a framework for computing physics observables in a controlled manner as a function of a well-defined set of parameters. Those, in turn, can thus be constrained by the experimental limits. Whereas lepton-flavor number violation in rare μ decays may well have the same source as neutrino oscillations, the rates generically are no longer suppressed by powers of neutrino masses. Roughly speaking there are two different supersymmetric sources of lepton-flavor number violation: flavor non-diagonal soft terms, and R-parity violation.

A mismatch in flavor space between the lepton and slepton mass matrices generates tree-level transitions between different leptonic generations, both in charged and neutral currents. For instance, if the mixing angle between the first two generations of sleptons is schematically indicated by $\theta_{\tilde{e}\tilde{\mu}}$,

$$B(\mu \to e\gamma) \simeq \frac{\alpha^3 \pi \theta_{\tilde{e}\tilde{\mu}}^2}{G_F^2 \tilde{m}^4}, \qquad (10)$$

where $\tilde{m}$ is a typical supersymmetric mass. Complete formulae for the rates of rare muon processes with the functional dependence on the different supersymmetric parameters can be found in Ref. [9]. As is apparent from Eq. (10), the rates for rare muon processes can be large in supersymmetric models, since $\theta_{\tilde{e}\tilde{\mu}}$ is not necessarily small.

For generic values of the soft supersymmetry-breaking mass scale $\tilde{m} \sim 1$ TeV, the experimental limits on rare muon decays transform in very stringent upper limits on $\theta_{\tilde{e}\tilde{\mu}}$. Therefore, it is common to invoke some universality condition or flavor symmetry that implies $\theta_{\tilde{e}\tilde{\mu}} = 0$ at the scale at which the soft terms are generated. If this scale is sufficiently low, loop corrections are small and rare muon processes are unobservable. However, if the scale of supersymmetry breaking is large (as in "supergravity" models), the soft-terms receive significant quantum corrections from high-energy flavor-violating interactions, and one generally expects more sizeable slepton mixing angles [10].

A possible mechanism comes from renormalizable Dirac neutrino Yukawa couplings y_v at energies larger than M_R [9, 11], the the "seesaw" scale [12]. In the basis in which the charged-lepton Yukawa coupling matrix is diagonal, loop corrections to the slepton mass matrix are not diagonal in flavor space:

$$\left(m_{\tilde{\ell}_L}^2\right)_{ij} \simeq -\frac{3m_0^2 + A_0^2}{8\pi^2} (y_v)_{ki}^* (y_v)_{kj} \ln \frac{M_{Pl}}{M_R}. \qquad (11)$$

Here m_0 and A_0 are the universal soft supersymmetry-breaking scalar mass and trilinear term, respectively, and it has been assumed that all three right-handed neutrinos have a common mass M_R. The contribution in Eq. (11) induces flavor-violating mixing angles for both sneutrinos and charged sleptons, and therefore rare muon processes are generated by loop diagrams involving charginos and neutralinos.

The experimental information on neutrino oscillation parameters is not sufficient to reconstruct the complete structure of the relevant mass matrices and it is, hence, necessary to rely on specific ansätze for them. It is, however, fair to say that these effects can induce $\mu \to e\gamma$ at rates close to the current experimental sensitivity, but precise relations to neutrino mixing parameters are very model dependent. It should also be emphasized that, because the magnitudes of neutrino Dirac Yukawa couplings are related to the right-handed neutrino mass scale M_R in seesaw models, $B(\mu \to e\gamma)$ is very sensitive to the right-handed neutrino scale: $B(\mu \to e\gamma) \propto M_R^2 [\ln(M_{Pl}/M_R)]^2$.

Finally, in these types of models, the decay $\mu^+ \to e^+ e^- e^+$ is usually dominated by an intermediate photon, and therefore Eq. (4) holds rather generally. This dominance is less certain in $\mu^- - e^-$ conversion, so there may be some moderate deviations from Eq. (5).

In supersymmetric models where R parity is not imposed as a discrete symmetry, one encounters renormalizable interactions that violate total lepton number and individual lepton numbers. At the scale of the muon mass, these interactions correspond to effective four-fermion operators of the kind discussed earlier. Therefore, depending on the field content of the operator, either $\mu^+ \to e^+ e^- e^+$ or $\mu^- -e^-$ conversion become the dominant rare muon process (see, e.g. [8] for more details).

Theories with extra spatial dimensions represent another possible solution to the problem of the large hierarchy between the Planck and Fermi mass scales. The hypothesis is that the universe possesses $1+n$ ($n > 3$) space-time dimensions, while the Standard Model particles are constrained to live on a 1+3-dimensional subspace. Gravity, which propagates in all dimensions, appears very weak to us either because its strength is diluted in a large compactified extra-dimensional space [13] or because it is localized away from us in spaces with non-factorizable geometries [14]. Since these theories assume that gravity becomes strongly coupled at an energy scale M_D comparable to the electroweak scale, the possibility of explaining the smallness of the neutrino masses via the classical seesaw mechanism is lost. Nevertheless, the small neutrino masses could now have an explanation based on geometrical arguments, similar to those that led to a justification of the small ratio M_W/M_{Pl}. For this to happen, one needs to assume the existence of right-handed neutrinos which, like the graviton, also propagate in the extra dimensions. If this is the case, their Yukawa interactions with the SM left-handed neutrinos are effectively suppressed by large geometrical factors [15], and therefore the SM neutrinos obtain, after electroweak symmetry breaking, a very small Dirac mass. When neutrino family mixing is included, one finds that the Kaluza–Klein modes of the right-handed neutrino mediate, at the loop level, flavor transitions in the charged sector [16, 17].

Particularly interesting are 'minimal models' in which all flavor transitions are described by only two free parameters, besides the observable neutrino masses and mixing angles. These two parameters are the 'fundamental' cut-off scale Λ (expected to be of the order of the weak scale, if these models are motivated by the hierarchy problem) and a dimensionless coefficient ε, which is currently constrained to be approximately less than 10^{-2} and whose expected value depends strongly on the neutrino Yukawa couplings and on details of the extra dimensional model (see Ref. [17] for details).

Under these conditions, the branching ratio for $\mu \to e\gamma$ is

$$B(\mu \to e\gamma) = \frac{3\alpha}{8\pi}\varepsilon^2 \left| U_{e2}U_{\mu 2}^* \frac{\Delta m_{\text{sun}}^2}{\Delta m_{\text{atm}}^2} + U_{e3}U_{\mu 3}^* \right|^2, \qquad (12)$$

where U is the leptonic mixing matrix and $\Delta m_{\text{sun,atm}}^2$ are the neutrino solar and atmospheric mass-squared differences, for a hierarchical neutrino mass spectrum.

Unlike $\ell_i \to \ell_j \gamma$ decays, the rates for $\mu \to eee$ and $\mu \to e$ conversion in nuclei are quite dependent on the unknown ultraviolet details of the models. Nevertheless, these rates can be predicted as a function of ε and Λ (see Refs. [16, 17] for complete expressions) and it turns out that they can be significantly enhanced with respect to the rate for $\mu \to e\gamma$ in some regions of parameter space, as in the case of SUSY with R-parity violation.

In spite of the fact that the one-loop effects that give rise to rare muon and tau processes can only be qualitatively estimated because of their cut-off dependency, in minimal models it is still true that their flavor structure is directly related to the physical neutrino mass matrix, and one is capable of predicting and relating the rates of rare charged and neutral lepton processes in terms of observable neutrino oscillation parameters. This is in sharp contrast to other cases of physics beyond the SM. In SUSY models, as pointed out earlier, the rates for rare muon processes are perturbatively calculable, but their relations with neutrino oscillations parameters are indirect and strongly model-dependent.

ACKNOWLEDGMENTS

I thank Alessandro Baldini for the invitation to present this overview talk at the 2003 Neutrino Factory Conference and the members of the "Physics with Low-Energy Muons at a Neutrino Factory Complex" working group at CERN for enlightening discussions. I am especially grateful to Gian Giudice and Kazuhiro Tobe, with whom almost all of the above was written. This work was supported by the US Department of Energy Contract DE-AC02-76CHO3000.

REFERENCES

1. J. Äystö et al., hep-ph/0109217.
2. S.T. Petcov, Sov. J. Nucl. Phys. **25** (1977) 340; S.M. Bilenky, S.T. Petcov and B. Pontecorvo, Phys. Lett. **B67** (1977) 309; T.P. Cheng and L.-F. Li in *Proceedings of the Coral Gables Conference*, 1977, Ed. Saul Perlmutter (Plenum, New York, 1977).
3. W. Marciano and A. Sanda, Phys. Lett. **B67** (1977) 303; B.W. Lee, S. Pakvasa, R. Shrock and H. Sugawara, Phys. Rev. Lett. **38** (1977) 937; B.W. Lee and R. Shrock, Phys. Rev. **D16** (1977) 1444.
4. For a complete, recent review, see Y. Kuno and Y. Okada, Rev. Mod. Phys. **73** (2001) 151.
5. A. Czarnecki, W. J. Marciano and K. Melnikov, in *Proc. Workshop on Physics at the First Muon Collider and at the Front End of the Muon Collider*, Eds. S.H. Geer and R. Raja (AIP Conference Proc. 435), p. 409 [hep-ph/9801218].
6. M. Raidal and A. Santamaria, Phys. Lett. **B421** (1998) 250.
7. L. Hall and M. Suzuki, Nucl. Phys. **B231** (1984) 419; V. Barger, G. F. Giudice and T. Han, Phys. Rev. **D40** (1989) 2987; H. Dreiner, hep-ph/9707435.
8. A. de Gouvêa, S. Lola and K. Tobe, Phys. Rev. **D63** (2001) 035004.
9. J. Hisano et al., Phys. Rev. **D53** (1996) 2442.
10. L.J. Hall, V.A. Kostelecky and S. Raby, Nucl. Phys. **B267** (1986) 415.
11. F. Borzumati and A. Masiero, Phys. Rev. Lett. **57** (1986) 961; J. Hisano et al., Phys. Lett. **B357** (1995) 579.
12. T. Yanagida, in *Proceedings of the Workshop on Unified Theory and Baryon Number of the Universe*, Tsukuba, Japan, 1979, O. Sawada and A. Sugamoto (KEK, Tsukuba, 1979), p. 95; M. Gell-Mann, P. Ramond and R. Slansky, in *Supergravity, Proceedings of the Workshop*, Stony Brook, New York, 1979, Eds. P. van Nieuwenhuizen and D. Freedman (North-Holland, Amsterdam, 1979).
13. N. Arkani-Hamed, S. Dimopoulos and G. Dvali, Phys. Lett. **B429** (1998) 263.
14. L. Randall and R. Sundrum, Phys. Rev. Lett. **83** (1999) 3370.
15. N. Arkani-Hamed et al., hep-ph/9811448; K.R. Dienes, E. Dudas and T. Gherghetta, Nucl. Phys. **B557** (1999) 25; Y. Grossman and M. Neubert, Phys. Lett. **B474** (2000) 361.
16. A.E. Faraggi and M. Pospelov, Phys. Lett. **B458** (1999) 237; A. Ioannisian and A. Pilaftsis, Phys. Rev. **D62** (2000) 066001; R. Kitano, Phys. Lett. **B481** (2000) 39; T.P. Cheng and L. Li, Phys. Lett. **B502** (2001) 152.
17. A. de Gouvêa, G.F. Giudice, A. Strumia and K. Tobe, Nucl. Phys. **B623** (2002) 395.

A Future Muon $(g-2)$ Experiment to $< \pm 0.1$ ppm at a High Flux Muon Facility

B. Lee Roberts

Department of Physics, Boston University, Boston, MA 02215

Abstract. Data collection for the muon $(g-2)$ experiment at the Brookhaven National Laboratory AGS, E821, was ended in 2001. Four of the five data sets have been published, which resulted in a precision of ± 0.7 parts per million (ppm) for a_{μ^+}. The final data set from 2001, when μ^- were stored, is being analyzed, with the result expected in early 2004. The final result will have a combined precision of $\sim \pm 0.5$ ppm. An upgraded experiment at the 100 Tera-proton AGS could achieve a precision of 0.2 to 0.1 ppm. At J-PARC or the front end of a neutrino factory one might be able to reach 0.06 ppm.

INTRODUCTION

The muon $(g-2)$ experiment at the Brookhaven National Laboratory AGS has been underway since the mid 1980s. The original motivation was to observe the electroweak contribution from virtual W and Z^0 bosons. The precision goal for the experiment was set at ± 0.35 parts per million (ppm), which was about one fifth of the lowest-order electroweak contribution. Because the muon anomalous moment arises from virtual radiative corrections, there was also the possibility of observing contributions from physics beyond the standard model such as supersymmetry in the few-hundred GeV range, as well as a sensitivity to muon substructure at the few TeV scale.[1]

The principle of the experiment is to measure the difference frequency between the spin precession and momentum precession for muons in a storage ring, and results with precisions of 13 ppm, 5 ppm, 1.3 ppm and 0.7 ppm have been published.[2, 3, 4, 5] The apparatus is now mostly documented in the literature,[6, 7, 8, 9, 10, 11, 12]

DIPOLE MOMENTS

The magnetic dipole moment associated with a charged spin-one-half particle is related to the anomalous magnetic moment a (anomaly) through the g-factor as defined by

$$\vec{\mu}_s = g_s(\frac{e}{2m})\vec{s}, \quad \mu = (1+a)\frac{e\hbar}{2m}, \quad a = \frac{g-2}{2}. \tag{1}$$

We understand that the large anomalies of the proton and neutron are a result of their internal structure, whereas the leptons e, μ and τ have anomalies which are expected to only arise from radiative corrections. The lowest-order radiative correction gives an anomaly of $\alpha/2\pi$, which is $0.0011614\cdots$ and dominates the anomaly of the leptons.

While a magnetic dipole moment (MDM) is allowed, an electric dipole moment (EDM) is forbidden by both parity and time reversal symmetries, which can be seen from the transformation properties of the Hamiltonian, $\mathscr{H} = -\vec{\mu}\cdot\vec{B} - \vec{d}\cdot\vec{E}$, under the three symmetries C, P and T. The quantity $\vec{\mu}\cdot\vec{B}$ is even, and $\vec{d}\cdot\vec{E}$ is odd under both P and T, so an EDM implies that both P and T are violated. Thus while the magnetic anomaly has a substantial standard model value, the expectation for the EDM is orders of magnitude below what is experimentally accessible. Any discrepancy between the measured and expected MDM or EDM would signify the presence of new physics.

A popular example of new physics is SUSY, which connects the EDM, MDM and the lepton-number violating conversion process $\mu \to e$ in the field of a nucleus, which is shown pictorially below in Fig. 1 My colleague Bill Morse will discuss EDMs in his talk at this meeting.

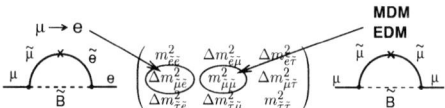

FIGURE 1. The supersymmetric contributions to the anomaly, and to $\mu \to e$ conversion, showing the relevant slepton mixing matrix elements. The MDM and EDM give the real and imaginary parts of the matrix element, respectively. (The × indicates a chirality flip)

THE EXPERIMENT

The difference frequency, which is the difference between the spin and momentum precession in the storage ring, $\omega_a = \omega_S - \omega_C$, is given by

$$\vec{\omega}_a = -\frac{e}{mc}\left[a_\mu \vec{B} - \left(a_\mu - \frac{1}{\gamma^2-1}\right)\vec{\beta}\times\vec{E}\right], \qquad (2)$$

where the electric field does not contribute to the spin motion for $\gamma = 29.3$. The storage ring is a weak focusing ring, and we ran with field indices of 0.142, 0.137 and 0.122 during our three main periods of data collection.

As can be seen from Eq. 2, both ω_a and the magnetic field B must be known to extract a value of a_μ from the experiment. The field is measured with NMR techniques, and has been shimmed to ± 1 ppm uniformity when averaged over azimuth. A contour map from one of the field maps is shown in Fig. 2. The B which appears in Eq. 2 is the field averaged over the muon distribution in the ring. With such a uniform field, only modest information is needed on the muon distribution, and in our data set from 2000 the systematic error from our knowledge of the muon distribution was ± 0.03 ppm.[5]

To monitor the magnetic field during data collection, 366 fixed NMR probes were placed around the ring and continuous readings from about 150 probes were used to track the field in time. About twice per week a trolley with 17 NMR probes was used to map the field in the storage ring. During muon data collection, the trolley is stored in a garage inside the vacuum chamber. The trolley probes were calibrated with a special spherical water probe, which provides a calibration to the free proton precession frequency ω_p.

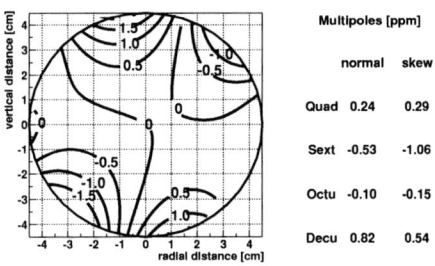

FIGURE 2. A contour plot of the magnetic field from the Y2000 run.

Positrons (electrons) from the parity violating decay $\mu^{+(-)} \to e^{+(-)} + \nu_e(\bar{\nu}_e) + \bar{\nu}_\mu(\nu_\mu)$ are detected in lead-scintillating-fiber calorimeters[12] where the energy and arrival time is measured. The highest energy positrons (electrons) carry the spin information, and the number of high-energy positrons (electrons) as a function of time is given by

$$N(t) = N_0(E) e^{-t/\gamma\tau} [1 + A(E) \sin(\omega_a t + \phi_a(E))], \quad (3)$$

where the energy threshold E is chosen to optimize the quantity NA^2. In the analysis of the data, many small effects such as coherent beam motion in the storage ring must be included.[4, 5]

At the end of E821, we will have collected about 9 billion high-energy decay positrons (or electrons), and our statistical error will be on the order of 0.5 ppm. To improve things further, both statistics and systematics must be improved. To accumulate the statistics needed for a substantial improvement would require a larger number of stored muons per unit time, which would require an improved beam line along with a new inflector magnet which has a larger aperture than the present one.[6] This increase in rates would mean that our understanding of pile-up, and of gain changes during the counting time would need to be significantly improved.

The key to any precision measurement is the systematic errors. The systematic errors from our 1999 and 2000 running periods[4, 5] are given below in Table 1. While we can see a way to improve things to the 0.05 ppm level, it will be very difficult, and much R&D will be necessary. Every aspect of the detectors, electronics, and the magnetic field measuring and monitoring system will have to be improved. A letter of intent has been submitted to J-PARC for an improved experiment.[14]

SUMMARY AND CONCLUSIONS

The muon $(g-2)$ experiment E821 at Brookhaven has achieved most of the goals set out twenty years ago when it began. The present experiment is still limited by statistics rather than systematic errors. Running was terminated before the statistical design goal was reached. After the last result is reported, there may well be compelling reasons to push further. We believe that the present technique can be pushed further, perhaps to the 0.06 ppm level. To go further than that would probably require a new idea such as the

TABLE 1. Systematic Errors from the 1999 and 2000 data sets.[4, 5] CBO stands for coherent betatron oscillations. The pitch correction comes from the vertical betatron oscillations, since $\vec{\beta} \cdot \vec{B} \neq 0$.

Syst. Error on ω_p	1999 (ppm)	2000 (ppm)	Syst. Error on ω_a	1999 (ppm)	2000 (ppm)
Inflector Fringe Field	0.20	-	Pile-Up	0.13	0.13
Calib. of trolley probes	0.20	0.15	AGS bgd.	0.10	0.01
Tracking B with time	0.15	0.10	Lost Muons	0.10	0.10
Measurement of B_0	0.10	0.10	Timing Shifts	0.10	0.02
μ-distribution	0.12	0.03	E-field/pitch	0.08	0.03
Absolute calibration	0.05	0.05	Fitting/Binning	0.07	0.06
Others[†]	0.15	0.10	CBO	0.05	0.21
			Beam debunching	0.04	0.04
			Gain Changes	0.02	0.13
Total for ω_p	0.4	0.24	Total for ω_a	0.3	0.31

one put forward to completely redesign the storage ring to run at higher momentum.[15] Certainly a core group of people from E821 are already thinking of how to do better,[14] both in the context of the existing storage ring design, and beyond.

ACKNOWLEDGMENTS

I want to thank my colleagues on E821 for many helpful discussions, and especially Jim Miller, Rob Carey, Paul Debevec, Dave Hertzog, Hugh Brown, Wuzheng Meng, Bill Morse, Yannis Semertzidis, Dave Warburton, Klaus Jungmann and Francis Farely for discussions on how to go to the next level.

REFERENCES

1. An overview of non-standard model physics is given by T. Kinoshita and W.J. Marciano in *Quantum Electrodynamics* (Directions in High Energy Physics, Vol. 7), ed. T. Kinoshita, (World Scientific, Singapore, 1990), p. 419.
2. R.M. Carey et al., Phys. Rev. Lett. **82**, 1632 (1999)
3. H.N. Brown et al., (Muon $(g-2)$ Collaboration), Phys. Rev. **D62**, 091101 (2000)
4. H.N. Brown, et al., (Muon $(g-2)$ Collaboration), Phys. Rev. Lett. **86** 2227 (2001)
5. G.W. Bennett, et al., (Muon $(g-2)$ Collaboration), Phys. Rev. Lett. **89**, 101804 (2002).
6. A. Yamamoto, et al., Nucl. Instrum. and Methods Phys. Res. **A491** 23-40 (2002).
7. G.T. Danby, et al., Nucl. Instr. and Methods Phys. Res. **A 457**, 151-174 (2001).
8. Efstratios Efstathiadis, et al., Nucl. Inst. and Methods Phys. Res. **A496** ,8-25 (2002).
9. Y.K. Semertzidis, Nucl. Instrum. Methods Phys. Res. **A503** 458-484 (2003)
10. S.I. Redin, et al., Nucl. Instrum. Methods Phys. Res. **A473**, 260-268, (2001).
11. R. Prigle, et al., Nucl. Inst. Methods Phys. Res. **A374** 118 (1996), and X. Fei, V. Hughes and R. Prigl, Nucl. Inst. Methods Phys. Res. **A394**, 349 (1997).
12. S.A. Sedykh et al., Nucl. Inst. Methods Phys. Res. **A455** 346 (2000)
13. W. Liu et al., Phys. Rev. Lett. **82**, 711 (1999).
14. J-PARC Letter of Intent L17, B.L. Roberts contact person.
15. F.J.M. Farley, Jul 2003, hep-ex/0307024.

Sensitive Measurement of the EDM of the Muon?

William Morse

Brookhaven National Lab

Abstract. A new sensitive method of measuring the EDM of particles with small magnetic anomalies $a = (g-2)/2$ is described. This is the written version of a talk given at the NuFact Conference, June 2003, Columbia University, N.Y.

INTRODUCTION

An electric dipole moment (EDM) of an elementary particle violates time reversal and parity invariance. The situation for a magnetic dipole moment (MDM) is shown in Fig. 1.

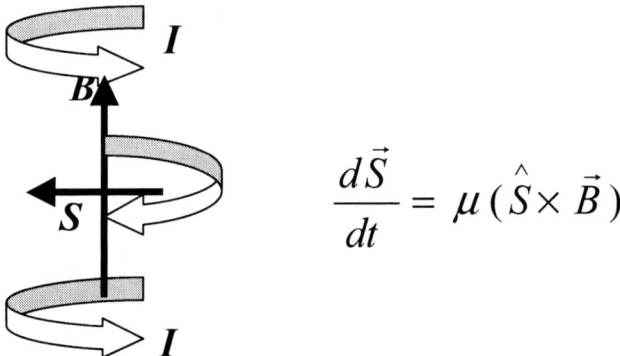

FIGURE 1. Precession of the magnetic dipole moment μ about the B field.

The magnetic field B is created by current distributions. The spin precesses about the magnetic field due to the torque created by the MDM. When time is reversed, the direction of the currents, and thus the magnetic field, reverses. Therefore, the MDM spin precession direction also reverses. The situation for an EDM is shown in Fig. 2. The electric field is created by charge distributions. The spin precesses about the electric field due to the torque created by the EDM. When time is reversed, the direction of the electric field does not reverse. The EDM spin precession does not reverse, unless the sign of d reverses. One comes to similar conclusions by requiring conservation of energy when time reverses: $H_\mu = \mu S \bullet B$, and $H_d = dS \bullet E$. Thus one can see whether time is flowing forward or backward by whether the EDM is aligned with or opposite to the spin, ie. the laws of physics are different when time goes forward compared to backward, unless $d = 0$.

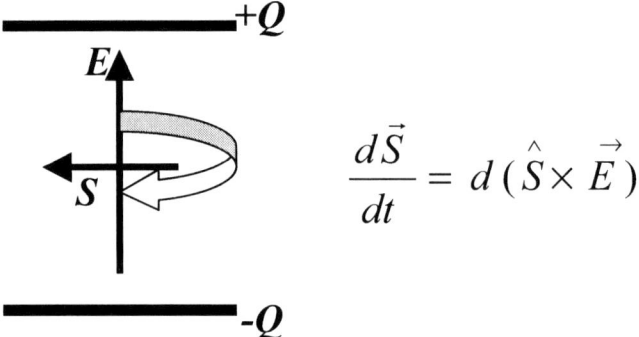

FIGURE 2. Precession of the electric dipole moment d about the E field.

A cartoon of a relativistic particle with charge e and velocity βc in a magnetic field which is directed out of the page is shown in Fig. 3. The origin of the electric dipole moment is a disparity between the charge distribution and the mass distribution. There is a torque on the spin due to the EDM which causes the spin to precess vertically:

$$\frac{d\vec{S}}{dt} = d\hat{S} \times \left(c\vec{\beta} \times \vec{B} \right) \quad (1)$$

This equation can also be derived in the particle's rest frame: $d\vec{S}/dt^* = d(\hat{S} \times \vec{E}^*)$, where $dt = \gamma dt^*$ and $\vec{E}^* = \gamma c \vec{\beta} \times \vec{B}$.

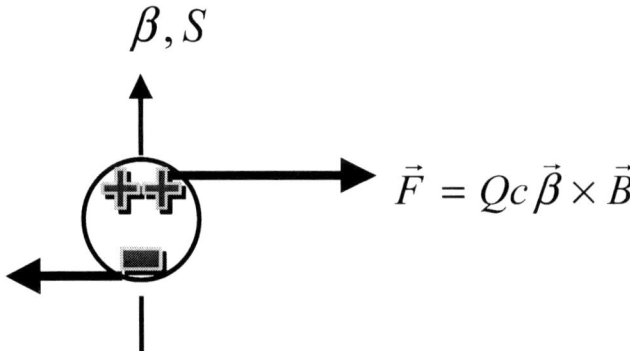

FIGURE 3. Cartoon of a relativistic particle in a magnetic field showing the torque on the spin due to an edm.

PRINCIPLE OF A SENSITIVE MUON EDM EXPERIMENT

The spin motion with respect to the momentum vector for the simple case where $\beta \cdot B = \beta \cdot E = 0$ is:

$$\omega_a = \omega_s - \omega_c = -\frac{e}{mc}\left[aB - \left(a - \frac{1}{\gamma^2 - 1}\right)\beta \times E\right] \quad (2)$$

where a is the anomalous magnetic moment of the muon [1]. In order to make a sensitive measurement of the vertical precession due to the electric dipole moment (see equ. 1), the effect of the horizontal precession due to the magnetic moment from equ. 2 must be largely canceled. This is possible with a radial electric field:

$$E_R = \frac{aBc\beta}{1-(1+a)\beta^2} \approx aBc\beta\gamma^2 \quad (3)$$

A letter of intent [2] was submitted to JPARC for an experiment to measure the edm of the muon to 10^{-24} e-cm in 10^7 seconds of running time in the PRISM II beamline. The parameters are given in Table 1.

TABLE 1. PRISM II Beamline and Muon EDM Storage Ring Parameters

Muon momentum	0.5 GeV/c
H and V acceptance of edm ring	800 π mm-mrad
Muon Polarization	60%
Number of muons per 10^7s	10^{17}
E_R	2 MV/m
B_0	0.25T
Bending radius	6.5m
Circumference	63m
$\beta_{x\,max}$	6.85m
$\beta_{y\,max}$	7.10m
dp/p	2%
$c\beta B_0$	75MV/m

A conservative value of $E_R \approx 2$ MV/m was chosen. The effective electric field from equ. 1, $c\beta B_0$, is then an astounding 75 MV/m. The main spin/beam dynamics systematic effect is when the radial electric field is not in a plane around the ring. This will also precess the spin into the vertical plane:

$$\frac{dS}{dt} = \frac{\mu E \theta_E}{\beta\gamma^2} \quad (4)$$

However, this effect can be canceled by injected the beam both clockwise and counter-clockwise. When injecting in different directions, the magnetic field must be reversed. The magnets will be laminated. The magnetic field will not be the same when the current is simply reversed, if only because the earth's magnetic field does not change sign; however, the magnetic field will be stabilized with NMR to much better accuracy than needed. The electric field remains in the same direction when injecting in different directions. Undoubtedly, the beam position, etc. will be different when injecting in different directions. However, the electric field is ideally pure dipole. Therefore, the electric field is the same independent of beam position, to first order. The beam position will be accurately monitored, and runs taken with the beam position different by an amount which is large compared to the position measurement accuracy. The magnetic forces are the same in both polarities, and the radial electric field electrodes will be supported independently of the magnets. The clockwise and counter-clockwise runs must be alternated to sample the temporal changes in the out of plane angle. The out of plane angle will be changing temporally due to small temperature, etc. changes. Mechanical devices [3] will monitor angular stability with better than 10^{-9} radian accuracy. If the injection direction is reversed every 10s during the run, for example, the knowledge of the temporal angular stability clockwise vs. counter-clockwise will be at the 10^{-9} radian/$\sqrt{10^6} \approx 10^{-12}$ radian level, and this systematic error will be at the $\approx 10^{-28}$ e-cm level, ie. well below the statistical error. Furthermore, polarized deuterons can be stored in the same ring: N, P, and A are all larger than for the muon. With lower velocity, they are much more sensitive to the electric field misalignment (see equ. 4) than the edm effect (see equ.1).

THEORETICAL IMPLICATIONS

In the Standard Model, T or CP violation comes from the complex phase in the CKM matrix. This leads to un-measurably small predictions for EDMs, as one needs interference between all three quark families. However, we know the Standard Model is not the whole story as it has recently been discovered that neutrinos have mass and there is large mixing between the flavor states. Some theoretical implications of this situation are discussed in ref. 4: "Enhanced Electric Dipole Moment of the Muon in the Presence of Large Neutrino Mixing". This paper predicts an EDM of the muon as large as 5×10^{-23} e-cm, while that of the electron is only about 10^{-28} e-cm, an order of magnitude below the present electron edm limit. Thus a sensitive measurement of the edm of the muon is an excellent way to find T violation which is not due to the Standard Model.

REFERENCES

1. G. Bennett et al., *Phys. Rev. Letters* **89**, 101804 (2002).
2. J-PARC Letter of Intent L22, *Search for a Permanent Muon Electric Dipole Moment at the 10^{-24} e cm Level*, Y. Kuno, J. Miller, Y. Semertzidis spokespersons.
3. S. Rescia, *Precise Measurements of Small Linear and Angular Displacements with Capacitance Methods*, http://www.bnl.gov/edm/papers/Sergio_Rescia_020118.pdf.
4. K.S. Babu, B. Dutta, and R.N. Mohapatra, *Phys. Rev. Letters* **85**, 5064-5608 (2000).

Status of the MEG experiment

Alessandro Baldini

INFN Pisa, Via F. Buonarroti 2, 56127 Pisa

Abstract. The phsysics motivations for this experiment are presented. A general description of the experiment and of the R&D status of each sub-detector is given. The sensitivity of the experiment and its time schedule are finally presented.

INTRODUCTION

The search for the $\mu \to e\gamma$ decay, which started back in 1948[1], has played a fundamental role in the construction of the Standard Model of elementary particles physics. The current best limit on the observation of this decay was given by the MEGA experiment[2] which established a 90% C.L. limit of $1.2\,10^{-11}$ for the banching ratio (BR) of the $\mu \to e\gamma$ decay with respect to the normal muon decay. The MEG experiment would like to improve this sensitivity by two orders of magnitude. Many theoretical models representing the most recent efforts to overcome the Standard Model predict observable decay rates for this experiment.

PHYSICS MOTIVATIONS

Grand unified suspersymmetric (SUSY-GUT) theories, owing to the large top quark mass, predict[3] the $\mu \to e\gamma$ decay to happen not much below the current experimental limit. In Figure 1 the SU(5) predictions for the $\mu \to e\gamma$ branching ratio as a function of the right handed selectron mass and for several values of $tan\beta$ are shown. Recent indications from the combined LEP experiments favor values of $tan\beta$ grater than 10. Predictions for SO(10) are even higher (about two orders of magnitude) than SU(5) predictions. In Figure 1 the current experimental limit and the sensitivity that the MEG experiment[4] aims at reaching are shown.

Another, independent, source for this decay might come from neutrino oscillations. After the recent KAMLAND results the large mixing angle solution, implying maximum mixing angle and a difference of $10^{-5}eV^2$ between $m_{\nu_e}^2$ and $m_{\nu_\mu}^2$, seems to represent the best solution for the so-called "solar neutrino problem". If a mechanism of the see-saw type is introduced in SUSY-GUT theories to reproduce the pattern of neutrino masses, sizeable contributions (of the same order of magnitude or even higher than the ones discussed above) to the $\mu \to e\gamma$ process take place[5]. These contributions add up to the previous ones, therefore making the $\mu \to e\gamma$ process an extremely sensitive probe of SUSY-GUT theories.

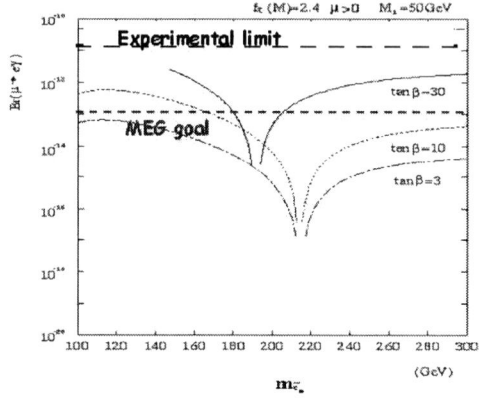

FIGURE 1. SUSY SU(5) predictions for the $\mu \to e\gamma$ decay

It must also be remarked that the BR of the $\mu \to e\gamma$ decay in the standard model due to neutrino oscillations would be completely unobservable ($BR \approx 10^{-54}$). The detection of $\mu \to e\gamma$ events would be a clear, unambiguous sign of physics beyond the standard model.

GENERAL LAYOUT OF THE MEG EXPERIMENT

A sketch of the experiment is presented in Fig.2. A surface muon beam with an intensity grater than $10^7 \mu/s$ will be stopped in a thin target. A magnetic spectrometer, composed of a superconducting magnet and 17 drift chambers azimuthally oriented, will be used for the measurement of the positrons trajectories. The magnetic field is non uniform, with a peak intensity of 1.27 T at the center and decreasing along the axis of the beam. This inhomogeneous field prevents positrons with a small longitudinal momentum from overcrowding the drift chambers. The central part of the magnet is very thin (0.2 X_0) because it is in the acceptance region of the electromagnetic calorimeter. The drift chambers spatial resolution needed for this experiment is approximately 300μ in the $r - \theta$ plane and along the beam direction. In order to reach this resolution along the beam direction cathodes segmentation is necessry.

Positrons timing will be measured at the end of their trajectory by an array of scintillators. A second (inner) layer of scintillating fibers read by APDs will be used for trigger purposes.

Photons will be detected by an innovative homogeneous electromagnetic calorimeter in which a total of about 800 photomultipliers will detect the light produced by photons initiated showers in about 800 liters of liquid Xenon. Liquid Xenon has optimal scintillation features:

- high light yield: 80% of NaI

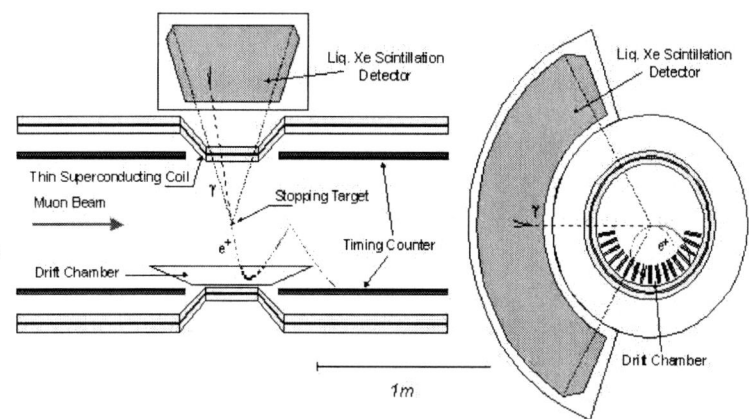

FIGURE 2. A sketch of the MEG detector

TABLE 1. Resolutions (FWHM) foreseen for the MEG experiment

$\frac{\Delta E_\gamma}{E_\gamma}$	$\frac{\Delta E_{e+}}{E_{e+}}$	$\Delta\theta_{e\gamma}$	Δt
4%	0.8%	19 mrad	0.15 ns

- fast emission: three components with exponential decay times of 4, 22 and 45 ns
- short radiation length: 2.77 cm

Two disadvantages are the the operating temperature ($\approx -100°C$) and the ultraviolet light emission spectrum.

The trigger of the experiment is based on an online digital reconstruction of the photon energy and direction, of the postron direction and of the relative timing between the two. Our estimates indicate that the final trigger should be about 20 Hz. All the signals (except for the scintillating fibers) will be read by a 2 GHz waveform digitizer based on an earlier development made at PSI[6], especially developed for this experiment.

The resolutions on the electron and photon momenta, obtainable in this experiment, are indicated in table1.

The two main backgrounds are:

- a prompt one due to the $\mu \to e\gamma\nu\bar{\nu}$ decay
- the one arising from the accidental coincidence of a photon and a positron with momenta equal, within the detector resolutions, to those expected for the $\mu \to e\gamma$ decay.

Our estimates indicate that the accidental background will be bigger than the prompt one by a factor ten. Since the accidental background is quadratically dependent on the muon rate while a possible $\mu \to e\gamma$ signal is obvioously linearly depending on it there would be no advantage in using a more intense beam.

STATUS OF THE R&D OF THE INDIVIDUAL SUB–DETECTORS

Several tests were made of the PSI beam. An electrostatic separator was used in order to eliminate the positron contamination in the μ^+ beam. The beam was then refocused and transported to a target similar to the one to be used in the final experiment. The measured rate was $4\,10^7 \mu/s$ with transverse dimensions of $\sim 5 \times 6 \, mm^2$. The beam rate could be increased by more than a factor 2 by improving some of the beam elements but this rate is already sufficient for our purposes.

The magnet of the spectrometer was already built and is already at PSI. Its field was measured and corresponds to the design characteristics.

Small prototypes (10 cm long) of the drift chambers were tested also in a magnetic field. A resolution of $\sim 400 \, \mu$ was obtained by means of the photocathodes segmentation method. A ful scale prototype is presently under construction and optimization studies of the photocathode structure in order to improve the position resolution are going on. The full prototype will be tested in the first half of next year.

Prototypes of the positron timing counters already reached the $100 ps$ FWHM resolution necessary for this detector.

A large prototype (100 l of liquid Xenon seen by 228 PMTs) is currently under test at PSI by means of monochromatic 55 MeV photons obtained by the $\pi^- p \to n\pi^0$ charge exchange raection followed by π^0 decay. Previous tests in Japan verified that the transparency properties of purified liquid Xenon do not deteriorate the homogeneity of this calorimeter.

SENSITIVITY AND TIME SCHEDULE

All our present measurements and simulations indicate that we should be a able to reach a single event sensitivity (SES: the branching ratio for which one event should be observed) to the $\mu \to e\gamma$ decay of 4×10^{-14} in about two years of data taking and that 0.5 (accidental) background events are correspondingly expected . In case no event is observed this SES corresponds to a 90% C.L. limit of 10^{-13} which represents an improvement of two orders of magnitude with respect to the present best experimental limit. We envisage two years of construction starting next year. The start of the data taking is therefore foreseen in 2006. The first results should be available in 2007. We believe it would be important to perform this experiment before the start of the LHC in order to address the SUSY searches for the experiments at this collider.

REFERENCES

1. E.P. Hinks and B. Pontecorvo, Phys. Rev. Lett. **73** (1948) 246
2. M. L. Brooks *et al.*, Phys. Rev. Lett. **83** (1999) 1521
3. R. Barbieri and L.J. Hall Phys. Lett. **B338** (1994) 212
4. The MEG experiment: search for the $\mu \to e\gamma$ experiment at PSI, September 2002, available on the web at the address: http://meg.psi.ch/docs/prop_infn/nproposal.pdf
5. J. Hisano and N. Nomura, Phys. Rev. **D59** (1999) 116005
6. C. Bronnimann *et al.*, Nucl. Instr. and Meth. A420 (1999) 264

Precise measurement of the positive muon lifetime at the RIKEN-RAL muon facility

D. Tomono[a], S.N. Nakamura[bc], Y. Matsuda[a], M. Iwasaki[a], G. Mason[d], K. Ishida[ac], T. Matsuzaki[a], I. Watanabe[a], S. Sakamoto[e] and K. Nagamine[c]

[a] *Muon Science Laboratory, RIKEN, 2-1 Hirosawa, Wako, Saitama, 351-0198, Japan*
[b] *Department of Physics, Tohoku University, Aoba, Aramaki, Aoba-ku, Sendai, 980-8587, Japan*
[c] *Meson Science Laboratory, High Energy Accelerator Research Organization, (KEK-MSL), 1-1 Oho, Tsukuba, Ibaraki, 305-0801, Japan*
[d] *Department of Physics and Astrophysics, University of Victoria, Victoria, BC V8W 2Y2, Canada*
[e] *High Intensity Accelerator Facility Development Center, Tokai Research Establishment, Japan Atomic Energy Research Institute, (JAERI), 2-4 Shirane, Tokai, Naka-gun, Ibaraki, Japan*

Abstract. The positive muon lifetime have been precisely measured at the RIKEN-RAL muon facility. We developed a new method to realize a higher data accumulation rate with a strong pulsed beam. The data taking have already been finished and about 10^{10} decay events were accumulated. The off-line analysis is now proceeding by comparing with results obtained by a Monte Carlo simulation.

INTRODUCTION

The lifetime of the positive muon (τ_μ) was measured with a strong pulsed beam at the RIKEN-RAL muon facility. The τ_μ is directly associated with the Fermi coupling constant (G_F), which is one of the fundamental parameters in the Standard Model (SM). In the framework of SM weak sector, at least three experimental input parameters are required to calculate any observables. These parameters are usually chosen to be, the fine structure constant (α 0.045 ppm), the mass of Z boson (M_Z 24 ppm), and the Fermi coupling constant (G_F 9 ppm, PDG average) [1].

The relationship between G_F and τ_μ is given by the following formula [2]:

$$\tau_\mu^{-1} = \frac{G_F^2 m_\mu^5}{192\pi^3} F\left(\frac{m_e}{m_\mu}\right)\left(1 + \frac{3}{5}\frac{m_\mu^2}{m_W^2}\right)(1+\Delta q), \tag{1}$$

where m_μ, m_e and m_W denote a muon, electron and W mass, respectively, and Δq denotes a higher order QED radiative correction term which was recently calculated up to 2-loop QED radiative correction [2]. At present, the τ_μ can be theoretically translated with an accuracy better than one ppm. The precise measurement of τ_μ gives the most precise value of G_F. We have observed $\sim 10^{10}$ muon decay events and our final goal is to improve the τ accuracy at the level of 10 ppm.

The muon lifetime measurement has a long history and the most accurate value of the τ_μ was obtained at TRIUMF in 1984 [3]. They determined $\tau_\mu = 2.19695 \pm 0.00006$ μs

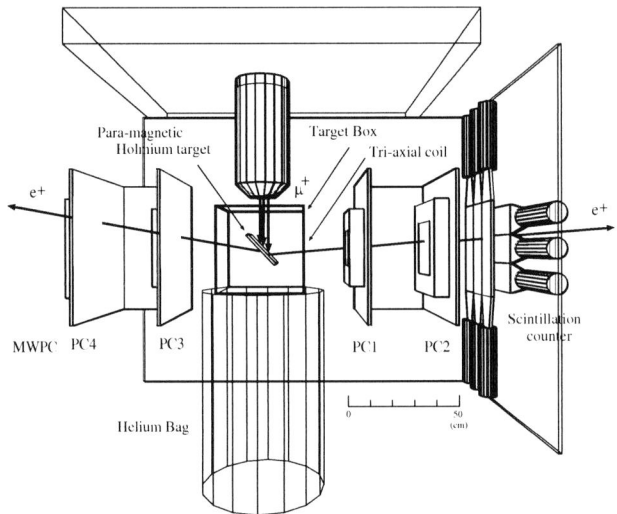

FIGURE 1. Schematic figure of our Set-up

(27 ppm) and the PDG average gives 2.19703 ± 0.00004 μs (18 ppm) [1]. However, no improvements have been made for more than 20 years.

In the previous measurements [3, 4] accuracy of τ_μ was limited, since they observed only a few muons in a time window in order to decrease a count-loss due to the pile-up. To improve the τ_μ accuracy within a realistic beam-time, it is essential to increase the data accumulation rate. Our novel approach realizes to observe 200 counts muon decay in a spill to increase the event rate drastically. In this case, the count-loss due to the pile-up becomes a source of the most severe systematic error. In order to suppress this effect, we considered following points: 1) A multi-wire proportional chamber (MWPC) was used as a main detector to observe decay-positron. It has a large segmentation and an event rate for each wire can be decreased about 1.0 counts/spill. 2) In the off-line analysis, we estimate residual count-loss mathematically and correct the observed muon decay time spectrum. This procedure is explained in the following section.

EXPERIMENTAL SET-UP

We have carried out the experiment at the RIKEN-RAL muon facility in the Port-2 area [5]. We used surface muons which have a momentum of 27 MeV/c at a repetition rate of 50 Hz and their spins are almost 100% polarized anti-parallel to the muon momentum. About $10^4 \sim 10^5$ muons per double pulses come to the Port-2 area.

Figure 1 shows the schematic figure of our set-up. Muons came into the target-box and stopped at the target. For the target, a para-magnetic holmium (Ho) was selected. Since Ho has a large fluctuation of the internal field, stopped muons lose their polarization exponentially with a time constant of $T_1 \sim 500$ nsec which is independent to the external

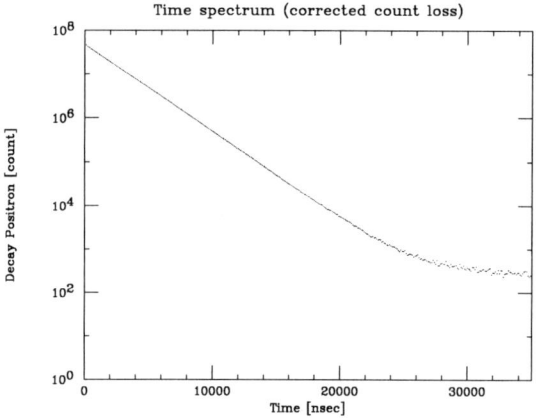

FIGURE 2. Muon decay time spectrum (Preliminary). This spectrum was corrected by scheme 1 and 2. This is not all data sets.

magnetic field. However, for the safety, the tri-axial magnetic coil system was installed around the target-box to cancel the residual magnetic field from the beam-line magnet and geomagnetism. Target region was filled with helium gas in order to minimize the number of stopped muons in the target region. A helium bag was installed to the downstream. Muons which did not stop in the target was guided to a beam dump.

Decay positrons were observed during a time window of 70 μs by four MWPCs. They were located to the both sides of the target in order to control the effect of the spin asymmetry. Small chambers (PC1,PC3) and large chambers (PC2, PC4) have x and y sense plane of 64 or 96 wires, respectively. In total, 192 segmentations are obtained even by using the single wire mode. Scintillation counters were also installed for a calibration purpose [6].

For the absolute time calibration, a very accurate clock system was developed. We used a latching memory module (LM) synchronized with a GPS time and frequency reference receiver (HP58503A). The GPS receiver contains a quartz oscillator which is continuously calibrated with atomic clocks on the satellites, and thus it provides an accurate clock signal (10MHz, $< 10^{-12}$) Therefore, our clock system has very accurate integral and differential linearity. The wire hit timings were also digitized with a long range multi-stop TDC (LeCroy TDC3377). These data were handled by four CAMAC systems in parallel at a rate of 3 Mbyte/sec [7].

Analysis and Current Status

Data analysis is now in progress. The count-loss correction is a major part for oure lifetime analysis. Time spectrum was distorted by the count-loss effect. We estimate the amount of this effect based on an assumption that the event rate follows the Poisson distribution. These are proceeded in the off-line analysis by comparing to the results with a Monte Carlo simulation. This correction enables us to realize less parameter fit. The procedures of the count-loss correction are briefly summarized as:

1. Input parameters for the correction should be known precisely. We required a fixed dead-time ($d = 200$ nsec) by using LM-GPS clock system and mean count rate per spill (m). These parameters determine a time region and an amount of the correction.
2. We assume that the event rate follows a Poisson distribution and decay can be written in the exponential at the no dead-time limit. Due to the count-loss, event distribution is also deviated from the Poisson and a observed mean value is reduced from the actual value of m to m_L. However, when we focus on events coming during the dead-time, they still follow the Poisson distribution with a mean value of $m(t, t+d)(= \frac{m}{\tau} \int_t^{t+d} e^{-t/\tau} dt)$ calculated by the exponential distribution. As a result, missing events during the dead-time can be corrected with their mean value of m and it is essential to find a relation between m_L and m. We are trying to test this scheme by the Monte Carlo simulation and apply this scheme to the real data sets.
3. When the exponential distribution is assumed, the reported τ_μ value is used to calculate the correction. In the last stage, we need to correct our data iteratively, and check it whether it is self-consistent. Then, we examine validity of this procedure whether consistent τ_μ values are obtained in any fitting regions and in the case of a different event-rate run.

A preliminary time spectrum is shown in Figure 2 which is corrected by taking steps 1 and 2. The correction procedure is successfully performed in this statistics. Analysis of all data and further Monte Carlo simulation are now in progress.

SUMMARY

A new method to measure the muon lifetime was developed. MWPCs were used as the decay positron detector, which has a large segmentation in order to minimize the count-loss effect. Data taking has already finished and analysis of the count-loss correction is in the last stage. The final result will be obtained in the near future.

REFERENCES

1. K. Hagiwara, et al., Particle Data Group, *Phys. Rev.* **D66** (2002) 010001.
2. T.van Rittbergen and R.G. Stuart, *Phys. Rev. Lett.* **82** (1999) 488.
3. K.L. Giovanetti, et al., *Phys. Rev.* **D29** (1984) 343.
4. G. Bardin, et al., *Phys. Lett.* **B137** (1984) 135.
5. T. Matsuzaki, et al., *Nucl. Instrum. Methods* **A465** (2001) 365.
6. S.N. Nakamura, et al., *RIKEN Review* No.20, p64 1999.
7. S.N. Nakamura and M. Iwasaki, *Nucl. Instrum. Methods* **A388** (1997) 1220.

Towards an improved determination of the Fermi coupling constant from the µLan experiment

C.J.G. Onderwater
on behalf of the µLan Collaboration[1]

Department of Physics, University of Illinois, 1110 W. Green Street, Urbana IL 71801, USA

Abstract. The status to date on the preparations for a measurement of the lifetime of the positive muon using the µLan detector is presented. Significant progress has been made towards the realization of the experiment, which utilizes a high intensity pulsed surface muon beam and a nearly 4π solid angle detection system. Construction of all major components is approaching completion and first data taking is underway.

µLAN EXPERIMENTAL GOAL AND CONCEPT

The goal of the µLan experiment is to reduce the uncertainty on the experimental determination of the positive muon lifetime by a factor of twenty to 1 ppm. From the muon lifetime the Fermi coupling constant G_F, which governs the strength of the weak interaction, can be extracted with a precision of 0.5 ppm, thanks to recent theoretical progress[2].

The principle behind the µLan experiment is very simple. A thin stopping target is charged with about twenty muons in a short burst lasting several microseconds, and is then monitored for muon decays for about ten muon lifetimes ($22\mu s$). This cycle is repeated until more than 10^{12} muon decays are registered.

The pulsed muon beam is created from the high intensity continuous πE3 surface muon beam at PSI using a new electrostatic kicker. Positrons from subsequent muon decay are registered using a highly segmented double layer scintillator shell, which covers a solid angle of nearly 4π. Each scintillator element is coupled to a fast photo-multiplier tube (PMT) and read out individually by precisely timed 500 MHz waveform digitizers (WFDs). From the recorded waveforms the muon decay times can be reconstructed. The muon lifetime is obtained by fitting the decay time distribution.

Minimization of systematic errors was the key consideration in the design of the experiment. Primary sources are multi-particle pile-up, muon spin precession, time variation of the PMT response or electronic threshold, and backgrounds. Pile up is reduced to a manageable level by segmenting the detector and the relatively low peak rate. The effect of spin precession is minimized using a depolarizing target, active dephasing of the incoming muons with a magnetic field and a highly symmetric detector geometry. Detector response and threshold variations are dealt with by recording the signal height on an event-by-event basis, which makes the detector self-calibrating. Backgrounds are reduced by requiring coincident hits between the inner and outer

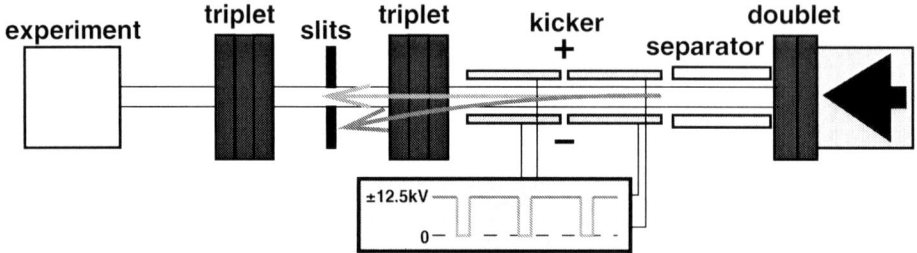

FIGURE 1. Schematic layout of the beamline extension of the standard πE3 setup. The beam goes from right to left.

scintillator elements.

MUON BEAM AND ELECTROSTATIC KICKER

The layout of the extension of the standard πE3 beamline at PSI is shown in fig. 1. Spatial limitations dictate that the separator and kicker be positioned back-to-back. The kicker and separator operate in the same (vertical) direction for optimal performance. The second half of the beamline consists of a triplet-slit-triplet combination to produce foci at the experimental target and slit, which blocks the beam when kicked.

Over the past years we have performed detailed simulations and several beamline tests in order to obtain and optimize a new beamline tune with sufficient intensity and extinction when kicked. In 2002, we assembled the setup described above and tested our new tune for flux, spot size and extinction with a non-switching magnetic equivalent of a proposed electric kicker. At a 15 MHz beam rate, we measured an extinction factor of 3×10^{-4}, which exceeds the design criterium of 10^{-3}. At full acceptance, the measured flux was as high as 50 MHz.

Based on these findings, the design of a fast electrostatic kicker was finalized and construction has since been completed. The device consists of two pairs of plates. Each pair is 75 cm in length, 20 cm wide and separated by a 15 cm gap. A 25 kV potential difference can be switched on and off with a rise-time of about 45 ns. In July 2003, the kicker was installed at PSI and basic functionality was confirmed. Further commissioning will take place in the fall of 2003.

STOPPING TARGET AND MAGNETIC FIELD

The muons arrive at the target highly polarized. A slow spin precession can therefore lead to a significant systematic error because of the strong relation of the decay positron emission direction to the muon spin. To alleviate the sensitivity, the target assembly is set up to maximally depolarize the stopped muons.

In a literature search, sulfur was found to significantly depolarize muons stopped in it. In the summer of 2002, we measured its depolarizing properties. The level of

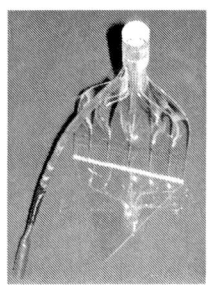

FIGURE 2. Left-to-right: triangular scintillator element with 90° light-guide and LED pulser; "hex-houses" and "penthouses"; the detector approaching completion.

residual polarization was obtained from observing the count rates in two detectors on opposite sides of the target assembly. It was found that sulfur has 11.5 times less residual polarization than silver, the standard polarization-preserving target.

To reduce the effect of remnant polarization, an external magnetic field of about 100 Gauss is applied to the target region to increase the spin precession rate. Moreover, when the strength of this field and the duration of the muon accumulation period are matched, dephasing further reduces the polarization.

Alternatively, we found that a permanently magnetized foil could combine both functions in one material. In the summer of 2003 we studied a sample of Arnokrome-3™[1] foil, with a remnant magnetic field of ~9 kG. No statistically significant polarization was observed, even when an additional external field was applied. This indicates complete scrambling of the incoming muon spin and full depolarization.

SCINTILLATOR DETECTOR

Several prototypes of the detector were studied over the last couple of years. In the final design, the active elements of the detector consist of two nested triangular scintillator tiles coupled to adiabatic light-guides that are attached through mirrored edges beveled at 45°. The tiles are 3 mm thick, with a 15 cm base and a 60° or 72° top angle. Five or six of these elements form pentagonal and hexagonal superstructures, which combine to a truncated icosahedron, or soccer-ball, geometry. Fig. 2 shows the final version of some of the sub-structures.

An extensive program was completed to optimize the performance of the scintillator elements. The initial icosahedron geometry of the detector was truncated to prevent particles hitting the tiles under relatively large angles. The dimensions of the outer tiles were adjusted to avoid particles simultaneously traversing the sides of the inner and outer layer. Both reduced the sensitivity to gain and threshold fluctuations. The orientation and configuration of the light-guides were chosen to minimize internal reflections and to

---

[1] Arnold Engineering Company, http://www.arnoldmagnetics.com/

avoid false coincidences between inner and outer tiles from Čerenkov-light generation in the light-guides. Several candidate photo-multiplier tubes were thoroughly tested for gain, efficiency and after-pulsing.

Two complete "hexhouses" and "penthouses" were built and tested *in situ* in the summer of 2002. They performed as expected. The design of the full detector was then finalized; construction of the full detector was recently completed.

READOUT ELECTRONICS AND DATA ACQUISITION SYSTEM

The PMTs of each scintillator element are individually sampled by waveform digitizers (WFDs), operating at 500 MHz with 8-bit resolution. The heart of these WFDs is a Field Programmable Gate Array (FPGA), which performs zero suppression and data formatting and guarantees essentially dead-time-free data collection. The principal challenge in the design is programming the FPGA to handle the high data rate produced by the flash ADC. In 2002 a single channel prototype was designed, built and tested as a proof-of-principle. Based on the success of this test, a four-channel board including all necessary peripherals was designed and tested. The final design is now nearly complete and the production of the full set of boards is expected to commence soon. :q

The DAQ is based on the PSI MIDAS data acquisition package and designed to handle a rate of 10^6 muon decays per second, corresponding to about 60 MBytes/s. It reads out and processes the data and also provides the control and monitoring of the high voltage, beamline and other experimental settings. It consists of a local network of eight frontend processors providing single VME crate readout and data processing, one backend processor, which assembles the single crate event fragments and performs low level analysis, and one mass storage processor. The complete setup was assembled and tested in July 2003 and is now in use at PSI for first data taking.

CONCLUSION

Work on most of the major component of the experiment has been completed. We are well underway to reach our goal to measure the muon lifetime with a precision of 1 ppm.

REFERENCES

1. R. Carey, W. Earle, A. Gafarov, B. Graf, M. Hance, M. Hare, E. Hazen, I. Logachenko, K. Lynch, Y. Matus, J. Miller, Q. Peng, L. Roberts, M. Saia, J. Wasserman, Boston University, Boston, USA; D. Chitwood, S. Clayton, P. Debevec, D. Hertzog, P. Kammel, B. Kiburg, G. Onderwater, C. Özben, C. Polly, A. Sharp, D. Webber, S. Williamson, University of Illinois at Urbana-Champaign, Urbana, USA; S. Cheekatamalla, M. Deka, S. Dhamija, T. Gorringe, M. Ojha, University of Kentucky, Lexington, USA; C. Arnold, K. Giovanetti, A. Werner, R. Wiita, James Madison University, Harrisonburg, USA; K. Crowe, F. Gray, B. Lauss, University of California, Berkeley, USA.
2. T. van Ritbergen, R.G. Stuart, Nucl.Phys. **B564**:343-390,2000

A PRECISION MEASUREMENT OF μ^+ DECAY

Peter Kitching (on behalf of the *TWIST* collaboration)

TRIUMF, 4004 Wesbrook Mall, Vancouver, B.C. V6T2A3, Canada

Abstract. *TWIST* (the TRIUMF Weak Interaction Symmetry Test) is an experiment to measure the energy and angular distributions of the positrons (e^+) emitted from the decay of polarized positive muons (μ^+). The distributions are described by four parameters ρ, η, δ, and ξ, widely referred to as the Michel parameters. The values of the four parameters are exactly predicted by the Standard Model; the goal of *TWIST* is to search for new physics that leads to deviations from the predictions by measuring ρ, δ, and $\mathcal{P}_\mu \xi$ to a few parts in 10^4. This represents an improvement of more than one order of magnitude over the existing precision.

INTRODUCTION

An intense beam of highly polarized surface muons is stopped in a thin planar target at the center of a precisely constructed, low-mass array of drift chambers (Fig. 1) at the centre of a large, uniform, 2-T solenoidal magnetic field. The polarization of the muon beam, $\vec{\mathcal{P}}_\mu$, is preserved to a high and predictable degree in the process. When the muons decay, the array of chambers records the tracks of the e^+ in the magnetic field, allowing the measurement of e^+ momentum for several thousand decays every second.

THE MICHEL PARAMETERS IN MUON DECAY

Muon decay has been described in terms of the most general local, derivative-free, lepton-number-conserving, four-fermion point interaction [1], which includes scalar(S), vector(V), and tensor(T) interactions among charged lepton spinors of definite (right(R) or left(L)) chirality. The matrix element depends on 10 complex amplitudes $g^\gamma_{\varepsilon\mu}$, where $\gamma =$ S,V or T and $\varepsilon, \mu = R$ or L. The elements g^T_{LL} and g^T_{RR} are zero, resulting in 19 independent real parameters. In the Standard Model(SM), $g^V_{LL} = 1$ and all others are zero ($V-A$). Constraints on the values of the coupling constants must be derived from observables.

Neglecting neutrino masses and radiative corrections, averaging over the polarization of the decay e^+, the differential decay rate of the positive muon is expressed as [2]

$$\frac{d^2\Gamma}{dx\, d\cos\theta} = \frac{1}{4} m_\mu W^4_{\mu e} G^2_F \sqrt{x^2 - x_0^2} \left\{ \mathcal{F}_{IS}(x) + \mathcal{P}_\mu \cos\theta \cdot \mathcal{F}_{AS}(x) \right\}$$

where $W_{\mu e} = \frac{m_\mu^2 + m_e^2}{2m_\mu}$, $x = \frac{E_e}{W_{\mu e}}$, $x_0 = \frac{m_e}{W_{\mu e}}$, $\mathcal{P}_\mu = |\vec{\mathcal{P}}_\mu|$, and $\cos\theta = \frac{\vec{\mathcal{P}}_\mu \bullet \vec{p}_e}{|\vec{\mathcal{P}}_\mu||\vec{p}_e|} \sim -\cos\theta_T$.

The isotropic and asymmetric, respectively, terms in the decay rate are given in terms of the Michel parameters ρ, η, δ, and ξ.

$$\mathscr{F}_{IS}(x) = x(1-x) + \frac{2}{9}\rho(4x^2 - 3x - x_0^2) + \eta x_0(1-x)$$

$$\mathscr{F}_{AS}(x) = \frac{1}{3}\xi\sqrt{x^2 - x_0^2}\left[1 - x + \frac{2}{3}\delta\left\{4x - 3 + \left(\sqrt{1-x_0^2} - 1\right)\right\}\right]$$

They can be expressed in terms of bilinear combinations of the coefficients $g^{\gamma}_{\varepsilon\mu}$. Notice that in the differential decay rate, ξ appears only in the product $\mathscr{P}_\mu\xi$, requiring that for ξ the polarization of the sample of decaying muons must also be well known. Increased precision in the measurements of the Michel parameters translates directly into more precise limits on the values of the coupling coefficients. Radiative corrections to the distribution are substantial, and must be incorporated. Corrections have been calculated which are adequate for measurements with a precision of 10^{-3}, and higher order estimates are in progress.

The Michel parameters have the values $\rho = \frac{3}{4}$, $\eta = 0$, $\delta = \frac{3}{4}$, and $\xi = 1$ in the SM. The Review of Particle Physics values for the best current measurements are: $\rho = 0.7518 \pm 0.0026$, $\eta = -0.007 \pm 0.013$, $\delta = 0.7486 \pm 0.0026 \pm 0.0028$, $\mathscr{P}_\mu\xi = 1.0027 \pm 0.0079 \pm 0.0030$, $\mathscr{P}_\mu\frac{\xi\delta}{\rho} > 0.99682$.

The goal of *TWIST* is to search for new physics (beyond the SM) through measurements of the Michel parameters. We aim to set new limits on the right-handed coupling of the muon in a model independent way, as well as to squeeze the parameter space for classes of extensions to the Standard Model, such as those invoking left-right symmetry.

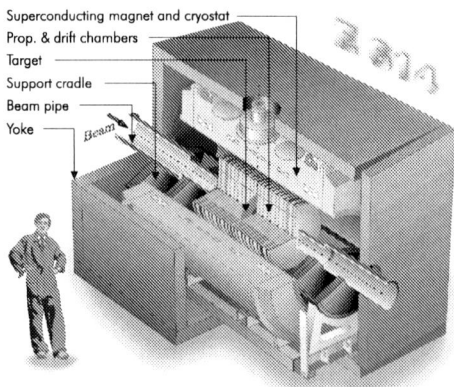

FIGURE 1. The *TWIST* spectrometer.

FIRST MEASUREMENTS

Data were taken in the fall of 2002, with the goal of determining two of the Michel parameters, ρ and δ, with a precision of better that one part in 10^3. While this level of statistical precision can be reached using only a few days of beam, the investigation of possible systematic uncertainties takes much longer. Altogether, approximately 6×10^9 events were recorded, comprising many separate measurements of the Michel parameters under different experimental conditions, in order to measure the effects of the changes. Data sets were obtained with different detector efficiencies, solenoidal fields, muon beam characterstics (polarization, stopping position, focus position), detector gas density, trigger rates, and backscattered event rates.

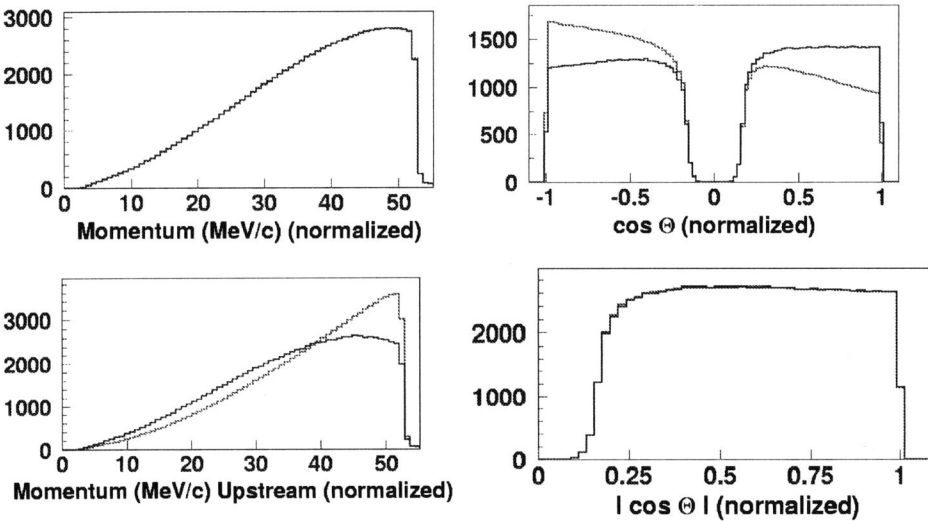

FIGURE 2. Comparison of momentum and angular distributions of highly polarized surface muons with cloud muons. The region near $\cos\theta_T = 0$ is not within the spectrometer acceptance. Fiducial cuts on the other variable have been applied to the histograms. Events with $0.3 < x < 0.95$ (15.8MeV/c $< p <$ 50.8MeV/c) and $0.5 < |\cos\theta_T| < 0.95$ are considered to be within the fiducial volume Bottom left: comparison of momentum distributions of highly polarized surface muons with cloud muons, for $\cos\theta < 0$, showing the expected effect of polarization on the energy distribution. Bottom right: comparison of the distributions in $|\cos\theta|$ (removing the polarization asymmetry) for the same two data sets, showing the restoration of symmetry within the angular region of measurement.

Figure 2 shows a comparison of decay distributions from highly polarized ($\mathscr{P} \sim -1$) surface muons with oppositely polarized ($\mathscr{P} \sim +0.3$) cloud muons. The effects of the difference in polarization are clearly seen. Notice that when the polarization asymmetry is removed, by plotting the distributions *vs* momentum or $|\cos\theta|$, the differences are small.

A critical task for *TWIST* is to test and validate both the analysis software and GEANT simulation programs. With the availability of the first data, this has begun. There are many comparisons which can be made. For example,agreement between the simulated and measured muon stopping distributions is impressive, showing that the energy loss models used in the simulation are so far adequate for *TWIST*. The symmetry of the detector allows one half to be used to independently define a positron track , while the other half can be used to analyze another part of the same track. This is accomplished by using muons which have been selected to stop near either end of the detector, so that the decay e^+ traverses both halves. Comparison of the two halves permits checking characteristics such as pattern recognition, reconstruction efficiency, and resolution in angle and momentum. Moreover, comparisons can also be made using the GEANT simulation under the same conditions, to help to validate the physics of simulated e^+ motion through the detector. Comparisons of the track information from the two halves of the detector are shown in Fig. 3. The simulation and real data show nearly identical behavior. Furthermore, if the data and GEANT predictions are studied differentially, the deviations are even smaller, and probably arise from a small difference in the properties of the incident muon beam.

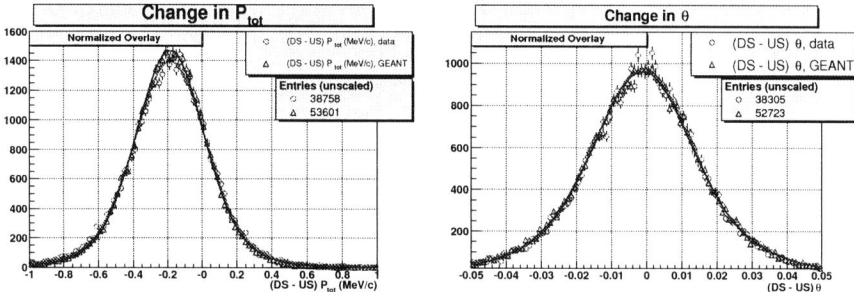

FIGURE 3. Distributions showing the differences in fit momentum (left) and angle (right), between two halves of the same track as determined by the upstream (US) and downstream (DS) halves of the *TWIST* detector. The graphs show both data and simulation. Notice the shift from zero in the momentum, corresponding to the energy loss (mainly in the mylar target foil) between the two halves of the track.

NEXT STEPS

The data will be analyzed in a way which is blind to the final results, in order to avoid possible human biases. A simulated spectrum generated with hidden Michel parameters different from the accepted values is generated, including radiative corrections up to next-to-leading order. Derivative spectra are also generated from the derivatives of the Michel spectrum to the parameters, and the experimental data are fit to a linear combination of these spectra. The linear nature of the Michel parametrization allows the corrections to the hidden parameters to be derived directly from the fitting coefficients. When the fits are completed, the hidden values are revealed and the results of the fit are used to calculate the true experimental values for the Michel parameters It is expected that the first blind analysis will be completed in 2003. Meanwhile data taking continues with the ultimate goal of measuring ρ, δ, and $\mathscr{P}_\mu \xi$ to a few parts in 10^4

ACKNOWLEDGMENTS

This research was supported in part by the Natural Sciences and Engineering Research Council and the National Research Council of Canada, and the U.S. Department of Energy

REFERENCES

1. Fetscher, W., and Gerber, H.-J., "Precision Measurements in Muon and Tau Decays," in *Precision Tests of the Standard Electroweak Model*, edited by Langacker, P., World Scientific, Singapore, 1995, vol. 14 of *Advanced Series in Directions in High Energy Physics*, pp. 657–705.
2. Michel, L., *Proc. Phys. Soc.*, **A63**, 514 (1950).

Measurement of the transverse polarization of the e^+ from the decay of polarized μ^+

W. Fetscher[*], K. Bodek[†], A. Budzanowski[**], N. Danneberg[*], C. Hilbes[*‡],
L. Jarczyk[†], K. Kirch[*‡], S. Kistryn[†], K. Köhler[*], J. Lang[*], A. Kozela[*], J. Smyrski[†], E. Stephan[§], A. Strzałkowski[†], A. von Allmen[*] and J. Zejma[†]

[*]*Institut für Teilchenphysik, ETH Zürich, CH-8093 Zürich, Switzerland*
[†]*Institute of Physics, Jagellonian University, Cracow, Poland*
[**]*H. Niewodniczanski Institute of Nuclear Physics, Cracow, Poland*
[‡]*present address: Paul Scherrer Institut, CH-5232 Villigen-PSI, Switzerland*
[§]*Institute of Physics, University of Silesia, Katowice, Poland*

Abstract. The transverse polarization components of the positrons from the decay of polarized muons have been measured at the Paul Scherrer Institute by annihilating in flight the decay positrons with polarized electrons. The use of a stroboscopic method greatly reduces systematic errors. The energy dependence of the transverse polarization component P_{T_1}, which lies in the plane spanned by muon spin and positron momentum, yields the low energy parameter η and thus an improved model-independent value of the Fermi coupling constant. A non-zero value of the transverse component P_{T_2}, which is perpendicular to the plane mentioned above, would be the first observation of time reversal violation in a purely leptonic decay. Part of the data (40×10^8 annihilation events) have been evaluated. Both P_{T_1} and P_{T_2}, averaged over the positrons' energy, are consistent with zero with a precision improved by a factor of three as compared with the previous measurement [1]. The preliminary results are $\langle P_{T_1} \rangle = (7 \pm 8) \times 10^{-3}$, $\langle P_{T_2} \rangle = (-1 \pm 8) \times 10^{-3}$ in full agreement with the standard model.

Muon decay, $\mu^+ \to \nu_\mu(e^+ \nu_e)$, as a purely leptonic decay, is one of the pillars on which the standard model rests. In fact, only a few years ago it has been shown that $V - A$, as one of the basic assumptions of the standard model, *follows* from the results of a selected set of muon decay experiments (including inverse muon decay)[2, 3]. The experimental limits obtained up to now, however, still allow for substantial contributions from non-standard couplings which differ in their spin structure from the $V - A$ interaction. The limits on these couplings can be efficiently improved by performing experiments with polarized muons and positrons. One interesting possibility is to measure the two transverse polarization components P_{T_1} and P_{T_2} as a function of the positron energy, since the standard model predicts for P_{T_1} only a tiny value and zero for the (P-odd and T-odd) component P_{T_2}. Thus any evidence for a substantial transverse polarization of the e^+ from polarized μ^+ decay would be evidence for physics beyond the standard model. The energy dependence of P_{T_1} offers the possibility to obtain the low energy parameter η without the suppression factor m_e/m_μ, which makes the determination of η from the electron energy spectrum extremely difficult, and the simultaneous measurement of the polarization component P_{T_2} allows one to test for time reversal invariance.

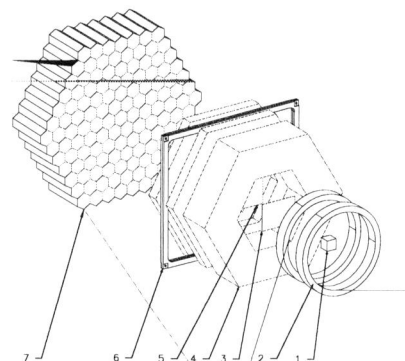

FIGURE 1. Experimental setup: 1 - Be target, 2 - spin precession magnet, 3 and 5 - plastic trigger counters, 4 - drift chamber (10 planes), 6- iron yoke of the magnetized Vacoflux foil, 7 - BGO calorimeter. Two additional drift chambers (2 planes each) sandwich the magnetized foil. An array of plastic veto counters (ANTI) in front of and cosmic trigger scintillators on top and below the BGO wall are not shown.

The experimental setup [4] is shown in Figure 1. A beam of highly polarized muons ($P_\mu \approx 91\%$) enters a beryllium stop target with bunches every 20 ns. The polarization of the stopped muons precesses in a homogeneous magnetic field with the same frequency ν as the accelerator RF. Thus every new muon bunch is added coherently with the muon spins pointing in the same direction as the polarization vector. Decay e^+ emitted parallel to the B-field are tracked by drift chambers and can annihilate with polarized e^- in a magnetized foil. The two annihilation quanta are then detected by a hexagonal array of 127 BGO crystals. A valid annihilation event requires a coincidence of two plastic scintillator counters in front of the magnetized foil with two separated clusters of BGO detectors and an anticoincidence with a plastic counter array in front of the BGO wall. A possible transverse polarization would be detected as a harmonic time dependence of the annihilation rate with frequency ν for a given detector pair. The time distribution of the annihilation events is given by

$$R(t) = F_{\text{res}}(t) \cdot F_{\mu\text{SR}}(E, P_\mu, \chi, \alpha, t) \cdot F_{\gamma\gamma}(E_3, E_4, \mathbf{P}_{e^+}, P_{e^-}, \psi, t) \tag{1}$$

Here, E designates the positron energy, χ and α the polar and azimuthal angles of emission of the e^+, respectively, E_3 and E_4 the energies of the annihilation quanta, $\mathbf{P}_{e^+}$ and P_{e^-} the polarizations of the e^+ and e^-, respectively, and ψ is the azimuthal angle of emission of the two annihilation quanta.

The origins of these three functions are quite different:

1. F_{res} is due to the differential nonlinearity of the TDC and to the time variation of the muon stop distribution, i.e the arrival of a new burst and the subsequent decay until the arrival of the next burst. This function depends neither on the positron's direction of emission (χ, α) nor on the sign of the electron's polarization P_{e^-}.

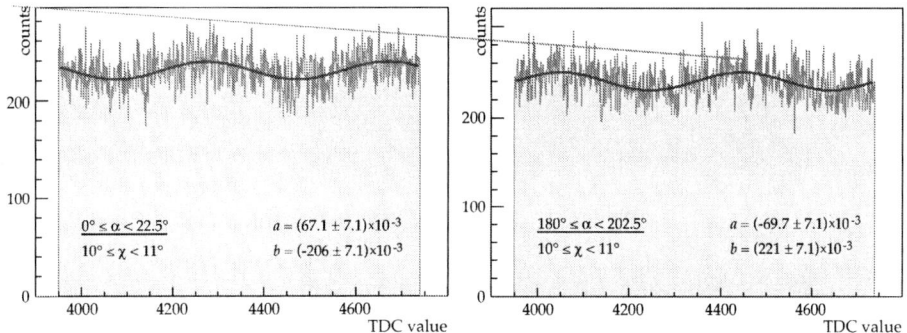

FIGURE 2. Annihilation event rate for positrons emitted into the same polar angle χ, but two opposite azimuthal angles α.

2. $F_{\mu SR}$ is due to the decay asymmetry of the emitted positrons: The accepted decay positrons are emitted close to $\chi = 90°$ and thus show a small remnant μSR effect which depends on the azimuthal angle of emission α of the positron.
3. $F_{\gamma\gamma}$ The effect due to a possible transverse polarization $\mathbf{P}_T$, in contrast, does not depend on α, but instead on the relative orientation of $\mathbf{P}_T$ and the electron polarization in the magnetized foil. This measurement yields the absolute value of $\mathbf{P}_T$.

In order to obtain the two transverse polarization components we have to first derive these three functions separately from our data. For this task we observe that we are only interested in a harmonic signal of frequency $\nu = 50$ MHz. Thus we describe our normalized time distribution by

$$R(t) = 1 + a\cos\omega t + b\sin\omega t \tag{2}$$

with the angular frequency $\omega = 2\pi\nu$, neglecting freqencies in higher order, and determine the Fourier amplitudes a and b. By integrating over the fiducial region of the magnetized foil and by averaging over the foil polarization we obtain exclusively a_{res} and b_{res}. These terms, being the same for all possible experimental arrangements, are then subtracted from the individually derived Fourier amplitudes.

The amplitudes $a_{\mu SR}$ and $b_{\mu SR}$ are obtained in function of the azimuthal angle α and averaged over the foil polarization. Fig. 2 shows the results for two angular bins 180° apart from each other. It turns out that this signal is not only a background which has to (and can be) subtracted, but is actually essential for the experiment because it fixes the position of the muon polarization vector in terms of the TDC channels and thus also determines the reference plane for the T - odd and the T - even transverse polarization components.

The third function, finally, is derived by integrating over α, thus eliminating the residual μSR effect, and by subtracting the Fourier coefficients obtained from measurements with negative electron polarization from those obtained with positive polarization, which

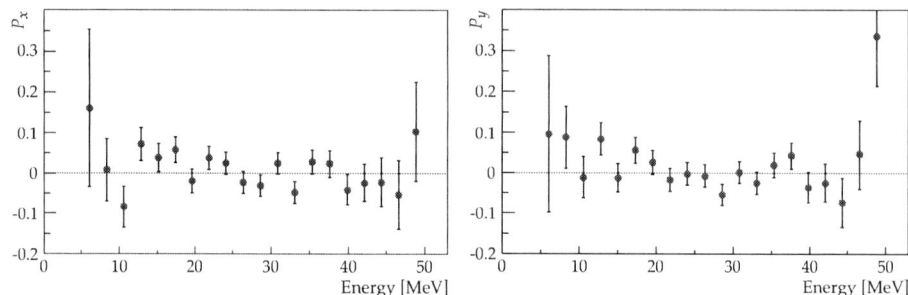

FIGURE 3. Transverse polarization components of the e^+ from μ^+ decay as a function of the e^+ energy at *the time of annihilation* (preliminary results). x is the horizontal, y the vertical axis of the apparatus.

eliminates the residual contribution. By properly taking account of the energies of the annihilation quanta and of their azimuthal orientation, one calculates the analyzing powers and derives the *absolute value* $|\mathbf{P}_T| = \left\{ P_{T_1}^2 + P_{T_2}^2 \right\}^{1/2}$ as a function of the positron energy. Together with the time information obtained from the μSR effect one finally arrives at the distribution of the two transverse polarization components shown in Fig. 3. We note that these are the components *at the time of annihilation*, while we are actually interested in their values at the time of muon decay. This analysis and its results for the decay parameters η, η'', α'/A and β'/A will be published shortly. Here we conclude with our (preliminary) results for the two transverse polarization components averaged over the positron energy:

$$\langle P_{T_1} \rangle = (7 \pm 8) \times 10^{-3} \qquad \langle P_{T_2} \rangle = (-1 \pm 8) \times 10^{-3} \qquad (3)$$

This result is an improvement by a factor of 3 as compared to the first measurement of P_T [1] and is in complete agreement with the predictions of the standard model.

ACKNOWLEDGMENTS

This project is supported in part by the Swiss National Science Foundation and by the Polish Committee for Scientific Research under Grant No. 2P03B05111.

REFERENCES

1. H. Burkard *et al.*, Phys. Lett. **160 B** (1985) 343.
2. W. Fetscher, H.-J. Gerber and K.F. Johnson, Phys. Lett. **173B** (1986) 102.
3. W. Fetscher and H.-J. Gerber, in *Precision Tests of the Standard Electroweak Model*, ed. P. Langacker, World Scientific, Singapore, 1995
4. I. Barnett *et al.*, Nucl. Instr. Meth. **A 455** (2000) 329.

Radioactive Muonic Atom Studies with Intense Muon Beams

P. Strasser[*], K. Nagamine[†*], T. Matsuzaki[*], K. Ishida[*], Y. Matsuda[*], K. Itahashi[*] and M. Iwasaki[*]

[*]Muon Science Laboratory, RIKEN (The Institute of Physical and Chemical Research), 2-1 Hirosawa, Wako-shi, Saitama 351-0198, Japan
[†]Meson Science Laboratory, Institute of Materials Structure Science, High Energy Accelerator Research Organization (KEK), 1-1 Oho, Tsukuba-shi, Ibaraki 305-0801, Japan

Abstract.
New intense muon beams with flux several orders of magnitude higher than at present muon facilities would allow many novel experimental studies that were until now statistically not feasible. The investigation of the nuclear properties of short-lived nuclei using muonic atom spectroscopy would become possible. In this paper, the recent progress of a feasibility study at RIKEN-RAL muon facility using the cold hydrogen film method to produce radioactive muonic atoms is being reported. The first experimental results have just been obtained with stable argon ions implanted in solid deuterium films.

INTRODUCTION

The study of muonic atoms has played since many years an important role in establishing and refining nuclear structure models, and has been successfully used to study stable nuclei. New intense muon beams with flux several orders of magnitude higher than at present muon facilities would allow many novel experimental studies that were until now statistically not feasible. The investigation of the nuclear properties of short-lived nuclei using muonic atom spectroscopy [1] would become possible. This would be a unique tool to increase our knowledge of the nuclear structure far from stability where new effects may be expected, in particular the nuclear charge distribution and the deformation properties of nuclei. Muonic X-ray measurements can yield very precise and absolute values for the charge radii and other ground state properties. It would usefully complement the knowledge obtained from electron scattering and laser spectroscopy, since in the past calibration data were used from muonic atom measurements with stable nuclei.

We proposed recently the cold hydrogen film method [2, 3] to extend muonic atom spectroscopy to the use of nuclear beams, including in the future radioactive isotope (RI) beams, and produce radioactive muonic atoms. This method would allow studies of unstable nuclei by means of the muonic X-ray method at facilities where both intense μ^- and RI beams would be available. The new neutrino factory concept to produce intense muon beams is very attractive to realize the proposed study, because the same driver beam could be used for second generation RI beam facilities. This would be a unique opportunity to combine massive amount of muons with very intense RI beams.

There are also several other experimental approaches that have been envisaged to combine muon with radioactive nuclei, including beam merging scenario and cyclotron trap [4]. Strong efforts are being reported towards possible formation and investigation of radioactive muonic atoms and even possibly radioactive antiprotonic atoms, which could be produced similarly and promise measurements of neutron distribution in nuclei.

COLD HYDROGEN FILM METHOD

The basic concept of the cold (solid) hydrogen film method is to stop both μ^- and nuclear beams simultaneously in a solid hydrogen (H_2/D_2) film, followed by the direct muon transfer reaction to higher Z nuclei to form radioactive muonic atoms [2, 3]. There are several advantages in using this method to produce and investigate radioactive muonic atoms. The system is a windowless target in vacuum, with a well-defined interaction region between μ^- and radioactive nuclei to set a compact detector system around the target. The H_2/D_2 film can easily be evaporated and replaced by a fresh one when needed. RI beam characteristics are not critical in terms of beam emittance, energy spread, and purity (if impurities are below a few percent). Also, the use of a pulsed μ^- beam combined with a magnetic confinement field around the target will help to reduce background noise by several orders of magnitude, and confine incoming μ^- within a small target area, while reducing μ-decay e^- associated background. One very important aspect to consider when using pulsed μ^- beams and measuring X-rays and γ-rays is that the counting rate per detector should be kept well below one event per pulse, otherwise the detector will be "blind" to delayed events. Large detector array with small individual solid angle will be required. It should be noted that we would benefit enormously from a machine that would produce muon beams of same intensities but with a much higher repetition rate such as 10 kHz. Each pulses would contain almost 3 orders of magnitude less muons making measurements more accessible.

RECENT PROGRESS AND FUTURE PLANS

As already reported in several short publications [5, 6, 7], an experimental program has been initiated at the RIKEN-RAL muon facility [8] to establish experimentally the feasibility of using the cold hydrogen film method to produce muonic atoms. An experimental setup (μA*) was constructed to implant stable ions in solid hydrogen films. The design and construction was described in detail in [9]. The first experimental results with implanted argon ions have just been obtained.

At first, a two-layer arrangement was used [2, 3]. The primary layer made of 0.5-mm H_2/D_2 is used to efficiently stop 27 MeV/c μ^- followed by pμ formation, pμ to dμ transfer and dμ emission with the help of the Ramsauer-Townsend effect in the diffusion of the dμ atoms, while the second layer made of pure D_2 (several μm thick) with implanted Ar ions is used to confine the production region of muonic argon (μAr) atoms to an optimized film thickness. The first targets with implanted Ar ions that were measured, only faint μAr X-rays could be detected. It was then realized that too many

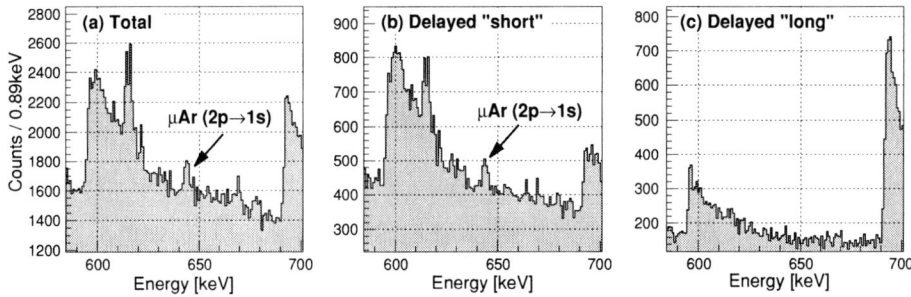

FIGURE 1. Germanium energy spectra measured with 0.5-mm $H_2/D_2 \oplus 7$-μm D_2(Ar): (a) total, (b) delayed events from +75 to +250 ns after each muon pulse and (c) delayed events from +250 ns to 32 μs after the second muon pulse, respectively.

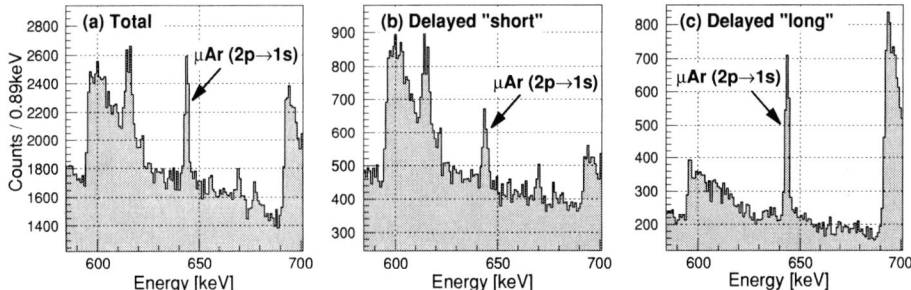

FIGURE 2. Germanium energy spectra measured with 0.5-mm D_2(Ar): (a) total, (b) delayed events from +75 to +250 ns after each muon pulse and (c) delayed events from +250 ns to 32 μs after the second muon pulse, respectively.

Ar ions were actually being implanted, and due to the large sputtering yield of solid D_2 [10], the region with implanted Ar ions was being removed by the argon beam a while later. According to Monte Carlo simulations performed with SRIM, the range of 33-keV Ar ions in solid D_2 is only about 250 nm with a range straggling of roughly 100 nm (FWHM). This sputtering yield was measured with our system by depositing a thin D_2 film (~ 1 μm) directly on the cryostat silver foil and measuring the time needed by the argon beam to remove it. When the ions reach the silver foil, the interaction between the beam and the metallic substrate produces an enhanced sputtering and the pressure in the vacuum chamber increases momentarily. A preliminary sputtering yield of 300–400 D_2 molecules per incoming Ar ions was extracted. This value is consistent with those published in ref. [10] for hydrogen ions with comparable energies.

The subsequent target was carefully made by taking into account this sputtering effect, and alternate implantation and D_2 deposition were used to built a D_2 layer with a larger thickness. The germanium energy spectra shown in Fig. 1 indicates very clearly μAr 2p→1s transition X-rays at 644 keV from the muon transfer transfer reaction. Around 10^{16} Ar ions per cm^2 were implanted in a deuterium thickness of 5-μm, corresponding to an average concentration of 500 ppm. At this concentration, all dμ atoms emitted from the primary 0.5-mm H_2/D_2 layer are expected to transfer very rapidly. The "short" delayed events in Fig. 1b are mainly due to the diffusion time of the $d\mu$ atoms before reaching the added D_2 layer. Almost no "long" delayed events can be seen in Fig. 1c, either because muon transfer already occurred or $d\mu$ atoms escaped into vacuum. The triangular peaks observed at 596 keV and 691 keV, respectively, result from inelastic neutron excitation of the Ge isotope nuclei within the detector itself [11]. The neutrons are produced following μ^- capture on nuclei and consist mainly of evaporation neutrons from de-excitation of the nucleus formed after the capture.

Later, a different target arrangement was investigated. A pure D_2 layer was used, and each argon implantation was separated from the next by depositing about 20 μm of D_2 to make a total D_2 layer thickness of 0.5 mm. Each implantation region had a thickness of only 0.1 μm with a local concentration of 1000 ppm. However, the average Ar concentration throughout the D_2 layer was around 5 ppm, which corresponds to approximately the same amount of Ar ions used in the previous target (i.e., $10^{16}/cm^2$). The germanium energy spectra shown in Fig. 2 clearly indicate that even with an inhomogeneous target very strong delayed events can be detected from the muon transfer reaction. Due to the strong reduction of the Bragg cross-section at low $d\mu$ collision energy in solid D_2 [12], the $d\mu$ atom mean-free-path is strongly increased to nearly 10–20 μm below $\sim$ 1 meV, resulting in a high $d\mu$ atom mobility and a very long diffusion length. Preliminary analysis shows that the disappearance rate of the 644-keV X-rays is consistent with an average Ar concentration of 5 ppm. At this concentration, most of the muon transfer occurs delayed, and the S/N ratio is greatly improved (see Fig. 2c).

This first test experiment was performed with a large amount of implanted Ar ions. Further measurements are planned with reduced Ar concentrations to study the muon transfer reaction and investigate the diffusion process of $d\mu$ atoms in solid. This effect will limit the layer thickness that can be used without being affect by the $d\mu$ atom loss.

As an intermediate step towards muonic X-ray spectroscopy with unstable nuclei, an experiment using long-lived isotopes is also under consideration. Radium isotopes are of strong interest since there are no stable isotopes for good measurements of nuclear parameters like the nuclear charge radius. These parameters would be urgently needed to exploit the full potential of the atom for atomic parity non-conservation studies [4].

REFERENCES

1. L. Schaller, Z. Phys. C 56 (1992) S48.
2. P. Strasser et al., Hyp. Int. 119 (1999) 317.
3. P. Strasser et al., Eur. Phys. J. A 13 (2002) 197.
4. K. P. Jungmann, Hyp. Int. 138 (2001) 463.
5. P. Strasser et al., Hyp. Int. 138 (2001) 497.
6. P. Strasser et al., J. of Phys. G 29 (2003) 2047.
7. P. Strasser et al., Nucl. Phys. A 722 (2003) 523c.
8. T. Matsuzaki et al., Nucl. Instr. and Meth. A 465 (2001) 365.
9. P. Strasser et al., Nucl. Instr. and Meth. A 460 (2001) 451.
10. B. Stenum et al., Nucl. Instr. and Meth. B 48 (1990) 530.
11. E. Gete, et al., Nucl. Instr. and Meth. A 388 (1997) 212.
12. A. Adamczak, Hyp. Int. 119 (1999) 23.

Recent Development of a point positive muon source at the RIKEN-RAL muon facility

Y. Matsuda*, P. Bakule*, P. Strasser*, K. Ishida*, T. Matsuzaki*,
M. Iwasaki*, Y. Miyake†, K. Shimomura†, S. Makimura† and
K. Nagamine†

Muon Science Laboratory, RIKEN, Wako, 351-0198, Japan
†*Meson Science Laboratory, KEK, Tsukuba, 305-0801, Japan*

Abstract. We report recent progress of slow muon beam line constructed at the RIKEN-RAL muon facility. The slow muons are muons which are re-accelerated from the one at thermal velocity. This beam is expected to have very small emittance compared with conventional muon beam, where muons are obtained from decays of pions at rest or on move. Especially, its smallness of energy spread will make it possible to apply μSR technique to the studies of thin films, multilayered systems, nanomaterials, and other objects of restricted dimensionality. Intense slow muon beams which can be realized at future neutrino factories will be a valuable tool for material science.

INTRODUCTION

The μSR technique has been successfully used for studies of condensed matter for long time. μSR stands for muon spin rotation, relaxation and/or resonance, depends on experimentalist's interests, but the principle is simple. When polarized muons are implanted to a sample, they lose energy very rapidly without depolarization and stop at preferred sites. When the muon decays, it emits positron preferentially along the spin direction of the muon. By observing positron's angular distribution and its change over a time after the muon beam is injected, we can establish local magnetic field distributions and its dynamic fluctuation at the muon implanted site. This technique complements ESR and NMR techniques for its frequency being in the middle of the two, but unlike those, it does not require external electro-magnetic field. This makes μSR technique as an unique and indispensable tool for studying, for example, magnetic systems, superconductors and conducting polymers.

One of the biggest limitation of this technique is that it requires bulk materials. Currently, the muon used for the study is formed from pion's decay at rest with energy of 4.1MeV. This means the beam penetrates the sample into a depth of 0.1mm~1mm, depends on the materials, with a wide distribution of about 20% of the stopping range.

On the other hand, we have seen very rapid development of studies of thin films, multilayers, nanomaterials, and generally objects of restricted dimentionality. These systems have attracted wide interests in science and industry, in mind of realization of new devices such as spin-FET and MRAM. In order to study such objects by means of μSR technique, we need a spin-polarized positive muon beam with tunable energies from a few eV to a few keV; we call it as slow muon beam.

Recently there has been considerable progress in generating such beam by moderating muon beam in thin layer of cyrosolids at PSI[1][2]. But it is not suited for pulsed muon beam facility like our RIKEN-RAL muon facility at ISIS, because the moderated beam's time structure preserve that of initial muon beam, which is then determined by that of proton beam. This means we will have two pulses of moderated muons coming every 20msec, each is approximately 80nsec wide and separated by about 330nsec. This poor timing resolution leads to poor resolution of muon's implanted depth and lesser sensitivity for dynamic fluctuation of magnetic field inside the sample. The J-PARC muon facility, currently constructed at Tokai in Japan, will also have similar time structure, and future proton drivers for neutrino factory and/or muon collider is also likely to be based on pulsed proton source.

Aiming to develop suitable scheme to generate slow muon beam at intense pulsed muon sources, we started to construct a slow muon beam line based on a different scheme, laser resonant ionization method, first proposed at KEK[3]. In this method, the muons are first injected to a hot tungsten film in which they lose their energy and are thermalized. The muons that diffuse close to the surface escape the film via thermonic emission of muonium with kinetic energy of about 0.2eV. This process is highly effective; several percent of initial muons are converted to muoniums[4]. The bound electron is then removed by subsequent laser ionization, leaving a free muon, which will be extracted and accelerated in a static electrical field. This scheme is best suited for pulsed muon source, because the time structure of the extracted beam is mainly determined by the size of ionization region and time structure of laser pulse, not that of initial beam pulse. Combined with accurate timing of laser irradiation, this would give us even better energy resolution of the slow muon beam compared with current PSI slow muon beam.

Nagamine has suggested that this scheme could also be considered as an alternative beam cooling scheme for future muon collider. Feasibility study of such use of slow muon beam is also in our scope.

RECENT PROGRESS AND FUTURE PERSPECTIVES

Our scheme is similar to that of laser ion sources for (un)stable nuclei, but differs at the frequency of laser; we use a pulsed Lyman-α (122nm) laser to resonantly excite muonium from 1S ground state to 2P state, and another 355nm laser to photo-ionize muonium from 2P state to unbound state. We have constructed a sophisticated laser system to generate Lyman-α wavelength to ionize muonium atoms efficiently. A detailed explanation of our laser system can be found elsewhere[5]. The first slow muon beam was successfully observed at the facility in 2001 at the rate of 0.03μ/sec from initial beam intensity of $5 \times 10^5 \mu$/sec as previously reported[6].

In order to improve the situation, we have introduced phase matching scheme at the generation stage of Lyman-α, where resonantly enhanced sum-difference frequency mixing process occurs in Kr gas. The negative phase mismatch of Kr gas for Lyman-α frequency is compensated by a positive phase mismatch of a buffer gas, like Ar. We also have placed a retractable retro-reflector for Lyman-α to increase energy density at

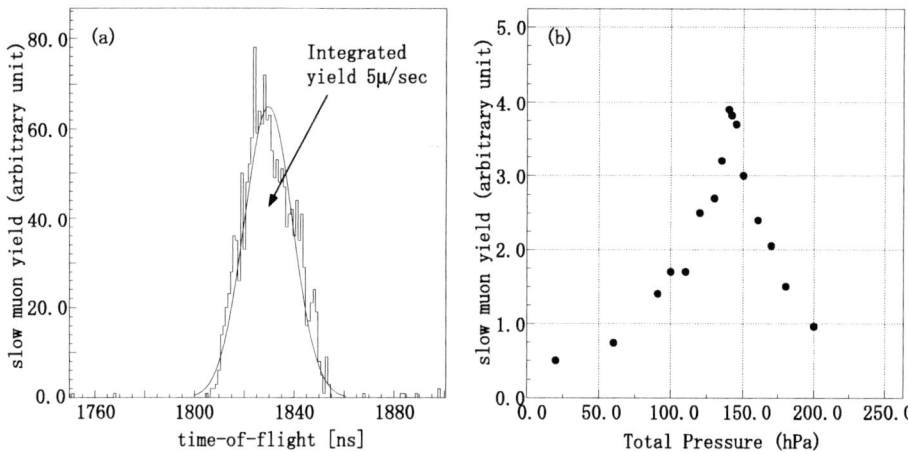

FIGURE 1. (a) Time-of-flight spectrum of slow muons is shown. The muons are accelerated to 7.5kV and transported to the target position which is about 3.9m away from the ionization position. (b) The yield of slow muon is plotted against different pressure of Kr-Ar mixture gas. The partial pressure of Kr gas is kept at 20hPa. The yield peaks at 140hPa, corresponding to mixing ratio Kr:Ar=1:6.

the ionizing region.

After a failed beam time at 2002, we realized that our muon beam line had not been tuned properly after construction of a new beam line which took place between the first beam time and the second beam time. The problem was addressed using a newly designed scintillator hodoscope as a beam profile monitor at the beginning of 2003.

In last beam time at April 2003, we observed about 5μ/sec slow muons, more than 100 times improvement over our first beam time. Figure 1(a) shows the time-of-flight spectrum of slow muons after being accelerated to 7.5kV and transported to a MCP detector which was placed approximately 3.9m away from the ionization region. The peak shows a Gaussian shape with FWHM of about 25nsec. The width is much shorter than that of initial muon beam, confirming an advantage of this scheme when it is applied to pulsed muon beam. The width is larger than that of laser pulse, though, because finite size of ionizing region gives different acceleration voltage and flight length to each slow muons.

Figure 1(b) shows the merit of phase matching in Kr-Ar mixture gas. We kept partial pressure of Kr at 20hPa, while changing partial pressure of compensating Ar gas from 0hPa to 180hPa. The yield of slow muon is plotted against total pressure of mixed gas. The yield peaked at total pressure of 140hPa, which corresponds to Kr:Ar ratio of 1:6 and agrees with theoretical calculation.

In next months, we are installing a 2-dimensional position sensitive MCP detector to the detector position, with which we will be able to measure the size of slow muon beam. Optimization of the size of the ionizing region and transport optics to minimize the size and the time-width of the slow muon beam will be carried out, as well as the measurement of beam emittance.

We are also working on the laser system; in fact, several difficulties in out laser system were limiting the power of Lyman-α light, by factor of 4 to 10 compared to specification. The 355nm laser will be also replaced with more powerful system. The signal showed almost linear dependence on energies of Lyman-α and 355nm pulses, suggesting we can expect significant improvement in slow muon yield.

Currently a design work of a new μSR spectrometer is underway. We will first measure the depth resolution of the slow muon beam by measuring change of signal from Au/SiO$_2$ composite while changing implantation energy. μSR signal from Au layer and SiO$_2$ layer will be clearly distinguished because of large difference of rotation frequency between free muon and muonium.

With improved intensity of slow muon beam, we then would go to studies of multi-layers which consists of alternative ferromagnetic and non-magnetic layers. The Giant Magnetic Resistance effect is first discovered in this system, and extensive efforts have been taken to understand the origin of this effect. Though we believe the nature of the interlayer exchange coupling between ferro-magnetic layers through a intermediate non-magnetic layer is well established theoretically and experimentally, we still lack the experimental data directly measuring induced electron spin in the non-magnetic layer. The μSR technique with slow muon beam will offer a unique tool for this purpose thanks to its high sensitivity to spin polarization and depth-selectivity.

Slow muon beam at the RIKEN-RAL muon facility has a drawback, though, which is limited statistics. When compared with intensities of muon beams currently available for bulk materials, it is clear that we need to improve current yield of a few/sec to a few thousands/sec. This can only realized with much higher intensity of initial muon beam, which will be available at future neutrino factories; indeed these factories will allow many novel experimental studies not only for neutrino physics.

ACKNOWLEDGMENTS

This research was partially supported by the Japanese Ministry of Education, Science, Sports, and Culture, Grant-in-Aid for Scientific Research (B), 2000, 11559019.

REFERENCES

1. Morenzoni, E., Kottmann, F., Maden, D., Matthias, B., Meyberg, M., Prokscha, T., Wutzke, T., and Zimmermann, U., *Physical Review Letter*, **72**, 2793–2796 (1994).
2. Morenzoni, E., Khasanov, R., Kuetkens, H., Prokscha, T., Suter, A., Garifianov, N., Gluckler, H., Birke, M., Forgan, E., Keller, H., Litterst, J., Niedermayer, C., and Nieuwenhuys, G., *Physica B*, **326**, 196–204 (2003).
3. Nagamine, K., Miyake, Y., Shimomura, K., Birrer, P., Marangos, J. P., IUwasaki, M., Strasser, P., and Kuga, T., *Physical Review Letter*, **74**, 4814–4817 (1995).
4. Mills Jr., A. P., Imazato, J., Saitoh, S., Uedono, A., Kawashita, Y., and Nagamine, K., *Physical Review Letter*, **56**, 1463–1466 (1986).
5. Bakule, P., Matsuda, Y., Miyake, Y., Strasser, P., Shimomura, K., Makimura, S., and Nagamine, K., *Apectrochimica Acta B*, **58**, 1019–1030 (2003).
6. Matsuda, Y., Bakule, P., Strasser, P., Matsuzaki, T., Ishida, K., Watanabe, I., Miyake, Y., Shimomura, K., Makimura, S., and Nagamine, K., *Physica B*, **326**, 217–221 (2003).

A Review of Target Developments in Europe

J. R. J. Bennett

CCLRC, Rutherford Appleton Laboratory, Chilton, Didcot, Oxon OX11 0QX, UK

Abstract. A review is given of neutrino factory targets and their developments in Europe.

INTRODUCTION

The European neutrino factory [1] is based on designs that provide proton beams of 2-30 GeV energy, pulsed at 50 Hz. The pulse is a few µs long and consists of a train of 1-2 ns long micro-pulses. The target absorbs an average power of 1 MW from the 4 MW (average) proton beam. A suitable beam dump absorbs the remaining 3 MW. The target will be 20 cm long and ~2 cm in diameter, giving an average power density in the target of 16 kW cm^{-3}. The energy density per pulse is 300 J cm^{-3}. The high power density creates severe problems in dissipating the heat and the short pulses produce thermal shocks due to the rapid expansion of the target material. These shocks can potentially exceed the mechanical strength of solid materials.

In addition the pions and muons created in the target must be collected in a 20 Tesla solenoidal field or a magnetic horn. This imposes strong restraints on the target and collector system which must ultimately be designed as a single entity.

Several targets which potentially can withstand the huge power density are currently being considered: a) a liquid metal (mercury) jet; b) a contained flowing liquid metal (mercury), c) a granular solid target, d) a tantalum rotating toroid, thermally radiating at ~2300 K.

THE LIQUID METAL JET

The proton beam is passed through the liquid metal jet. Figure 1 shows a jet curving under the influence of gravity and the beam passing tangentially through the jet. The jet will disintegrate with each proton pulse and then reform in time for the next pulse. Thus, shock waves are not a problem. The target can dissipate at least 10 MW of power and hence there is plenty in reserve for future developments. The "target" is easily removed by draining the system into a suitable container, thus reducing activity in the target area when maintenance is required. There are no beam entry or exit windows on the target itself. Mercury is a good candidate for the fluid since it is liquid at room temperature. It would be possible to distil the mercury during operation to remove impurities, including some of the radioactive products.

Thus, the mercury target has important advantages, but precautions must be taken to ensure that the mercury does not contaminate other parts of the machine and is not

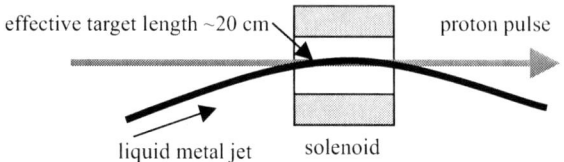

FIGURE 1. Schematic diagram of the liquid metal jet target.

released into the environment. Windows will be required both sides of the target area. The mercury is expelled from the jet at considerable velocity on being hit by the beam and the interior of the target chamber may need protection.

Ravn [2] has suggested the use of a hollow nozzle which would allow the beam to pass through the centre of the nozzle and into the jet. It also has applications for high power converter targets for radioactive beams. An alternative to the liquid metal jet is the liquid metal curtain [3,4] where the fluid is dropped down through the beam in a strip, 20 cm in length. A variation is to send the jet up though the beam and then allow it to fall back under gravity. The fluid curtain is not being pursued at present.

Tests already made in the USA and Europe have shown that the jet can be successfully achieved in the intense magnetic field of the collection solenoid [5] and a high intensity proton beam has been successfully used on a mercury jet [6]. The magnetic field improved the jet formation and provided damping. Thus, the scheme appears viable.

However, the combination of full mercury flow and velocity in the full magnetic field with a proton beam of the required intensity has still to be demonstrated. There are only two facilities in Europe capable of delivering suitable high power intensity pulsed beams: ISIS at RAL, UK and ISOLDE (and other parts of the accelerator complex), CERN, Geneva. Several proposals have been put forward to do in-beam tests at these facilities but considerable expenditure is needed to set up the equipment and to provide the beam time against competition with other established programmes.

THE CONTAINED LIQUID METAL TARGET

The contained fluid target is shown in Fig. 2. The target requires beam entry and exit windows. A counter flow form is also possible similar to that proposed for the high power neutron targets of the SNS [7]. It is likely that the contained liquid metal target design will suffer from the same shock problems and erosion that is being encountered with the SNS target design. These problems are actively being pursued for the SNS and when resolved could generate a viable target design for neutrino factories.

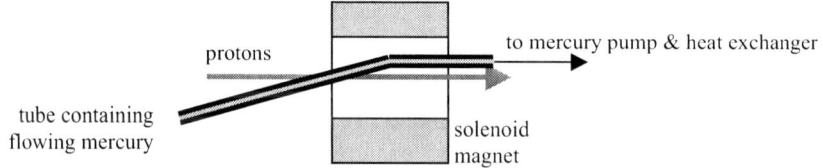

FIGURE 2. Schematic diagram of a contained liquid metal target.

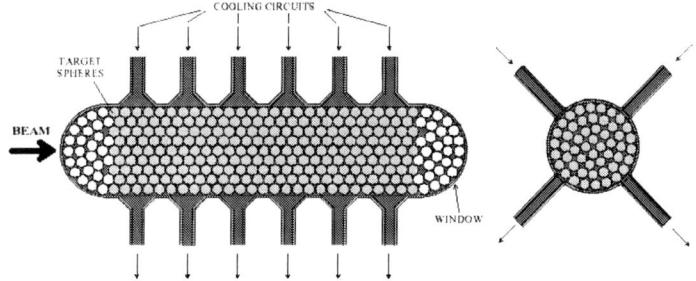

FIGURE 3. Schematic sections of the granular target

THE GRANULAR TARGET

Thermal shock is a problem when the length of the pulse is shorter than the time taken for the speed of sound in the target to cross its largest dimension [8]. Thus by making the target from small pieces of solid material the shock problem is removed. This is the principle of the design by Sievers [9], shown in Fig. 3. The granules of tantalum are cooled by flowing helium gas. It has been proposed to make 4 such targets within magnetic horns as collectors, steering the proton beam into each target on successive pulses. Only 250 kW will be dissipated in each channel, increasing the target life. Finally the pions are funneled into a single decay channel [10].

THE LARGE SOLID TARGET

A radiation-cooled tantalum rotating toroid (Figure 4) has been proposed [11] (based on a water-cooled rotating band by King [12]). It operates at high temperature (2300 K) to remove the heat by thermal radiation to the water cooled surroundings. The toroid will be levitated, driven and guided by electromagnetic forces and can dissipate tens of megawatts of power depending on its size and rotational speed.

The target may suffer from shock damage. However, comparing these targets with the pbar target at FNAL [13] and positron target at SLAC [14], which operate at up to two orders of magnitude higher energy density, it is considered that they will survive the shocks for at least 10^9 pulses to give an operational life of one year.

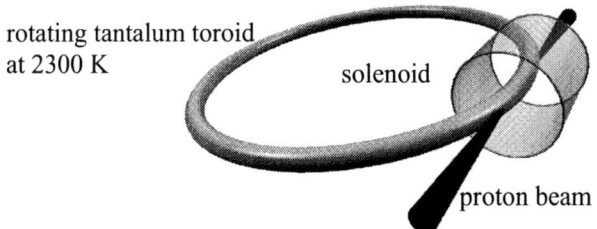

FIGURE 4. Schematic diagram of the rotating toroid

TARGET DEVELOPMENTS

At present there is little funding for development of neutrino factory targets. Pulsed electron beam shock tests [15] on thin tantalum foils will be continued by RAL. However, a small award has been made by the EU towards continuing networking. A proposal has been made to the UK funding agency, PPARC, to fund studies of a radiation cooled rotating toroid. This would include: measurements of the mechanical properties of tantalum at high temperatures under shock conditions; modeling of the shock to predict the effect of different geometries and the pulse lifetime; levitation, drive and stabilisation modeling; in-beam tests at ISOLDE ($\sim 10^5$ pulses) and lifetime tests on ISIS. A proposal is to be made to the EU for a design study, in which it will be proposed that mercury jets, granular targets and toroids be studied.

REFERENCES

1. Haseroth, H.D., Neutrino Factory R&D in Europe, see the proceedings of this conference.
2. Ravn, H. L., the ISOLDE Collaboration and the Neutrino-Factory Working Group, *Nucl. Instr. and Meth. B* **204** 197–204 (2003).
3. Johson, C., *Targetry Workshop*, LBNL, 1999. http://cdj.home.cern.ch/cdj/jet-targ/jet-targ/index.htm
4. Sievers, P., Moving and Stationary High-power Targets for Neutrino Factories, *Workshop on High-power Targetry for Future Accelerators*, Ronkonkoma, Long Island, N. Y., September 2003, http://www.cap.bnl.gov/mumu/conf/target-030908/tuesday_pm/Sievers5.pdf
5. Kirk, H. et al., in: Lucas, P. & Webber, S. (Eds.), *Proc. 2001 Particle Accelerator Conf.*, Chicago, IL, June 2001, Vol. 2, p. 1535.
6. Fabich, A., Lettry, J., Experimental observation of photon induced shocks and magneto-fluid-dynamics in liquid metal, *Proc. Int. Workshop NuFACT01*, Tsukuba, Japan, May 2001, http://psux1.kek.jp/~nufact01/ and *Nucl. Instr. and Meth. A*, **503** 336-339 (2003).
7. Status of the Spallation Neutron Source, Norbert Holtkamp for the SNS collaboration, EPAC 2003, http://warrior.lbl.gov:7778/pacfiles/papers/MONDAY/AM_ORAL/MOAL003/MOAL003.PDF
8. Sievers, P., Elastic Stress Waves in Matter due to Rapid Heating by an Intense High-energy Particle Beam, CERN Internal Report, LAB.II/BT/74-2, 26 June 1974.
9. Sievers, P., *Nuclear Instruments and Methods A* **503** 344-347 (2003).
10. Autin, B., Sievers, P., Verdier, A., Meot, F., Funnelling Pions and Muons, *Proceedings of this Workshop*.
11. Bennett, J. R. J., Densham, C. J., Drumm, P. V., A High Power Radiation Cooled Target for a Neutrino Factory, European Particle Accelerator Conference 2000, http://accelconf.web.cern.ch/accelconf/e00/PAPERS/MOP2A06.pdf
12. King, B. J., Moser, S. S., Weggel, R. J., Mokhov, N. V., "A Cupronickel Rotating Band Pion Production Target for Muon Colliders", in *Proceedings of the 1999 IEEE Particle Accelerator Conference*, New York City, NY, U.S.A., 29 March-2 April, 1999, 3041-3043, IEEE 99CH36366.
13. O'Day, S., Bieniosek, K., Anderson, K., "New Target Results from the Antiproton Source", in *Proceedings of the 1993 IEEE Particle Accelerator Conference*, Washington, DC, 17-20 May, 3096-3098, IEEE, 1993.
14. Stein, W., Sunwoo, A., Bharadwaj, V. K., Schultz, D. C., Sheppard, J. C., Thermal Shock Structural Analyses of a Positron Target, in *Proceedings of the Particle Accelerator Conference 2001*, Chicago, 2001, http://www-project.slac.stanford.edu/lc/local/systems/Injector/Talks%20and%20 Papers/PAC2001/Thermal%20Shock%20WPAH015.pdf
15. Paul Drumm and Chris Densham, Material Testing with Electron Beams for Neutrino Factory Targets, Proc. 2001 Particle Accelerator Conference, Chicago, 2001 http://accelconf.web.cern.ch/AccelConf/p01/PAPERS/TPAH157.PDF

MECO Production Target Development

J.L.Popp[1]

University of California, Irvine
Dept. of Physics and Astronomy, 4129 Reines Hall
Irvine, CA 92697-4575

Abstract. Production target design for the Muon-Electron Conversion Experiment has undergone significant evolution recently. Limited progress in producing a suitable radiation-cooled target has spurred interest in a water-cooling system with drastically lower operating temperatures. This system results in less than about five percent reduction in stopped muon yield compared to a similar radiation cooled unit. The theoretical and experimental research surrounding target design is discussed.

INTRODUCTION

MECO, the **M**uon-**E**lectron **CO**nversion experiment, is a search for direct evidence of coherent conversion of a muon into an electron in the field of a nucleus, $\mu^- N(A,Z) \to e^- N(A,Z)$, sensitive to one event for 5×10^{16} muon captures $\mu^- N(A,Z) \to \nu_\mu N(A,Z-1)$. Such an observation would have a profound effect on our understanding of elementary particle physics and cosmology. In fact, even a null result would still set an upper limit on the branching fraction approximately 10^{-4} times lower than the current one, a sensitive test of the Standard Model. An overview of this project can be found in [1–3].

The production target is nominally the size and shape of a pencil, and constructed out of a high-density metal; it lies in an axially-graded solenoidal magnetic field, designed to collect pions over a large solid angle. In MECO, a high-intensity proton beam, with energy $\sim 7.5\,\mathrm{GeV}$, strikes the target end-on to produce pions which decay into muons; the beamline then filters out unwanted particles and guides negative muons to the stopping target where, after slowing down through ionizing collisions, many quickly become bound in Coulomb states about target nuclei. To produce some 10^{17} stopped muons in the planned 30-week running time MECO must aim 4×10^{13} protons/sec at the production target. The Alternating Gradient Synchrotron (AGS) at Brookhaven National Laboratory (BNL) is especially suited to provide the required beam energy and intensity, as well as pulsed microsecond time structure with high beam extinction between pulses, a feature crucial to rejecting prompt backgrounds.

The target heating scenario is dictated by the macroscopic time structure of the beam; namely, the AGS cycle time of one second and duty factor 0.5. For one candidate target material, tungsten, a 16 cm long, 4.0 mm radius rod weighs about 155 g and receives an instantaneous power of 10 kW, i.e., a time average of 5 kW.

---

[1] On behalf of the MECO Collaboration

Critical to achieving a high muon yield, we must collide protons on a target with as large an atomic number as possible. Yield, as well as cooling, depends on target geometry. More generally, muon yield is sensitive to material anywhere in the clear bore of the production solenoid. Hence, the target, cooling system, and mechanical support must be compact and introduce little additional mass beyond that of the target, thus minimizing pion reabsorption. The target mounting must be sufficiently insensitive to vibrations to maintain good positioning. Furthermore, temperature must be controlled to avoid operating too close to the melting point and avoid thermal stress levels that may lead to mechanical failure or geometrical distortion.

Two cooling strategies have received significant study: radiation and forced convection with water as the coolant. Radiation cooling is attractive because it provides a high muon yield, is simple to operate and virtually maintenance-free. On the other hand, high operating temperatures lead to evaporation of some of the target material, requiring strategies to control contamination of the rest of the muon beam line and detector area, and require good control of thermal stresses. Convective cooling makes drastically lower operating temperatures possible, extending the possibility of achieving even greater muon beam intensities. An essential requirement for such a system is to maintain a non-boiling condition, since gas formation would impede flow through small water channels. This type of cooling provides more latitude in selecting target materials with higher densities and thermal conductivities, thus raising the possibility of reduced thermal stresses and an increased useful lifetime of the target. However, this target cooling system requires greater maintenance, and significantly more system design consideration. Protocols for coolant storage and disposal would be required.

RADIATION COOLING

For a radiatively-cooled target, refractory metals are often the materials of choice. Tungsten was selected for MECO because it has the highest melting point, ~ 3683 K, and thermal conductivity of all pure-metal refractories; in addition, its yield strength is better than most materials at high temperatures. The power distribution along the length of the MECO target is peaked within the first two to four centimeters and slowly decays over the remainder, a feature reflected in the temperature distribution shown in Figure 1 for an early target design. The radial power distribution in this figure was approximated with a Gaussian. Note that the maximum core temperature near the power deposition peak is 3231 K with a surface temperature there of 3113,K.

Thermal stress analysis of the radiation-cooled single solid cylinder has shown that this geometry is mechanically unstable. Redesign efforts aimed at reducing operating temperature and thermal stresses have sought to add high emissivity coatings, divide the target into thin slices & space them out along the beam, and adding to these smaller pieces slots both along & perpendicular to the target axis. The results of this investigation has produced a significantly reduced operating temperature, with the maximum between 1977-2032 K on spill. This target consists of 2.0 mm slices, spaced by 8.0 mm along the beam, assuming the surface is treated to obtain an emissivity of 0.9. Unfortunately, this modification of a single rod makes the length of the target unacceptable, due to

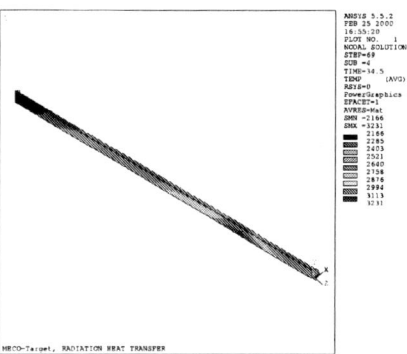

FIGURE 1. On-spill temperature distribution for radiatively-cooled tungsten target with radius 4.0 mm and 16 cm length. Temperatures are in Kelvins. ANSYS calculation courtesy of C. Pai, BNL.

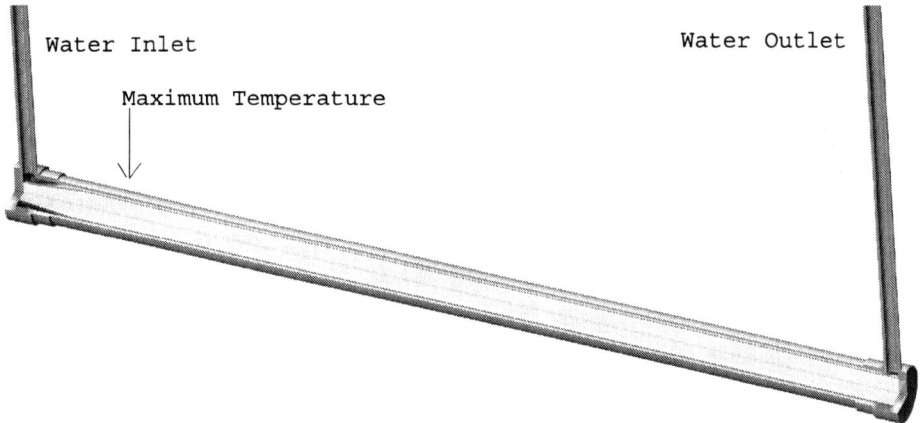

FIGURE 2. The MECO target and coolant channel. The beam strikes the target on the left, parallel to the rod axis. CAD drawing courtesy of B. Christensen, UCI.

the increased complexity of the supports and greater probability for pion reabsorption. Research on radiation cooling continues.

FORCED CONVECTIVE COOLING

A water-cooled target has been designed to cool using a high velocity water stream surrounding a cylindrical target in an annular coolant channel. The current design, shown in Figure 2, uses a gold or platinum rod, 16.0 cm long with a 3.0 mm radius. The gap size of the cooling channel is 0.3 mm. With an inlet water temperature of 20 C and a volumetric flow rate of one gallon per minute the maximum target surface temperature (at the outlet end) can be held at 71 C. The results of our design calculations are shown

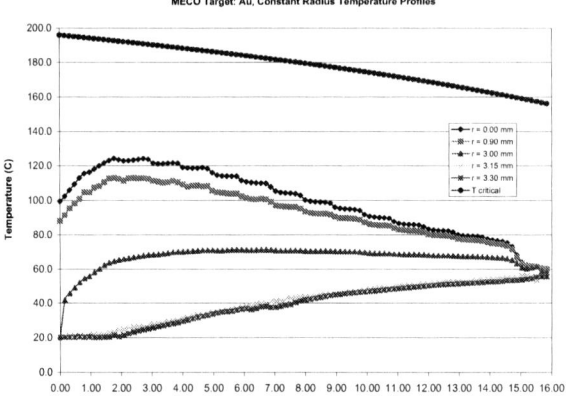

FIGURE 3. Water-cooled target and coolant temperature at fixed radii as a function of position along the length. These results are for 9.5 kW steady state conditions.

in Figure 3. Target surface temperature is shown at $r = 3.0$ mm. The top curve labeled T_{critical} gives the critical temperature for boiling assuming 25 cm inlet & outlet pipes.

In circular ducts the transition from stable laminar flow to fully-developed turbulent flow occurs over the range of Reynolds numbers Re: 2100-4000. The water flow velocity in the MECO target is above 10.5 m/s, leading to local values of Re: 12000-24500. Our calculations suggest that the pressure drop from inlet to outlet is below 130 psi, a fact confirmed in flow tests performed on a target much like that shown in Figure 2. A slight taper was introduced to reduce the pressure drop to about 110 psi. The water containment shell, endcaps, inlet, and outlet pipes are made of a high strength titanium alloy. The tube surrounding the target rod has a wall thickness of 0.5 mm. The inlet & outlet pipes are 25 cm long with inner and outer diameters 2.1 mm and 3.2 mm, respectively. Measurements at the above flow rate show that the pressure drop across each pipe is 67 psi.

Cooling tests are underway using induction heating. The first round of tests aimed at evaluating the performance of our test setup, induction coil, and water containment shell have been successfully completed. This first test was able to investigate a retricted set of target parameters. The next series of tests will aim for investigating and confirming our design calculations for a broader set of flow parameters; particularly in regards to channel size and flow rate. We have also devised methods for obtaining greater control over position in making target surface temperature measurements.

REFERENCES

1. Yamin, P., "The Meco Experiment," in *NUFACT'03: 5th International Workshop on Neutrino Factories and Superbeams*, edited by A. Para, AIP, New York, NY, 2004, pp. 212–213.
2. Molzon, W., *Nuclear Physics B Proceedings Supplements*, **111**, 188–193 (2002).
3. Popp, J. L., *Nuclear Instruments and Methods in Physics Research A*, **472**, 354–358 (2001).

Targetry R&D for PRISM Project

K. Yoshimura*, H. Ohnishi[†*], T. Nakamoto*, A. Yamamoto*, Y. Ajima*, M. Aoki**, N. Fukasawa[‡*], K. Ishibashi[†], Y. Kuno**, T. Miura*, K. Nakahara**, N. Nosaka**, M. Numajiri*, T. Ogitsu*, A. Sato**, A. Yamanoi*, B. Autin[§] and Peter Sievers[§]

*High Energy Accelerator Research Organization (KEK), Tsukbua 305-0801, Japan
[†]Kyushu University, Fukuoka 812-8581, Japan
**Department of Physics, Osaka University, Toyonaka 560-0043, Japan
[‡]Tokyo University of Science, Noda 278-8510, Japan
[§]CERN, CH-1211 Geneve 23, Switzerland

Abstract. A conceptual design of the targetry for the high intensity slow muon source is presented. Based on simulation studies, critical issues are discussed and a baseline design is determined. We have started R&D works on superconducting solenoids as well as conducting targets, which is an alternative option.

INTRODUCTION

PRISM (Phase Rotation Intense Slow Muon source) is a dedicated secondary muon beam with high intensity and narrow energy spread[1]. Figure 1 shows the schematic layout of a FFAG-based PRISM facility, which comprises a targetry section, a matching section, and a FFAG phase rotator. The goals of intensity and energy spread are 10^{11}–$10^{12}\mu$/sec and less than 5%, respectively.

The targetry section, which is composed of a pion production target and a superconducting capture solenoid, is one of the critical parts in the PRISM project. The components suffer heavy radiation and heat caused by high power proton beams. It is a technical challenge to develop the system which can be operated under such a environment. In this paper, we present the baseline design of the targetry for PRISM based on simulation studies. R&D activities on superconducting solenoids are introduced. A conducting target, which is now considered as an alternative option, will be briefly presented.

TARGETRY FOR PRISM

We studied pion production and capture by using various simulation code, such as MARS, GEANT3, and FLUKA. PRISM would provide a low energy muon beam ($p = 68$MeV/c) which is suitable for stopped muon experiments. We survey the parameters such that the pion yield become maximum in the low energy range (50 MeV/$c < p < 140$ MeV/c). The results are as follows:

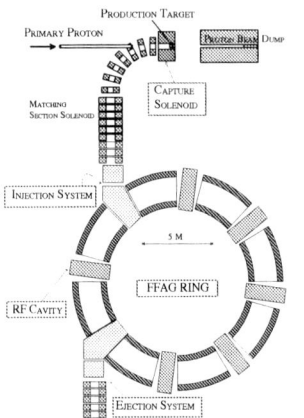

FIGURE 1. Schematic layout of the FFAG-based PRISM

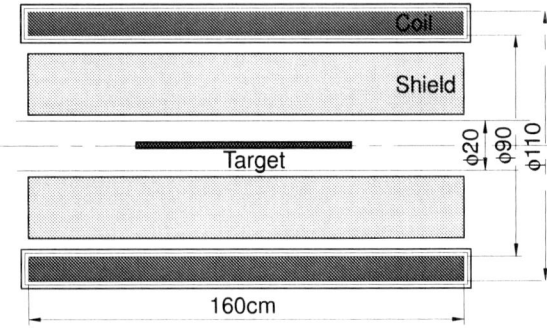

FIGURE 2. Baseline design of the Targetry for PRISM

TABLE 1. Targetry properties for the baseline option

Parameters	Values
Magnetic field (B)	6 Tesla
Warm bore radius	10 cm
Total length	160 cm
Target	Graphite L=2λ=80 cm
Shield	Tungsten 25 cm
Coil Inner radius (IR)	450 cm
Coil Thickness	10 cm
Cooling power	500 W

Backward capture

In the low energy range, pions are produced almost symmetrically in forward- and backward- direction. To avoid high energy pion background, we adopt backwards pion capture as shown in Fig.1.

Target material and dimension

To collect slow pions, heavy material with high Z is better than light material. For example, tungsten gives twice yield as graphite. However, heat deposit and thermal shock may cause damage of the heavy metal target. According to the past R&D, graphite target is considered to survive up to 1.5 MW. At present stage, we choose graphite target as a baseline option and continue to pursue heavy targets for upgrade options.

Superconducting Solenoid

Simulation study shows that 25 cm-thick tungsten shield is necessary to reduce heat deposit in the superconducting magnet. That implies that the coil must have a large bore with 80 cm in diameter . It is also suggested that higher magnetic field is

FIGURE 3. PRISM R&D Model Magnet

FIGURE 4. Beam test to measure heat load at KEK 12 GeV PS

desired since yield is roughly proportional to the field strength. Simple calculation shows that the stored energy of such a 12 Tesla magnet reaches 190 MJ as compared to 20 MJ in case of 6 Tesla magnet. Compromising on various condition, e.g, costs and maintenance, we choose 6 Tesla magnet as a baseline option and expect the future R&D progress for upgrade over 10 tesla.

Taking into account above consideration, we have determined the baseline(minimum) design parameters for the PRISM targetry as shown in Fig.2 and Table.1.

R&D WORKS

R&D works on the targetry for the PRISM have already been started as follows:

Model Magnet

As a prototype of capture solenoid, we have developed a hybrid coil which produced 10.6 T. It is composed of three coils, i.e., High TC, Nb3Sn, and NbTi from inside to outside. All coils are conductive cooled by GM cryocooler. Fig.3 shows a photograph of the magnet.

Beam test for measurement Heat load test

Since design parameters strongly depend on the calculations of heat deposit in the super conducting solenoid, we should carefully evaluate of the calculations. We have carried out the beam experiment to measure heat load directly for calibrate simulation code (Fig.4). Detailed description is presented in other paper in this proceedings[2]. The preliminary result shows that MARS simulations well reproduce the direct measurements with a level of 20% - 30%.

Further R&D plans

Now we have started to investigate cooling scheme for radiation heat load of 500 W. Following two methods are now in consideration; Active forced flow cooling and indirect convection cooling. A preliminary cooling test have been started and we plat to build half-sized R&D coil for the further study.

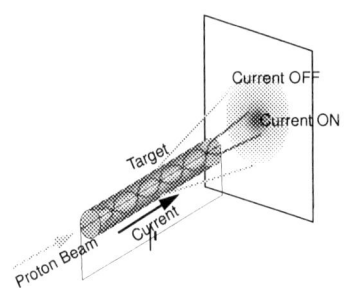

FIGURE 5. Principle of conducting target

FIGURE 6. Test Mercury loop for R&D of conducting target

CONDUCTING TARGET

Basic concept of the conducting pulsed target is schematically shown in Fig.5 [3]. By applying strong pulsed current to the target, produced secondary particles travel inside the target, confined by the toroidal focusing field. As a result, a low emittance beam comes out from the end of the target. Advantages of this scheme are as follows: 1) Beam has low emittance at production as compared to solenoidal capture, 2) No need for additional capture device, e.g., superconducting solenoid. 3) Large space outside the target, which can be used for cooling or mechanical structure. Simple calculations show that if we apply 2 MA pulsed-current, we obtain more pion yield than that in case of solenoidal capturing with 20 Tesla.

Although technical issues, such as energy deposit and electromagnetic forces by the pulsed current, are challenging, previous work on antiproton target and lithium lens may help. We have started R&D by using test mercury loop and target containers (Fig.6). We will make proof of principle experiment and study electromagnetic force as well as erosion and pitching effect in further R&D.

SUMMARY

We have studied targetry method for the PRISM project and determined the baseline design parameters. The next step is to manufacture R&D model coil to study how to cool the heat load of 500 W. As an alternative option, conducting pulsed target has been studied. A proof of principle experiment is prepared by using the test mercury loop.

ACKNOWLEDGMENTS

K.Y was supported by a Grant-in-Aid for Scientific Research from Japanese Ministry of Education, Culture, Sports, Science, and Technology.

REFERENCES

1. K. Yoshimura, Nucl. Instr. and Meth. **A 503** (2003) 254 (See http://www-prism.kek.jp).
2. Ohnishi et. al, presented at this workshop.
3. B. Autin, S. Gilardoni, and P. Sievers, Nucl. Instr. and Meth. **A 503** (2003) 348.

Preliminary Optimization of the Pion Capture and Decay Channel

K. Paul* and C. Johnstone[†]

*University of Illinois at Urbana-Champaign, Urbana, IL
[†]Fermi National Accelerator Laboratory, Batavia, IL

Abstract. A preliminary study of the pion decay channel in the US neutrino factory design has been completed, assuming a 1 MW beam of 16 GeV protons incident on the graphite target described in the first neutrino factory feasibility study. Preliminary data suggests that improvements in adiabaticity and increased field strength in the decay channel can increase muon yields as much as 20%, and possible more. The data also suggests that the muon beam can possibly be captured into a smaller emittance if a higher field strength is used in the decay channel solenoids.

INTRODUCTION

A good deal of time has passed since the second neutrino factory feasibility study (FS2) [1] in which the design of the pion capture and decay channel was last considered. While the final design and full optimization of the channel cannot be completed without knowing the details of the upstream and downstream elements, it is useful to investigate the channel's optimizable properties to determine if there is any room for improvement, and if so, then for what reasons and by how much. This article will serve as a *preliminary* analysis of the channel's general properties in the hopes of uncovering a means of extracting greater muon yields and possibly capturing the muon beam into a smaller emittance from inception.

CHANNEL DESIGN

For this study, we have returned to the graphite target design introduced in the first feasibility study (FS1) [2]. This is an 80-cm long, 1.5-cm diameter graphite target suspended in a 20-T capture solenoid at and angle of 100 mrad with respect to the solenoid central axis. It is assumed that a 1 MW, 16 GeV beam of protons is incident on the target and aligned with the center of the 100 mrad tilted target. This is a conservative target design, and considering more liberal designs involving mercury jets and higher energy proton beams will also improve muon yield. We hope to obtain desirable muon yields from the graphite target design after optimization of the capture and decay channel. No modification of the FS1 target design is considered.

The capture solenoid field strength, B_0, is chosen to capture pions of a given transverse momentum,

$$p_T < eB_0 \left(\frac{R_0}{2}\right), \qquad (1)$$

where R_0 is the radius of the beampipe within the capture solenoid. From the design of the downstream components, such as the phase rotation system and the buncher, the final aperture of the beampipe and the solenoidal field strength are determined, fixing the magnetic flux at the end of the decay channel,

$$\Phi_f = B_f (\pi R_f^2) = 0.35 \text{ Wb}, \qquad (2)$$

where the final solenoidal field strength is $B_f = 1.25$ T and the final radius of the beampipe is $R_f = 30$ cm. By transporting the beam adiabatically from the target to the end of the decay channel, the magnetic flux must be conserved. Hence, the chosen maximum captured transverse momentum fixes the beampipe radius at the target,

$$p_T < eB_0 \left(\frac{R_0}{2}\right) = \frac{e\Phi_f}{2\pi R_0}. \qquad (3)$$

The pion transverse momentum spectrum peaks near $p_T \approx 150$ MeV/c, and dropped to less than half its maximum at $p_T = 225$ MeV/c. Thus, we see that $p_T < 225$ MeV/c is a reasonable cutoff, which then fixes the beampipe radius at the target to $R_0 = 7.5$ cm, and due to conservation of flux, it also fixes the field strength of the capture solenoid to $B_0 = 20$ T.

To adiabatically match the beam from the high-field capture region at the target to the low-field decay channel, we simultaneously vary the field strength and beampipe radius with arclength, s, downstream of the channel such that the contained magnetic flux remains constant, $B_0 R_0^2 = B(s) R(s)^2$. Adiabaticity demands that the change in the field be slow such that the Larmor radius of the particles orbit, $a = \frac{p_\perp}{eB}$, satisfies

$$a \ll R_B = \frac{1}{\kappa_B}, \quad \text{and} \quad a \ll B\left(\frac{\partial B}{\partial s}\right)^{-1}, \qquad (4)$$

where R_B and κ_B are the radius of curvature and the curvature of the magnetic field lines, respectively. Hence, we can satisfy both conditions by choosing an appropriate *radius function*, $R(s)$, and determining the on-axis field strength from conservation of flux,

$$B(s) = B_0 \left(\frac{R_0}{R(s)}\right)^2. \qquad (5)$$

The motion of the particles in such adiabatic matching sections are most easily understood in terms of the adiabatic invariants Ba^2 and $\frac{p_\perp^2}{B}$ [3]. As the particles travel from a high-field region into a low-field region, their Larmor radius increases, and their transverse momentum decreases. Due to conservation of kinetic energy in the magnetic field, a decreasing transverse momentum indicates an increasing longitudinal momentum. Hence, the beam divergence decreases at the cost of increasing spot size.

We consider a radius function, $R(s)$, defined over the arclength range $s = s_1$ to $s = s_2$, describing the radius of the beampipe with a number of different tunable parameters that, in principle, the effectiveness of the matching section can depend upon. We consider four main constraints on the radius function,

$$R(s_1) \equiv R_1, \quad \frac{\partial R}{\partial s}(s_1) \equiv \lambda_1, \tag{6}$$

$$R(s_2) \equiv R_2, \quad \frac{\partial R}{\partial s}(s_2) \equiv \lambda_2, \tag{7}$$

where the constraints on the first derivatives are chosen to assure that the direction of the magnetic field matches at the upstream and downstream ends of the matching section. Then, we demand that the radius function monotonically increases over the range from $s = s_1$ to $s = s_2$. This means that a valid form for the radius function is

$$R(s) = (\rho_0 + \rho_1 s + \rho_2 s^2 + \rho_3 s^3)^{\frac{1}{\chi}} \tag{8}$$

where χ is chosen to ensure monotonicity while minimizing the curvature, κ_R, and the normalized field gradient, $\frac{1}{B}\frac{\partial B}{\partial s}$. For fixed values of ρ_i, this means choosing χ as close to unity as possible.

By choosing the field strength of decay channel, we fix the radius function at upstream and downstream ends of the matching section, R_1 and R_2, respectively. Since the downstream end of the matching section must match into a long, straight solenoid, we fix $\lambda_2 = 0$, but the match at the upstream end is less obvious. In this preliminary study, we fix the direction of the field lines at the upstream end such that the field points away from the upstream end of the target, or that $\lambda_1 = \frac{R_2}{\ell_{target}} = 0.09375$, where $\ell_{target} = 80$ cm is the length of the graphite target. To fully understand the properties of these adiabatic matching sections, we need to vary this parameter, and this will be done in future analysis.

We consider a 50-m long capture and decay channel. If the decay channel field strength is greater than 1.25 T, then an additional matching section is required at the downstream end of the decay channel to match into a 1.25 T field strength. The design of this matching section is similar to the upstream matching section, except that $\lambda_1 = 0$ since we assume to be matching two long, straight solenoids.

The reason for increasing the field strength in the decay channel is easy to understand in terms of the maximum transverse momentum cutoff. In a 1.25 T decay channel, the contained transverse momentum has $p_T < 28$ MeV/c. However, during pion decay, the resulting muon will be given an average transverse momentum kick of

$$\langle p_T \rangle \approx \frac{1}{2} m_\pi \left(1 - \frac{m_\mu^2}{m_\pi^2}\right) \langle \sin^2 \theta \rangle \approx 15 \text{ MeV/c}, \tag{9}$$

where θ is the angle between the pion and muon momenta. Thus, pions near the edge of the channel's contained phase space can decay into muons that that escape. By increasing the field strength in the decay channel to 5 T, we increase the contained transverse

TABLE 1. Muon yields at the end of a 50 m capture and decay channel, equal to the total number of muons with total momentum $100 < p < 350$ MeV/c divided by the number of incident protons.

	1.25 T		5 T	
	SHORT 240 cm	LONG 720 cm	SHORT 240 cm	LONG 720 cm
μ^+/P	0.163	0.181	0.192	0.193
μ^-/P	0.154	0.170	0.183	0.185
$\mu^\pm/P$	0.317	0.352	0.375	0.378

momentum to $p_T < 56$ MeV/c, holding on to more muon daughter products. In terms of beam parameters, the RMS divergence of the muon beam is equal to the RMS divergence of the pion beam and the RMS angular spread from pion decay added quadratically. The RMS angular spread from pion decay is independent of the decay channel field strength, and thus by increasing the decay channel field strength, the relative effect on the divergence due to pion decay diminishes.

PRELIMINARY RESULTS

In FS1, the design of the adiabatic matching section is similar to the design discussed in the previous section. The FS1 design fixes $\chi = 2$, $\rho_2 = 0$, and $\rho_3 = 0$, and does not consider the constraints that match the direction of the magnetic field at the ends of the matching section. The overall length of the FS1 adiabatic section is 300 cm, and the field strength of the decay channel is 1.25 T. In this article, we consider a short matching section, $s_2 - s_1 = 240$ cm, and a long matching section, $s_2 - s_1 = 720$ cm, abiding by the constraints discussed in the previous section. Hence, we consider the short matching section with the 1.25 T decay channel approximately equivalent to the FS1 design. To study the effects of decay channel field strength on the yields, we again consider a short, 240 cm, and a long, 720 cm, matching section with a 5 T decay channel. After the 5 T decay channel, then, we insert a 3-m long matching section taking the beam from 5 T to 1.25 T. Yields were then computed for all four possible designs using a full MARS [4] simulation of the target region and the capture and decay channel.

The total number of muons per incident proton at the downstream end of the decay channel (50 m) for the total momentum range $100 < p < 350$ MeV/c are given in Table 1. One can immediately see the effects of lengthening the matching section when the decay channel field strength is low. The gain in yield is 11% when the length of the section is tripled. When the field strength is high, the difference in adiabatic effects is negligible, suggesting that a high-field decay channel can be made shorter without significant loss. However, while the yields in both the short and long sections are equivalent, they are still 18% greater than the short, low-field (FS1) configuration.

The RMS transverse emittances, computed from the RMS radial and angular distributions of the muon beam at the downstream end of the decay channel, are shown in

TABLE 2. Absolute RMS transverse emittances at the end of a 50 m capture and decay channel, assuming equivalent horizontal and vertical distributions.

	1.25 T		5 T	
	SHORT 240 cm	LONG 720 cm	SHORT 240 cm	LONG 720 cm
100 MeV/c	1.78π cm rad	1.76π cm rad	1.42π cm rad	1.30π cm rad
200 MeV/c	1.36π cm rad	1.38π cm rad	1.30π cm rad	1.30π cm rad
300 MeV/c	$.941\pi$ cm rad	$.911\pi$ cm rad	$.936\pi$ cm rad	$.908\pi$ cm rad

Table 2. Emittances are computed for muons with various total momenta in the ranges $p \pm 20$ MeV/c for $p = 100, 200, 300$ MeV/c. One can immediately see that the high-field decay channel captures the lower momentum muons into a smaller emittance. Adiabaticity may have a small effect on the emittance (less than 10%), but a definite statement cannot be made.

CONCLUSIONS

Certainly, the preliminary results shown in this article suggest a more thorough investigation of the parameters involved in the design of the pion capture and decay channel. While lengthening the channel improves yields because of its obvious effect on adiabaticity, there are many more parameters in the design of the adiabatic sections that must be varied in order to precision determine the contributing factors to particle loss. A thorough analysis of the effects of all design parameters needs to be done for various decay channel field strengths.

Additionally, the full analysis should be compared to the FS2 adiabatic section design, as well. While a direct comparison between the quoted yields in FS1 can be made to the yields quoted in this article, such a comparison cannot be made to FS2 yields because of the difference in target material and the difference in proton beam energy. For this reason, a more delicate comparison should be made, and this work is also in progress.

However, the preliminary data suggests that it may be possible to get $\sim 20\%$ improvement in yields or greater with a more detailed optimization. Such an improvement would be significant, and it would be worth considering the implications of such an optimization on the design of downstream elements and even on the choice of target.

REFERENCES

1. S. Ozaki et al., BNL-52623.
2. N. Holtkamp et al., SLAC-REPRINT-2000-054.
3. J.D. Jackson. *Classical Electrodynamics, 2nd Ed.* John Wiley & Sons, Inc. New York-London, 1975.
4. N.V. Mokhov, "The Mars Code System User's Guide", Fermilab-FN-628 (1995). O.E. Krivosheev, N.V. Mokhov, "MARS Code Status", Proc. Monte Carlo 2000 Conf., p. 943, Lisbon, October 23-26, 2000; Fermilab-Conf-00/181 (2000); N.V. Mokhov, "Status of MARS Code", Fermilab-Conf-03/053 (2003); http://www-ap.fnal.gov/MARS/.

Horn R&D for 2002-2003

S. Gilardoni[*][†], G. Grawer[*], G. Maire[*], J.-M. Maugain[*], S. Rangod[*], F. Voelker[*]

[*]*CERN, CH 1211 Geneve 23, Switzerland*
[†]*Université de Genève, Département de Physique Nucléaire et Corpuscolaire, 24 Quai Ernest-Ansermet 1211 Genève 4, Switzerland*

Abstract. A further step in the horn R&D at CERN has been achieved with the construction of a new power supply, which pulses the horn prototype at 100 kA. This paper reviews the steps leading, during the last year, to this goal, and the first mechanical measurements of the horn.

HORN R&D GOALS FOR THE PAST YEAR

The construction of the first inner prototype (see figure 1) for a Neutrino Factory has been concluded in 2002 (see ref. [1]), and in parallel the design and commissioning of a test power supply has been carried out. Those two elements allowed the start of mechanical and electrical tests, which are to be considered as necessary steps towards the validation of the horn device as focusing system for the Neutrino Factory.

FIGURE 1. Horn first prototype in the test area. Lateral view (left, from ref. [2]) and back view (right) with the electric connections and the water-cooling pipes.

First Horn Power Supply Prototype

According to the CERN Neutrino Factory baseline design [1], the magnetic horn has to be pulsed with a peak current of 300 kA with a repetition rate of 50 Hz. The current pulses has been chosen to be ≈100 μs long to limit the maximum voltage on the horn below 6 kV, and the pulse length is kept longer than the proton pulse length of 3.2 μs. The first step towards the final power supply design consists of the realization of a prototype that can deliver a peak current of 30 kA at 1 Hz pulse frequency with a pulse length of ≈100 μs. Even if the current is only one tenth of the final one, this prototype is sufficient for the mechanical measurements of the horn, as described in the following, and for testing of crucial components for the final design, like the high voltage switches. The outline of the prototype, in fact, is such that it can be scaled up to the final 300 kA power supply.

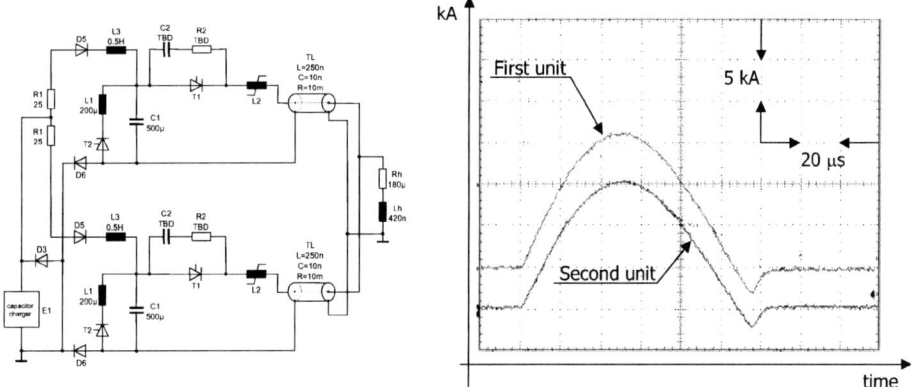

FIGURE 2. Power supply circuit (left) and oscilloscope measurement (right) of the current from the two units.

This prototype consists of two twin units that operate in parallel and deliver each a current of 15 kA. Figure 2 shows on the left the electric circuit scheme and on the right the oscilloscope measurements of the currents (red and blue lines) from the two units. Each of these is composed of two cupboards, one for the capacitors, which are charged by a common high voltage charger, and one for the power electronic components. The magnetic horn is connected to the power supply by 10 parallel 10 m long coaxial cables (see figure 1, right).

A further upgrade of the existing power supply, which involved the substitution of the capacitor charging high voltage and a part of the circuit, led the maximum total current to 100 kA with a pulse length kept around ≈100 μs. The repetition rate is reduced to 0.5 Hz after a failure due to undersized resistances. This power supply will be used to repeat and improve the mechanical measurements described in the following.

Mechanical Measurements

The knowledge of the horn vibration eigenfrequencies is necessary to avoid resonant excitations induced by the power supply pulse during normal operation. Moreover every vibration has to be completely dumped after 20 ms (50 Hz), otherwise an unwanted pile up of mechanical stresses could limit the lifetime of the conductors. Two different methods have been applied to measure the mechanical vibrations. The first one uses a directional microphone to record the sound produced by the conductors excited by the current. Mechanical vibrations are very often source of measurable sound and the horn conductors under the current passage behave like a massive membrane of a percussion instrument. The Fourier analysis of the sound spectrum reveals the mechanical vibration eigenfrequencies. This method has been validated by measuring the eigenfrequencies of the CNGS horn (see ref. [4] for more details) and the results were in good agreement with the data collected by an accelerometer, with a maximum disagreement of about 10%.

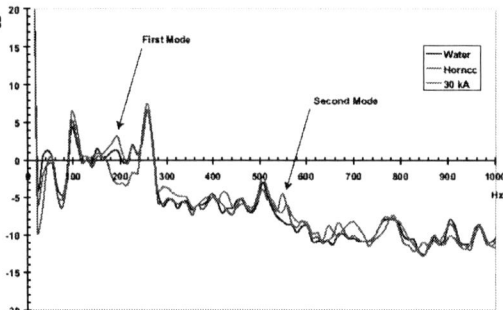 

FIGURE 3. Fourier transformation of the sound produced by the horn (left). The three lines represent the background from the water cooling and the power supply (blue line), the background from the power supply alone (violet line) and the final measurement of the horn (blue line). The right picture is a measurement setup with a laser vibrometer.

The measurements for the Neutrino Factory horn, fed by the 30 kA power supply, show the first two vibration modes (see figure 3 for the sound spectrum), the first at 193.7 Hz and the second at 549.1 Hz.

However, there are two main limitations to this method: first, if the background noise is too high it becomes difficult to isolate the signal. This is the case encountered for the horn powered by 30 kA, where the noise induced by the power supply itself and the water-cooling system hid the horn vibration. To reject these source of background, the measurements where taken with the water cooling active, but without any current (blue line on figure 3, left), then without the water and horn current at 30 kA but with the power supply in short circuit (violet line on figure 3, left), and at the end the horn connected to the supply and the cooling working to get the final measurement (green line on figure 3, left). The second limitation of the method comes from the lack of information about the absolute displacement of the conductors induced by the current. The time behavior of the sound signal is proportional to the displacement velocity, the integration of which to obtain the displacement would require the knowledge of the proportionality constant via a precise calibration. The latter turned out to be too complicated. For these reason, a laser vibrometer has provided a second set of measurement. In this case the velocity is measured by the Doppler shift of the laser light frequency induced by the vibration of the horn surface. The displacement, instead, is measured by the difference between the phase of the laser light reflected by the horn conductor and a reference laser light. A preliminary measurement shows a good agreement with the microphone measurements (see figure 3 on the right for an example of the experimental setup).

Conclusions

The testing foreseen for the past Neutrino Factory year (june 2002- june 2003) comes to a conclusion with the measurements of the horn eigenfrequencies and the upgrade to the horn power supply to 100 kA, with a repetition rate of 0.5 Hz and a pulse length of ≈ 100 μs. The program for the next year will cover further measurements of the horn mechanical stability with the laser vibrometer and a more detailed study of the power supply to reach the final design which foresees 300 kA, 100 μs pulse length at 50 Hz.

ACKNOWLEDGMENTS

The authors are grateful to J. Lettry and A. Fabich for their help during the mechanical measurements.
S. Gilardoni is supported by the CERN Doctoral Student Program.

REFERENCES

1. S. Gilardoni et al., "Status of a magnetic horn for a neutrino factory" in *NuFact02- The 4th International Workshop on Neutrino Factories*, edited by K. Long and R. Edgecock, Journal of Physics G, Nuclear and Particle Physics, vol. 29, n. 8, 2003, pp. 1801-1805.
2. Courtesy of R. Wilfinger
3. S.Gilardoni et al., "Updated Results of the Horn Study For the Neutrino Factory", *CERN-NUFACT-Note-129*, available from http://slap.web.cern.ch/slap/NuFact/NuFact/NFNotes.html
4. S.Gilardoni, "Horn Vibration Acoustic Measurements", *CERN-NUFACT-Note-126*, available from http://slap.web.cern.ch/slap/NuFact/NuFact/NFNotes.html

Optimization of the transmission in an alternating gradient muon collection channel

B. Autin*, F. Lemuet†, F. Méot† and A. Verdier*

*CERN, Geneva
†CEA, Saclay

Abstract. This paper reports on recent results regarding the optimization of the transmission in a muon production channel based on quadrupole focusing technique.

Introduction

The alternating gradient pion decay channel ("AG" in the following) is comprised of a 10 m long upstream section with very large aperture that funnels pion beams from four horns into a single 20 m long FODO decay section with 80 cm aperture (Fig. 1). Descriptions of the system and first stages of its optimization have been subject to earlier publications one can refer to for more details [1,2].

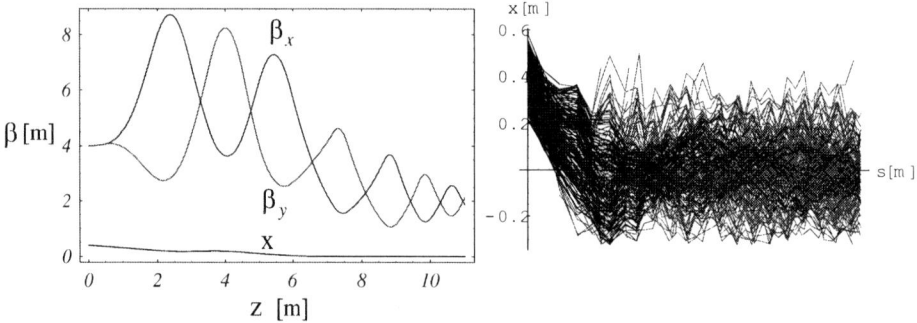

FIGURE 1. Left : β variations in the funnel and central trajectory (x) that starts from the horn center 40 cm off-axis and ends at zero on the FODO channel axis. Right : π (blue) and μ (red) trajectories over the 30 meters of the AG channel.

The main goal in the optimization is to maximize the muon transmission. For that purpose ray-tracing methods are used, that on the one hand permit the use of realistic models of magnetic fields in the large aperture magnets traversed by the beam and on the other hand allow accounting for $\pi \to \mu$ decay based on Monte Carlo technique.

Compared to most recent results [2] the transmission efficiency in the AG channel has been sensibly increased, thanks mostly to the optimization of magnetic fields along the channel, element by element. For the sake of comparison, optimization of muon transmission through a 30 m long, 80 cm aperture solenoidal channel is also reported.

Working hypothesis

Fig. 2 shows the MARS pion beam conditions at horn exit [3] as used in the present transmission optimizations. Note that time at origin is set to zero for all pions (analysis of proton bunch length effects is addressed in an analytical way in [4], detailed tracking studies are postponed to further investigations, numerical estimates in the solenoid case can be found in [5]).

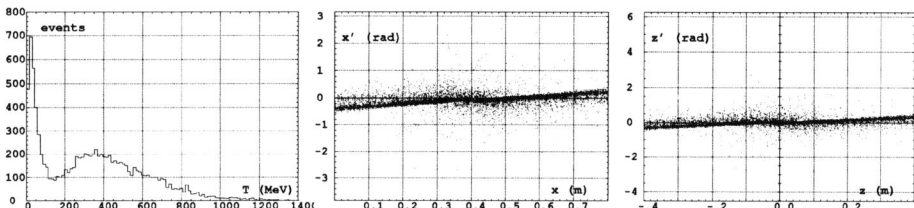

FIGURE 2. MARS distributions at horn exit - channel entrance. Left : kinetic energy, middle and right : xx' and zz' phase spaces.

Preliminary tracking results

Both the AG and the solenoid channels feature acceptance much smaller than the beam initial emittance shown in Fig. 2, this is illustrated in Fig. 3 that displays the subset of the initial coordinates of the pions yielding muons that make it to the end of

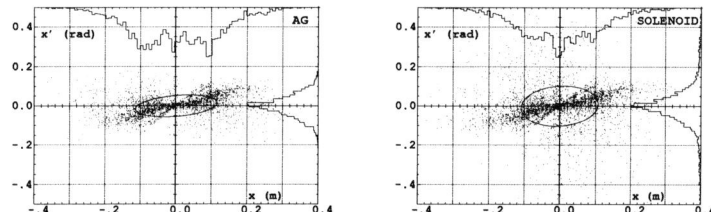

FIGURE 3. Initial vertical phase space of pions yielding muons that make it to the end of the 30 m channel (the horizontal phase space is similar).

the channel. It can be observed that the solenoid transverse acceptance is wider than the AG one, this is a property clearly established by the transmission simulations, however these need to take into account the longitudinal acceptance (of the RF bucket) which has a strong effect on transmission as made clear in Fig. 4 : the time extent of the muon beam at the exit of the solenoid (about $[0.1, > 0.4]$ μs) is much wider than in the AG case (about $[0.1, 0.11]$ μs).

A strong cause of transverse loss in the AG channel is the momentum spread in the beam (Fig. 2-left), by far larger than the momentum acceptance of the AG optics. A consequence is that the energy spread of the transmitted muons is strongly reduced compared to the initial pions' one as can be observed by comparison of the projected time spectra in Fig. 4-left with the initial one in Fig. 2-left.

339

Transmission calculation and optimization

The transmission efficiency through the channel is defined as

$$\frac{\text{Number of muons in a given 6-D acceptance at channel exit}}{\text{Number of pions at channel entrance}}$$

Comments on this formula are in order :
- the pion beam at channel entrance is that of Fig. 2, and the "Number of pions at channel entrance" is $5\ 10^4$, this number is associated with 10^6 p.o.t.
- the "Number of muons in a given 6-D acceptance at channel exit" is maximized using an automatic procedure that optimizes, for each one of the 3 phase spaces, the shape (i.e., the parameters α, β) of the acceptance ellipse

$$\gamma(y - y_c)^2 + 2\alpha(y - y_c)(y' - y'_c) + \beta(y' - y'_c)^2 = \varepsilon_y/\pi$$

($\gamma = (1 + \alpha^2)/\beta$; y stands for x, z, l for respectively the (x, x'), (z, z') and $(time, energy)$ motions), and the positioning (y_c, y'_c) of that ellipse ; the only parameter being fixed in that procedure is the acceptance ε_y/π.

This is illustrated in Fig. 4-right that shows the various so-optimized longitudinal phase space ellipses with surfaces $\varepsilon_l = 0.1$, 0.5, 1 and $2\ \pi$ eV.s.

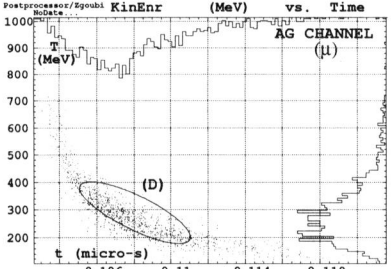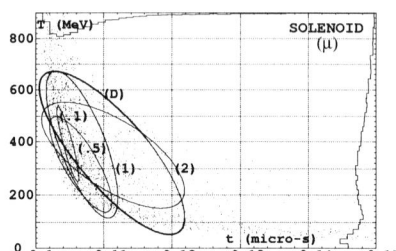

FIGURE 4. Time-energy phase space of the muon beam at AG and solenoid channel ends, and *rms* ellipses (D). Optimum longitudinal ellipses at downstream end are shown for the solenoid case.

Maximization of the transmission in the AG channel has been obtained by adjusting the value of the magnetic fields in all the magnets, one by one. This has the advantage of accounting for the variation of the average rigidity of the $(\pi + \mu)$ beam along the channel due to the decay [2,4]. The same kind of adjustment has been performed on the solenoid field. It comes out that for both types of channels, the optimum value of the fields, yielding maximum transmission, sensibly depends on the 6-D acceptance at channel end. It is found for instance that in the case of the solenoid, if the longitudinal acceptance is taken infinite the optimum field is 4.8 T about whereas it is in the range $1.5 - 2$ T with limited longitudinal acceptance. In the case of the AG channel the optimum field is rather high, of the order of 2 T at pole tip in the 40 cm radius quadrupoles of the FODO cells. Relaxing on transmission efficiency by about 30% has to be conceded if one wants to utilize warm magnets with lower field.

Optimization results

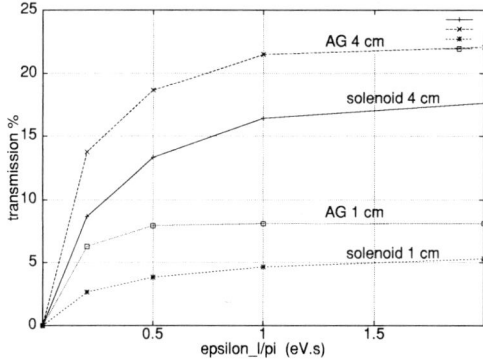

FIGURE 5. Transmissions versus longitudinal acceptance ε_l/π, in the AG and in the solenoid channels, for $\varepsilon_{x,z}/\pi = 1$ cm and $\varepsilon_{x,z}/\pi = 4$ cm transverse acceptances at exit.

Fig. 5 summarizes transmission results. In both non-cooling case (1 cm transverse acceptance) and cooling case (4 cm transverse acceptance), it can be seen that the AG has slightly larger transmission than the solenoid channel. One key reason for that is the wide initial momentum byte of the beam, which causes large beam lengthening and hence larger surface in longitudinal phase space at solenoid exit. AG channel with lower field yields transmissions (not shown here) comparable to that of the solenoid.

Conclusion

For the case of 40 cm inner radius, the 6-D transmission efficiencies of the AG channel and of the solenoid channel are comparable. Starting with a pion distribution generated with the MARS program for a proton energy of 2.2 GeV, about 8% of the muons produced arrive in emittances of $0.01\,\pi$ m.rad transverse and $0.5\,\pi$ eV.s longitudinal. For 10^{16} p.o.t./s this yields $4\,10^{13}$ muons/s, i.e. 40% of the nominal value.

These emittances can probably be transmitted into the RLA and the decay ring ; how the beam will be accelerated to its final energy is still to be studied. This opens the possibility of proposing a NuFact design without cooling, the latter being considered as an upgrade of the project.

With transverse and longitudinal coolings, emittances of $0.04\,\pi$ m.rad transverse and $1\,\pi$ eV.s longitudinal can be considered, yielding more than twice as much muons/s. This has to be weighted against losses in the cooling channel.

REFERENCES

1. Muon collection in an AG channel, B. Autin, F. Méot, A. Verdier, Proc. NuFact02.
2. Efficiency Of An Alternating Gradient Muon Collection Channel, B. Autin, F. Méot, A. Verdier, CERN NuFact Note 128 (2003).
3. S. Gilardoni and J. Pasternak, , private communication, CERN (2002).
4. Time-energy densities in $\pi \rightarrow \mu$ decay, B. Autin and F. Méot, these proceedings.
5. 3. Tab. 5.2, Section 5.6, in Study Of A European Neutrino Factory Complex, CERN NUFACT Note 122 (2002).

High Intensity Surface Muon Beam Using A Large Acceptance Axial Focusing Channel

H. Miyadera[1,2], K. Nagamine[1], K. Shimomura[1], K. Nishiyama[1,2], H. Tanaka[3,4], Y. Ikedo[1] and K. Ishida[1,3]

[1]*KEK (High Energy Accelerator Research Organization), Oho, Tsukuba, Ibaraki, Japan*
[2]*Graduate School of Science, University of Tokyo, Hongo, Bunkyo, Tokyo, Japan*
[3]*RIKEN (The Institute of Physical and Chemical Research), Wako, Saitama, Japan*
[4]*Graduate School of Science, Nagoya University, Furoh, Chikusa, Nagoya, Japan*

Abstract. A large acceptance surface muon channel "Dai Omega" was successfully constructed at KEK-MSL. At Dai Omega, four large aperture superconducting coils were utilized for the muon collection with solid angle acceptance of 1sr. As a result, the muon intensity of $4 \times 10^5 \mu^+$/sec has been realized at KEK-NML (500MeV, 5µA). A µSR detection system was developed, and the event rate of 36M/hour was achieved. As for the particle separation, axially symmetric electric separator was installed at the inside of 4th coil.

INTRODUCTION

Surface muon beam is the μ^+ emitted from π^+ decay at rest near the surface of the production target, and has been utilized in various fields of physics. One important application of surface muon beam is the generation of ultra-slow muon beam using the resonance laser ionization of thermal muonium [1] from hot tungsten foil. This slow muon beam is applicable as a small emittance narrow time width muon source for the future muon colliders and neutrino factories. However, the generation efficiency of slow muon is relatively small ($\sim 10^{-4}$), so that the high intensity source of surface muon beam is essential for the applicable use. Towards the high intensity muon beam, JPARC and other projects are planned. Along with the upgrade of primary proton beam, the improvement of muon channel is also essential to realize the higher intensity muon beam; this paper will describe the first working model of large acceptance muon channel.

AXIALLY SYMMETRIC MUON CHANNEL "DAI OMEGA"

Conventional muon channels consist of axially asymmetric magnetic components: magnetic dipoles and magnetic quadruples to transport the muon beam of a defined momentum. A cross field separator is commonly used in order to remove the contaminated positrons and electrons from the muon beam. In these systems, solid angle acceptances of the channels are restricted around 50msr. One of the profitable methods to improve the acceptance of surface muon channel is the use of large aperture coils placed near the production target. Axially symmetric magnetic field of coils can be used for axial focusing beam transport which is effective to improve the luminosity of the beam.

The first working model of axially symmetric muon channel, a large solid angle axial focusing superconducting surface muon channel "Dai Omega", was constructed at KEK Meson Science Laboratory II [2]. As shown in Figure 1, Dai Omega transports surface muon beam from a point source to a point focus on the symmetry axis of the coils.

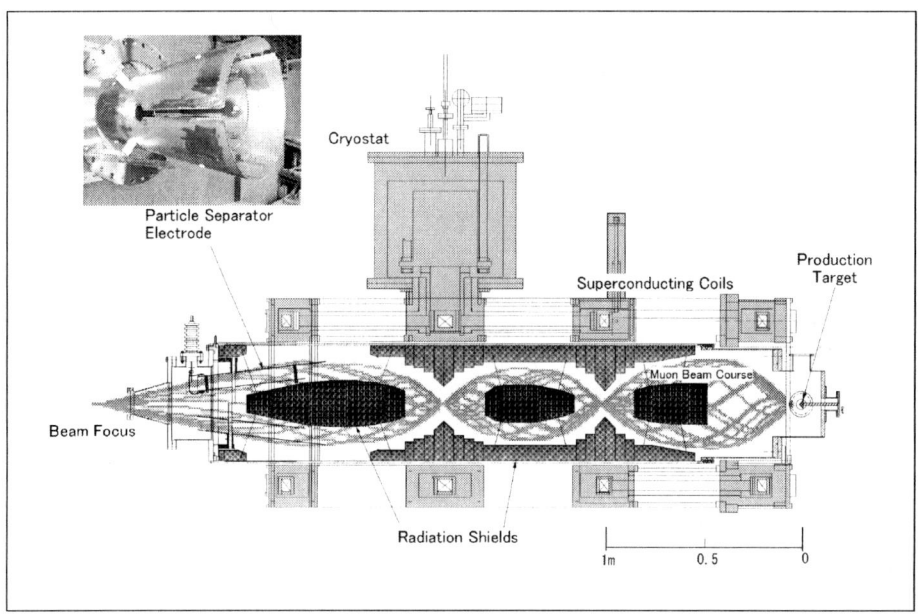

FIGURE 1. Schematic view of Dai Omega. Muon beam is focused after three Larmor periods of their helical orbit. Thus, the beam course of Dai Omega forms two intermediate focus points.

Table 1 shows the calculated solid angle acceptance of Dai Omega. The acceptance was above 1sr for the central momentum (30MeV/c for surface muon), which was 20 times larger than the conventional muon channels. Since the first beam of Dai Omega at December 2001, several beam study experiments were performed. The measured muon intensity was $4\times10^5 \mu^+/s$, which was the world most intense source of pulsed muon as shown in Table 2.

TABLE 1. Momentum acceptance of Dai Omega

Momentum [MeV/c]	Acceptance [msr]
31	754
30	1382
29	565
28	314
27	188

TABLE 2. World pulsed muon facilities

	Proton Beam	Target Thickness	μ^+/pulse
Dai Omega	500 MeV 5μA	4mm	2×10^4
KEK-MSL I	500 MeV 5μA	30mm	1.3×10^3
RIKEN-RAL	800 MeV 200μA	10mm	$\sim 2\times10^4$

Surface muon beam has been commonly used for μSR (Muon Spin Rotation / Relaxation / Resonance) spectroscopy. A new μSR detection system was designed and installed at Dai Omega [3], and the event detection rate of 36 Million/hour was achieved. The installed μSR detector consists of 128 coincidence counters (64 counter pairs each for upstream and downstream) to avoid the pile up loss of the event. As a whole, the detector covers 5% of the total solid angle of the emitted μ^+-decay e^+ from the sample target. Each scintillation counter is connected to 4x4 multi anode photomultiplier through the optical fiber; this combination makes the total detection system compact, inexpensive and low power consuming, which is applicable to more intense pulsed surface muon beam line planned at coming JPARC.

AXIALLY SYMMETRIC SEPARATOR FOR DAI OMEGA

Dai Omega transports μ^+ beam of 30MeV/c to the final focus with momentum acceptance of 6%, however charged particle such as e^+, e^-, π^+, π^- and crowd μ^- of the same momentum range are transported, too. To remove these beam contaminations, axially symmetric particle separator (shown in Figure 1) was designed and installed at the inside of 4th coil (downstream).

Because of charge mass ratio deference, electric field produces flight angle differences between μ^+ and other charged particles, thus it is possible to focus only μ^+ if the suitable condition is applied. The length of electrodes are 65cm (outside) and 55cm (inside), and the electric field of 30-60kV is applied on 7cm gap. The size and the shape of the electrodes were optimized by SIMION 7, Monte Carlo simulation. In order to compensate the applied electric field, the current of 4th coil was reduced to 30-50%.

As for the efficiency test of the separator, a plastic scintillation counter (1cm×1cm, 1mm thickness) was placed at the focus of Dai Omega, and the time spectrum of the beam was read by a photomultiplier. The observed time spectrum using this "Beam monitor" is shown in Figure 2. The prompt noise is mainly due to the e^+ and e^- which are produced by the decay of π^0 mesons ($\pi^0 \rightarrow \gamma\gamma$, $\gamma \rightarrow e^+e^-$) at the proton impact of production target. Because of the TOF differences, such prompt e^+ and e^- arrive at the focus point 50ns earlier than the μ^+. As shown in Figure 2, the prompt noise height was reduced to 1/10 by applying voltage of ±40kV on the electrodes. There also exist delay noise components, especially e^+ due to the μ^+ decay at the inside of production targets. Such decay e^+ are unpleasant for the μSR spectroscopy because they simulate the muon life time and distort the μSR spectrum. The distortion of μSR asymmetry time spectrum, which was observed before the installation of separator, was greatly improved by the installation of the separator.

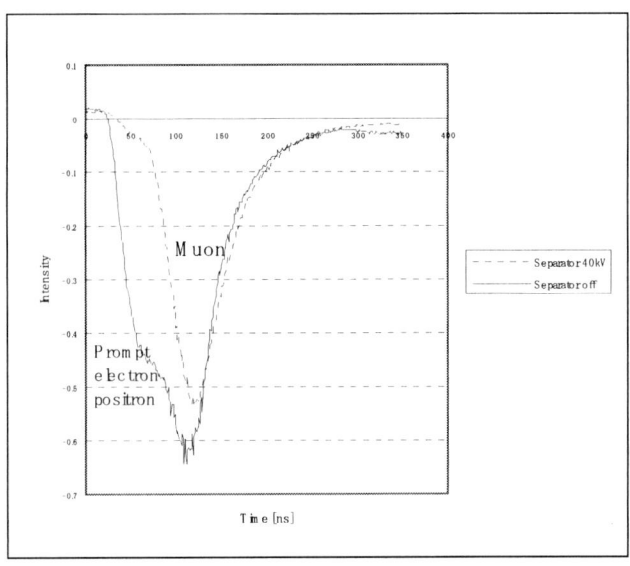

FIGURE 2. Prompt and muon signal observed by beam Monitor.

SUMMARY AND FUTURE PROSPECT

Axially symmetric surface muon channel "Dai Omega" was successfully installed at KEK-MSL II. Muon intensity of $4 \times 10^5 \mu^+/s$ has been achieved at KEK-NML (500MeV, 5μA). Axially symmetric particle separator was installed, and contaminated e^+ and e^- was declined by one order. Dai Omega is a milestone for the large acceptance curved solenoidal surface muon channel schemed at JPARC [4]. Dai Omega is also a pilot experiment for the future neutrino factory: a large acceptance coil may also be used as the pion injection part of a decay muon channel [5].

ACKNOWLEDGMENTS

We appreciate MEXT (Ministry of Education, Culture, Sports, Science and Technology) for supporting arrangement of COE budget. The Dai Omega project is supported by Grant-in-Aid for Scientific Research by JSPS (Japan Society for the Promotion of Science). H.M was supported by JSPS.

REFERENCES

1. Nagamine, K., et. al., Phys..Rev. Lett. **74**, 4811-4814 (1995).
2. Miyadera, H., et. al., Physica **B326**, 265-269 (2003).
3. Tanaka, H., et. al. Nucl. Instr. and Meth. **A** (submitted).
4. Shimomura, K., et. al., (following paper).
5. Nagamine, K., J. Phys. **G29**, 2031-2038 (2003).

Super Omega – new concepts of super intense surface muon beam

K. Shimomura, K. Ishida*, H. Miyadera[#], K. Nagamine

Meson Science Laboratory, Institute of Material Structure Science, KEK
1-1 Oho, Tsukuba, Ibaraki 305-0801, Japan
*Muon Science Laboratory, RIKEN (The Institute of Physical and Chemical Research)
Wako, Saitama 351-0198, Japan
[#]Department of Physics, GraduateSchool of Science, University of Tokyo
Hongo, Bunkyo, Tokyo, 113-0033, Japan

Abstract. A design study for new concepts of the axial focusing muon channel for the world highest intense surface muon beam is presented, which will be useful for the feasibility studies of the new scheme of the neutrino factory.

Recently, inspired by the progress of muon science (like ultra slow muon generation/muon catalyzed fusion/muon rare decay etc.) [1] and the design study of the muon collider /neutrino factory, several new concepts of the intense muon beam channel are extensively studied.

In KEK-MSL, a large sold angle axial focusing surface muon channel (Dai Omega) was successfully constructed [2]. This channel consists of four sets of superconducting coil and transports the surface muon with an acceptance of more than 1000 msr. In PSI, to provide the highest intensity of the surface muon beam for the generation of the low energy muon, the new μE4 beam line is now under construction by using two normal solenoids as the first focusing elements [3]. The residual magnetic field at the production target is less than 10G in this case. And also in BNL, for the search of the rare process μ^-N to e^-N, the new muon beam line is now under development. As the major part of this system, the transport solenoid (a set of solenoids and sections of toroidal) is designed to transport low energy negative muon with helical trajectories to the stopping target [4].

On the other hand, the Muon Science Facility will be constructed at the Materials and Life Science Facility in J-PARC, which is enable us to promote the excellent futures of muon science. 3 GeV; 333 μA, 1 MW beam will be sent to the Materials and Life Science Facility. At the period of Phase I, one carbon target of 2 cm thickness will be installed for the production of intense pulsed pion and muon beams. Two types of the conventional muon channel are now designing for the simultaneous use, namely, superconducting muon channel and surface muon channel [5].

By combining the progress of these muon beam technologies, we propose the world highest intensity surface nuon channel called; Super Omega, which is planed to install the Muon Science Facility. Super Omega consists of three parts; 1) a double normal conducting solenoid lens, 2) a bending superconducting solenoid and 3) four axial symmetric superconducting coils.

1) A double normal conducting solenoid with 56 cm diameter and 100cm length for each will be installed at the position of 60 cm from the carbon target, which can accept muons in a solid angle of 400msr. Their typical applied fields are from 3 to 4 kG. It should be noted the residual magnetic field around the production target is about 10G, therefore no serious effect on the primary proton beam is expected.

2) A curved superconducting solenoid consists of 10 pieces of superconducting coils which size is same as the currently used one at the decay muon channel in KEK-MSL, TRIUMF and RIKEN/RAL. The applied magnetic filed is 50kG and surface muon beam is confined within the radius of 5cm therefore transport without any significant loss. This solenoid bending is also helpful to reduce neutron and photon background

3) The final part, the existing Dai Omega will be installed for the collection of the divergent muon beam from the solenoid. By applying a pulsed electric-kicker system, the background charged particle like positron and electron are eliminated.

The schematic view of Super Omega is shown in Fig.1.

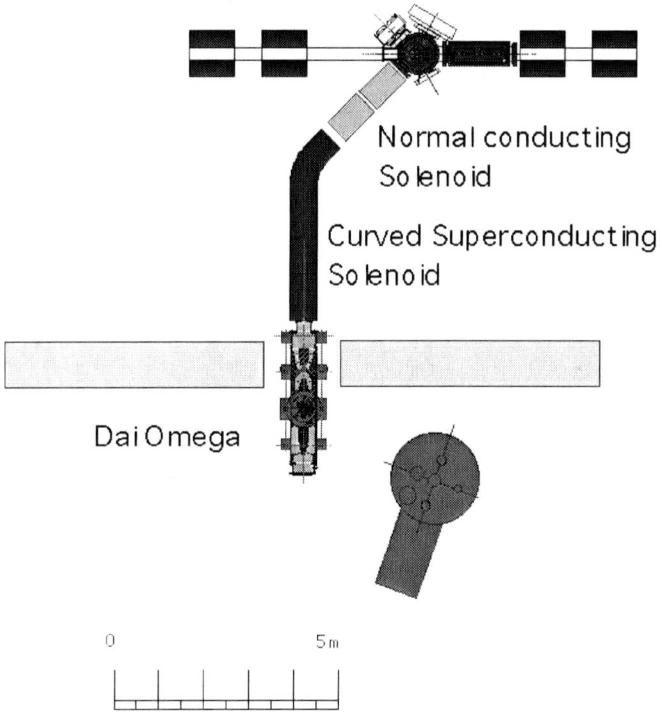

FIGURE 1. Conceptual design of super omega.

In Fig.2, typical muon beam trajectories along with the path are shown. In this configuration, beam is drifted about 2cm, due to the vertical magnetic field to the beam axis at the bending solenoid. This drift effect is also helpful to the suppression of the negative muons and pions. From the Monte Carlo Simulation for the surface muon yield on the target by Ishida [6], we can obtain the expected surface muon intensity 2 to 5×10^8/sec with momentum bite 4 to 10 % at this condition.

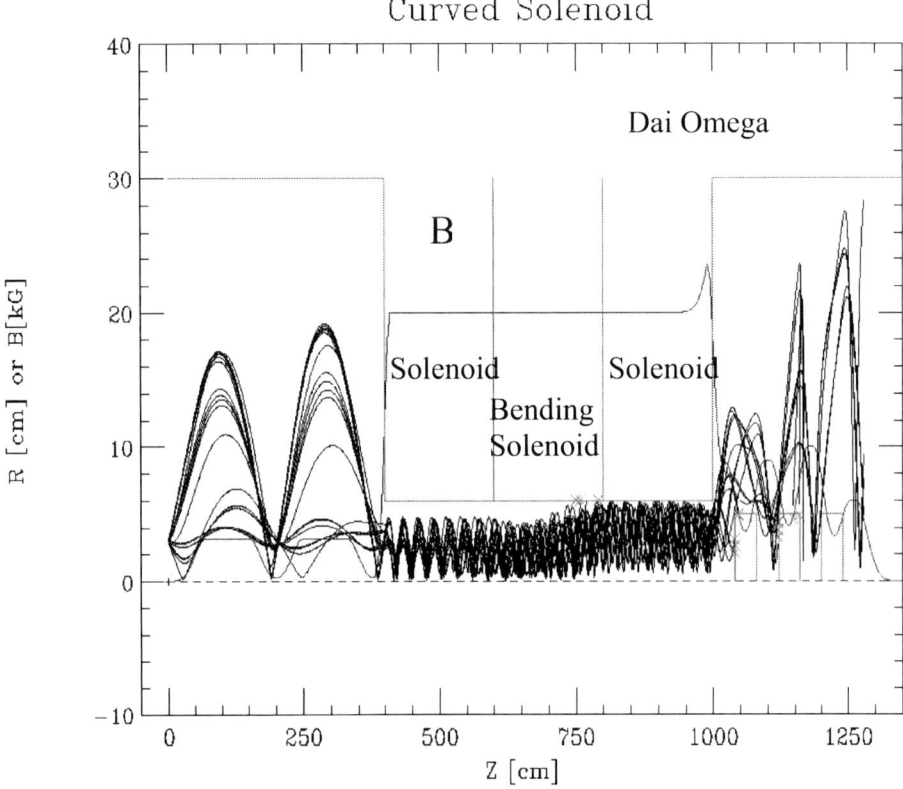

Figure 2. Typical muon beam trajectory.

The intense ($\sim 10^9$ p/s) and high-quality 4 MeV μ^+ beam generated from the Super-Omega channel will be used for the various types of advanced muon experiments. Some distinguished examples are summarized as follows.
1) Production of ultra-slow μ^+ beam and its application to surface science As demonstrated in the series of pioneering experiments at KEK-MSL, laser resonant-ionization of thermal muonium from the hot tungsten can be used to generate ultra-slow (down to 0.2eV) intense (conversion factor 10^{-3}) μ^+ beam[7]. The use of Super-Omega will produce intense ($>10^6$ p/s) ultra-slow μ^+ beam. The intense ultra-slow μ^+ beam can be used not only for further beam developments as described below but also for the surface science studies by using µSR methods with high spin polarization ($\geq 50\%$). Some representative surface science studies will be as follows; (i) surface magnetism (static and dynamic) in transition metals; (ii) electric state of H-like impurity on surface; (iii) electric conductivity on surfaces of synthesized materials such as nano-technology devices by using the

labeled-electron method.
2) Muon micro-beam production for life science application By using intense and cooled μ^+ beam as described above, the μ^+ micro-beam with a spot size of below 1 mm diameter can be realized by acceleration up to 100 keV. These produced μ^+ beam can be used for the electron-transfer property measurements in wide variety of protein crystals [8] which are produced in a size of below (mm)^3. Muon life science studies may become widely popular.
3) Injector for μ^+ Acceleration

Some further advanced technical developments will be done for the intense and cooled μ^+ beam. By considering such a ultra-slow μ^+ beam as an ion-source of μ^+, one can inject into post-accelerator up to further higher energy, say 100 MeV. Some R & D studies will be performed at Super-Omega facility, in particular, checking a muon loss before reaching relativistic energy.
4) New scheme of neutrino factory

Once successfully accelerated up to 10 GeV, one can store such muon beam to generate intense and high-quality neutrino ($\bar{\nu}_\mu$, ν_e) beam. The performance test of post-acceleration mentioned above will provide a clear information regarding feasibility[9].

So far, the present Super-Omega project has no schedule to be funded. The budget request will be done through various routes of short-term grant from either governmental or non-governmental agencies, in relation to life science, nano-technology and environment science. In any case, installation of the front-end double coil with a gate vale must be completed before spring of '07.

ACKNOWLEDGMENTS

Stimulating discussions with K. Nishiyama, Y. Miyake, K. Fukuchi and S. Makimura are greatly acknowledged.

REFERENCES

1. K. Ngamine, Introductory Muon Science, Cambrige, 2003.
2. H. Miyadera, K. Nagamine, K. Shimomura, K. Nishiyama, H. Tanaka, K. Fukuchi, S. Makimura and K. Ishida, *Physica B 326 (2003), 265.*
3. http://lmu.web.psi.ch/lem/newmue4/.
4. http://meco.ps.uci.edu/.
5. Y. Miyake et. al, *Physica 326 (2003) 255.*
   Detailed information is also available from http://j-parc.jp/MatLife/en/facilities/index.html.
6. K.Ishida, *KEK Proceedings 2001-1, 207.*
7. K. Nagamine et. al, *Phys. Rev. Lett. 74, (1995) 4811*
8. K. Nagamine et. al, *Physica 326 (2003) 255.*
9. K. Nagamine, *Nucl. Inst. Meth. A326 (2003) 414.*

Looking for strangeness with neutrino-nucleon scattering

W. M. Alberico*, S.M. Bilenky[†] and C. Maieron**

*Dipartimento di Fisica Teorica, Università di Torino and INFN, Sezione di Torino, Via P. Giuria 1, I-10125 Torino, Italy
[†]Joint Institute for Nuclear Research, Dubna, Russia
**Dipartimento di Fisica e Astronomia, Università di Catania and INFN, Sezione di Catania, Via S. Sofia 64, I-95123 Catania, Italy

Abstract. The possibility to determine the axial strange form factor of the nucleon from neutrino scattering experiments is studied. The existing experimental information is shortly mentioned and several related observables which could be measured in the near future at new neutrino facilities are discussed.

INTRODUCTION

The measurement of the cross-sections for neutral current (NC) neutrino (antineutrino) nucleon elastic scattering

$$\nu_\mu(\bar{\nu}_\mu) + N \longrightarrow \nu_\mu(\bar{\nu}_\mu) + N \tag{1}$$

has been indicated as a key tool for the determination of the so-called strange form factor of the nucleon, namely the matrix element of the (isoscalar) strange axial current:

$$\langle p,s|\bar{S}\gamma^\alpha\gamma^5 S|p,s\rangle = 2Ms^\alpha g_A^s.$$

Here $S, \bar{S}$ are the strange quark field operators, $|p,s\rangle$ is the state vector of a nucleon with momentum p and spin s. The NC of the nucleon which intervenes in the process (1) is:

$$J_\alpha^Z = V_\alpha^3 + A_\alpha^3 - 2\sin^2\theta_W J_\alpha^{em} - \frac{1}{2}V_\alpha^s - \frac{1}{2}A_\alpha^s. \tag{2}$$

It contains the customary vector and axial isovector components, which usually dominate over the isoscalar ones, the latter being associated with the strange quark component of the sea (though heavier quarks could come into play as well).

In order to disentangle the tiny effect (the present estimates are of the order of a few %) of the strange form factors, various observables have been suggested, all of them being *ratios* of cross sections: these quantities have the advantage of minimizing the uncertainties connected with, e.g., the determination of the ν-flux and/or the influence of the nuclear medium, when ν ($\bar{\nu}$) are scattered off nuclei[1].

Let us thus consider the following quantities:

1. NC over CC ratio:

$$R_{NC/CC}(Q^2) = \frac{(d\sigma/dQ^2)_v^{NC}}{(d\sigma/dQ^2)_v^{CC}} \tag{3}$$

2. Neutrino-antineutrino asymmetry:

$$\mathscr{A}(Q^2) = \frac{\left(\frac{d\sigma}{dQ^2}\right)_v^{NC} - \left(\frac{d\sigma}{dQ^2}\right)_{\bar{v}}^{NC}}{\left(\frac{d\sigma}{dQ^2}\right)_v^{CC} - \left(\frac{d\sigma}{dQ^2}\right)_{\bar{v}}^{CC}} \tag{4}$$

The cross sections in the denominators refer to the charged current (CC) processes:

$$\begin{aligned} v_\mu + n &\longrightarrow \mu^- + p, \\ \bar{v}_\mu + p &\longrightarrow \mu^+ + n \end{aligned} \tag{5}$$

and can be determined with higher accuracy, since all particles in the final state can be detected, while the final v ($\bar{v}$) in the NC process (1) is not observed.

The elastic NC v ($\bar{v}$)-nucleon scattering can be written in the form:

$$\left(\frac{d\sigma}{dQ^2}\right)_{v(\bar{v})}^{NC} = \frac{G_F^2}{2\pi}\left[\frac{1}{2}y^2(G_M^{NC})^2 + \left(1-y-\frac{M}{2E}y\right)\frac{(G_E^{NC})^2 + \frac{E}{2M}y(G_M^{NC})^2}{1+\frac{E}{2M}y} + \right.$$

$$\left. + \left(\frac{1}{2}y^2 + 1 - y + \frac{M}{2E}y\right)(G_A^{NC})^2 \pm 2y\left(1-\frac{1}{2}y\right)G_M^{NC}G_A^{NC}\right]. \tag{6}$$

In the above formula $y = p \cdot q/p \cdot k = Q^2/2p \cdot k$, E is the v($\bar{v}$) energy in the laboratory system, p (k) the initial nucleon (neutrino) four-momentum, M is the nucleon mass and $Q^2 = -q^2$ the square of the four-momentum transfer. Moreover G_E^{NC}, G_M^{NC}, G_A^{NC} are the electric, magnetic and axial weak NC form factors of the nucleon, all of them containing an isoscalar strange component. In particular

$$G_A^{NC;p(n)}(Q^2) = \pm\frac{1}{2}G_A(Q^2) - \frac{1}{2}G_A^s(Q^2), \tag{7}$$

where the strange axial form factor can be parameterized with the usual dipole form $G_A^s(Q^2) = g_A^s/1 + Q^2/M_A^2$, with $g_A^s = G_A^s(0)$.

From Equation (6) it is clear that to the NC scattering process several unknown quantities contribute: in particular there are three (electric, magnetic and axial) strange form factors, two of which (G_M^s and G_A^s) can produce contributions of similar size. Moreover it has been pointed out[2] that the present uncertainty on the axial cutoff mass, $M_A = 1.032 \pm 0.036$ GeV, allows one to obtain equally good fits to the elastic $v - N$ scattering cross sections with values of $|g_A^s|$ ranging from 0 to 0.25.

THE $\nu - \bar{\nu}$ ASYMMETRY

From the explicit evaluation of the NC and CC ν ($\bar{\nu}$)-nucleon cross sections, one can express the neutrino-antineutrino asymmetry as follows[3]:

$$\mathscr{A}_{p(n)} = \frac{1}{4}\left(\pm 1 - \frac{G_A^s}{G_A}\right)\left(\pm 1 - 2\sin^2\theta_W \frac{G_M^{p(n)}}{G_M^3} - \frac{1}{2}\frac{G_M^s}{G_M^3}\right), \tag{8}$$

where the $+$ ($-$) sign refer to proton (neutron) respectively. Taking into account only terms which linearly depend on the strange form factors:

$$\mathscr{A}_{p(n)} = \mathscr{A}_{p(n)}^0 \mp \frac{1}{8}\frac{G_M^s}{G_M^3} \mp \frac{G_A^s}{G_A}\mathscr{A}_{p(n)}^0 \tag{9}$$

we find out that any deviation with respect to the (known) term,

$$\mathscr{A}_{p(n)}^0 = \frac{1}{4}\left(1 \mp 2\sin^2\theta_W \frac{G_M^{p(n)}}{G_M^3}\right), \tag{10}$$

must be ascribed to a non-vanishing contribution of G_M^s and/or G_A^s.

Measurements of the asymmetry (8) is a quite demanding task from the experimental point of view, since it requires both ν and $\bar{\nu}$ beams of comparable intensity. An indirect "experimental" value of the this asymmetry with flux-averaged cross sections was extracted [4] from the data of the BNL-734 experiment [2]. We refer the reader to ref. [4] for the details of the analysis. The main conclusion, however, was that the present experimental uncertainty is compatible with any value of g_A^s, in the range $0 \geq g_A^s \geq -0.12$.

From Equation (9), the interference between the magnetic (G_M^s) and axial (G_A^s) strange form factors is evident: should they have the same sign, then their effects on the asymmetry get enhanced. The opposite is true, however, if they have opposite sign.

FUTURE PERSPECTIVES

We consider here the ratio of NC to CC elastic $\nu - p$ scattering cross sections: the information on the strange form factors one can extract from this quantity is not free from ambiguities, however it deserves to be carefully considered. It was recently proposed [5] to use the high intensity Booster neutrino beam at Fermilab, to measure ν-nucleon CC quasi-elastic and NC elastic scattering, with neutrino energies in the $0.5 \div 1.0$ GeV range. This kinematical conditions appear to be quite interesting to analyze the ratio $R_{NC/CC}(Q^2)$, Equation (3). From a throughout analysis we have performed on this quantity, we can summarize the following outcomes:

1. It is sensitive to g_A^s, but not much affected by the cutoff mass of the axial form factors, assumed in the above quoted dipole form.
2. The interference between axial and vector strange form factors (in particular the magnetic strange one) can hinder the effect of g_A^s alone. However G_M^s is under

investigation also with polarized electron-proton scattering experiments [6] and one can hope to have complementary information from this source.
3. The sensitivity to the flux is negligible, because it is largely eliminated in the ratio of cross sections.
4. The same argument applies to nuclear medium effects: indeed a large fraction of processes would occur on ^{12}C, where nucleons are bound and subject to final state interactions. These sizeably reduce the single cross sections, but their net effect on the ratio can be safely neglected.

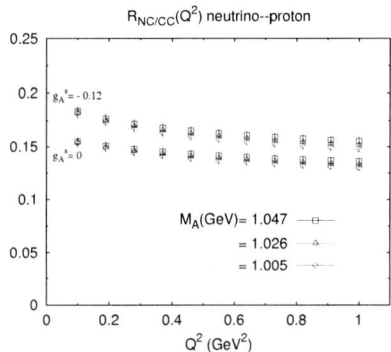

FIGURE 1. Plot of the ratio $R_{NC/CC}(Q^2)$, obtained with the neutrino cross sections averaged over the ν spectrum, for different choices of M_A and g_A^s.

To illustrate some of the above points, we show in fig. 1 the NC/CC ratio, for different choices of the axial cutoff mass M_A and of the strange axial constant g_A^s, as indicated. We have assumed that this ratio could be measured with a 5% accuracy, represented by the small "error band" plotted for each calculated point. We can see that, for the moderate Q^2 values represented here, the sensitivity of this ratio to G_A^s is large enough to allow a precise determination of it.

We conclude by observing that $\bar{\nu}$ scattering, if feasible, would offer relevant and complementary information on the strange form factors of the nucleon, and, eventually, would allow the determination of the neutrino asymmetry.

REFERENCES

1. For a recent review on this subject see:
   W.M. Alberico, S.M. Bilenky and C. Maieron, *Phys. Rept.*, **358**, 227 (2002).
2. L.A. Ahrens, *et al.*, *Phys. Rev.*, **D35**, 785 (1997).
3. W.M. Alberico, S.M. Bilenky, C. Giunti and C. Maieron, *Z. Phys.*, **C70**, 463 (1996).
4. W.M. Alberico, *et al.*, *Nucl. Phys.*, **A651**, 277 (1999).
5. See the contribution by R. Tayloe, in these Proceedings.
6. R. Hasty, *et al.* [SAMPLE Collaboration], *Science*, **290**, 2117 (2000).

Neutrino Scattering on the Nucleon and Determining Parton Distribution Functions

Yoshiyuki Miyachi

Department of Physics, Tokyo Institute of Technology, Oookayama 2-12-1, Meguro-ku, Tokyo, 152-8551, Japan

Abstract. Nucleon structure has been investigated using deep inelastic lepton scattering for more than 30 years. Precise measurements of the spin averaged structure functions using unpolarized DIS help understanding of the nucleon structure in the frame work of QCD inspired quark parton model. Recent experimental developments on the polarized nulceon target have shed light on "Spin Puzzle". Present understanding of the nucleon structure based on the charged lepton DIS scattering experiments and future prospects with the neutrino factory are addressed.

INTRODUCTION

Deep Inelastic Scattering (DIS) is the most basic tool to investigate the structure of nucleon. Four momentum square of the virtual photon emitted from the incoming lepton, Q^2, and total mass of the hadron system are large enough comparing with nucleon mass, so that the virtual photon can dissolve the nucleon into asymptotically free point-like partons which are quarks and gluons.

Inclusive measurement of DIS where only the scattered lepton is detected in the final state can provide information on structure functions of the nucleon which commonly written as $F_i(x, Q^2)$ ($i = 1, 2, 3$) for spin averaged and $g_i(x, Q^2)$ ($i = 1, 2, 3, 4, 5$) for spin dependent part [1]. Those structure functions can be decomposed into parton distributions which can be regarded as a number density and helicity distribution of quarks and gluon which carries momentum fraction x of the parent nucleon at scale Q^2: $q(x, Q^2)$, $\Delta q(x, Q^2)$, $G(x, Q^2)$ and $\Delta G(x, Q^2)$.

Through electro-magnetic interaction which is sensitive to electric charge square of the partons, $F_{1,2}(x, Q^2)$ and $g_{1,2}(x, Q^2)$ can be determined. On the contrary neutrinos interact by exchanging charged or neutral weak boson. It allows to measure $F_3(x, Q^2)$ and $g_{3,4,5}(x, Q^2)$, and therefore quark and anti-quark distributions can be separated.

In this report, I discuss the present understanding of nucleon structure from deep inelastic scattering of both charged and neutral leptons. Some prospect of nucleon spin physics in the future neutrino factory where the relevant scale is about $1 < Q^2 < 20\,\text{GeV}^2$ are also presented.

nified approach for modelling neutrino and ctron nucleon scattering cross sections from very high Q^2 to $Q^2 = 0$

Arie Bodek* and Un-ki Yang[†]

*rtment of Physics and Astronomy, University of Rochester, Rochester, New York 14618, USA
[†]Enrico Fermi Institute, University of Chicago, Chicago, Illinois 60637, USA

act. We present the results of a new scaling variable, ξ_w in modelling neutrino- and on-nucleon scattering cross sections using effective leading order PDFs. Our model desribes ep inelastic scattering charged lepton-nucleon scattering data including resonance data A/NMC/BCDMS/SLAC/JLab) from very high Q^2 to very low Q^2 (down to photo-productin), as well as CCFR neutrino data. Non-perturbative QCD effects at low Q^2 region turn out to l described by this new scaling variable. Our model is currently used for neutrino oscillation ments at few GeV region.

previous communication [1] a modified scaling variable x_w was used, and the leading order PDFs was modified such that the PDFs describe both high energy energy e/μ data. In order to describe low energy data down to the photopro- limit ($Q^2 = 0$), and account for both target mass and higher twist effects, the g modifications of the GRV94 LO PDFs are need:

increased the d/u ratio at high x as described in our previous analysis [2]. tead of the scaling variable x we used the scaling variable $x_w = (Q^2 +)/(2M\nu + A)$ (or $=x(Q^2+B)/(Q^2+Ax)$). This modification was used in early fits SLAC data [9]. The parameter A provides for an approximate way to include h target mass and higher twist effects at high x, and the parameter B allows the to be used all the way down to the photoproduction limit ($Q^2=0$).

addition as was done in earlier non-QCD based fits [10] to low energy data, we ltiplied all PDFs by a factor $K=Q^2/(Q^2+C)$. This was done in order to describe Q^2 data in the photoproduction limit, where F_2 is related to the photoproduction ss section according to

$$\sigma(\gamma p) = \frac{4\pi^2 \alpha_{EM}}{Q^2} F_2 = \frac{0.112\, mb\, GeV^2}{Q^2} F_2$$

ally, we froze the evolution of the GRV94 PDFs at a value of $Q^2 = 0.24$ (for < 0.24), because GRV94 PDFs are only valid down to $Q^2 = 0.23\, GeV^2$.

r analyses, the measured structure functions were corrected for the BCDMS tic error shift and for the relative normalizations between the SLAC, BCDMS

LEPTON SCATTERING OFF UNPOLARIZED NUCLEON TARGET

Structure function measured in the unpolarized charged lepton DIS can be written in quark parton model (QPM) as $F_2(x) = x\sum_q e_q^2(q(x) + \bar{q}(x))$. Such measurements of proton structure functions have been carried out on various laboratories intensively [2, 3, 4, 5, 6, 7]. HERA e-p collider experiments (H1 and ZEUS) extended the accessible kinematic area [8, 9].

QCD inspired QPM, involving the gluon distribution in the framework, describes well the observed large Q^2 scale violation in the structure function among those experiments. This tells us that at very high energy scale, anti-quarks and gluon give non-negligible contribution to the nucleon structure. Indeed a half of the nucleon momentum is carried by the gluons.

The structure function of neutron which is an iso-spin partner of proton has been also measured. The difference between proton and neutron structure functions which is known as Gottfried sum rule [10] indicated a possible flavor symmetry violation in the light sea quark [11, 12, 13, 14].

Due to the difficulty of neutrino beam handling there are rather weak constrains from neutral lepton scattering so far. Using heavy targets, structure function $F_3(x)$ to which quark and anti-quark distributions give contributions with opposite signs also has been measured [15, 16, 17, 18]. On the contrast to F_2 structure function, the statistical accuracy of F_3 is still rather limited.

Based on all the available experimental results on the spin averaged structure functions, several sets of unpolarized parton distributions have been extracted [19, 20, 21].

LEPTON SCATTERING OFF POLARIZED NUCLEON TARGET

Since EMC reported its surprising results that quarks carry only a small part of nucleon spin [22, 23], the spin of nucleon has come into the spotlight. This can be decomposed into sum of quark spins $\Delta\Sigma$, gluon spin ΔG and orbital angular momentum L_z as $\frac{1}{2} = \frac{1}{2}\Delta\Sigma + \Delta G + L_z$. Inspired by that result, several experimental attempts have been made. in order to determine the spin dependent structure function $g_1(x) = \frac{1}{2}\sum_q e_q^2(\Delta q(x) + \Delta\bar{q}(x))$ [24, 25, 26, 27, 28]. The difference of first moments, $\int_0^1 dx g_1^p(x) - g_1^n(x)$, has been regarded as a clean test of QCD since it has a simple relation with the axial charge of the nucleon which is known as Bjorken sum rule [29]. All the available experimental results from the inclusive measurements of polarized DIS at CERN, DESY and SLAC confirm Bjorken sum rule. The helicity distribution functions Δq and ΔG were extracted with similar way applied to the unpolarized PDF [30, 31, 32, 33]. Information on the gluon distribution comes from the scale dependence of $g_1(x, Q^2)$. However the kinematic coverage at present only gives very marginal constrain on ΔG. In order to obtain further experimental constrain on ΔG, COMPASS at CERN has been taking data since 2002. In a different approach, RHIC-spin program at BNL is running using polarized protons.

Semi-inclusive measurement which requires the hadron detection in the final state as well as the scattered electron allows flavor decomposition of quark distributions.

Such measurements have been carried out by SMC [34] and HERMES [35]. Recently quark helicity distributions for 5 different flavors were extracted [36]. The anti-quark distributions were nearly zero. Such information also put additional constrain on the QCD based helicity distribution extraction [37]. However for concrete statement, one needs further measurements.

In order to explore the last remained element, L_z, there are novel distributions proposed. On the same footing as $\Delta q(x)$, transverse parton distribution functions were defined [38]. Recently single spin azimuthal asymmetries of semi-inclusive π and K cross section have been reported [39]. Measurement of exclusive γ or meson production was proposed as a probe of parton total angler momentum in the framework of 'generalized parton distribution function (GPD)' [40, 41]. Single beam spin asymmetry measurements of deeply virtual compton scattering have been carried out [42, 43]. Succeeding experiments are expected to provide further information on those new distributions to study the orbital angular moment.

SPIN PHYSICS AT FUTURE NEUTRINO FACTORIES

One of the most important characteristics of neutrino beam is 100 % beam polarization by its nature. The left handed neutrino can select one state of partons, then the spin dependent structure functions g_5 becomes accessible. Quark and anti-quark distributions form g_5 with different signs as for F_3. Exchanging charged current enables further flavor selection like $g_5^{W^-}(x) = (-\Delta u(x) + \Delta \bar{d}(x) + \Delta \bar{s}(x) - \Delta c(x))$. However it has not been measured up to now, due to the technical difficulties of neutrino scattering experiment off polarized nucleon. It was stated that the inclusive measurement of neutrino DIS is dominated by scattering off u quark [44]. In order to enhance sensitivity to the specific quark flavor, the semi-inclusive measurement is required [45]. Combination of hadron identification and weak current interaction allows to have maximum sensitivity to specific flavor information.

Because of the high beam intensity expected in the neutrino factory, one can use nucleon target instead of nucleus target for the first time. Developments of the possible polarized active nucleon target are important for the wide energy band neutrino beam line. As stated above, the hadron identification also becomes more important for the flavor tagging measurement at the neutrino factory. The HERMES spectrometer which can identify π, K, p from 2 to 15 GeV/c [46] can be a good test bench to start considering the spectrometer for the neutrino factory.

SUMMARY

With charged or neutral lepton beams DIS experiments have revealed the rich structure of nucleon both for the spin averaged and polarized cases. Present progress of understanding helicity distributions for sea quarks from the semi-inclusive DIS measurement suggests the rich physics to be explored with neutrino factories.

REFERENCES

1. Hagiwara, K., et al., *Phys. Rev.*, **D66**, 010001 (2002).
2. Benvenuti, A. C., et al., *Phys. Lett.*, **B223**, 485 (1989).
3. Aubert, J. J., et al., *Nucl. Phys.*, **B259**, 189 (1985).
4. Allasia, D., et al., *Z. Phys.*, **C28**, 321 (1985).
5. Adams, M. R., et al., *Phys. Rev.*, **D54**, 3006–3056 (1996).
6. Arneodo, M., et al., *Nucl. Phys.*, **B483**, 3–43 (1997).
7. Whitlow, L. W., Riordan, E. M., Dasu, S., Rock, S., and Bodek, A (1992).
8. Adloff, C., et al., *Nucl. Phys.*, **B497**, 3–30 (1997).
9. Derrick, M., et al., *Z. Phys.*, **C72**, 399–424 (1996).
10. Gottfried, K., *Phys. Rev. Lett.*, **18**, 1174 (1967).
11. Arneodo, M., et al., *Phys. Rev.*, **D50**, 1–3 (1994).
12. Baldit, A., et al., *Phys. Lett.*, **B332**, 244–250 (1994).
13. Hawker, E. A., et al., *Phys. Rev. Lett.*, **80**, 3715–3718 (1998).
14. Miller, C. A., *Nucl. Phys. Proc. Suppl.*, **79**, 146–148 (1999).
15. Romosan, A., et al., *Phys. Rev. Lett.*, **78**, 2912–2915 (1997).
16. Marage, P., et al., *Z. Phys.*, **C49**, 385–394 (1991).
17. Berge, J. P., et al., *Z. Phys.*, **C49**, 187–224 (1991).
18. Sidorov, A. V., et al., *Eur. Phys. J.*, **C10**, 405–408 (1999).
19. Pumplin, J., et al., *JHEP*, **07**, 012 (2002).
20. Gluck, M., Reya, E., and Vogt, A., *Eur. Phys. J.*, **C5**, 461–470 (1998).
21. Martin, A. D., Roberts, R. G., Stirling, W. J., and Thorne, R. S., *Eur.*
22. Ashman, J., et al., *Phys. Lett.*, **B206**, 364 (1988).
23. Ashman, J., et al., *Nucl. Phys.*, **B328**, 1 (1989).
24. Anthony, P. L., et al., *Phys. Rev.*, **D54**, 6620–6650 (1996).
25. Abe, K., et al., *Phys. Lett.*, **B404**, 377–382 (1997).
26. Adeva, B., et al., *Phys. Rev.*, **D58**, 112001 (1998).
27. Airapetian, A., et al., *Phys. Lett.*, **B442**, 484–492 (1998).
28. Anthony, P. L., et al., *Phys. Lett.*, **B493**, 19–28 (2000).
29. Bjorken, J. D., *Phys. Rev.*, **148**, 1467–1478 (1966).
30. Leader, E., Sidorov, A. V., and Stamenov, D. B., *Eur. Phys. J.*, **C23**, 4
31. Gehrmann, T., and Stirling, W. J., *Phys. Rev.*, **D53**, 6100–6109 (1996)
32. Blumlein, J., and Bottcher, H., *Nucl. Phys.*, **B636**, 225–263 (2002).
33. Goto, Y., et al., *Phys. Rev.*, **D62**, 034017 (2000).
34. Adeva, B., et al., *Phys. Lett.*, **B369**, 93–100 (1996).
35. Ackerstaff, K., et al., *Phys. Lett.*, **B464**, 123–134 (1999).
36. Airapetian, A., et al., hep-ex/0307064 (2003).
37. de Florian, D., and Sassot, R., *Phys. Rev.*, **D62**, 094025 (2000).
38. Barone, V., Drago, A., and Ratcliffe, P. G., *Phys. Rept.*, **359**, 1–168 (2
39. Airapetian, A., et al., *Phys. Lett.*, **B562**, 182–192 (2003).
40. Ji, X.-D., *Phys. Rev.*, **D55**, 7114–7125 (1997).
41. Goeke, K., Polyakov, M. V., and Vanderhaeghen, M., *Prog. Part. Nuc*
42. Airapetian, A., et al., *Phys. Rev. Lett.*, **87**, 182001 (2001).
43. Stepanyan, S., et al., *Phys. Rev. Lett.*, **87**, 182002 (2001).
44. Forte, S., Mangano, M. L., and Ridolfi, G., *Nucl. Phys.*, **B602**, 585–6
45. Sudoh, K., hep-ph/0212246 (2002).
46. Akopov, N., et al., *Nucl. Instrum. Meth.*, **A479**, 511–530 (2002).

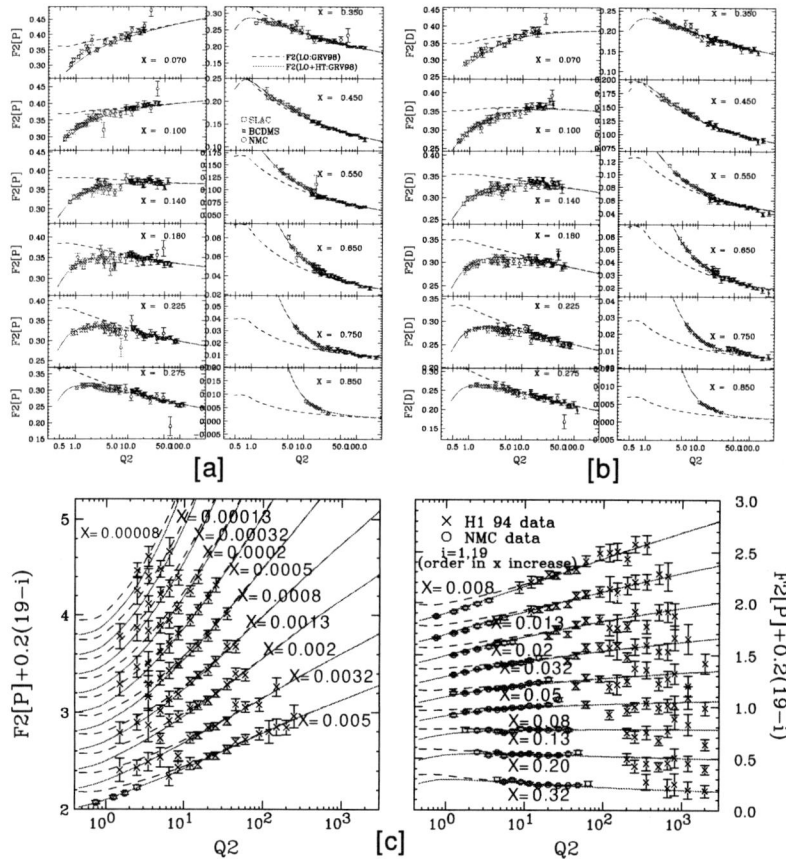

FIGURE 1. Electron and muon F_2 data (SLAC, BCDMS, NMC, H1 94) used in our GRV98 ξ_w fit compared to the predictions of the unmodified GRV98 PDFs (LO, dashed line) and the modified GRV98 PDFs fits (LO+HT, solid line); [a] for F_2 proton, [b] for F_2 deuteron, and [c] for the H1 and NMC proton data at low x.

and NMC data [2, 3]. The deuterium data were corrected for nuclear binding effects [2, 3].

In this publication we update our previous studies [8], which were done with a new improved scaling variable ξ_w, and fit for modifications to the recent GRV98 LO PDFs such that the PDFs describe both high energy and low energy electron/muon data. We now also include NMC and H1 94 data at lower x. Here we freeze the evolution of the GRV98 PDFs at a value of $Q^2 = 0.8$ (for $Q^2 < 0.8$), because GRV98 PDFs are only valid down to $Q^2 = 0.8$ GeV^2. In addition, we use different photoproduction limit multiplicative factors for valence and sea. Our proposed new scaling variable is based on the following derivation. Using energy momentum conservation, it can be

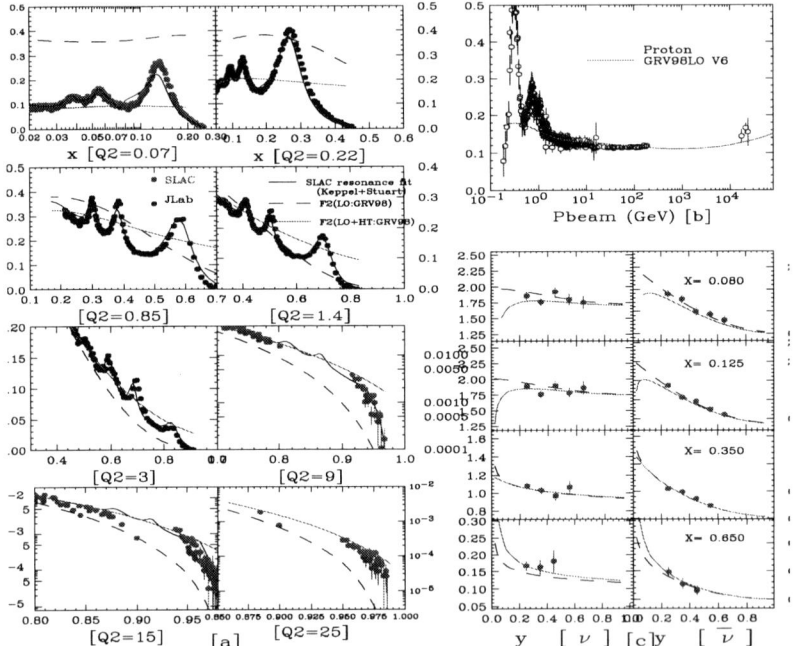

FIGURE 2. Comparisons to data not included in the fit. (a) Comparison of SLAC and JLab (electron) F_{2p} data the resonance region (or fits to these data) and the predictions of the GRV98 PDFs with (LO+HT, solid) and without (LO, dashed) our modifications. (b) Comparison of photoproduction data on protons to predictions using our modified GRV98 PDFs. (c) Comparison of representative CCFR ν_μ and $\bar{\nu}_\mu$ on iron at 55 GeV and the predictions of the GRV98 PDFs with (LO+HT, solid) and without (LO, dashed) our modifications.

shown that the factional momentum $\xi = (p_z + p_0)/(P_z + P_0)$ carried by a quark of four-mometum p in a proton target of mass M and four-momentum P is given by $\xi = xQ'^2/[0.5Q^2(1 + [1 + (2Mx)^2/Q^2]^{1/2})]$, where $2Q'^2 = [Q^2 + M_f^2 - M_i^2] + [(Q^2 + M_f^2 - M_i^2)^2 + 4Q^2(M_i^2 + P_T^2)]^{1/2}$.

Here M_i is the initial quark mass with average initial transverse momentum P_T and M_f is the mass of the quark in the final state. The above expression for ξ was previously derived [6] for the case of $P_T = 0$. Assuming $M_i = 0$ we use instead: $\xi_w = x(Q^2 + B + M_f^2)/(0.5Q^2(1 + [1 + (2Mx)^2/Q^2]^{1/2}) + Ax)$.

Here M_f=0, except for charm-production processes in neutrino scattering for which M_f=1.5 GeV. For ξ_w the parameter A is expected to be much smaller than for x_w since now it only accounts for the higher order (dynamic higher twist) QCD terms in the form of an enhanced target mass term (the effects of the proton target mass are already taken into account using the exact form in the denominator of ξ_w). The parameter B accounts for the initial state quark transverse momentum and final state quark effective ΔM_f^2 (originating from multi-gluon emission by quarks).

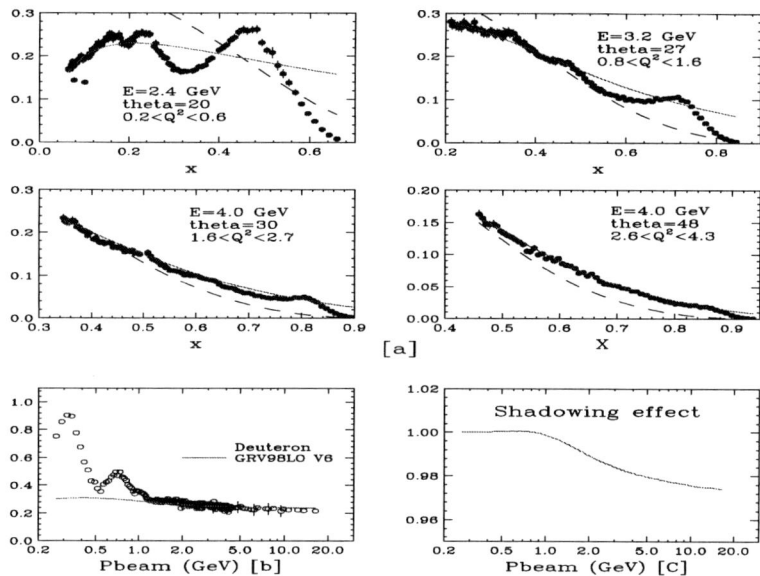

FIGURE 3. Comparisons to data on deutrerium which were not included in our GRV98 ξ_w fit. (a) Comparison of SLAC and JLab (electron) F_{2d} resonance data and the predictions of the GRV98 PDFs with (LO+HT, solid) and without (LO, dashed) our modifications. (b) Comparison of photoproduction data on deuterium to predictions using our modified GRV98 PDFs (including shadowing corrections). (c) The shadowing corrections that were applied to the PDFs for predicting the photoproduction cross section on deuterium.

Using closure considerations [11] (e.g. the Gottfried sum rule) it can be shown that the scaling prediction for the *valence* quark part of F_2 at low Q^2 should be multiplied by the factor $K=[1-G_D^2(Q^2)][1+M(Q^2)]$ where $G_D = 1/(1+Q^2/0.71)^2$ is the proton elastic form factor, and $M(Q^2)$ is related to the magnetic elastic form factors of the proton and neutron. At low Q^2, $[1-G_D^2(Q^2)]$ is approximately $Q^2/(Q^2+C)$ with $C = 0.71/4 = 0.178$ In order to satisfy the Adler Sum rule [12] we add the function $M(Q^2)$ to account for terms from the magnetic and axial elastic form factors of the nucleon. Therefore, we try a more general form $K_{valence}=[1-G_D^2(Q^2)][Q^2+C_{2v}]/[Q^2+C_{1v}]$, and $K_{sea}=Q^2/(Q^2+C_{sea})$. Using this form with the GRV98 PDFs (and now also including the very low x NMC and H1 94 data in the fit) we find A=0.419, B=0.223, and C_{1v}=0.544, C_{2v}=0.431, and C_{sea}=0.380 (all in GeV^2, χ^2 = 1235/1200 DOF). With these modifications, the GRV98 PDFs must also be multiplied by N=1.011 to normalize to the SLAC F_{2p} data. The fit (Figure 1) yields the following normalizations relative to the SLAC F_{2p} data ($SLAC_D$=0.986, $BCDMS_P$=0.964, $BCDMS_D$=0.984, NMC_P=1.00, NMC_D=0.993, $H1_P$=0.977, and BCDMS systematic error shift of 1.7).(Note, since the GRV98 PDFs do not include the charm sea, for $Q^2 > 0.8$ GeV^2 we also include charm production using the photon-gluon fusion model in order to fit the very high ν HERA data. This is not needed for any of the low energy comparisons but is only needed to describe the highest ν HERA electro and photoproduction data).

Comparisons of predictions using these modified GRV98 PDFs to other data which were *not included* in the fit is shown in Figures 2 and 3. From duality [14] considerations, with the ξ_w scaling variable, the modified GRV98 PDFs should also provide a reasonable description of the average value of F_2 in the resonance region. Figures 2(a) and 3(a) show a comparison between resonance data (from SLAC and Jefferson Lab, or parametrizations of these data [15]) on protons and deuterons versus the predictions with the standard GRV98 PDFs (LO) and with our modified GRV98 PDFs (LO+HT). The modified GRVB98 PDFs are in good agreement with SLAC and JLab resonance data down to $Q^2 = 0.07$ (although resonance data were not included in our fits). There is also very good agreement of the predictions of our modified GRV98 in the $Q^2 = 0$ limit with photoproduction data on protons and deuterons as shown in Figure 2(b) and 3(b). In predicting the photoproduction cross sections on deuterium, we have applied shadowing corrections [16] as shown in Figure 3(c). We also compare the *predictions* with our modified GRV98 PDFs (LO+HT) to a few representative high energy CCFR ν_μ and $\bar{\nu}_\mu$ charged-current differential cross sections [4, 13] on iron (neutrino data were not included in our fit). In this comparison we use the PDFs to obtain F_2 and xF_3 and correct for nuclear effects in iron [1]. The structure function $2xF_1$ is obtained by using the R_{world} fit from reference [5]. There is very good agreement of our *predictions* with these neutrino data on iron at 55 GeV (assuming that vector and axial structure functions are the same). We are currently working on further corrections to account for the fact that at low energies, the vector and axial structure functions are different.

REFERENCES

1. A. Bodek and U. K. Yang, (hep-ex/0203009) Nucl.Phys.Proc.Suppl.112:70-76,2002
2. U. K. Yang and A. Bodek, Phys. Rev. Lett. **82**, 2467 (1999).
3. U. K. Yang and A. Bodek, Eur. Phys. J. C**13**, 241 (2000).
4. U. K. Yang, Ph.D. thesis, Univ. of Rochester, UR-1583 (2001).
5. L. W. Whitlow et al. (SLAC-MIT), Phys. Lett. B**282**, 433 (1995); A. C. Benvenuti et al. (BCDMS), Phys. Lett. B**237**, 592 (1990); M. Arneodo et al. (NMC), Nucl. Phys. B**483**, 3 (1997).
6. H. Georgi and H. D. Politzer, Phys. Rev. D**14**, 1829 (1976); R. Barbieri et al., Phys. Lett. B**64**, 171 (1976), and Nucl. Phys. B**117**, 50 (1976); J. Pestieau and J. Urias, Phys.Rev.D**8**, 1552 (1973)
7. A.L. Kataev et al., Phys. Lett. B**417**, 374 (1998), and also hep-ph/0106221; J. Bluemlein and A. Tkabladze, Nucl. Phys. B**553**, 427 (1999).
8. A. Bodek, U. K. Yang, hep-ex/0210024, J. Phys. G. Nucl. Part. Phys. 29, 1 (2003); A. Bodek and U. K. Yang, To be published in Proceeding of NUINT02 2nd Workshop on Neutrino - Nucleus Interactions in the Few GeV Region (NuInt01), Irvine CA ,2002; A. Bodek and U. K. Yang, hep-ex/0301036
9. A. Bodek et al., Phys. Rev. D**20**, 1471 (1979).
10. A. Donnachie and P. V. Landshoff, Z. Phys. C **61**, 139 (1994); B. T. Fleming et al.(CCFR), Phys. Rev. Lett. **86**, 5430 (2001).
11. S. Stein et al., Phys. Rev. D**12**, 1884 (1975); K. Gottfried, Phys. Rev. Lett. **18**, 1174 (1967).
12. S. Adler, Phys. Rev. **143**, 1144 (1966); F. Gillman, Phys. Rev. **167**, 1365 (1968).
13. U. K. Yang et al.(CCFR), Phys. Rev. Lett. **87**, 251802 (2001).
14. E. D. Bloom and F. J. Gilman, Phys. Rev. Lett. **25**, 1140 (1970).
15. C. Keppel, Proc. of the Workshop on Exclusive Processes at High P_T, Newport News, VA, May (2002).]
16. Badelek and Kwiecinski, Nucl. Phys. B**370**, 278 (1992).

Near Liquid Argon TPC Detectors for Near Future

Franco Sergiampietri

INFN-Sezione di Pisa, Via Buonarroti 2, Ed. B, I-56127 Pisa, Italy
and
UCLA, Department of Physics & Astronomy, University of California, Los Angeles, CA 90095-1547, USA

Abstract. As demonstrated by the events collected during the surface test of the first ICARUS module, detectors based on the technique of liquid argon TPC appear as the ideal candidates for neutrino detectors of the second generation. Their low energy threshold, combined with their high-resolution imaging and calorimetric response allow for detailed kinematical reconstruction of neutrino-induced events. A reduced scale detector, based on this technique, is described as "near" detector for near future neutrino beams.

INTRODUCTION

The working principle of a liquid argon time projection chamber (TPC) can be described as follows. Primary ionization electrons, generated in a liquid argon volume by ionizing particles, are drifted by an electric field toward an anode plane made of a set of (two or three) parallel successive wire planes. Wires are oriented along different angles on each plane. Wires are suitably dc biased to get a kind of "transparency" of the first crossed planes (induction planes) for the drifting electrons that are finally collected on the last plane (collection plane). Electric signals on each wire, detected and amplified by low noise amplifiers, are continuously time sampled and acquired. The event reconstruction is based on the time correlation of the time samples from the different wires allowing a 3-dimensional reconstruction of the original ionization tracks.

The detector is continuously sensitive, while the data acquisition can be operated continuously or triggered by:
- the scintillation light produced in liquid argon by ionizing particles and detected by photomultipliers working in the liquid;
- selective analog sum of the wire signals
- beam spill synchropulse, when the detector is on a pulsed beam line.

Much of what we know at present on this technique results from the research activity made during more than 20 years for the ICARUS experiment [1].

Events recorded during a first test run on one of the ICARUS-T600 detector semi modules widely feature the detection and imaging potential of this technology.

The performances attainable by this technique can be summarized as follows:
- homogeneous response over the full active volume

- events with 3-d bubble-chamber imaging quality
- calorimetric information
- excellent kinematical reconstruction
- heavy/minimum ionization particle discrimination (dE/dx, range)
- ideal for low/moderate rate experiments
- easy to extend to active masses of several ktons.

Liquid argon is chosen as ionizable medium due to its rather high density ($\rho = 1.4$ g/cm^3, $X_0 = 14.2\ cm$, $\lambda_I = 84\ cm$, $dE/dx_{min} = 2.1\ MeV/cm$), its abundance in the atmosphere ($\sim 1\% \equiv 17\ ktons/km^3$), the consequent relatively low cost ($\sim 800\ \$/ton$), its high dielectric strength and its zero electronegativity. Its use is favored also by the low ionization energy (23.6 eV/electron-ion pair) and by the easiness of its liquefaction and thermal stabilization by means of liquid nitrogen.

The above described features made liquid argon the candidate medium for sampling calorimetry and for high density TPC's.

The granularity in liquid argon TPC's is related to the wire pitch p_w and to the sampling time interval δt_s. With $p_w = 3\ mm$ and $\delta t_s = 400\ ns$, events are displayed by an array of 3-d pixels ~ 1 mm rms in transverse size.

This kind of detector deals with primary ionization signals and a critical part is the optimum matching between read-out wires and front-end (low noise) electronics.

The energy threshold is mainly given by the signal-to-noise ratio S/N at the preamplifier input. With the ICARUS electronics, the equivalent noise charge (ENC) can be expressed by $ENC \approx 145 e^- + 3 e^-/pF$, resulting of the order of 900 e^- rms. The minimum track signal on a wire (track in a plane parallel to the wire plane and with direction orthogonal to the wire orientation, with length = p_w = 3 mm) is $dE_{min} = 0.64$ MeV equivalent to $\approx 15 \cdot 10^3\ e^-$, with $S/N \approx 16$. Energy releases over 250 KeV are then clearly distinguished from the noise.

Intrinsic limits for this kind of detector are due to the long electron drift times (1.6 ms/m with electric field $E = 0.5\ kV/cm$) and to the poor imaging on the induction wires for tracks aligned along the drift direction.

A 40 ton NEAR DETECTOR FOR FINeSE AND NuMI

Beyond what was made up to now for the ICARUS experiment, improvements, systematic calibrations and implementations can be planned for a diffused use of the liquid argon TPC technique. Among these, outfitting such a detector by a magnetic field is of special interest especially in view of neutrino factory beams and, in general, as development of "electronic bubble chamber" detectors. Based on this concept, a study line has been started during these last years to design intermediate and very large mass magnetized neutrino detectors [2, 3].

In this paper a relatively small detector, Mini-LANNDD T40, is described, that could be motivated *a)* by physics programs in experiments like FINeSE [4] (as main neutrino interaction detector) and/or along the NuMI beam line (as near neutrino detector), *b)* to promote this kind of detection technique with improved performances for what is concerning power and LN_2 consumption, safety and eventual operation in magnetic field.

The detector is structured as a cylindrical volume with horizontal axis, filled by liquid argon (see Fig. 1). A vertical wire chamber (item *1*) is aligned along the cylinder axis and splits the volume into two drift regions (left and right).

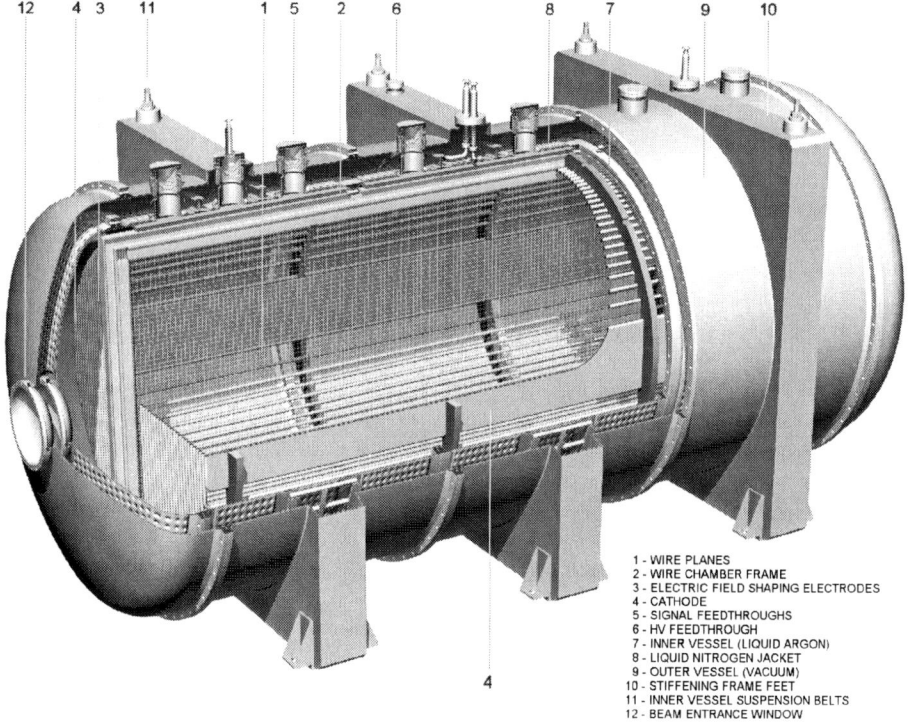

FIGURE 1. Mini-LANNDD T40: cutaway view.

A horizontal and uniform electric field is generated for the electron drift into each region by a cathode (*4*) and a system of field shaping electrodes (*3*). The chamber is built of two specular sets of wire planes. A central, grounded wire plane screens the left and the right sets one from the other. An eventual grid plane can be mounted in front of each readout wire plane pairs to ensure position independent signals on the wires.

Signals from wires are brought to the front-end electronics, outside the cryostat, through twisted pair cables and low voltage feedthrough flanges (*5*). Two high voltage feedthroughs (*6*) are used to bias the cathodes and the field shaping electrodes.

The cryostat, vacuum insulated and liquid nitrogen cooled, is built of an inner vessel and an outer vessel, with vacuum in between.

The inner vessel is surrounded by a dimpled surface jacket, welded to it and refillable with LN_2. Three square frames are part of the outer vessel. They are used as bases for it and as supporting frames for the inner vessel. Wrapping the inner vessel and the LN_2 jacket by super-insulation layers and suspending them to the outer vessel frame by three steel belts (*11*) minimize the heat input.

TABLE 1. Mini-LANNDD T40 - Parameters

Active liquid argon sizes	$\varnothing = 2.4\ m,\ L=6.0\ m$
Active liquid argon mass	38.5 Ton
Number of drift regions	2
Max drift lengths	$2 \times 1.2\ m$
Maximum required high voltage	60 kV
Number of cathode planes	2
Number of wire chambers	1
Number of readout wire planes	4
Orientation of readout wires	0°, 90
Number of readout wires	5632
Heat Input	
a) Radiation (with $w_r = 1\ watt/m^2$)	72 W
b) Conduction (cables + mech. supports)	240 W
Total	312 W
Equivalent liquid nitrogen consumption	0.2 m^3/d
Dipole magnet (warm, $B = 0.6\ T$)	
Copper weight	60.2 Ton
Power @ 35 kA	2.5 MW
Iron yoke	900 Ton

The main parameters of the Mini-LANNDD T40 are reported in Table 1. The low nitrogen consumption allows operating the detector with a reduced size, refillable nitrogen storage dewar.

The possibility of operating liquid argon TPC's in magnetic field is, by far, the most interesting implementation of this kind of detector. This feature allows easy muon sign discrimination and momentum evaluation. At beam momenta lower than 5 GeV/c, with a magnetic field intensity of 0.6-1.0 T, calculations indicate that the sign discrimination for $e^{\pm}$ that initiated an electromagnetic shower becomes possible too. The main parameters for a possible configuration with a dipole magnet are indicated at the end of Table 1.

REFERENCES

1. ICARUS Collaboration, "A first 600 ton ICARUS Detector Installed at the Gran Sasso Laboratory", Addendum to proposal, LNGS-95/10, 1995. For complete list of references to ICARUS publications, see http://pcnometh4.cern.ch/publications.html.
2. D. B. Cline, J. G. Learned, K. McDonald and F. Sergiampietri, "LANNDD-a massive liquid argon detector for proton decay, supernova and solar neutrino studies and a neutrino factory detector", Nucl. Instr. and Meth., A 503 (2003) 136.
3. D. B. Cline, S. Otwinowski and F. Sergiampietri, "Mini-LANNDD: A very sensitive neutrino detector to measure sin sup 2(2 theta sub 13)", J. Phys. G29 (2003) 1893-1898
4. D. H. Potterveld et al., "FINeSE, Fermilab Intense Neutrino Scattering Experiment", Expression of Interest presented to FNAL PAC 11/15/02

First Results from E158 Measuring Parity Violation in Moller Scattering

Imran Younus[1]

Department of Physics, Syracuse University, Syracuse, NY 13244

Abstract. SLAC E158 is an experiment to measure the parity nonconserving asymmetry in Moller scattering. Longitudinally polarized 48 GeV electrons are scattered off unpolarized (atomic) electrons in a liquid hydrogen target with an average Q^2 of 0.027 GeV^2. The asymmetry in this process is proportional to $(\frac{1}{4} - \sin^2\theta_W)$, where $\sin^2\theta_W$ gives the weak mixing angle. The preliminary results give $A_{PV} = -151.9 \pm 29.0(stat) \pm 32.5(syst)$ parts per billion. This in turn gives $\sin^2\theta_W^{eff} = 0.2371 \pm 0.0025 \pm 0.0027$, which is consistant with the standard model prediction (0.2386 ± 0.0006).

INTRODUCTION

Over the last three decades, the Standard Model has evolved as the most successful theory describing the electroweak interactions. It has been verified experimentally with an ever-increasing accuracy. The experiments studying the weak interactions at high energies (at the $Z°$ resonance) have measured the electroweak parameters, such as $\sin^2\theta_W$, with spectacular precision, and have established the correctness of the Standard Model. However, at lower energies, tests of the electroweak theory in the weak neutral current sector are less sensitive by more than an order of magnitude. Equally precise measurements away from the $Z°$ resonance are crutial for a complete test or might provide an insight to the new physics beyond the Standard Model.

The SLAC E158 experiment[1, 2] is measuring the weak mixing angle at low Q^2 by observing the parity-violating asymmetries (A_{PV}) in polarized Moller scattering ($e^-e^- \rightarrow e^-e^-$). This asymmetry arises from the interference between the electromagnetic and weak neutral currents, and has not been measured before. The goal of the experiment is to measure the asymmetry with a precision of 8% which would in turn give the most precise measurement to date of $\sin^2\theta_W$ at low Q^2. Such a measurement of $\sin^2\theta_W$, when compared with the measurements at the $Z°$ pole, would establish the Q^2 dependance of the weak coupling angle at a singnificance of $\sim 8\sigma$.

The E158 results are sensitive to several new physics scenarios at the TeV scale. A deviation from the prediction of $\sin^2\theta_W$ can indicate the presence of additional neutral current effects such as Z' bosons in the mass range of 600 to 900 GeV. E158 is also capable of setting a limit to the electron compositness scale of up to 10 to 14 TeV, thus providing a complementary approach to high-energy collider experiments' searches.

---

[1] Speaker. For the SLAC E158 Collaboration.

PARITY VIOLATING MOLLER SCATTERING

The Moller scattering proceeds via both t- and u-channel processes. The parity-violating asymmetry in scattering of polarized electrons off unpolarized target is defined as

$$A_{PV} = \frac{\sigma_R - \sigma_L}{\sigma_R + \sigma_L}, \qquad (1)$$

where $\sigma_R(\sigma_L)$ is the scattering cross-section from incident right(left) handed electrons. The interference between the tree level γ and Z° amplitudes gives rise to the standard model asymmetry given by[3]

$$A_{PV} = -m_e E \frac{G_F}{\sqrt{2}\pi\alpha} \frac{4\sin^2\theta_{cm}}{(3+\cos^2\theta_{cm})^2}(1-4\sin^2\theta_W). \qquad (2)$$

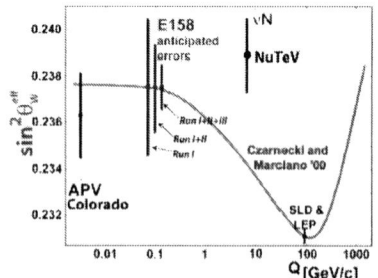

FIGURE 1. Predicted running of $\sin^2\theta_W$ from the precision measurement at Z° resonance.

The simple functional form of the asymmetry suggests Moller scattering to be an excellent probe of the weak mixing angle. At the tree level, for $Q^2 \approx 0.03$ GeV^2, the asymmetry $A_{PV} \approx -280 \times 10^{-9}$. The electroweak radiative corrections decrease this asymmetry by almost 40%[4]. The beam polarization of 80% and some dilution from the background processes bring the raw experimental asymmetry down to approximately 130 parts per billion. The experiment goal is to measure this small asymmetry with a precision of 8%, which would determine the $\sin^2\theta_W$ at low Q^2 with $\sigma(\sin^2\theta_W) < 0.001$.

The 40% reduction in the A_{PV} due to the radiative corrections corresponds to a 3% increase in $\sin^2\theta_W$ from its value obtained at $Q^2 = M_Z^2$ in measurements obtained by SLC and LEP[5]. This Q^2 dependance of $\sin^2\theta_W$ is shown with solid curve in fig. 1[6].

DESCRIPTION OF THE EXPERIMENT

A high intensity longitudinaly polarized electron beam ($3.5-6 \times 10^{11}$ elec./pulse; 300 ns pulses at 120 Hz) is scattered off atomic electrons in an unpolarized liquid hydrogen target. The helicity of the electron beam is flipped psudo-randomly on a pulse-by-pulse basis, and the pulse pair flux asymmetry is measured:

$$A_{measured} = (F_R - F_L)/(F_R + F_L), \qquad (3)$$

where $F_R(F_L)$ is the detected flux of scattered right(left) handed electrons. This asymmetry is then corrected for the differences between the right and left handed beam properties: $A_{raw} = A_{measured} - \sum \alpha_i \Delta x_i$, where Δx_i stands for the charge, energy, position and angle differences between the right and left handed pulses. The correlation coefficients α_i are determined experimentally. The final physics asymmetry is obtained by subtracting the backgrounds and normalizing with beam polarization. The background processes mainly consist of elastic and inelestic electron-proton scattering, real and virtual photo-production of pions and synchrotron radiation.

Apparatus

E158 uses 150 cm long liquid hydrogen target (0.15 r.l.) kept at 17.5 K. The pulse-by-pulse density fluctuations in the target are maintained at a level of <50 parts per million. The scattered electrons are transported through a spectrometer consisting of 3 dipole and 4 quadrupole magnets, and an acceptance collimator (several other collimators are used to suppress different backgrounds). The dipole chicane (figure 2) is placed downstream of the target for suppression of soft particles created in the target. The acceptance collimator passes the Moller electrons with momenta in the range 13-25 GeV and eP electrons with momenta of ~ 48 GeV. The quadrupole magnets downstream of the collimator then seperates the Moller and eP fluxes by focusing the Moller electrons on a smaller ring 60 m downstream of the target where the detector is placed.

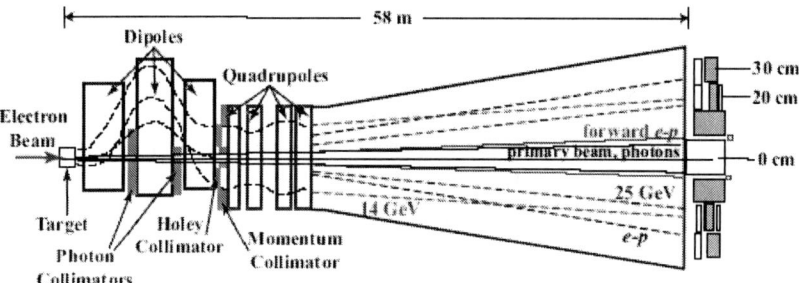

FIGURE 2. Schematic of E158 Spectrometer.

The electromagnetic calorimeter surrounds the beam axis in the radii range 15.0-23.5 cm and has an energy resolution of $\sim 10\%$. It consists of layers of quartz fibers interspersed between copper plates with a total thickness of 15 r.l. The fibers make an angle of $45°$ with the beam axis to maximize the Cerenkov light output. The response of the detector is integrated over the duration of the pulse to determine the flux of scattered Moller electrons $(3 - 4.5 \times 10^7 \, e^-/\text{pulse})$.

To measure the elastic and inelastic eP backgrounds, a calorimeter is placed at a larger radius from the beam axis surrounding the Moller detecor. The eP detecor is similar in construction to that of the Moller detector but have 10 times lesser quartz density. Additionaly, a pion detecor is placed behind the Moller calorimeter to measure the pion contribution to the Moller signal.

In addition to the above mentioned detectors, a movable profile detector may be placed in front of the main calorimeter to scan the radial and azimuthal distribution of the flux at the face of the Moller detector. It consists of a quartz Cerenkov detector with PMT readout and can be remotely positioned at any r and ϕ. This scanner is used for collimator alignments, background determination, and Q^2 measurement etc.

Beam Systematics

A critical requirement of the experiment is to keep the beam related systematics at the level of few parts per billion. The rapid, psudo-random flipping of the beam helicity on a pulse to pulse basis minimizes the effects of slow drifts in spectrometer acceptance, pedestals or calibrations. However, the correlated pulse to pulse differences in the beam characteristics (intensity, energy, position and angle) can induce false asymmetries. These effects are corrected for by precisely measuring the beam intensity and positions for every pulse. Additional cancellation of systematic effects come from the passive reversal of beam helicity. This is done by running the beam at two different energies, 45 and 48 GeV which correspond to integer number of g-2 precessions of electron spin in the 24.5° bending of the beam before it enters the experimental area. Another passive reversal of the beam helicity is achieved by inverting the helicity of the laser light at the source using a half-wave plate.

PRELIMINARY RESULTS

The preliminary result is $A_{PV} = -151.9 \pm 29.0(stat) \pm 32.5(syst)$ parts per billion. The value of $\sin^2 \theta_W^{eff} = 0.2371 \pm 0.0025 \pm 0.0027$ at a Q^2 of 0.027 GeV^2. The standard model prediction at this Q^2 value is 0.2386 ± 0.0006[4]. To compare these results with the other experiments, we convert this value to the renormalized mixing angle defined by $\overline{MS}$ (modified minimal subtraction) renormalization scheme: $\sin^2 \theta_W(m_Z)_{\overline{MS}} = 0.2296 \pm 0.0038$. Figure 3 gives this comparison.

Thus, E158 has observed for the first time the parity violation in the Moller scattering. With these results, the limit on Z' mass can be set at the level of 400-500 GeV, and the limit on Λ_{LL} (electron compositness scale) can be set to 3-4 TeV (90% c.l.).

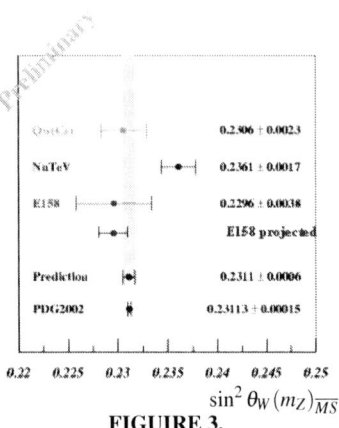

FIGUIRE 3.

REFERENCES

1. K. S. Kumar, et al., *Mod. Phy. Lett. A*, **10**, 2979–2992 (1995).
2. P. A. Souder, et al., *SLAC-Proposal-E-158* (1997).
3. Derman, E., and Marciano, W. J., *Annals of Physics*, **121**, 147–180 (1979).
4. Czarnecki, A., and Marciano, W. J., *Physical Review D*, **53**, 1066–1072 (1996).
5. A combination of preliminary electroweak measurements and constraints on the standard model, Tech. Rep. CERN-EP/2001-98 (2001).
6. Czarnecki, A., and Marciano, W. J., *Int. J. Mod. Phys. A*, pp. 2365–2375 (2000).

DIS-Parity: Measuring $\sin^2(\theta_W)$ with Parity Violating Deep Inelastic Scattering

P.E. Reimer

Argonne National Laboratory, Argonne, IL 60439
Representing the SLAC and Jefferson Laboratory DIS-Parity collaborations.

Abstract. Deep inelastic scattering (DIS) parity violation, arising from the interference between the electromagnetic and weak neutral currents, offers a unique opportunity to study the Electro-weak Standard Model. In the Standard Model, this parity violation measures $\sin^2(\theta_W)$. An experiment is presented to measure the DIS parity violating asymmetry on a deuterium target at a Q^2 near that of the NuTeV measurement (which shows a three standard deviation discrepancy with the Standard Model). The experiment's sensitivity to the Standard Model couplings C_{1q} and C_{2q} is also discussed.

The Standard Model has proven remarkably successful in describing a wide variety of phenomena associated with the electro-weak interaction. Despite this success, there is a slowly emerging view that the Standard Model is incomplete, motivating an enormous effort to find physics not included in the Standard Model. Precision electro-weak measurements offer excellent window through which to look for these extensions.

One such measurement is parity violation in polarized electron-deuterium ($\vec{e}$-d) deep inelastic scattering (DIS). Parity violation arises from an interference between the photon and Z exchange between the electron and the quarks in the deuterium. Within the Standard Model, the asymmetry is sensitive to the Standard Model parameter $\sin^2(\theta_W)$. Extensions to the Standard Model may modify these couplings, changing the apparent value of $\sin^2(\theta_W)$ extracted from the asymmetry measurement.

The Standard Model predicts that $\sin^2(\theta_W)$ will vary, or "run," as a function of Q^2, as illustrated in Fig. 1 [1]. Multiple measurements at the Z-pole have accurately determined $\sin^2(\theta_W)$, but measurements at other values of Q^2 are few. Atomic parity violation in Cesium has measured $\sin^2(\theta_W)$ at very low Q^2 and appears to agree with the Standard Model [2, 3]. The Fermilab NuTeV collaboration recently reported results based on neutrino and antineutrino DIS on an iron target in which a three standard deviation discrepancy with Standard Model predictions was observed [4]. This anomaly has been interpreted as physics beyond the Standard Model [5]; however, conventional explanations, including nuclear effects in the iron target [6] and QCD effects [5], may account for a substantial fraction of the difference. Two other experiments are approved that will measure $\sin^2(\theta_W)$ in the next decade. The SLAC E-158 experiment, which measures parity violation in Møller scattering to determine the electron's weak charge, will complete data collection in the summer of 2003 [7]. The Jefferson Laboratory Q-Weak experiment will measure parity violation in electron-proton elastic scattering to determine the weak charge of the proton. Both experiments will be at $Q^2 \approx 0.03$ GeV^2. In the Standard Model, all of these experiments are sensitive to $\sin^2(\theta_W)$, as shown in

Fig. 1. In the context of extensions to the Standard Model, however, these experiments are sensitive to different couplings, and as such, different Standard Model extensions.

Parity violation in $\vec{e}$-d DIS was originally measured by Prescott et al. [8] in the late 1970's. This experiment provided conclusive evidence for the electro-weak unification model of Weinberg, Salam and Glashow. Now, almost 30 years later with vast improvements in polarized beams and polarimetry, a parity violating DIS experiment ("DIS-Parity") can once again be used to confront the Standard Model.

The asymmetry, A_d, for scattering longitudinally polarized electrons from an unpolarized deuterium target, assuming charge symmetry, is given by [9]

$$A_d = \frac{\sigma_L - \sigma_R}{\sigma_L + \sigma_R} = -\left(\frac{3G_F Q^2}{\pi\alpha 2\sqrt{2}}\right) \frac{2C_{1u} - C_{1d}[1+R_s(x)] + Y(2C_{2u} - C_{2d})R_v(x)}{5 + R_s(x)}. \quad (1)$$

The $C_{1u(d)}$ represents the axial neutral current-electron coupling times the vector neutral current-$u(d)$ quark coupling. The $C_{2u(d)}$ are the vector-electron coupling times the axial-$u(d)$ quark coupling. Within the Standard Model, these are given by

$$C_{1u} = -\tfrac{1}{2} + \tfrac{4}{3}\sin^2\theta_W \approx -0.192, \quad C_{1d} = \tfrac{1}{2} - \tfrac{2}{3}\sin^2\theta_W \approx 0.346,$$
$$C_{2u} = -\tfrac{1}{2} + 2\sin^2\theta_W \approx -0.038, \quad C_{2d} = \tfrac{1}{2} - 2\sin^2\theta_W \approx 0.038,$$

with $\sin^2(\theta_W) \approx 0.231$. The kinematic variable Y is defined as $Y = \frac{1-(1-y)^2}{1+(1-y)^2 - y^2 R/(1+R)}$, with $y = (E - E')/E$, where E (E') is the energy of the incident (scattered) electron. The ratio of longitudinal to transverse cross sections, $R = \sigma_L/\sigma_T \approx 0.2$, is experimentally well determined [10]. The ratios $R_s(x)$ and $R_v(x)$ depend on the quark distributions:

$$R_s(x) = \frac{s(x) + \bar{s}(x)}{u(x) + \bar{u}(x) + d(x) + \bar{d}(x)} \quad \text{and} \quad R_v(x) = \frac{u_v(x) + d_v(x)}{u(x) + \bar{u}(x) + d(x) + \bar{d}(x)}.$$

At large x, where sea quark contributions vanish, $R_v \sim 1$ and $R_s \sim 0$, the asymmetry simplifies to

$$A_d \approx 10^{-4} Q^2 \left[\frac{3}{2}(1+Y) - \left(\frac{10}{3} + 6Y\right)\sin^2(\theta_W)\right].$$

The dependence of the asymmetry on Y allows for the extraction of $2C_{1u} - C_{1d}$ and $2C_{2u} - C_{2d}$ separately. The sensitivity to the $2C_{2u} - C_{2d}$ makes the experiment quite unique. Since "new" physics may affect each of these coefficients in a different way, the ability to separate these coefficients will provide a valuable tool to understand the "new" physics [11] if a deviation from the Standard Model is observed.

The kinematics necessary to do this experiment are well matched to the beam available in SLAC End Station A (ESA). To ensure scattering off of quarks, the experiment must take place at well above the resonance region. Higher twist contributions to the asymmetry must also be small. These drive the experiment to require high Q^2 and W (the invariant mass of the final state hadrons). Other considerations include, minimizing pion contamination and electromagnetic radiative corrections ($E'/E > 0.2$) and minimizing uncertainties due to sea quarks ($x > 0.3$). To maximize the sensitivity to $\sin^2(\theta_W)$, Y must be as large as possible. All of these conditions can be readily met at SLAC ESA or, possibly, at Jefferson Laboratory after the 12 GeV upgrade.

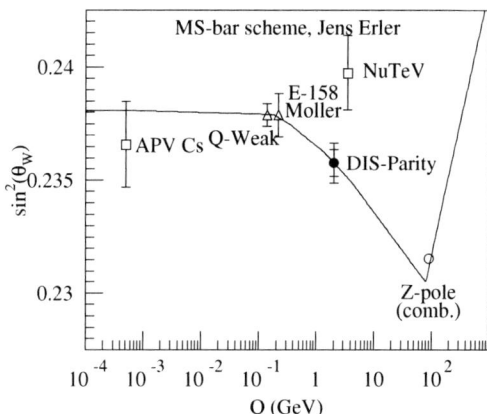

FIGURE 1. The running of $\sin^2(\theta_W)$ in the $\overline{MS}$ scheme [1], showing existing measurements from Cs Atomic Parity Violation (APV Cs) [2, 3], NuTeV vDIS [4], and a combined Z-pole measurement [1]. The proposed measurements of Q_{weak} (proton) [12], SLAC E-158 Møller scattering [7] are shown with their expected uncertainties, as well as the SLAC measurement discussed here (labeled "DIS-Parity").

At SLAC ESA, a spectrometer composed of two identical arms, each at 12° using a lead glass array for electron detection has been designed for this experiment [13]. Given the large asymmetry $A_d \approx 2 \times 10^{-3}$ at $Q^2 \approx 20$ GeV^2, it can be measured in a counting experiment, with flash ADC's reading the lead-glass array, and statistical uncertainty of $\delta A_d/A_d = 0.006$ can easily be achieved. The largest systematic uncertainties are the beam polarization ($\delta A_d/A_d = 0.003$), determination of Q^2 ($\delta A_d/A_d = 0.003$) and Electromagnetic radiative corrections ($\delta A_d/A_d = 0.003$). Overall, the absolute uncertainty, $\delta \sin^2(\theta_W) = \pm 0.0006$ (stat) ± 0.0006 (expt. syst.) ± 0.0004 (theor. syst.).

The experiment is also possible at Jefferson Laboratory after the 12 GeV upgrade using either Hall A or Hall C's (upgraded) standard spectrometers. With the high beam current, a statistical uncertainty of $\delta A_d/A_d = 0.005$ is easily achievable. The dominate systematic uncertainties would be the polarization measurement and the determination of Q^2, contributing 0.01 and 0.005, respectively, to $\delta A_d/A_d$. For an 11 GeV measurement, effects of Higher Twist on the asymmetry are possibly large, and relatively unknown. To help understand these, additional, equally precise, measurements will be needed at lower Q^2 and at different x values. The absolute uncertainty of a measurement at Jefferson Laboratory is expected to be $\delta \sin^2(\theta_W) = \pm 0.0006$ (stat.) ± 0014 (expt. syst.) ± 0.0004 (theor. syst.) *plus uncertainties due to Higher Twist contributions.*

In order elucidate any contributions from non-Standard Model physics, it is important to determine the couplings C_{1q} and C_{2q}. As shown in Eq. 1 and in Fig. 2, DIS-Parity has sensitivity to both of these couplings. Using the measurements from atomic parity violation and Q-Weak to constrain the C_{1q} couplings, allows the DIS-Parity to set vastly improved limits on the C_{2q} couplings.

A measurement of parity violation in electron-deuterium DIS has been outline here. The asymmetry arises from an interference between the photon exchange and the Z

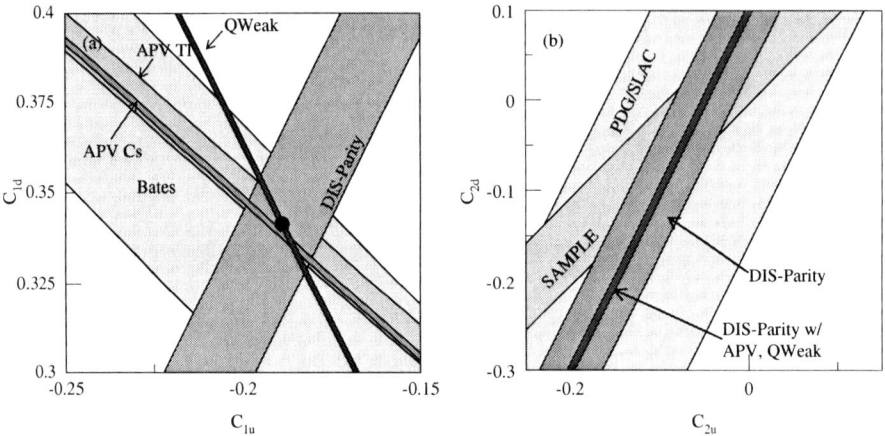

FIGURE 2. The sensitivity of DIS-Parity at SLAC to (a) C_{1d} vs. C_{1u} and (b) C_{2d} vs. C_{2u}. Similar sensitivity is obtained with the Jefferson Laboratory experiment. Also shown in (a) are the Cs [2, 3] and Tl [14] atomic parity violation results, results from Bates [15], and the sensitivity of the approved Q-Weak experiment [12]. Shown in (b) are results based on the original SLAC experiment [16] and the SAMPLE experiment [17, 18]. The shaded bands indicate one standard deviation limits on the parameters.

exchange. Within the Standard Model, this measurement is sensitive to the $\sin^2(\theta_W)$.

This work is supported by the U.S. Department of Energy, Nuclear Physics Division, under contract No. W-31-109-ENG-38 and by the U.S. National Science Foundation.

REFERENCES

1. Erler, J. (2002), Private Communication.
2. Bennett, S. C., and Wieman, C. E., *Phys. Rev. Lett.*, **82**, 2484–2487 (1999).
3. Milstein, A. I., Sushkov, O. P., and Terekhov, I. S. (2002), hep-ph/0212072.
4. Zeller, G. P., et al., *Phys. Rev. Lett.*, **88**, 091802 (2002).
5. Davidson, S., Forte, S., Gambino, P., Rius, N., and Strumia, A., *JHEP*, **02**, 037 (2002).
6. Miller, G. A., and Thomas, A. W. (2002), hep-ex/0204007.
7. Carr, R., et al. (1997), SLAC-PROPOSAL-E-158.
8. Prescott, C. Y., et al., *Phys. Lett.*, **B84**, 524 (1979).
9. Cahn, R. N., and Gilman, F. J., *Phys. Rev.*, **D17**, 1313 (1978).
10. Abe, K., et al., *Phys. Lett.*, **B452**, 194–200 (1999).
11. Ramsey-Musolf, M. J., *Phys. Rev.*, **C60**, 015501 (1999).
12. Carlini, R., et al., The Q_{weak} experiment: A search for new physics at the TeV scale via a measurement of the proton's weak charge (2001), Jefferson Laboratory Proposal.
13. Arrington, J. R., et al., DIS-Parity: Search for new physics through parity violation in deep inelastic electron scattering (2003), SLAC Letter of Intent.
14. Vetter, P. A., et al., *Phys. Rev. Lett.*, **74**, 2658–2661 (1995).
15. Souder, P. A., et al., *Phys. Rev. Lett.*, **65**, 694–697 (1990).
16. Hagiwara, K., et al., *Phys. Rev.*, **D66**, 010001 (2002).
17. Hasty, R., et al., *Science*, **290**, 2117 (2000).
18. Beise, E. (2003), Private Communication.

Low Energy Neutrino Cross Sections

G.P. Zeller

Columbia University, Department of Physics, New York, NY 10027

Present atmospheric and accelerator based neutrino oscillation experiments operate at low neutrino energies ($E_\nu \sim 1$ GeV) to access the relevant regions of oscillation parameter space. As such, they require precise knowledge of the cross sections for neutrino-nucleon interactions in the sub-to-few GeV range. At these energies, neutrinos predominantly interact via quasi-elastic (QE) or single pion production processes, which historically have not been as well studied as the deep inelastic scattering reactions that dominate at higher energies.

Data on low energy neutrino cross sections come mainly from bubble chamber, spark chamber, and emulsion experiments that collected their data decades ago. Despite relatively poor statistics and large neutrino flux uncertainties, these measurements provide an important and necessary constraint on Monte Carlo models in present use. The following sections discuss the current status of QE, resonant single pion, coherent pion, and single kaon production cross section measurements at low energy.

QUASI-ELASTIC SCATTERING

At low energy, neutrinos predominantly interact quasi-elastically ($\nu_\mu n \to \mu^- p$). Over the years, quasi-elastic processes have been studied extensively in bubble chamber experiments at ANL, BNL, CERN, FNAL, and Serpukhov (see Figures for references). The bulk of this data came from light targets and had limited precision due to large neutrino flux uncertainties. Figure 1 compares a collection of this data to a theoretical calculation from the NUANCE Monte Carlo [1]. NUANCE, while different in some regards from other neutrino generators, was chosen because it is representative of the available model predictions [2]. As can be seen from these plots, there is sizable spread in the available experimental data at any given energy and no data on heavy targets below $E_\nu \sim 1$ GeV. At best, this cross section is known to $\sim 15 - 20\%$ based on this data.

RESONANT AND COHERENT SINGLE PION PRODUCTION

The dominant means of single pion production in low energy neutrino interactions arises through the excitation of baryon resonances (Δ, N, etc.) which decay to nucleon-pion final states. There are seven known resonant 1π reaction channels, three CC ($\nu_\mu p \to \mu^- p \pi^+, \nu_\mu n \to \mu^- p \pi^0, \nu_\mu n \to \mu^- n \pi^+$) and four NC ($\nu_\mu p \to \nu_\mu p \pi^0, \nu_\mu p \to \nu_\mu n \pi^+, \nu_\mu n \to \nu_\mu n \pi^0, \nu_\mu n \to \nu_\mu p \pi^-$). With a few excep-

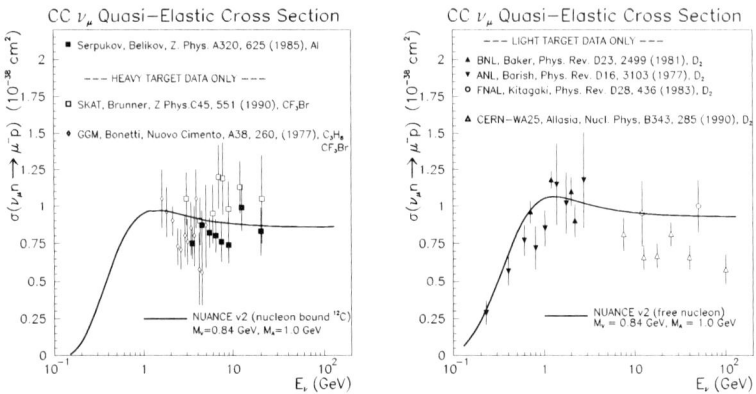

FIGURE 1. QE cross section measurements on both heavy (left) and light (right) targets.

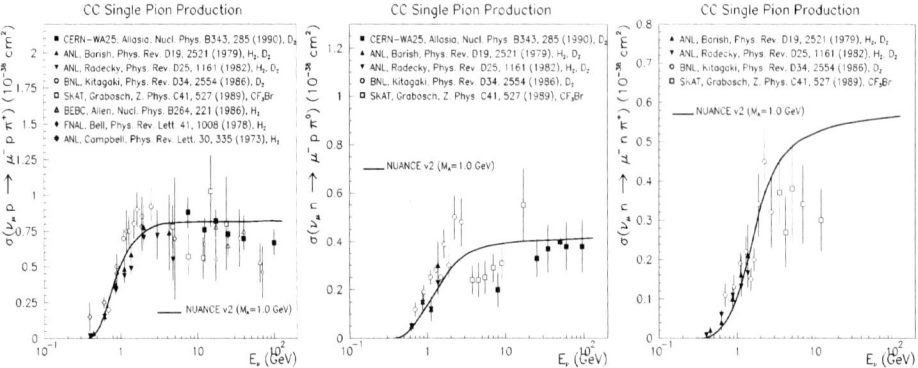

FIGURE 2. CC 1π cross section data; also, Rein-Sehgal-based [3] prediction from NUANCE [1].

tions, most of the measurements of CC 1π cross sections come from D_2 or H_2 bubble chamber experiments (Figure 2). As in the quasi-elastic case, there is no data on single pion production cross sections on heavy targets below $E_\nu = 1$ GeV.

The data on NC single pion production is even more limited. Almost all of this data exists in the form of NC/CC ratios (e.g., see the partial review in [2]). Data on *absolute* NC 1π cross sections is extremely sparse. While the ANL 12 ft D_2 bubble chamber experiment [4] reported a cross section for the NC $1\pi^-$ channel, $\nu_\mu n \to \nu_\mu p \pi^-$, the only measurements of the remaining NC 1π cross sections come from a recent reanalysis of Gargamelle propane-freon bubble chamber data [5] (Figure 3). Overall, the CC and NC resonant 1π cross sections are known to $\sim 20-40\%$ from this data.

In addition to resonance production, neutrinos can also coherently produce single pion final states. In this case, the neutrino coherently scatters from the entire nucleus,

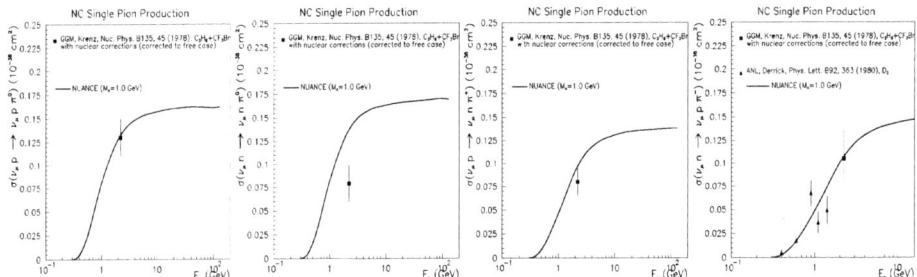

FIGURE 3. NC single π^0 and $\pi^\pm$ cross section data with same Monte Carlo as in Figure 2.

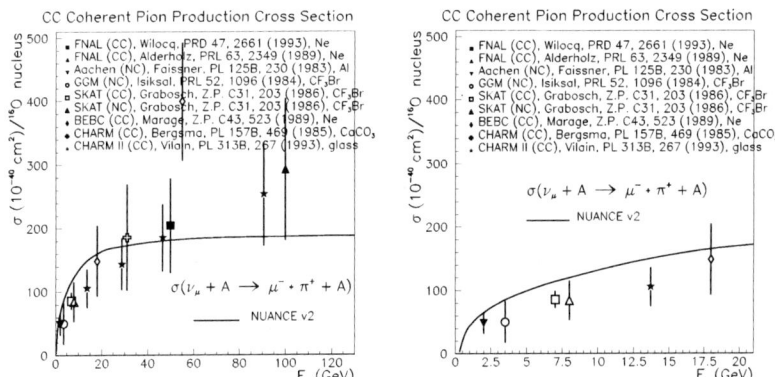

FIGURE 4. Coherent 1π production data. NC and CC data are displayed together after rescaling the NC data $\sigma_{NC}^{coh} = 1/2 \sigma_{CC}^{coh}$; data from various targets have been corrected to ^{16}O assuming $A^{1/3}$ scaling [8].

transferring negligible energy to the target nucleus (A). The result is a distinctly forward-scattered single pion. The cross sections for both NC and CC coherent pion production processes ($\nu_\mu A \to \nu_\mu A \pi^0, \nu_\mu A \to \mu^- A \pi^+$) are predicted to be small, but have been measured in a number of neutrino experiments [6] and over a wide variety of energies (Figure 4). Given the lack of coherent pion data below 2 GeV and the large variation in model predictions at these energies [7], present oscillation experiments typically assign a 100% uncertainty to this particular mode of single pion production.

SINGLE KAON PRODUCTION

Atmospheric neutrino production of single kaon final states holds particular importance as the dominant background to proton decay searches in the SUSY-GUT preferred mode: $p \to \nu K^+$. There are few predictive theoretical models [9] and little experimental data on strange particle production. In fact, there is no data at energies near threshold which are most relevant for proton decay searches. Figure 5 shows results from the only two

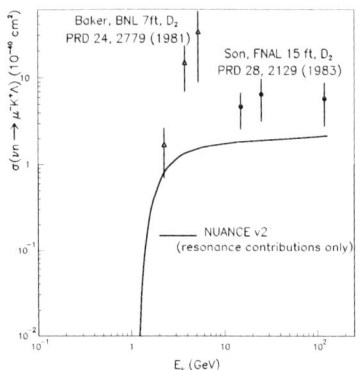

FIGURE 5. Measurements of the CC associated production cross section, $\sigma(\nu_\mu n \to \mu^- K^+ \Lambda^0)$.

experiments which have published cross sections on the dominant associated production channel at atmospheric neutrino energies, $\nu_\mu n \to \mu^- K^+ \Lambda^0$. Both measurements are based on the observation of only a handful of events.

CONCLUSIONS AND FUTURE PROSPECTS

While this collection of low energy neutrino data is generally limited in its precision and availability, new data are already on the way. Both the K2K near detector and MiniBooNE experiments have to date collected at least an order of magnitude more low energy neutrino events than previous experiments. Better constraints are already being obtained by both of these efforts [10]. In addition, proposals for future experiments to further improve this knowledge are also currently being considered [11]. As neutrino oscillation experiments strive for greater precision, they will rely even more heavily on our understanding of low energy neutrino interactions; hence, improvements in this knowledge will carry mounting importance as the neutrino field moves forward.

REFERENCES

1. D. Casper, Nucl. Phys. Proc. Suppl. **112**, 161 (2002).
2. G. P. Zeller, NuInt02 proceedings, to be published in Nucl. Phys. Proc. Suppl., *hep-ex/0312061*.
3. D. Rein and L. M. Sehgal, Annals Phys **133**, 79 (1981).
4. M. Derrick *et al.*, Phys. Lett. **B92**, 363 (1980).
5. E. Hawker, NuInt02 proceedings, to be published in Nucl. Phys. Proc. Suppl.
6. P. Vilain *et al.*, Phys. Lett. **B313**, 267 (1993).
7. E.A. Paschos, A.V. Kartavtsev *hep-ph/0309148*; J. Marteau *et al.*, *hep-ph/9906449*; B.Z. Kopeliovich, P. Marage, Int. J. Mod. Phys. **A8**, 1513 (1993); D. Rein, L.M. Sehgal, Nucl. Phys. **B223**, 29 (1983).
8. D. Rein and L. M. Sehgal, Nucl. Phys. **B223**, 29 (1983).
9. R. Shrock, Phys. Rev. **D12**, 2049 (1975); A.A. Amer, Phys. Rev. **D18**, 2290 (1978).
10. see talks by B. Louis, C. Moore, and C. Walter, these proceedings.
11. L. Bugel *et al.*, FINeSSE, *hep-ex/0402007*; D. Drakoulakos *et al.*, MINERvA, *hep-ex/0405002*.

Working Group 2: *Neutrino Scattering Physics*

B. T. Fleming

Fermi National Accelerator Laboratory, Batavia, IL 60510

Abstract. Presentations from the neutrino scattering talks from working group two (WG2), summarized here, were presented at the Neutrino Factory Workshop, 2003.

INTRODUCTION

Recent results in neutrino oscillation physics coupled with new intense neutrino sources and precision detection techniques have rekindled interest in conventional neutrino scattering physics. Measurements using neutrinos as probes of nucleon structure and as searches for new physics, as well as neutrino cross section measurements, can be performed with increased precision and sensitivity. Puzzling results, such as the NuTeV anomaly continue to be explored. Finally, new high precision detection techniques to address these new measurements and other open questions are being studied. Talks on these subjects presented in WG2 at the NuFACT03 workshop are summarized here.

NEUTRINOS AS PROBES OF NUCLEON STRUCTURE

Strangeness

For many years, experiments have worked to measure and understand the spin structure of the nucleon. This fundamental property can shed light on quark and gluon substructure of the nucleon and is crucial for dark matter searches. Y. Miyachi presented the latest results from the HERMES experiment on the spin carried by the sea quark distributions. This newer analysis tags the flavor of the outgoing hadron in addition to the charged lepton in order to untangle the spin contribution from strange quarks from the that carried by the other light sea quarks. In contrast to previous measurements from charged lepton experiments indicating the spin carried by the strange sea is negative [1], these new results show the spin carried by the sea quarks is consistent with zero.

Measuring the spin carried by the strange quarks using neutrino scattering can address this discrepancy. R. Tayloe presented an experiment in the proposal stage, the FINeSSE experiment [2] designed to to this by using neutrinos as a unique and clean probe of nucleon structure and thus resolving this modern day spin-crisis. Neutrino-proton elastic scattering is sensitive to the strange part of the axial form factor which, at $Q^2 = 0$, is equal to Δs.

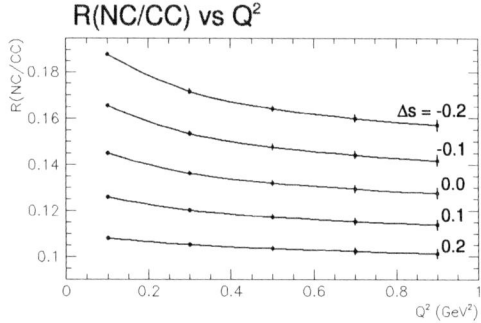

FIGURE 1. The NC/CC ratio for different values of Δs as will be measured by the FINeSSE experiment. Error bars are statistical only.

Unfortunately, uncertainties in the neutrino flux make a precision absolute cross section measurement impossible. Measuring the ratio of $v - p$ to charged current quasi elastic scattering makes this measurement possible, although is slightly less sensitive to Δs. The neutrino energy spectrum for 8 GeV POT, as it is in the Booster neutrino beamline, is ideal for this measurement as it is low in Q^2, but above the region where nuclear effects begin to contribute. The FINeSSE experiment is designed to measure $v - p$ elastic scatters at low Q^2 as well as charged current quasi elastic scattering necessary for the normalization. This ratio will be determined with a 6% total error down to $Q^2 = 0.2 GeV$, making it a definitive measurement of Δs. Fig. 1 shows the NC/CC ratio versus Q^2 for different values of Δs.

Theoretical uncertainties in low energy neutrino scattering generate large systematic errors. Measurements involving ratios reduce these uncertainties, but still different models for these ratios must be tested to ensure errors are negligible. W. Alberico presented studies of nuclear model errors in the ratio of neutral current to charged current neutrino scattering. She showed that for several different models, these troublesome nuclear effects cancel in the ratio measurement.

Deep Inelastic Scattering and the NuTeV anomaly

Neutrino-nucleus scattering at high energies is another excellent probe of nucleon structure and the weak interaction. Relating these well understood cross sections at high energies to those at lower energies where resonance scattering and DIS meet is challenging. U. Yang presented a unified approach for modeling neutrino scattering cross sections and electron scattering cross sections at high and low Q^2 [3].

High statistics precision deep inelastic scattering can provide insight into the weak interaction. In particular, the weak mixing angle can be extracted from neutrino and antineutrino scattering using the Paschos-Wolfenstein technique. The NuTeV experiment, taking advantage of this technique, recently measured a 3.0 σ discrepancy compared to existing non-neutrino measurements and predictions from the Standard Model, as

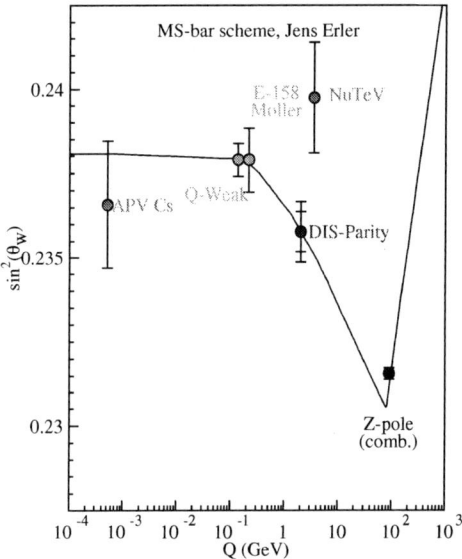

FIGURE 2. Measurements of $\sin^2\theta_W$ versus Q^2 along with the Standard Model prediction of the mixing angle dependence indicated by the solid line. The DIS-parity data point indicates the size of the expected error bars only. The NuTeV result disagrees with the Standard Model at the 2.6 σ level

presented by J. Yu [4]. $\sin^2\theta_W$ can also be measured via Atomic Parity Violation, Moller scattering and DIS-parity scattering. P. Reimer presented new results from E-158, a Moller scattering experiment from SLAC. This electron based measurement is consistent with the Standard Model. I. Younus presented sensitivity to $\sin^2\theta_W$ from a SLAC DIS-parity experiment in the proposal stage. The NuTeV result, SLAC's E-158 Moller scattering result, and expected sensitivity for the SLAC DIS-parity result (shown in agreement with the SM prediction) are shown in Fig. 2.

The NuTeV experiment also has preliminary results on a measurement of the structure function, $F_2(x, Q^2)$ over a broad range of x and Q^2 values. B. Bernstein presented these results, in agreement with previous $v - Fe$ scattering measurements from the CDHSW and the CCFR experiment.

NEUTRINO CROSS SECTION MEASUREMENTS

Improved neutrino cross section measurements are crucial to future neutrino oscillation measurements as well as probes of our understanding of neutrino interactions. Atmospheric neutrino and long-baseline oscillation experiments need improved measurements in the not well-understood 0.1-10 GeV region. S. Zeller presented an overview of how well we understand neutrino cross sections focusing on recent results from the MiniBooNE experiment. C. Walter continued this discussion with recent results from the K2K experiment. K. McFarland completed this session with an overview of future dedicated neutrino scattering physics measurements. These efforts include the MINERvA experiment, in the proposal stage to run at Fermilab on the NuMI beamline, designed to study the DIS to resonance region, exclusive final states in single pion production, and the A dependence of neutrino-nucleon scattering [5]. The FINeSSE experiment, also in the proposal stage at Fermilab will use the lower energy, cleaner neutrino spectrum of the Booster neutrino beamline to study interactions below the deep inelastic scattering turn on. In particular, FINeSSE will measure the strange part of the axial form factor and extract Δs as well as quasi-elastic scattering and single pion production.

Neutrino cross sections at even lower energies are crucial for understanding the supernova process. The SNS facility, under construction, will be a copious source of low energy neutrinos produced from muon decay at rest. Y. Efremenko presented a program of low energy neutrino cross section physics, to be performed at the SNS facility.

Neutrino mass and mixing parameters in general have a big impact on Big Bang nucleosynthesis and cosmology. K. Abazajian presented work on understanding the impact of these mixings on the early universe [6].

NEW INTERACTIONS IN NEUTRINO SCATTERING PHYSICS

Our improved understanding of the physics of the neutrino sector is opening doorways to look for new physics in the neutrino sector and beyond. In particular, searches for GeV scale sterile neutrinos – neutrinos which do not interact via the weak interaction – could lead to new physics. J. Formaggio presented the physics of these GeV scale steriles, event signatures and present limits. He also outlined improved sensitivities possible at a neutrino factory. Along these same lines, J. Sato presented studies on searching for lepton flavor violation in long baseline experiments.

In the Standard Model, massive charged particles have intrinsic magnetic moments by virtue of their spin. If we introduce neutrino mass into the theory, an effective neutrino magnetic moment can arise via one-loop radiative corrections. How neutrino mass fits into the Standard Model, be it via a standard Dirac mechanism, SUSY, or something else, affects the size and origin of this magnetic moment. Thus, we can use a neutrino magnetic moment measurement to tell us how neutrinos fit into the larger theory. B. Fleming presented studies on improved sensitivities to muon neutrino magnetic moment at existing and future experiments.

PRECISION DETECTION TECHNIQUES IN NEUTRINO SCATTERING

The new initiatives described above bring with them new detection techniques. These include optically isolated plastic scintillator stacks, such as the K2K 'SciBar' detector and improvements to this, as well as open volume liquid scintillator detectors read out with a grid of wave length shifting fiber.

Perhaps the most tantalizing technology in neutrino physics is liquid argon TPC detectors. These combine the spatial accuracy of bubble chambers of the past with

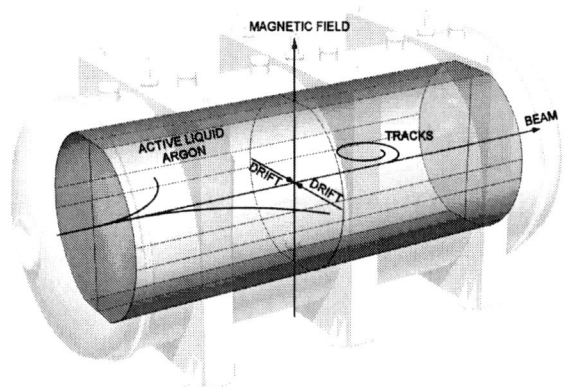

FIGURE 3. The MiniLANNDD 40 ton Liquid argon TPC with wire chamber planes situated down the length of the detector. [7]

active readout, necessary for modern day neutrino experiments. This technology has come to maturity over the last couple of decades and new ideas for both large and small detectors are being considered. F. Sergiampietri presented designs for small, medium, and large Liquid Argon TPCs to be used in conventional neutrino scattering experiments and neutrino oscillation experiments alike. Fig. 3 shows the design for a 40 ton Liquid Argon TPC.

ACKNOWLEDGMENTS

I would like to thank all the speakers who participated in these sessions. Their presentations, new ideas, and lively discussion made these sessions enjoyable and interesting for all. I would also like to thank the organizers of the NuFACT 03 workshop for the opportunity to organize and participate in this parallel session, and for their efforts in hosting such a successful workshop.

REFERENCES

1. D. Adams *et al.*, Phys. Rev. **D56**, 5330 (1997).
2. L. Bugel *et al.* [FINeSSE Collaboration], arXiv:hep-ex/0402007.
3. A. Bodek and U. K. Yang, arXiv:hep-ex/0308007.

4. G. P. Zeller *et al.* [NuTeV Collaboration], Phys. Rev. Lett. **88**, 091802 (2002) [Erratum-ibid. **90**, 239902 (2003)] [arXiv:hep-ex/0110059].
5. [Minerva Collaboration], "Proposal to perform a high-statistics neutrino scattering experiment using a fine-grained detector in the NuMI beam," arXiv:hep-ex/0405002.
6. arXiv:hep-ph/0312163.
7. D. B. Cline, Y. H. Seo and F. Sergiampietri, arXiv:astro-ph/0301545.

WORKING GROUP 3

Tetra Muon Cooling Ring

S. Kahn, R. Fernow

Brookhaven National Laboratory, Upton, NY 11973

V. Balbekov, R. Raja, Z. Usubov

Fermi National Accelerator Laboratory, Batavia, IL 60510

Abstract. Recent simulations have shown that muon cooling rings can effectively reduce both longitudinal and transverse emittance. The muon collaboration is investigating several varieties of muon cooling rings. This study looks at the first of these ring cooling scenarios that was proposed by V. Balbekov. This simulation of this ring shows significant cooling in the hardedge field approximation. We discuss the status of using realistic fields in the tetra simulation.

INTRODUCTION

Muon cooling rings have shown promise in simulation studies as a technique for reducing muon beam emittance. In muon ring coolers, the muons pass through absorber material followed by acceleration repeatedly for a number of turns. This can produce cooling if there is sufficient focusing at the absorber. Currently there are three different ring coolers that are being investigated by the collaboration: 1) The Tetra four sided solenoid cooling ring, 2) the RFoFo cooling ring, and 3) dipole edge focused cooling rings. This paper will primarily discuss the simulations that have been performed for the Tetra ring. The Tetra ring, which was originally proposed by Valeri Balbekov [1], was the first of the muon cooling ring designs to show significant phase space cooling. The Tetra ring consists of long straight sections with a large solenoid magnet surrounding RF cavities and a liquid hydrogen absorber to provide transverse cooling. The field in the long solenoid varies from 2 T at the ends to 5 T in the center where the absorber is placed, so as to reduce the beta function at the absorber. The RF cavities are adjusted to replace the energy lost in the absorber. In the corner region between the two straight sections there is a short straight section with two opposite field solenoids sandwiched between two 45° dipole bend magnets. The bend magnets are combined function dipoles with a field index of 0.5 to provide focusing in both planes. Figure 1 shows the layout of the cooling ring, along with the principle parameters that describe it. The solenoid field flips sign in each of the solenoid channels, as indicated in the figure. The field flip occurs in the center of the short solenoid channel. A LiH wedge absorber is positioned near the field flip position.

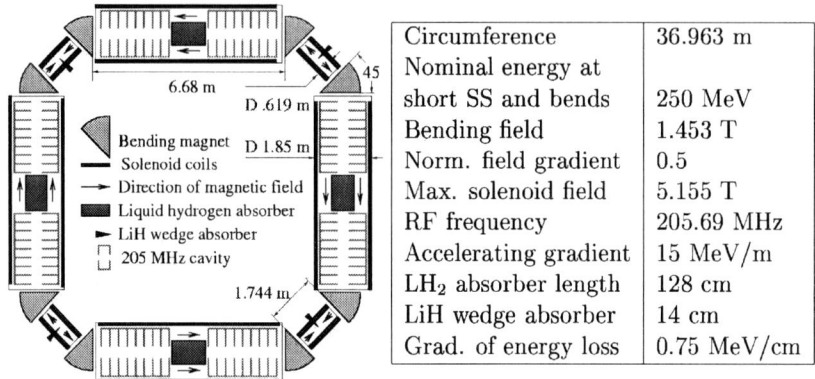

FIGURE 1. Layout of the Tetra muon cooling ring and a table of the principle parameters describing the ring.

Simulations of the Tetra ring have been carried out by Balbekov using his own simulation program [2], by Fernow using ICOOL [3], and by S. Kahn, R. Raja and Z. Usubov using GEANT. The ICOOL simulation study shows significant cooling with merit factors up to 103 after 15 turns. (The merit factor, which is the muon transmission divided by the emittance reduction, is a measure used to compare various cooling schemes.) As promising as these studies have been, there are problems with the implementation. First of all, the studies that have been performed have used hardedge field descriptions, which ignore fringing fields. Secondly, there is no space between the ring elements, which could cause the solenoid and dipole fields to couple and would make it difficult to bring in RF and cryogenic services. Also there is no reasonable approach yet for injection to or extraction from the ring.

HARDEDGE MODEL

Figure 2 shows the transmission, transverse emittance, and 6D emittance as a function of turn number for this ring simulated in ICOOL using a hardedge description of the field. The figure also shows the cooling merit factor. The cooling ring lattice is highly optimized to provide this performance. In an attempt to add a 15 cm gap on each side of the wedge bend magnet, matching coils were added symmetrically to the ends of the adjacent solenoid magnets to compensate and to match into the bending magnets such that $\alpha=0$. The matching coil current was varied to maximize the transmission and minimize the final emittance while keeping the focusing of the solenoids, $\int B_s^2 ds$, and the value of B_s at the absorber unchanged. Inserting this 15 cm gap reduces the transmission by 80% and increases the transverse emittance by a factor of 1.5. The longitudinal emittance is unchanged. The orientation angle of the

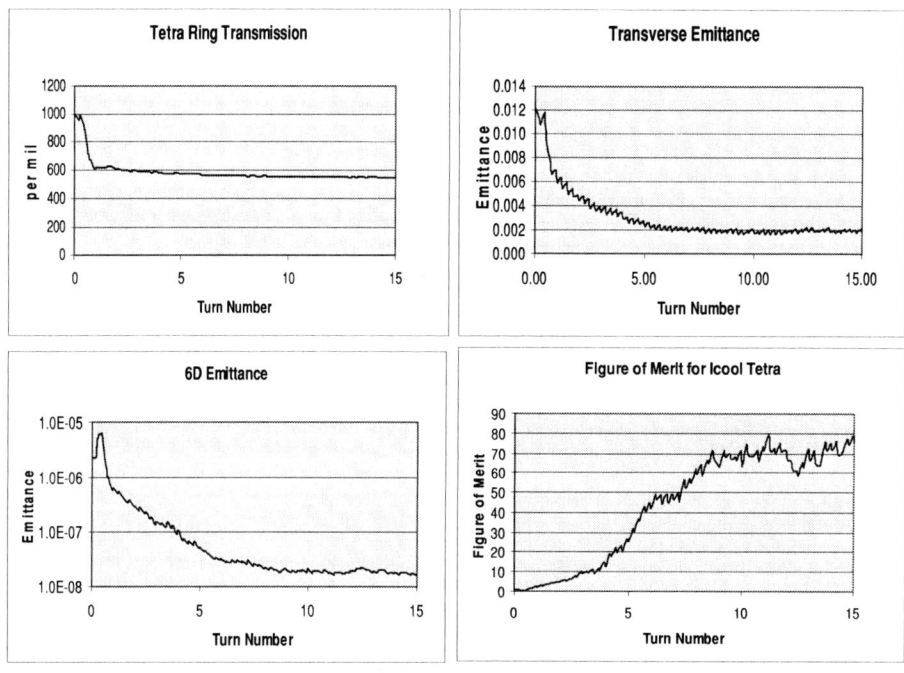

Figure 2: Transmission, transverse emittance, 6D emittance and figure of merit for the Tetra cooling ring with hardedge fields. Emittances are in meters and the transmission is given in parts per mil.

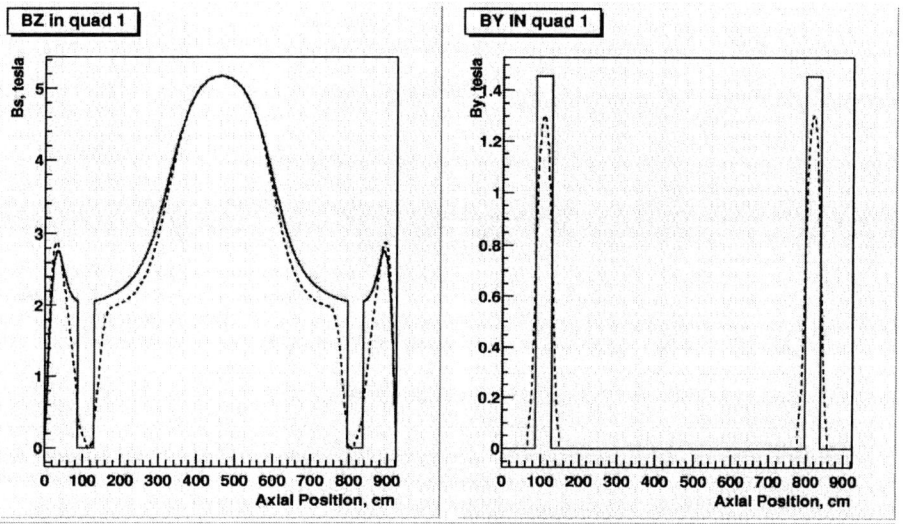

Figure 3: A comparison of a realistic field description (dotted) to the hardedge description (solid). The figure shows the axial component (left) and the vertical component (right) of the field for one quadrant of the ring.

wedge absorber should be adjusted when varying the fields, however the results are not strongly sensitive to this.

REALISTIC FIELDS

The hardedge field description does not satisfy Maxwell's equations and there is a concern that fringing fields may alter the cooling efficiency of the ring. An attempt to make a realizable design of a magnet system for the Tetra cooling ring was reported at the PAC01 Accelerator Conference [4]. In this report the 45° bend magnets are 1.45 T iron C magnets with the poles shaped to give the field index of 0.5. The pole region is highly saturated, but the magnet does provide the proper field index. Field clamps on the ends of the magnet are provided to reduce the fringing field from the dipole magnet. In order to minimize the fringe field from the large aperture of the long solenoid magnet, iron plates (with extra coils at the inner radius) are placed on the ends of the solenoid to return the flux. A similar arrangement is made for the short field flip solenoid. The resulting field from this more realistic magnet system is compared to the hardedge field in figure 3. The figure shows both the axial and the vertical component of the field for one quadrant of the tetra lattice. This field is currently being implemented in the GEANT model of this ring.

As another step toward simulating the tetra ring with realistic fields, the hard edge solenoids (with fictitious mirror plates) in the ICOOL model were replaced with an elliptic integral representation that produces a fringe field that extends into the dipole magnet region. The bend magnet in this study is still represented by the hard edge description. The solenoid fringe field in this approximation does not follow the curved reference path in the bend magnet. The emittances, transmission and merit factors obtained by replacing the hard edge solenoid by this more realistic representation remain essentially unchanged.

ACKNOWLEDGMENTS

This work was performed with the support of the US DOE under Contract No. DE-AC02-98CH10886.

REFERENCES

1. V. Balbekov et al., *Muon Ring Cooler for the MUCOOL experiment*, Proceedings 2001 Particle Accelerator Conference, Chicago, IL, 2001, pp. 3867-9.
2. V. Balbekov, *Ring Cooler Progress,* Muon Collaboration Note MUC-NOTE-246 (2002).
3. R. Fernow, *Hardedge ICOOL Model of the Balbekov Square Cooling Ring,* Muon Collaboration Note MUC-NOTE-258 (2002).
4. S. Kahn et al., *Design of a Magnet System for a Muon Cooling Ring Experiment,* Proceedings 2001 Particle Accelerator Conference, Chicago, IL, 2001, pp. 3239-41.

RFOFO Cooling Ring: Simulation Results

J.S. Berg, R.C. Fernow, J.C. Gallardo, and R.B. Palmer

Physics Department, Brookhaven National Laboratory, Upton, NY 11973

Abstract. Practical cooling rings could lead to lower cost or improved performance in neutrino factory or muon collider designs. The ring modeled here uses realistic 3-dimensional fields and includes such "real-world" effects as windows on the absorbers and RF cavities and leaving empty lattice cells for injection and extraction. The ring increases the density of muons in a fixed acceptance volume by a factor of 4.2.

INTRODUCTION

Designs for neutrino factories [1] and muon colliders [2] use ionization cooling to reduce the emittance of the muon beam prior to acceleration. Current baseline designs make use of linear channels that only cool the beam in transverse phase space. However, there has been considerable progress over the past two years in achieving 6-dimensional ionization cooling in cooling rings [3]. At present the most realistic modeling has been done for the RFOFO ring [4-7]. Alternating polarity solenoids provide transverse focusing. The bending field is provided by alternately tipping the axis of the solenoids above and below the orbital midplane. A short cell length is used to obtain a small beta function with a reasonable value of the solenoid field strength. Wedge-shaped absorbers are placed in the beam path at locations where the solenoidal field changes direction. Most of the lattice cell is filled by RF cavities to restore the energy lost in the absorbers.

MODELING THE RING

The RFOFO ring was modeled using the simulation code ICOOL [8]. A layout of three cells of the ring is shown in Fig. 1 and a summary of ring properties are given in Table 1.

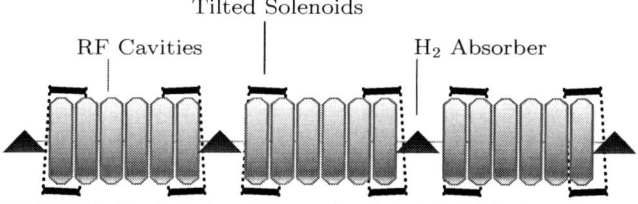

FIGURE 1. Vertical view of three cells of the RFOFO lattice.

The particle motion is centered on a 33 m circumference circle. The centers of the solenoids are displaced radially outwards from this circle by 10 cm to reduce the on-axis radial field. The 3-dimensional magnetic field from the ring of tipped solenoids was calculated in an independent code by summing the fields from a system of current sheets [9,10]. The resultant field components were shown to satisfy Maxwell's equations to a high level of accuracy and agreed well with independent calculations [11,12]. The RF cavities were modeled using cylindrical pillboxes running in the TM010 mode. The liquid hydrogen absorbers have a wedge shape and are located in dispersive regions in order to decrease the momentum spread in the beam.

TABLE 1. RFOFO ring parameters.

circumference	33	m
cells	12	
coil tilt angle	3	degrees
min. transverse beta function	38	cm
max. dispersion function	8	cm
wedge opening angle	100	degrees
RF frequency	201.25	MHz
peak RF gradient	12	MV/m
RF phase from 0-crossing	25	degrees

SIMULATION RESULTS

We first examine the performance of an "ideal" ring, ignoring the effects due to windows on the absorbers and RF cavities and leaving empty space for injection. Since the magnetic field has a small radial component on-axis, the closed orbits are non-planar. Fig. 2 shows the transverse motion of the closed orbit for a 200 MeV/c muon.

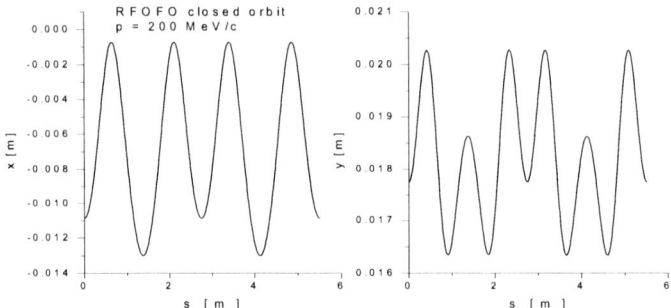

FIGURE 2. Transverse position of a closed orbit along two cells of the system axis. y: vertical, x: radial position.

Note that the closed orbit is offset by 11 mm in x and by 18 mm in y at the beginning of the cell. Along the cell x varies by ±6 mm and y varies by ±2 mm. For the simulations we use a Gaussian input beam with normalized transverse emittance of 12 mm and normalized longitudinal emittance of 18 mm. The initial beam had a correlation between the axial momentum and the transverse amplitude to minimize the tendency for the particles in the bunch to spread out longitudinally in the solenoidal field. The correlation causes the average axial momentum to be larger than the reference momentum of 203 MeV/c. The ring has a momentum acceptance from 160 to 260 MeV/c. Fig. 3 shows the radial and longitudinal phase space after 1 and 15 turns.

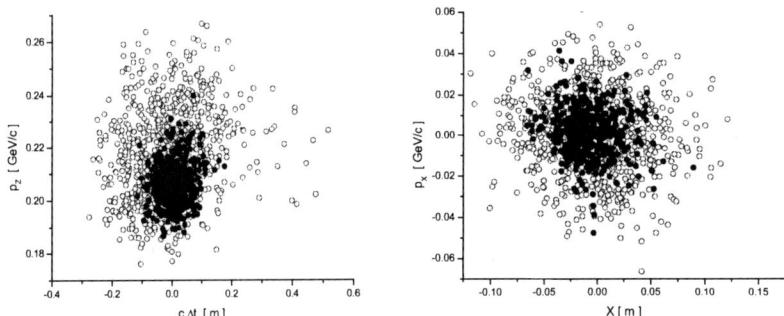

FIGURE 3. Longitudinal phase space (left) and radial phase space (right) after 1 turn (open circles) and 15 turns (closed circles).

The reduction of phase space area can be seen clearly in both distributions. The vertical transverse distribution is similar to the radial one because of the solenoids.

We now consider the effects on performance of including windows for the absorbers and RF cavities and leaving empty cells for injection. We refer to this as the "practical" ring design. We enclose the liquid hydrogen absorbers with 360 μm aluminum windows. The RF cavities have tapered 20 μm thick beryllium windows on the two ends and tapered 70 μm thick beryllium windows between the interior cells. Two adjacent cells have the absorber and RF cavities removed in order to leave space for the injection kicker, but the magnetic periodicity is maintained. Fig. 4 shows the evolution of the emittances and transmission for the practical ring design.

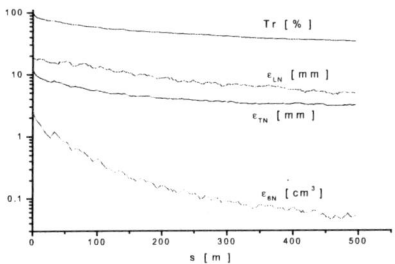

FIGURE 4. Normalized emittances and transmission with decay as a function of accumulated distance in the practical cooling ring. One turn is 33 m.

The performance of the practical ring is summarized in Table 2.

TABLE 2. Performance of the practical RFOFO cooling ring.

ε_{TN}	3.1	mm
ε_{LN}	4.8	mm
ε_{6N}	0.05	cm^3
Tr	33	%
M	18	
D	4.2	

Two common merit factors for cooling rings listed in Table 2 are

$$M = \frac{\varepsilon_{6N}(initial)}{\varepsilon_{6N}(final)} Tr, \quad D = \frac{N(s)/V}{N(0)/V}$$

where N is the number of muons and V is a fixed 6-dimensional acceptance volume. The corresponding factors for the ideal ring are M=112 and D=8.9.

ACKNOWLEDGMENTS

This work was supported by the U.S. Department of Energy under contract DE-AC02-98CH10886.

REFERENCES

1. Alsharo'a, M. et al, Recent progress in neutrino factory and muon collider research within the Muon Collaboration, *Phys. Rev. Special Topics-Accelerators and Beams* **6**, 081001-1-52 (2003).
2. Ankenbrandt, C.M. et al, Status of muon collider research and development and future plans, *Phys. Rev. Special Topics-Accelerators and Beams* **2**, 081001-1-73 (1999).
3. Palmer, R.B., Ring coolers, *J. Phys.G* **29**, 1577-1583 (2003).
4. Berg, J.S., Fernow, R.C., and Palmer, R.B., RFOFO ring cooler, *J. Phys.G* **29**, 1657-1659 (2003).
5. Fernow, R.C., Berg, J.S., Gallardo, J.C., and Palmer, R.B., Muon cooling in the RFOFO ring cooler, submitted to the proceedings of the 2003 Particle Accelerator Conference, Portand, OR.
6. Berg, J.S., Fernow, R.C., and Palmer, R.B., An alternating solenoid focused ionization cooling ring, MUC-NOTE-COOL-THEORY-239, Mar. 2002. This series of technical notes can be found at (http://www-mucool.fnal.gov/mcnotes/).
7. Fernow, R.C., Berg, J.S., Gallardo, J.C. and Palmer, R.B., Muon cooling in the RFOFO ring, MUC-NOTE-COOL-THEORY-273, Apr. 2003.
8. Fernow, R.C., ICOOL: a simulation code for ionization cooling of muon beams, Proc. 1999 Particle Accelerator Conference, p. 3020-3022.
9. Fernow, R.C., and Gallardo, J.C., Realistic on-axis fields for the RFOFO cooling ring, MUC-NOTE-COOL-THEORY-265, Nov. 2002.
10. Fernow, R.C., and Gallardo, J.C., Calculation of RFOFO fields using the off-axis expansion in ICOOL, MUC-NOTE-COOL-THEORY-268, Jan. 2003.
11. Balbekov, V., Simulation of RFOFO ring cooler with tilted solenoids, MUC-NOTE-COOL-THEORY-264, Nov. 2002.
12. Bracker, S., Magnetic field maps for the RFOFO muon cooling ring, MUC-NOTE-COOL-THEORY-271, Mar. 2003.

A Muon Ring Cooler with Lithium Lenses

Yasuo Fukui, David Cline, Alper Garren

Physics and Astronomy, UCLA, CA 90095, USA

Harold Kirk

Brookhaven National Laboratory, Upton, NY 11973, USA

Abstract. We designed a muon cooling ring with straight inserts with Lithium lenses and injection and extraction, and we demonstrated the transverse phase space cooling of muon beam in the ring with a tracking simulation code with hard-edged magnetic elements. The muon coling ring can provide the final transverse muon cooling for a Higgs Factory, a low energy version of a $\mu^+\mu^-$ collider.

MUON COOLING WITH LITHIUM LENSES

The ionization cooling is one of the most promising methods to reduce the 6 dimensional phase space of muon beam, where both transverse and longitudinal momenta are reduced due to the energy loss in absorbers, and only longitudinal components of the muon momenta are restored through the accelerating fields of RF cavities. The multiple Coulomb scattering contributes to heat the transverse phase space. And the dE/dx straggling through absorbers contributes to heat the longitudinal phase space. Wedge absorbers in dispersive region in bending cells perform the emittance exchange between the longitudinal emittance and the horizontal emittance with normalized transverse phase space above 1 mm·rad [1]

Lithium lens is an active focusing and energy absorbing element. With β at 1 cm with high current density Lithium lenses, the equilibrium normalized transverse emittance can be at around 100 mm·mrad, which is necessary for a $\mu^+\mu^-$ collider. [2,3]

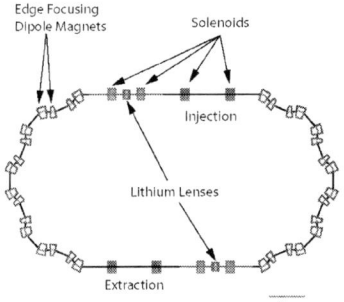

FIGURE 1. A schematic diagram of a muon cooling ring with Lithium lenses in straight sections.

In the muon cooling ring with Lithium lenses, Lithium lenses with the β function as low as 1 cm is placed in a straight section with matching solenoid magnets. Figure 1 shows a schematic view of a muon cooling ring with Lithium lenses. Circumference is 47.0 m, the straight section is 8.9 m long each, and the radius of the bending section is 4.6 m.

Muon Cooling in Straight Channels with Lithium Lenses

We designed muon cooling rings with Lithium lenses which are made of 2 matching higher β Lithium lenses sandwiching the central lower β Lithium lens. β at the inner 22 cm long Lithium lens is 1.0 cm . The matching Lithium rod with the length of 6.3 cm each, which sandwich the central Lithium lens, has an equilibrium β at 4.0 cm, which swings the β function from the β at 16 cm at the outer end to the β at 1 cm at the inner end of the matching Lithium lens. The solenoids have 6 Tesla B_z field where the B_z direction of solenoids is opposite to each other, and each solenoid is 0.41 m long. Figure 2 shows a schematic diagram of a Lithium lens and straight section which is made of 2 matching solenoids and a set of Lithium lenses. Figure 3 shows the β as a function of z in the Lithium lens and matching cells with solenoids.

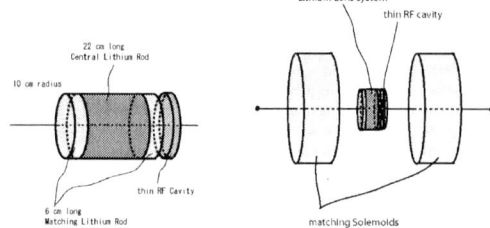

FIGURE 2. A schematic diagram of a Lithium lens(left) and straight section(right)

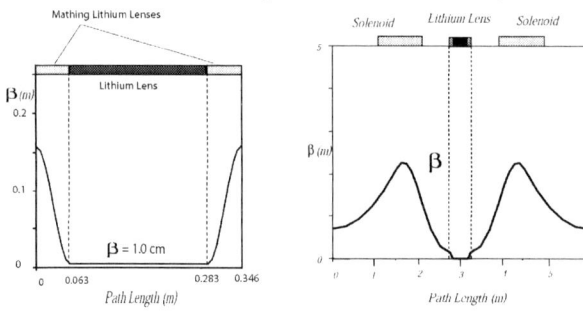

FIGURE 3. β as a function of s in the Lithium lens(left) and matching cells with solenoids(right)

In order to study the muon beam dynamics only through a Lithium lens and matching solenoid lattices which sandwich the Lithium lens, we performed tracking simulation with ICOOL tracking code[4]. Original model was designed by using the SYNCH [5] which generates the input date for the tracking code ICOOL.

Figure 4 shows the normalized transverse emittance (left), the longitudinal emittance, Δp/p, and Δz (center) and the muon transmission (right) as a function of z through 50 sets of 4.0 m straight sections. In this simulation, the loss of muon p_z due to the dE/dx energy loss through the Lithium lens is recovered through a thin RF cavity by adding average p_z kick. The equilibrium normalized transverse emittance is around 0.3 mm·rad.

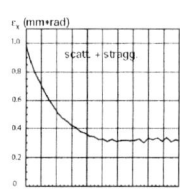 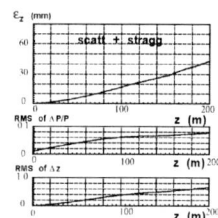 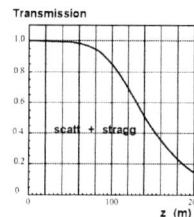

FIGURE 4 Normalized transverse emittance vs. z (left), normalized longitudinal emittance vs.z, $\Delta p/p$ vs. z, Δz vs. z (center), and muon transmission vs. z (right) through 50 sets of straight sections.

Muon cooling in a Cooling Ring with Lithium lenses

We designed a 45 degree bending cell by using two sets of zero-gradient dipole magnets with edge focusing. Figure 5 shows β function vs. z and the dispersion D vs. z in a 45 degree bending cell (left) and in an half of the muon cooling ring with Lithium lenses in straight sections (right). Table 1 lists parameters of the muon cooling ring with two Lithium lenses in straight channels.

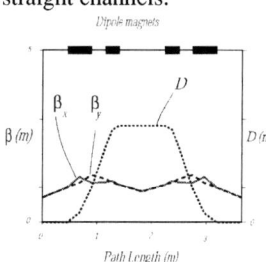 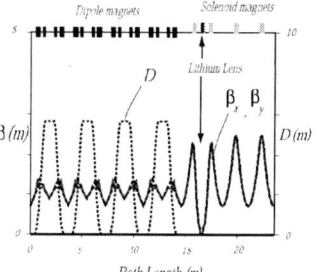

FIGURE 5 β vs. z and the dispersion D vs. z in a 45 degree bending cell (left) and in an half of the muon cooling ring with Lithium lenses in straight sections (right).

TABLE 1. Parameters of a muon cooling ring with Lithium lenses.

Parameters	Values
muon momentum	250 MeV/c
Circumference/ straight section	47.0 m / 8.9 m (x2)
dE/dx energy loss	35 MeV/turn
length of Lithium lens	0.345 m (x 2)
lowest/highest β in Lithium lens	1.0 cm / 16 cm
dipole magnetic field	1.8, -1.8 Tesla
dipole bend angles	54.9, -32.4 degree
dipole edge angles	18/12, 6/7 degree
ring tunes	15.50/15.50
chromaticity, momentum compaction	-10.04/-8.77 , -0.079

Figure 6 (left figure) shows the development of the normalized transverse emittance as a function of z through 8 turns of the muon cooling ring. In this simulation, the loss of muon p_z due to the dE/dx energy loss through the Lithium lens is recovered through a thin RFcavity by adding average p_z kick. The figure indicates the transverse cooling in the muon cooling ring with Lithium lenses with the normalized transverse emittance down to around 0.3 mm·rad. The middle and right figures show the development of the normalized longitudinal emittance, $\Delta p/p$ and Δz as a function of z, and the transmission as a function of z, respectively.

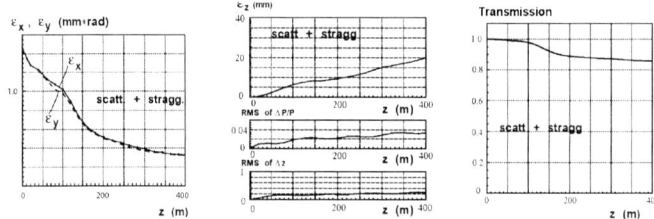

FIGURE 6 Normalized transverse emittance vs. z (left), normalized longitudinal emittance vs.z, Δp/p vs. z, Δz vs. z (center), and muon transmission vs. z (right) through 8 turns of the muon cooling ring.

Conclusion

We designed a muon cooling ring with 35 cm long Lithium lenses in straight channels with β at 1 cm. Bending cells have zero-gradient dipole magnets with edge focusing. With tracking simulation, we demonstrated the normalized transverse phase space cooling down to around 0.3 mm·rad, which is required for the final muon cooling in a Higgs Factory, a lower energy version of a $\mu^+\mu^-$ collider.

ACKNOWLEDGMENTS

The authors wish to thank R. B. Palmer, R. Fernow, A Sessler, M. Zisman, J. Wurtele, D. Neuffer, and V. Balbekov, for useful discussions, and their help and encouragement.

REFERENCES

1. H.G. Kirk, D. B. Cline, Y. Fukui, A. A. Garren, "Progress Toward A Muon Ring Cooler" *SNOWMASS-2001-M101*, Jun 2001. 5pp., Prepared for APS / DPF / DPB Summer Study on the Future of Particle Physics (Snowmass 2001),Snowmass, Colorado, 30 Jun – 21 Jul 2001.
2. C. M. Ankenbrandt, et., al., the NuMu Collaboration, "Status of Neutrino Factory and Muon Collider Reasearch and Developmemt and Future Plans", *Phys. rev. ST Accel. Beams* 2 (1999) 081001
3. Mohammad M. Alshar et al., "Recent progress in neutrino factory and muon collider research within the Muon Collaboration" *Phys. Rev. ST Accel. Beams* 6, 081001 (2003)
4. R. Fernow, "A Simulation Code for Ionization Cooling of Muon Beams", Part. Accel. Conf., Edts., A. Luccio and W. MacKay, Proc. 1999, p. 3020.
5. A. Garren, A. S. Kenney, E. D. Courant, A. D. Russel, and M. J. Syphers. "SYNCH A Compueter System for Synchrotron Design and Orbit Analysis, User's Guide", SSCL-MAN-0030, 1993

Features Of A Muon Cooling Ring For A Neutrino Factory

S J Brooks, M R Harold, C R Prior and G H Rees

Rutherford Appleton Laboratory, UK

Abstract. A low energy muon cooling ring is considered for a Neutrino Factory as it may lead to a large cost reduction of the facility.

1. INTRODUCTION

The ring must be able to handle, at injection, the large muon transverse and longitudinal emittances that remain after proton target irradiation, pion decay to muons in a solenoid focusing channel and subsequent muon bunch rotation in a number of 31.4 MHz radio frequency (rf) cavities. A suitable mean kinetic energy of the muons is 180 MeV. The ring requires compact bending and focusing, space for the injection, rf and extraction components, and suitable locations for ionizing cooling elements. Cooling is introduced in the transverse and longitudinal planes by the use of ionizing absorbers at low β-value lattice positions. For the transverse plane, parallel edged absorbers are needed at positions of zero disperion while, for the longitudinal plane, wedged shaped elements are used in finite dispersion regions.

Injection and extraction are demanding due to the large muon transverse emittances and the need for a compact ring to limit the muon decay. A free length of at least 4 m is needed for injection and a separate 4 m for extraction, and these have to be included within a ring circumference of < 45 m. A single muon bunch is injected for each cooling cycle, but it may divide depending on the choice of rf system. Frequencies considered have been 88 MHz (harmonic number, h = 14) and 201.14 MHz (h = 32), and the latter is favoured, with three bunches forming in the ring.

2. NEUTRINO FACTORY FRONT END

Ahead of the ring, the front end consists of an ~ 4 MW proton beam incident on a high power target, a superconducting solenoid pion decay channel, and a solenoid channel with rf cavities excited to give longitudinal muon bunch rotation. In an earlier study[1], the decay channel was ahead of a bending magnet chicane. Here, the chicane has been replaced by the rf bunch rotation system and cooling ring. The Muonplayer code[1,2] now has a more flexible optimisation routine for adjusting the decay channel solenoids, and the predicted muon yields have increased.

The decay channel length is ~ 30 m, as before, but it is now optimized for muon energies of 180 ± 70 MeV. In the channel, pion-muon bunches shear in longitudinal phase space, increasing in bunch extent from 1 to ~ 1.7 ns rms on reaching the bunch rotation cavities. These have a frequency of 31.4 MHz and include a 0.9 m long solenoid lens within their nose cone. There are 30, 1.4 m cavities, each adding a peak voltage of 2.25 MV, giving an upright bunch orientation of energies 180 ± 23 MeV. After the cavities is a short solenoid section for matching to the adjacent cooling ring.

3. COOLING RING LATTICE

A racetrack is proposed for the lattice as shown in Figure 1 for the h = 32 rf system. The two long straight sections each contain 4 m free spaces for injection or extraction, 8 solenoids and 18, 201.14 MHz cavities, of length 0.316 m. In each 180° arc there are 4 focusing solenoids and 4, 45° weak focusing, gradient bending magnets. The latter are used in pairs, with a small unit spacing and a common excitation coil. The spaces are introduced for longitudinal cooling elements. Solenoids are placed at the points S of Figure 1 and transverse cooling elements at the points marked A.

Focusing is made with solenoids not quadrupoles. They allow simpler matching from the front end; they are more easily included in rf cavities; they give equal focusing for the transverse planes, and focus the large emittance beams more strongly over the energy range. The symmetry also allows simpler superconducting technology for the solenoid fields of up to 5 T. The bending sections may be of dipole magnets, combined function units or tilted quadrupoles[3]. The chosen arc has 4 combined function magnets and has been found superior to an arc using a chicane[1]. Weak focusing, not zero gradient, dipoles are preferred as they allow equal transverse plane focusing without magnet end shaping. Tilted quadrupoles have also been considered for the arcs but use in a racetrack ring is more complex than in the high periodicity lattice[3] as transverse dispersions in the long straight sections are difficult to minimize. The asymmetric, tilted quadrupole, off-axis fields are also a concern.

Trace3D and Parmila have first been used to design the lattice, approximating the effects of solenoid fields, rf cavities and gradient magnets. Arc solenoid and gradient unit parameters are adjusted for an achromatic arc with 2π transverse betatron phase shifts and with dual betatron waists, of β-values ~ 0.72 m, at each gradient magnet pair mid-point. All 4 solenoids assume a 0.4 m length and an ~ 2.295 T hard-edged field, and with opposite field polarities in the central 2 units. Each 45° bending magnet needs an effective length of 0.63 m, a bending radius of 0.80214 m, zero degree entry and exit angles and an n value of ~ 0.66 for vertical focusing.

In the long straight sections, solenoids are adjusted to give betatron waists at the absorbers A and at injection and extraction centre points. The β-values are 0.067 m at the absorbers and ~ 2.2 m at the centre points. Solenoids at the ends of the straight are adjusted for β-matching at the ends of the arcs. The unnormalised transverse emittances assumed are 10,000 (π) mm mr. The long straights have zero dispersion due to opposed fields in the two central arc solenoids and the 2π arc transverse betatron phase shifts. Transverse dispersions are maximum at the arc centre.

Injection and extraction ring matching sections are identical. There is a solenoid next to a betatron absorber straight (A), and this is the last (first for extraction) of a set of four equal solenoids, of alternating polarity, which provide 2π transverse betatron phase shifts. In the second (third) solenoid cell, there is a dipole magnet to bend the muon beam through an angle equal to that of a kicker (in the fourth injection or first extraction cell). Solenoids up/downstream of the 2π section allow betatron matching.

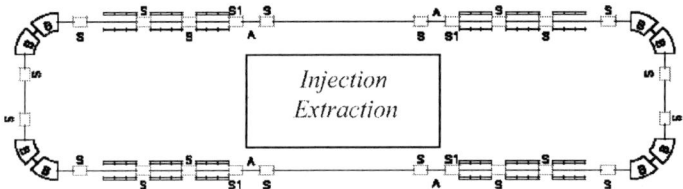

Figure 1: Muon Racetrack Cooling Ring

4. PARMILA TRACKING

Muon emittances and energy spread from pion decay are approximated, and muons are then tracked through the decay channel, the bunch rotation channel, the output matching section and the first three turns in the ring. The energy spread reduces from ± 70 MeV to ± 23 MeV due to bunch rotation, and further energy changes result in changing betatron envelopes. At first is assumed a 36 MV peak rf voltage per turn, a 22.3° synchronous phase (ring notation) and energy loss compensation, but no damping absorbers. With absorbers, the voltage is raised to ~ 200 MV. At injection, the single muon bunch is ~ 15 ns and, after 3 turns, it splits into 3 shorter bunches. Survival will improve from the observed 51% on adding longitudinal dampers. For the rf voltage assumed, the synchrotron tune is ~ 0.21, rising to ~ 0.5 at full voltage. A code is to be developed to give more realistic trapping efficiencies and ring survival.

5. OPERA3D MAGNET SIMULATIONS

A half straight section and arc have been simulated and single muons tracked using Opera3D. In the arc, gradient magnet n values have been reduced from 0.66 to 0.5 over the inner radii. The 0.4 m hard edged arc solenoids are changed to units 0.25 m long, of 0.44 m inside and 0.48 m outside radii. Each has a surrounding iron cylinder, of μ = 500, length = 0.26 m, and inside and outside radii 0.5, 0.8 m. Iron is also added near the 2 opposed field solenoids at the arc centre so that fields fall smoothly to zero. Alternating solenoid polarities are retained through the ring. Iron is not used in the straight section as it would interfere with injection and extraction. The 2 solenoids (of the 4), furthest from the arc, require fields ~ 5 T, suggesting superconducting units. Tracking shows some non-linearity, with a β_{min} of 0.13 m (cf with 0.067 m in Trace3D). Modifications are needed for improved dispersion down to 130 MeV.

6. INJECTION AND EXTRACTION KICKER MAGNETS

Conventional kickers are inadequate for injection and extraction. Instead, kickers with magnetic amplifier pulser systems[4] are proposed, allowing over an order of magnitude increase in peak pulse power. Due to the high fields of adjacent ring solenoids the kicker is designed without magnetic materials, so adding to the pulse current requirements. The air cored kickers are either induction linac modules with distributed input feed lines, or matched, push-pull transmission line kickers of low characteristic impedance. For the latter, electric and magnetic fields both contribute.

7. IONIZATION DAMPING ABSORBERS

The emittance damping per section due to ionizing absorbers may be expressed:

$$\alpha = J U_0 / (\beta^2 \gamma E_0)$$

where J is a damping partition number, as defined for synchrotron radiation rings,
U_0 is the mean muon ionization energy loss in the section,
E_0 is the muon rest mass energy and β, γ are the muon relativistic factors.

For the longitudinal plane, a wedge shaped absorber is assumed with an energy loss as a function of radial position, dU/dx. The relevant damping parameter, J_ε, for an absorber placed at a lattice position with a horizontal dispersion function, D_x, is:

$$J_\varepsilon = (dU/(dp/p)) / U_0 = D_x (dU/dx) / U_0$$

For a linear wedge absorber, of radial extent $\pm X_0$ and wedge angles $\pm \theta/2$, the central scattering length is $2X_0 \tan \theta/2$, while the partition number, J_ε, is given by (D_x / X_0). For a parabolic wedge absorber of the same extent, but of half the central scattering length, U_0 is halved, but J_ε is approximately doubled to $(2D_x / X_0)$. The coupled, transverse, partition numbers are: $J_h = J_v \cong 1 - (J_\varepsilon / 2)$. Two cases are to be considered:

1. Four, parallel edged absorbers at the low β positions in the long straights, of a total energy loss ~ 40 MeV, with four linear wedge absorbers in the arcs, of total energy loss ~ 20 MeV, with $D_x = 0.3$ m, $X_0 = 0.15$ m, $\theta \sim 4°$ and $J_\varepsilon = 2$.

2. As in 1, but with an energy loss of ~ 50 MeV in the long straights, where $J_h = J_v = 1$, and ~ 10 MeV in parabolic absorbers in the arcs, where $J_\varepsilon = 4$ and $J_h = J_v = -1$.

Emittance damping per turn is the same for 1. and 2. in all 3 planes, at ~ 1/6, so 18 ring turns are ~ 3 damping times. Case 2 is considered as its beam sizes should be less due to the smaller arc energy losses. Graphite is proposed for the transverse damping absorbers, and graphite or liquid hydrogen for the wedge absorbers.

8. FUTURE WORK

An accurate code is to be developed to allow more realistic evaluations. Opera3D field maps will be obtained and included in the tracking code. For the ring, longitudinal and transverse damping absorbers and rf cavity systems will be included and muon beam damping will be assessed over intervals of up to 20 ring revolutions.

REFERENCES

1. G H Rees et al, *Muon front-end chicane and acceleration*, Proceedings NuFact02.
2. S J Brooks, *Muonplayer tracking code*, Rutherford Appleton Laboratory
3. J S Berg et al, *RFOFO Ring Cooler*, Proceedings of NuFact02.
4. L L Reginato, LBL, and R B Palmer, BNL, *Private communications*.

Frictional Cooling of Protons

R. Galea

Nevis Laboratories, Columbia University, New York, NY, 10533, USA

Abstract. Muon cooling is the main technological obstacle in the building of a muon collider. A muon cooling scheme based on Frictional Cooling holds promise in overcoming this obstacle. An experiment designed to demonstrate the Frictional Cooling concept using protons was undertaken. Although the results were inconclusive in the observation of cooling, the data allowed for the qualification of detailed simulations which are used to simulate the performance of a Muon Collider.

INTRODUCTION

The basic idea of Frictional Cooling[1] is to bring the charged particles into a kinetic energy range where the energy loss per unit distance increases with kinetic energy. A constant accelerating force can be applied to the charged particles, which compensates for the energy loss, resulting in an equilibrium energy. The desired condition can be met for proton kinetic energies below a $\sim$ 100 KeV or proton kinetic energies beyond about 1 GeV. At the high energy end, the change in dT/dx with energy is only logarithmic, whereas it is approximately proportional to the speed at low energies. Below the dT/dx peak, protons are too slow to ionize the atoms in the stopping medium.

Despite some reported evidence for Frictional Cooling[1], it is clear that much more information is needed if this is to become a useful technique for phase space reduction of beams.

An experiment was performed using protons in order to demonstrate the behavior of charged particles in this low energy regime. This experiment had several goals: demonstrating Frictional Cooling; benchmarking the simulations, and employing many of the experimental components, detectors, etc., which would be needed in future experimentation with muons.

Using protons also simplifies the experiment considerably, as they are easily produced and are stable particles. A time of flight experiment was devised employing start and stop detectors, an electric field and a gas cell with thin entrance and exit windows.

TIME OF FLIGHT (TOF) EXPERIMENT SETUP

The experiment was performed at RARAF[1]. RARAF has a 4 MeV Van de Graaff accelerator, which provided a beam of diatomic Hydrogen with one electron stripped

---

[1] RAdiological Research Accelerator Facility

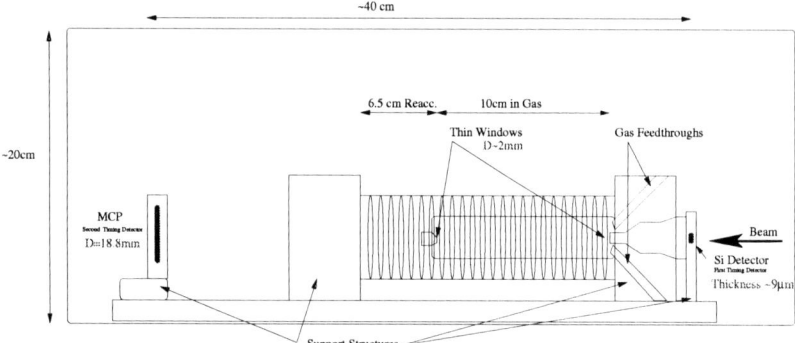

FIGURE 1. Schematic of the RARAF TOF experiment setup. The H_2^+ beam comes in from the right and breaks up inside the first timing detector. The protons then pass through the gas cell which is surrounded by an accelerating grid. Those protons which survive through the exit window of the gas cell are reaccelerated by accelerating grid, which extends beyond the gas cell, and drift toward the second timing detector.

(H_2^+) in the energy range of 1.3 – 1.6 MeV. Figure 1 shows the experimental setup. The proton beam was first collimated through a 1 mm hole separating the RARAF beam line from the experimental section. This collimator, while reducing the delivered current to the experiment, also acted as a baffle protecting the beam line vacuum from degradation as a result of Helium gas which leaked from the gas cell.

The H_2^+ breaks up in the first trigger detector resulting in an effective initial beam of protons with energies in the range of 650 – 800 KeV. The first detector consisted of a planar totally depleted silicon surface barrier detector. The silicon detector acted to degrade the beam energy, adding energy spread, while also providing the trigger.

The protons then entered the gas cell through a thin Carbon window, which was supported on a nickel grid with a 55% open area ratio. The windows were inspected and it was discovered that in a few grid spacings the carbon was perforated. The windows were epoxied between two precision washers and the sandwich arrangement was further epoxied to window holders, designed and built at Nevis Laboratories.

The accelerating grid consisted of 30 copper rings with thin (1 mm) Teflon separators. The rings were connected in series by a resistor chain resulting in a uniform accelerating field of 60 KV/m. The gas cell was shorter than the accelerating grid which provided for a short reacceleration field for protons exiting the gas cell. Those protons which were not sufficiently degraded by the silicon and were not in the energy range suitable for stopping were minimally affected by the small reacceleration field.

The second timing detector was a Micro Channel Plate (MCP) detector. There were two plates with a total potential difference across them of 1600 V.

ANALYSIS

The nominal H_2^+ beam energy for the analysis was 1.44 MeV. Since the breakup of the H_2^+ is not simulated the first step was to determine the kinetic energy of the incoming protons. This was possible using the calibration runs which were taken without the gas cell and accelerating grid structures installed in the beam line. From the calibration one can extract not only the T0 offset needed to reconstruct the TDC measurements but also the slope of the distribution. The slope of the distribution is representative of a velocity. The slope of 0.5 cm/ns represents the velocity for a proton with the most probable kinetic energy after the silicon detector (136 KeV). The incoming energy was varied in the simulations such that the output agreed with the values observed in the data. The transmitted proton energy is a strong function of the incoming energy. An initial proton energy of 721 KeV is needed to produce a most probable transmitted energy of 136 KeV.

The next step was to determine the effect of the gas cell windows. With the gas cell and accelerating grid in the beam line, a data run was taken without flowing Helium gas in the cell and leaving the grid off. In this way the only difference in the TOF distribution resulted from the extra energy loss of the protons in the entrance and exit windows of the gas cell. This data was compared to the simulation of various window thicknesses. An effective carbon window thickness of 300 nm reproduced the data. The apparent thickness of 300 nm was more than an order of magnitude larger than what was expected. The effect of this thick window was to change the expected TOF distributions by adding an effective lower threshold of proton energy to get through the system. Hence, no protons which would result in a TOF greater than 400 ns were expected to penetrate the exit window. This greatly reduced our chances of observing cooled protons.

Finally the pressure of the gas had to be determined from the data since the pressure gauge was not precise. After adding the gas the MC was tuned to extract the pressure of the Helium gas inside the gas cell. The MC was fit to the data under these conditions and the probable pressure of Helium gas was found to be between 0.055 and 0.06 atm. This was in line with our readings from the pressure gauge.

Cooled protons were searched for in TOF distributions of data runs with the gas on and the accelerating grid ramped up to produce a field of 60 KV/m. The background subtracted data is shown in Fig. 2 for the nominal beam energy.

The MC was fit to the main peak in the data, which corresponds to the protons which do not achieve the equilibrium energy. The MC expectation was then calculated by integrating the number of events over a time window. Data at other H_2^+ beam energies were also taken and are summarized in Table 1. We note that the MC expectations yield a very small number of cooled protons. This is in large part due to the effective thickness of the carbon windows. The data is consistent with no observation of cooled protons, but also compatible with the expectation from the simulations within the statistical errors.

CONCLUSIONS

This experiment was performed to study Frictional Cooling using protons. The number of cooled protons observed under various conditions was consistent with zero within large statistical errors. This result was explained by the small acceptance of the system and the large exit window thickness of our gas cell.

However, the data allowed for the tuning of unproven simulation code. The experimental experience and the refined simulations will be used in the design and implemen-

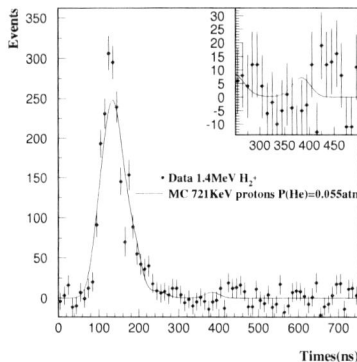

FIGURE 2. Measured time distribution after subtraction of the background. The solid curve is a prediction from our MC simulation with 300 nm thick Carbon entrance and exit windows.

TABLE 1. Results of data runs. Note that the MC expectation does not include the possible variation of the MCP efficiency with energy. MC expectation is given assuming 300 nm thick Carbon entrance and exit gas cell windows.

H_2^+ Beam Energy (MeV)	Fitted Proton Beam Energy (KeV)	Fitted He Pressure (atm)	$\Sigma_{250\ ns}^{750\ ns}$ Events (MC exp.)
1.3	710	0.055	15 ± 31 (2)
1.44	721	0.055	64 ± 82 (55)
1.5	745	0.06	31 ± 22 (0.1)
1.6	760	0.06	185 ± 176 (1)

tation of a future experiments.

ACKNOWLEDGMENTS

This research was performed in collaboration with Professor Allen Caldwell. This work was funded through an NSF grant number NSF PHY01-04619, Subaward Number 39517-6653. Special thanks to S. Schlenstedt and H. Abramowicz for their contributions to our simulation efforts. We owe special gratitude to the RARAF staff and especially S. Marino, for their efforts to deliver beam to the experiment.

REFERENCES

1. M. Muhlbauer et al., *Nucl.Phys.Proc.Suppl.* **51A**, 135-142 (1996).

"High Frequency" Buncher and Phase Rotation

David Neuffer

Fermilab
PO Box 500, Batavia IL 60510

Abstract. A scenario for capture, bunching and phase-energy rotation of µ's from a proton source is explored. It requires a drift section, a bunching section and a φ-δE rotation section. The rf frequency changes along the transport in order to form the µ's into a train of equal-energy bunches suitable for cooling and acceleration. Optimization and variations are discussed. The concept can operate in a wide range of rf frequencies and bunch train lengths. It also can simultaneously capture positive and negative muons.

Introduction

For a neutrino factory, short, intense bunches of protons are focused onto a target to produce pions, that decay into muons, which are then accelerated into a high-energy storage ring, where their decays provide beams of high-energy neutrinos. [1, 2] The challenge is to collect and accelerate as many muons as possible. The pions (and resulting muons) are initially produced within a short bunch length and a broad energy spread, much larger than the acceptance of any accelerator. In this paper we discuss a method of reducing that energy spread, providing bunched beam suitable for cooling and acceleration.

Overview of transport

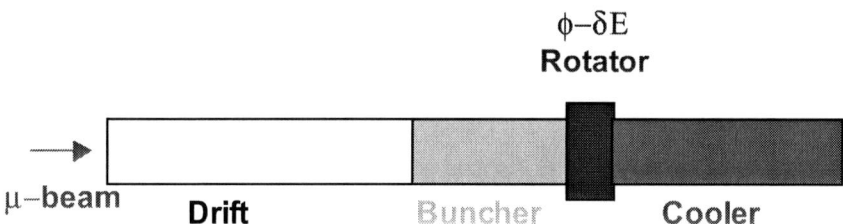

FIGURE 1. Schematic view of the components of the system, showing an initial drift, the varying frequency buncher, and the phase-energy (φ-δE) rotator leading into a cooling section. π's would be produced by protons on a target at the beginning of the drift, decay to µ's in the drift, while lengthening in phase the buncher and φ-δE rotator form the µ's into a string of bunches matched into the cooler.

In this method, the muons first drift from the production target, lengthening into a long bunch with a high-energy "head" and a low-energy "tail". (see Fig. 1) Then, the beam is transported though an "adiabatic buncher", a section of rf cavities that gradually increase in gradient and decrease in frequency (from ~300 to ~200MHz, in our initial example). The rf wavelength is fixed by requiring that reference particles at fixed energies remain separated by an integer number of wavelengths. This forms the beam into a string of bunches of differing energies. Following the buncher, the beam is transported through an "rf rotator" section that performs a phase-energy rotation that aligns the bunches to (nearly) equal central energies, suitable for injection into a fixed-frequency ~200 MHz cooling system.[3]

Example: Capture Into 200 MHz For A Neutrino Factory

To illustrate the method and its components, we discuss its application to a reference problem of forming a muon bunch with large energy spread into a long string of bunches matched into ~200MHz rf, and we present simulations of the process, tracking the phase–energy motion. Simulations of this example are presented in fig. 2.

Drift + Adiabatic Buncher

We set an initial reference kinetic energy $T_0 = 125$MeV. With $m_\mu c^2 = 105.66$ MeV, we find $1/\beta_0 = 1.12497$, where $\beta = v/c$. The rf frequency and phase are set so that the reference particle passes through at zero phase. We next determine a bunch spacing by setting an interval in $1/\beta$. A bunch timing spacing of ~200MHz ($\lambda_{rf} = 1.5$m) at the end of the drift + buncher is set if:

$$L_{tot}\left(\frac{1}{\beta_1} - \frac{1}{\beta_0}\right) = L_{tot}\delta\left(\tfrac{1}{\beta}\right) = \lambda_{rf} = 1.5\text{m} .$$

Here L_{tot} is the total distance from the target to the end of the drift + buncher section. If $L_{tot} = 150$m, then $\delta(1/\beta) = 0.01$. If the 200 MHz rf cavity at $z = 150$m is set to have zero phase when the reference particle passes through it, then the other test particles which differ in the parameter $(1/\beta)$ by integer multiples of 0.01 will also pass at zero phase. The complete buncher contains a string of rf cavities starting from an initial position $z = z_0$, continuing up to $z = L_{tot}$. The reference particles of the bunches remain at zero phase if the wavelength of each cavity is given by:

$$\lambda_{rf}(z) = \delta\left(\tfrac{1}{\beta}\right)z ,$$

and the reference particle remains at zero phase.

For the initial reference example, we start with a drift with a length of $z_0 = 90$m, during which the muons develop a position-energy correlation, and follow it with a 60m long rf buncher. At $\delta(1/\beta) = 0.01$, the frequency of the rf begins at 333MHz ($\lambda_{rf} = 0.9$m) and reduces to 200 MHz along the buncher. The rf gradient is increased gradually along the buncher, and the beam is "adiabatically" captured into a string of bunches, each of them centered about test particle positions with energies determined by the $\delta(1/\beta)$ spacing: $1/\beta_n = 1/\beta_0 + n \delta(1/\beta)$.

Fig. 2 shows simulation results of the initial example. Fig. 2A shows beam as produced at the target, Fig. 2B shows beam after the end of the drift, and Fig. 2C shows the beam at the end of the buncher. The beam is formed into a train of different energy bunches.

"Vernier" rf Rotator

In the transport immediately following the buncher, the rf is used to form the string of bunches into a string of bunches with (approximately) equal central energies In this initial example, we keep the initial reference particle (T_0 = 125MeV) and choose as a second reference particle the test particle at n = 10. This particle has $1/\beta_{10} = 1/\beta_0 + 10\,\delta(1/\beta)$ = 1.22497, or T_{10} = 77.281 MeV. This test particle initially trails the first reference particle by $\Delta ct = 10\lambda_{rf}$ = 15m. For vernier bunching, the first reference particle remains at zero phase, but the rf wavelength is changed to place the second particle at an acceleration phase. This is done by setting $\lambda_{rf} = \Delta ct/10.1$, which then places the second reference particle (which is the center of bunch 10) at at a phase of ϕ_{10} = 36°. $\delta\lambda/N\lambda_{rf}$, defined as the vernier offset parameter, is 0.1/10 for this case. Through the length of the vernier rotator, the rf wavelength is changed to maintain this phase, which means the reference particle energy changes following:

$$T_{10}(z_R) = T_{10}(0) + e\,E_{rf}\sin(\phi_{10})z_R\;,$$

where z_R indicates distance within the rotator. Other test particles (centering other bunches) would show proportional changes ($\phi_n \cong \phi_{10}\,n/10$). If E_{rf} = 10 MV/m and a rf rotator insert length of $z_{R,final}$ = 7.8m (where $T_{10} \cong T_0$) are chosen, the bunches are aligned with nearly equal energies. Fig. 2D shows the beam at the end of a 8.8m, 10MV/m ϕ-δE rotator.

At the end of the rotator, the muon bunches are aligned in energy, and can be matched into a constant-frequency cooler. The rf frequency of this cooler should be rematched to place all bunches at the same phase, which means that the rf spacing between the reference particles should become an integer number of wave lengths. ($\Delta ct = 10\lambda_{rf}$ from bunch 0 to bunch 10.)

Comments on example

In this initial example, we have arbitrarily set many system parameters, and these parameters can be greatly varied in other optimizations. We list some of these key parameters to invite consideration of variations:
1. *Drift:* The key parameter is the length of the section, z_0, which was arbitrarily set to 90m.
2. *Buncher*: The length of the section ($L_B = L_{tot} - z_0$), the bunching gradient $E_{rf}(z)$, the reference particle energy T_0, and $\delta(1/\beta)$ bunch spacing can be varied.
3. *ϕ-δE rotation:* The length and rf voltage of the phase rotation section ($z_{R,final}$ and E_{rf}) are the key parameters. Also the reference particle energies (T_0, T_N) and the vernier parameter $\delta\lambda/(N\lambda_{rf})$ can be changed.
4. *Cooling System*: The match into the following cooling/acceleration systems determines the effectiveness of the system. The key parameters of the cooling system are the rf frequency, the rf voltage, and the absorber energy loss rate, which set the longitudinal dynamics, and the transverse focusing, which determines the transverse cooling limits. All of these parameters can be varied to improve performance.

In the initial example we have separated the adiabatic buncher and the ϕ-δE rotation into separated consecutive systems. It is also possible to combine these functions into a single system, or to design a more gradual transition between the two functions.

In the above discussion we have considered only one sign of μ's in the capture and phase rotation. The initial target produces both μ^+ and μ^- in nearly equal amounts, and half-way between each of the stable phases for one sign of μ's in the buncher and ϕ-δE rotator, there is a stable phase for the opposite sign. The same system obtains strings of μ-bunches of both signs, at similar intensities. This is unlike the induction linac system of study-II, which can only capture one sign. [2]

3-D Simulation results and future studies

Scenarios were initially generated using a 1-D longitudinal-motion simulation code. Muon motion is highly nonparaxial and transverse motion must be included to obtain accurate evaluation. Also the focusing and accelerating fields can have large nonlinear components with signicant effects. Therefore, simulations in 3-D codes have been initiated.

Simucool, Geant4, and ICOOL simulations

Initial 3-D simulations were obtained using A. Van Ginneken's simulation code SIMUCOOL. [4] Initial π-distributions were generated using the MARS particle production simulation code, [5] where production in a mercury target by a 24 GeV proton beam was assumed. A 1.25T solenoidal field provided transverse focusing. The code SIMUCOOL was designed to track large numbers of particles, and that property was used in optimization, particularly for the φ-δE rotator. In these simulations it was noted that muon capture was somewhat improved by varying the rf frequency within the φ-δE rotation, and the large statistics was used to obtain a vernier-based optimum. The SIMUCOOL simulations were then verified using the transport and cooling simulation code ICOOL.[6]

Simulations were also performed using Geant4. [7] Geant4 is designed to provide accurate and detailed 3-D simulations of particle motion through realistically determined electromagnetic fields that include all particle-material interactions. In the simulations, the magnetic fields are determined by current coils in the 3-D geometry (designed to obtain ~1.25T solenoidal fields.), and acceleration fields are determined by the electromagnetic fields in multicell pillbox cavities. The simulations showed successful bunching and phase rotation, in agreement with 1-D simulations. [8, 9]

Matching into Cooling and Acceleration Systems

An accurate measure of the buncher-rotator effectiveness requires matching into the following cooling and acceleration systems, and calculations of the resulting complete system performance. A properly matched cooling channel has not yet been developed. A 120m long cooling system was designed for the Study II neutrino factory [10] and simulated within ICOOL. To obtain an initial estimate of the possibility of matching into a cooling system, we simply took the output beam from the ICOOL simulation of the buncher-rotator and inserted it into an ICOOL representation of the Study II cooling channel. Transverse motion, synchrotron motion and transverse-longitudinal correlations were not properly matched. For example, the emittance of the beam exiting the buncher-rotator was $\varepsilon_T = 0.02$m normalized, while the acceptance of the cooling channel was $\varepsilon_T = 0.012$m. (A matched cooler would start with initial cells that would cool ε_T from 0.02m to 0.012m.)

Even with the gross mismatches, the ICOOL simulation results were encouraging. [11] ~40% of the μ's are lost in the initial few cells of the cooling channel, due to the mismatch in acceptances. Following that initial loss, the cooling and losses are very similar to those obtained with the Study 2 beam. An initial example obtained ~0.22 cooled μ^+/p at the end of the Study II cooling channel. This was only ~5% less than that obtained in parallel ICOOL simulations (starting from the same π^+'s) of Study-II.

Practical considerations

The first simulations used rf systems that changed frequency every meter (30—60 different frequencies). It is economical to reduce the number of different frequencies, and particularly the number of different cavity sizes that need to be built. Elvira and Keuss studied an example and found that ~10 different frequencies provided bunching and phase rotation only slightly worse than a 60-frequency case. (20 frequencies was actually slightly, but not significantly, better.) [8] 1-D simulations show no deterioration in capture unless the number of different frequencies becomes less than ~12.

Some initial estimates of the possible cost of the system were obtained, with assistance from A. Moretti and M. Green. The ~150m long system could cost less than ~$1/4^{th}$ the expected cost of the 440m long induction linac phase rotation and buncher system of the Study-II ν-factory.

Variations and future directions

The initial example was developed in order to match the Study II neutrino factory design, which uses a long train (~100m long) of ~200 MHz bunches to match into a ~200 MHz cooling and acceleration design. The same procedure can be readily adapted to obtain shorter bunch trains, and to obtain bunches at other frequencies (50, 100, 200, or 400 MHz, etc.). 1-D simulations of these variants have been developed.) Also the central energy of 125 MeV could be moved to higher or lower energy, depending upon scenario optimization. The general utility indicates that the method could be used to develop μ-bunches for any neutrino factory scenario, including the CERN and JHF neutrino factory scenarios.

The method may also be adaptable to the somewhat different requirements of a μ^+-μ^- Collider. For a high-luminosity collider, we require a small number of both μ^+ and μ^- bunches. The method does produce both μ^+ and μ^- bunches, but would tend to produce a large number of bunches. Scenarios with a more limited number of initial bunches, plus some bunch combination in the cooling process, may be suited to a collider; appropriate methods could be developed.

Acknowledgments

I thank R. Palmer and A. Van Ginneken for important contributions in the development of this concept.

REFERENCES

1. N. Holtkamp and D. Finley, eds., A Feasibility Study of a Neutrino Source based on a Muon Storage Ring, Fermilab-Pub-00/108-E (2000).
2. S. Ozaki, R. Palmer, M. S. Zisman, Editors, Feasibility Study-II of a Muon-Based Neutrino Source, BNL-52623, June 2001. http://www.cap.bnl.gov/mumu/studyII/FS2-report.html.
3. D. Neuffer and A. Van Ginneken, Proc. 2001 Particle Acc. Conf., Chicago, p. 2029 (2001).
4. A. Van Ginneken, "Adiabatic and Vernier Bunching in Decay Channel", Mucool 220, October 2001.
5. N. Mokhov, "Particle Production and Radiation Environment at a Neutrino Factory Target Station", Fermilab-Conf-01/134, Proc. 2001 Particle Accelerator Conference, Chicago, P. 745 (2001), see http://www-ap.fnal.gov/MARS/.
6. R. Fernow, "ICOOL, a Simulation Code for Ionization Cooling of Muon Beams", Proc. 1999 Particle Accelerator Conference, New York, p. 3020 (1999), see http://pubweb.bnl.gov/people/fernow/.
7. V. Daniel Elvira, P. Lebrun, P. Spentzouris, "Beam Simulation Tools for GEANT4 (BT-v1.0) User's Guide", MuCOOL Note 260, October 2002.

8. D. Elvira, "High Frequency Adiabatic Buncher", MuCool 255, September, 2002.
9. N. Keuss and D. Elvira, "Fixed-frequency Phase Rotation for the Neutrino Factory (Geant4 Simulation)", MuCOOL Note 254, September, 2002.
10. R. C. Fernow, J. C. Gallardo, R. Palmer, P. Lebrun, "Ionization Cooling scenario for a Neutrino Factory", Proc. 2001 Part. Acc. Conf., Chicago, p. 3861, 2001.
11. D. Neuffer, "Exploration of the "High-Frequency" Buncher Concept", MuCOOL 269, March 2003.

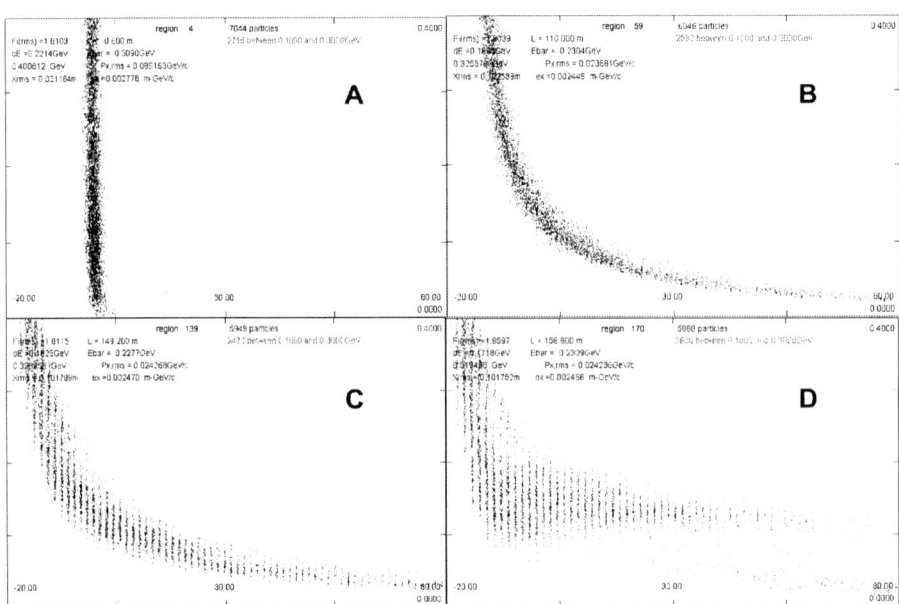

Figure 2: ICOOL simulation results of the buncher and phase rotation, at the parameters of the example described in the text. Each figure shows the A: π's and μ's as produced at the end of a 0.6m long target. B: μ's at z=110m; 90m drift + 20 m of initial buncher. C: μ's at z=150m, the end of the buncher. The beam has been formed into a string of ~200MHz bunches at different energies. D: At z= 158.80m after φ-δE rotation; the bunches are aligned into nearly equal energies. In each plot the vertical axis is kinetic energy (0 to 0.4 GeV) and the horizontal axis is longitudinal position with respect to the reference particle at 125 MeV (-20 to 80m).

Propagation of a Large-emittance Muon Beam through a Straight, Quadrupole-based Precooling Channel [1]

M. Berz*, K. Makino† and C. J. Johnstone**

*Department of Physics and Astronomy, Michigan State University, East Lansing, MI 48824, USA
†Department of Physics, University of Illinois at Urbana-Champaign, Urbana, IL 61801, USA
**Fermi National Accelerator Laboratory, Batavia, IL 60510, USA

Abstract. A straight quadrupole channel[1, 2] that was originally developed to precool the transverse dimensions of a large-emittance muon beam has demonstrated efficient cooling across a large momentum range, from about 150 to over 500 MeV/c[1]. A full simulation of the channel has been performed and previously reported[1, 2] using the transfer-map based code COSY INFINITY[4], which includes representative fringe field effects for large-aperture quadrupoles. In this paper, particle distributions derived from upstream systems designed for either a Neutrino Factory or Muon Collider[3] are matched and propagated through this channel to ultimately determine its feasibility as a precooling stage for either facility.

INTRODUCTION

In the U.S. studies for a Neutrino Factory and Muon Collider[3], the initial cooling channel designs utilized solenoids due to the ability to use their fringe-fields to simultaneously provide a focus in both transverse planes over a large momentum acceptance. However, a major drawback remained in that not only are the fringe fields strongly nonlinear, but also the accelerating rf cavities had to be located at large beam dimensions and, in addition, housed inside the solenoidal coils. This configuration resulted in enormous meter-scale solenoidal apertures. In the quadrupole precooler studied here, absorbers and rf cavities can be removed from the component apertures allowing only beam conditions to regulate component apertures. As shown in the layout in Figure 1, the cooling channel is based on repetitive FODO cell optics using alternating-gradient short quadrupoles, cavities, and absorbers without nesting of components[1, 2]. To accomplish a baseline design for the linear optics and also prevent the quadrupole element from being entirely determined in terms of its fringe fields, the full aperture is constrained to be equal to the length, which for this design is 60cm; larger apertures are possible and limited overlaying of consecutive components does not compromise the design nor result in superconducting-strength fields. The inclusion and study of various fringe-field representations on the performance of the channel is imperative, and the

---

[1] The work was supported by the US Department of Energy, the Illinois Consortium for Accelerator Research, and the National Science Foundation.

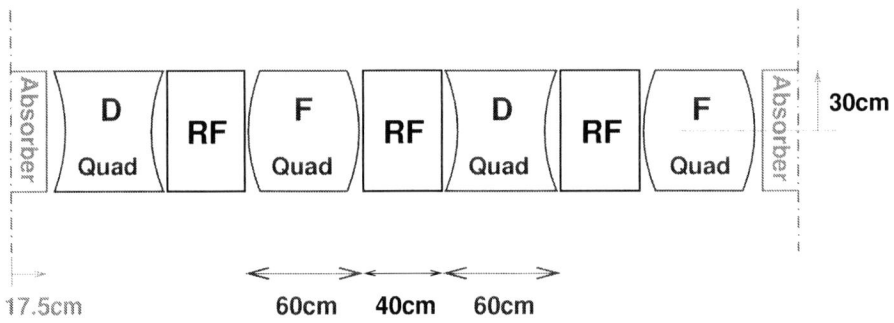

FIGURE 1. The layout of one straight quadrupole cooling cell.

extensive choices available for fringe field modeling represent one of strengths of the code COSY INFINITY[4], and provide a formal basis for the conclusions derived in the analysis of this channel. Further, tools in COSY INFINITY allow re-optimization of parameters in the presence of fringe fields, resulting in a close approximation of the desired linear-design behavior.

In this paper, first a matching section to the upstream buncher/phase rotator system[5, 6, 7] will be presented. Then, muons derived and tracked from the production target through the various upstream stages (targetry, collection, decay, bunching and phase rotation) are pushed through the matching section and the quadrupole precooler using the transfer map method. The final cooled muon distribution and losses are thus determined.

FROM THE BUNCHER SECTION TO THE QUAD PRECOOLER

The high frequency adiabatic buncher[5, 6] followed by the phase rotator[7] uses 1.25 Tesla solenoids for the beam containment. This system was arbitrarily ended[7] in an infinitely-long solenoidal field. The first task was to start at this point, but, in order to match to the quadrupole precooler as designed, the field had to be gradually decreased from 1.25Tesla to 0.86Tesla with a correspondingly increase in aperture. This change took place over a length of 7m with the increase in aperture scaled to maintain constant, continuous acceptance. The solenoidal fringe field is modeled by a kick approximation accounting for the angular momentum transfer, and immediately following is a short, 10cm-long, horizontally-defocusing quadrupole ($k = -8$), required to complete the

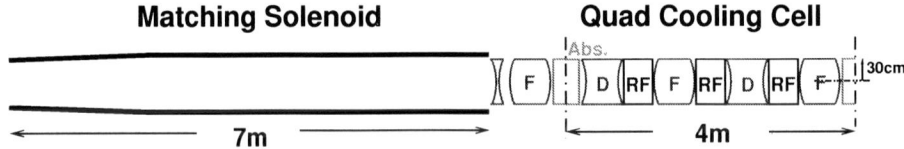

FIGURE 2. The layout of the matching magnets to the first straight quadrupole cooling cell.

match to the FODO optics of the cooling cell. Although, in principle, the optical match was done at a beam momentum of 200 MeV/c, this matching approach was found to be reasonably achromatic and valid over most of the momentum range of the muons. After 11cm of drift, the cooling channel begins with a horizontally-focusing quadrupole. The layout of the matching section followed by the beginning of the cooling channel is shown in Figure 2. All elements starting with the matching section were modeled in the code COSY INFINITY[4]. A thick solenoidal element was used to describe the 7m long matching solenoid in the simulation.

PROCESSING THE PARTICLE DISTRIBUTION

The individual state coordinates of thousands of muons at the exit of the phase rotator were supplied by V. D. Elvira[6, 7] for three different high frequency adiabatic buncher schemes[5]. We chose a set of particles from the 20-frequency scheme for this work. The 10-frequency scheme showed less resolved bunch structure and the 60-frequency scheme is considered not practical. The particles exhibited a Gaussian distribution in x and y within ± 0.3m, and p_x and p_y within ± 73MeV/c. The longitudinal distribution is tremendous, as can be seen from Figure 3(A), where the full time-of-flight of captured muons t ranges from 5.5×10^{-6}sec to 1.26×10^{-6}sec and momentum p_z from 46MeV/c to 1020MeV/c. Although the majority of the muons lies in the range from 150MeV/c to 400MeV/c, there is still a significant number up to 1GeV/c.

Since the particle data was given in ICOOL[8] format and did not supply the conventional or natural parameterization required for beam description and manipulation (relative to reference energy, reference time or frequency, reference trajectory, etc.), first the data had to undergo a coordinate transformation and, even further, pre-processing, in order to extract the necessary information from the raw distributions. Practically this amounted to an optimization procedure in which off-reference deviations were minimized for the raw data set to empirically determine the optimal reference parameters. The transformation to COSY coordinates $\{x, a, y, b, l, \delta_K\}$ is given as

$$a = p_x/p_0, \quad b = p_y/p_0, \quad l = -(t-t_0)v_0\gamma/(1+\gamma), \quad \delta_K = (K-K_0)/K_0, \quad (1)$$

where x, y and l are in meters, and a, b and the relative deviation δ_K of the kinetic energy K are unit-less[4]. After the coordinate transformation, all the transverse particle coordinates are within ± 0.3, of the reference trajectory, corresponding to the central axis of the solenoid. Longitudinally, l ranges from -122m to 15.3m relative to the reference particle as defined for the entire bunch train, and δ_K ranges from -0.93 to 4.67, though the majority of the muons are contained in the range from -0.5 to 1, as shown in Figure 3(B). In the coordinate transformation, the defined reference values $p_0 = 245$MeV/c, $K_0 = 161.2$MeV and $t_0 = 6.3389 \times 10^{-6}$sec are used.

In order to track efficiently a large number of particles using high-order transfer maps[9], it is important to confine the distribution to a numerically small range, specifically less than 1. While the transverse distributions pose no problems in this regard, the longitudinal distributions required extensive pre-processing. For this purpose, the particles were sorted into smaller ensembles; each ensemble with a relatively small energy spread and time-of-flight for the particles within the ensemble. The sorting of the original distribution by energy is shown in Figure 3(C). These defined groups now have manageable internal spreads in δ_K. To solve the problem of the spread in l, the bunched structure of the distribution was used to subdivide the train into smaller time intervals. Through a careful analysis of the deviation in time-of-flight over the entire bunch train, it became clear that the bunch frequency of 179.0MHz combined with a 52% offset from the initial ICOOL time variable produced the smallest time-of-flight distribution per bunch, as shown in Figure 3(D). After the described pre-processing, the l distribution in each ensemble is less than ± 1, and, for the most part, less than ± 0.5. In each small group, the average energy is specified as the reference energy, and the 179.0MHz frequency re-sets the zero for the reference time-of-flight on a per bunch basis. For each group, we compute the transfer maps of the matching section and the straight quadrupole cooling cell which are a function of energy. The transfer map of the matching section is then applied to the particle distribution in each group followed by successive applications of the transfer map of one quadrupole cooling cell, repeating and varying the number of cells to check the cooling status of the distribution.

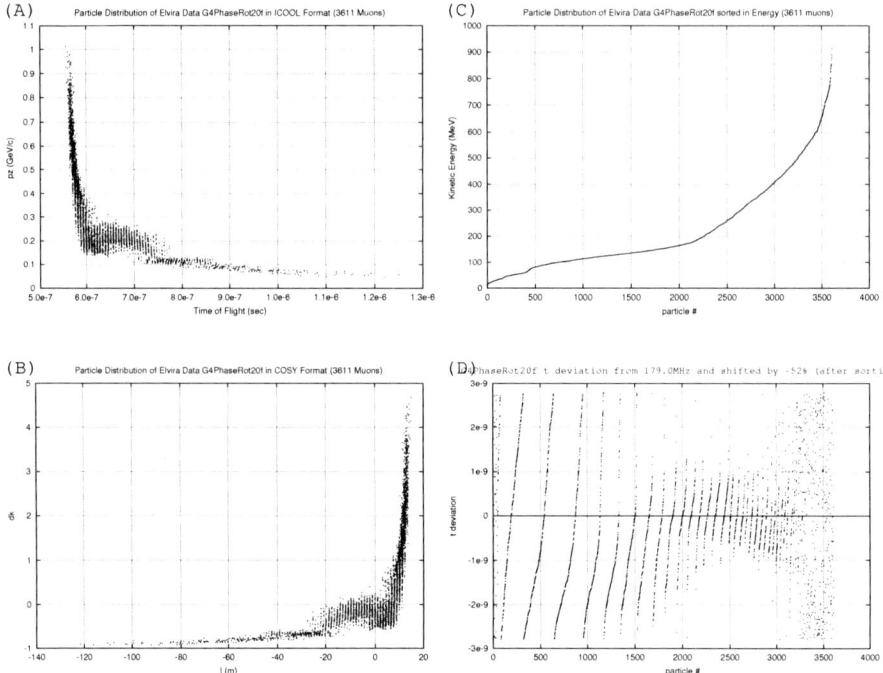

FIGURE 3. Distribution of muons off the phase rotator after the buncher. Shown are (A) time-of-flight t versus p_z, the data prepared in ICOOL format, (B) l versus δ_K, the data transformed to COSY coordinates, (C) all the muons sorted in energy, and (D) deviation in time-of-flight from the central bunched time-of-flight with frequency 179.0MHz and further 52% shifted in bunches.

CONCLUSION

The preliminary simulation of the matching section through the quadrupole precooler using large-emittance muon distributions derived from upstream systems shows a transmission and effective cooling of about 60% over most of the momentum range above 180MeV/c for 10 or more quadrupole cells. Further work will include the determination of optimal frequencies, optimization of the matching section, and adjustments to other relevant parameters based on a quantitative measure of cooling which reflects both emittance reduction and overall transmission.

REFERENCES

1. Johnstone, C. J., Berz, M., Errede, D., and Makino, K., Muon beam ionization cooling in a linear quadrupole channel, to appear in *Nuclear Instruments and Methods* (2003).
2. Makino, K., Berz, M., Errede, D., and Johnstone, C. J., High order map treatment of superimposed cavities, absorbers, and magnetic multipole and solenoid fields, to appear in *Nuclear Instruments and Methods* (2003).
3. Ozaki, S., et al., Feasibility study-II of a muon-based neutrino source, Tech. Rep. 52623, Muon Collider Collaboration, BNL (2001), see http://www.cap.bnl.gov/mumu/studyii/FS2-report.html.
4. Berz, M., and Makino, K., COSY INFINITY Version 8.1 - user's guide and reference manual, Tech. Rep. MSUHEP-20704, Department of Physics and Astronomy, Michigan State University, East Lansing, MI 48824 (2001), see also http://cosy.pa.msu.edu.
5. Neuffer, D., Exploration of the "high-frequency" buncher concept, Neutrino Factory/Muon Collider Notes MUC-NOTE-DECAY_CHANNEL-0269 (2003), see http://www-mucool.fnal.gov/notes/notes.html.
6. Elvira, V. D., High frequency adiabatic buncher for the neutrino factory (GEANT4 simulation), Neutrino Factory/Muon Collider Notes MUC-NOTE-COOL-THEORY-0253 (2002), see http://www-mucool.fnal.gov/notes/notes.html.
7. Keuss, N., and Elvira, V. D., Fixed frequency phase rotator for the neutrino factory (GEANT4 simulation), Neutrino Factory/Muon Collider Notes MUC-NOTE-COOL-THEORY-0254 (2002), see http://www-mucool.fnal.gov/notes/notes.html.
8. ICOOL at FNAL, http://www-pat.fnal.gov/muSim/icool.html/. Fernow, R. C., "ICOOL: a simulation code for ionization cooling of muon beams," in *Proc. 1999 Particle Accelerator Conference, New York*, p. 3020.
9. Berz, M., *Modern Map Methods in Particle Beam Physics*, Academic Press, San Diego, 1999.

Tetra Cooler Ring Simulation in COSY INFINITY

K. Makino* and M. Berz[†]

*Department of Physics, University of Illinois at Urbana-Champaign, Urbana, IL 61801, USA
[†]Department of Physics and Astronomy, Michigan State University, East Lansing, MI 48824, USA

Abstract. A tetra ring muon beam cooler designed by V. Balbekov[1] consists of large aperture magnets. Because the original design is based on the hard edge model, the realistic field description has to be considered in the simulations. Using the code COSY INFINITY, which can treat realistic fields, we studied the dynamics of muon beam in the tetra cooler ring.

INTRODUCTION

A tetra ring muon beam cooler designed by V. Balbekov[1] has large aperture long straight solenoid sections and short straight solenoid sections connected via inhomogeneous bending magnets. The original design uses the hard edge model for all the magnets, and a linear kick approximation was applied to represent the edge field effect of solenoids[2, 3]. Because of the large size of the magnet aperture, further simulations with realistic magnetic fields were considered crucial[4, 3].

The beam dynamics code COSY INFINITY[5] is based on the high order transfer map method, and there are various algorithms to compute the fields on the fly parallel to the system transfer map computation[6, 7], allowing the treatment of realistic fields. We used the code to represent the fields of the magnets in the tetra cooler ring to compare with the hard edge model, and studied the dynamics through the ring with various field models.

FIELDS IN STRAIGHT SOLENOID SECTIONS

The tetra cooler ring has eight straight sections, consisting of four long solenoid sections and four short solenoid sections, and inhomogeneous bending magnets are placed between. In the middle of 1.74m short section with 28cm radius coils, the longitudinal field direction flips. The coils of the long sections are 6.68m long each and the inner radius is 81cm[1].

The hard edge model of solenoids in the original design[1] has a longitudinal field profile with the assumption of infinitely long extension of coils, and a linear kick approximation was applied to account for the end field effects[2, 3]. With the original coil design parameters[1], however in reality, the field has a long tail of fringe field fall-off as shown in the top pictures in Figure 1. For instance, at the position 4m longitudinally away from the coil of the short section, B_z remains $5 \cdot 10^{-4}$ Tesla. In the long section with much larger aperture, the remaining field continues with high value as $4 \cdot 10^{-3}$ Tesla at the position 10m away from the coil.

Using the algorithm of the DA (Differential Algebraic) PDE (partial differential equation) solver in the code COSY INFINITY[5, 6, 7], the 3D fields of short and long solenoid sections of the tetra ring were computed as shown in Figure 1. While we show the field distribution for the demonstration, the actual 3D field computation necessary for the transfer map computation in the code happens as it is needed in parallel to the transfer map computation itself, differing from the need to prepare the 3D field data of a realistic field model ahead of time as normally performed in other beam dynamics codes.

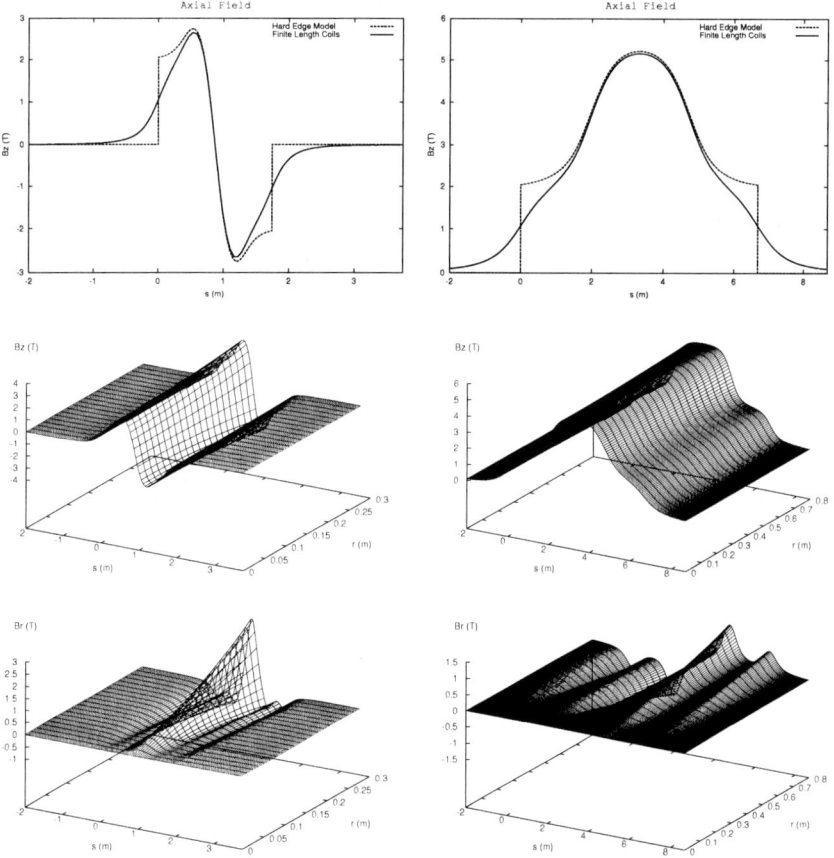

FIGURE 1. The magnetic fields of the short (left) and the long (right) solenoid sections of the tetra cooler ring. The top pictures show the axial field profile of the hard edge model and the realistic model with end fields. The 3D field distribution of the realistic model is shown for B_z (middle) and B_r (bottom).

We had earlier computed the transfer maps in the short and the long sections of the tetra ring using different field models. Refer to [3] for the details. The hard edge model assuming infinitely long coils and the hard edge model assuming finitely long coils were compared with the realistic model with correct fringe field, and the validity of the linear kick approximation was checked. While the short section showed good agreement between the realistic model and the hard edge model with finitely long coils, a similar comparison for the long section showed a big disagreement, indicating that the kick approximation is only valid when the solenoid aperture is small.

BEAM DYNAMICS IN VARIOUS FIELD MODELS

With those different field models for the solenoid sections, we performed the stability analysis of the ring by monitoring the traces of linear part of transfer maps as functions of energy as shown in Figure 2. In the result in Figure 2, we used the hard edge model for the bending magnets as in the original design simulation[1]. As it is expected, the simpler the solenoid model is, the more stable the system behaves. The crudest model, namely the hard edge model assuming infinitely long coils, is almost linearly stable because the absolute value of traces is almost less than 2, but sometimes it gets close to linear instabilities, particularly at the design energy of $E_{tot} = 250$ MeV. Observed also is that the hard edge model with finitely long coils is close to a linear instability at the design energy.

The beam dynamics is studied with tracking pictures for those models as shown in the top pictures in Figure 3. When realistic fields are considered for the bending magnets, the system gets unstable as shown in the bottom pictures. For this simulation, the fringe field model of default fall-off in the code COSY INFINITY[5] is applied.

To conclude, we observed that there are strong instabilities in those magnetic field

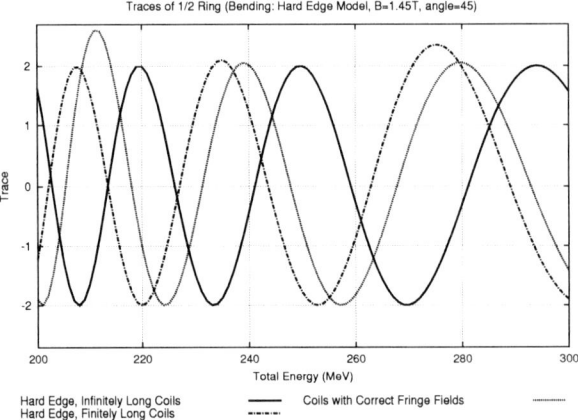

FIGURE 2. Traces of a half of the ring as functions of beam energy. Different solenoid models are compared; i.e. the hard edge model with infinitely long coils, the hard edge model with finitely long coils, and the realistic model with correct fringe fields. The hard edge bending magnets are used.

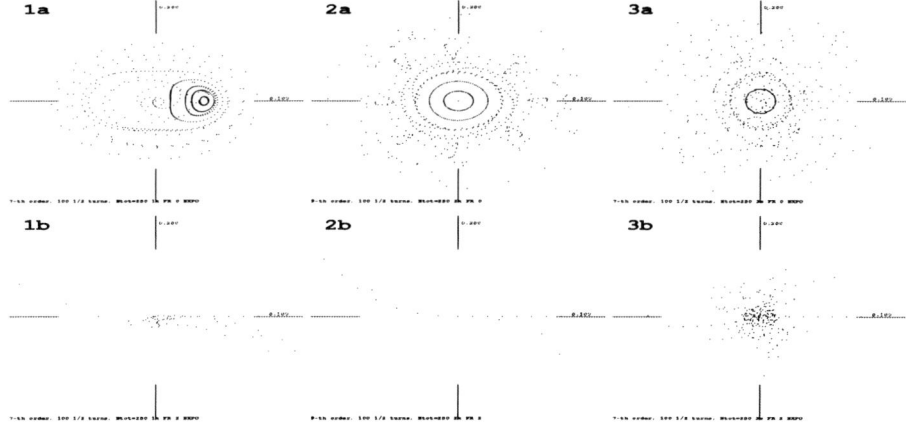

FIGURE 3. Tracking 50 revolutions in the ring at $E_{tot} = 250$ MeV. Different magnet models are compared for solenoid sections; i.e. the hard edge model with infinitely long coils (1), the hard edge model with finitely long coils (2), and the realistic model with correct fringe fields (3). Further for the bending magnets, the hard edge model (a) and the realistic model with fringe fields (b) are compared.

models studied. In particular, the fringe field effects have a dramatic impact on the performance of the ring.

ACKNOWLEDGMENTS

We would like to thank Carol J. Johnstone for many discussions. The work was supported by the Illinois Consortium for Accelerator Research, the US Department of Energy, and the National Science Foundation.

REFERENCES

1. Balbekov, V., Ring cooler progress, The Muon Collider and Neutrino Factory Notes MUC-NOTE-COOL-THEORY-0246, (2002), see http://www-mucool.fnal.gov/notes/notes.html.
2. Balbekov, V., Private communication (2002).
3. Maidana, C. O., Berz, M., and Makino, K., "Muon Beam Ring Cooler Simulations Using COSY INFINITY," to appear in *Seventh International Computational Accelerator Physics Conference*, 2003.
4. Kahn, S., Hanson, G., Schwandt, P., Balbekov, V., Raja, R., and Usubov, Z., "Design of a Magnet System for a Muon Cooling Ring Experiment," in *2001 Particle Accelerator Conference*, IEEE, 2001, p. 3239.
5. Berz, M., and Makino, K., COSY INFINITY Version 8.1 - user's guide and reference manual, Tech. Rep. MSUHEP-20704, Department of Physics and Astronomy, Michigan State University, East Lansing, MI 48824 (2001), see also http://cosy.pa.msu.edu.
6. Berz, M., *Modern Map Methods in Particle Beam Physics*, Academic Press, San Diego, 1999.
7. Makino, K., and Berz, M., "Solenoid Elements in COSY INFINITY," to appear in *Seventh International Computational Accelerator Physics Conference*, 2003.

ISIS as a Proton Driver for a Neutrino Factory

Christopher R. Prior

*ASTeC Intense Beams Group, CCLRC Rutherford Appleton Laboratory,
Chilton, Didcot, Oxfordshire OX11 0QX, United Kingdom*

Abstract. The paper describes plans to develop the ISIS accelerators into a high intensity, multi-megawatt, proton driver for a range of fixed-target studies. Possibilities include use in a Neutrino Factory and an advanced spallation neutron source. A phased upgrade would allow intermediate accelerator R&D on pressing problems such as nanosecond proton bunch compression, the pion target and the muon capture channel. A possible bunch compression experiment using the existing ISIS synchrotron is also outlined.

INTRODUCTION

At a time when funding and other limitations have considerably reduced effort elsewhere on Neutrino Factory R&D, the work programme based at the Rutherford Appleton Laboratory is currently the most active in Europe. The aim of the UK group is to produce a self-consistent end-to-end design of the accelerator complex and carry out prototyping and experimental tests in those areas that are regarded as most critical.

Two high intensity proton drivers were developed in the period 1999-2000 [1, 2] and studies have since covered optimisation of the pion capture and decay channel [3], attempts to transport the muon beam without cooling [4] and, more recently, development of a muon cooling ring [5]. Funding is presently being pursued in order to carry out the full programme of work, including those aspects not considered to date, such as a system of re-circulating linacs or FFAG rings for muon acceleration, and a storage ring to hold the muons while they decay.

In view of the extremely high cost of a Neutrino Factory (NF) complex, it is important to ask if any existing machines around the world can be used for relevant experiments and if any future machine upgrades might be linked to Neutrino Factory R&D. One such possibility is the ISIS spallation neutron source, which has been running successfully for more than 15 years and at 160 kW of proton beam power remains the most intense source of its kind in the world. Plans are being implemented to install a radio frequency quadrupole (RFQ) in the linac early in 2004, and addition of a dual harmonic RF system to the synchrotron should increase the power to about 240 kW. Ideas are also being developed to raise the beam power initially to 1 MW and to 4–5 MW in the longer term. The 1 MW option is based on an increase in proton energy to 3 GeV by means of a second synchrotron using ISIS as a booster; details were presented in [6]. Of particular interest is an alternative option of accelerating the beam in this new ring to the higher energy of 8 GeV at reduced frequency, which would allow the machine to be used as a test bed for the nanosecond bunch compression needed for the NF proton driver. There would be opportunities for target R&D and for experimental studies of the pion decay channel. The cost of these proposals is relatively modest compared with a completely new facility.

In the longer term, a combination of two such rings with a new synchrotron booster (replacing the existing ISIS) would give attractive options for both the neutron and neutrino communities.

ISIS BUNCH COMPRESSION EXPERIMENT

For the immediate future, a bunch compression experiment has been devised that could be carried out on ISIS during periods allocated to normal machine physics. The ISIS synchrotron takes a 70.44 MeV, H^- beam from the linac, creates a proton beam via charge exchange injection and accelerates the accumulated beam to 800 MeV. The repetition rate is 50 Hz and the total number of particles in the ring is $\sim 2.5 \times 10^{13}$, divided equally into two bunches. At extraction each bunch is about 100 ns ($\sim$20 ns rms) in duration. The total RF voltage available in the ring is 150 kV (Figure 1), which is insufficient to compress a full bunch significantly further. Therefore either fewer particles have to be injected or particles need to be deliberately lost at an early stage in the cycle to create a 'clean' beam for the compression tests. A theoretical study of the second option allows us to explore the level of intensity one might hope to achieve in the final beam.

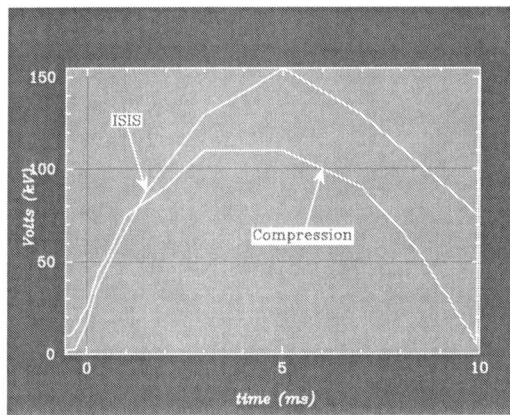

FIGURE 1. Bunch Compression RF programme compared with normal operating RF

The procedure, which for practical purposes is applied to only one test pulse in 50, is shown in Figure 2. The RF voltages are progressively lowered so as to reduce non-adiabatically the stable region of longitudinal phase space and trigger proton beam loss. After about 6 ms (two-thirds of the way through the 10 ms acceleration period), a core of particles will have been trapped and all of the unstable particles will have been removed by the ISIS collimation system. Controlled, adiabatic, lowering of the RF can then be used to stretch the bunch in phase and decrease its momentum spread without further beam loss. Between 9.93 and 9.96 ms, the full 150 kV of available RF power is applied as fast as possible, then held for 40 μs until extraction. The bunch rotates rapidly in phase space, producing the nanosecond duration required. The tracking studies illustrated in Figure 2 predict that 37.5% of the particles originally injected will remain in the bunch; the peak current will exceed 100 A; and

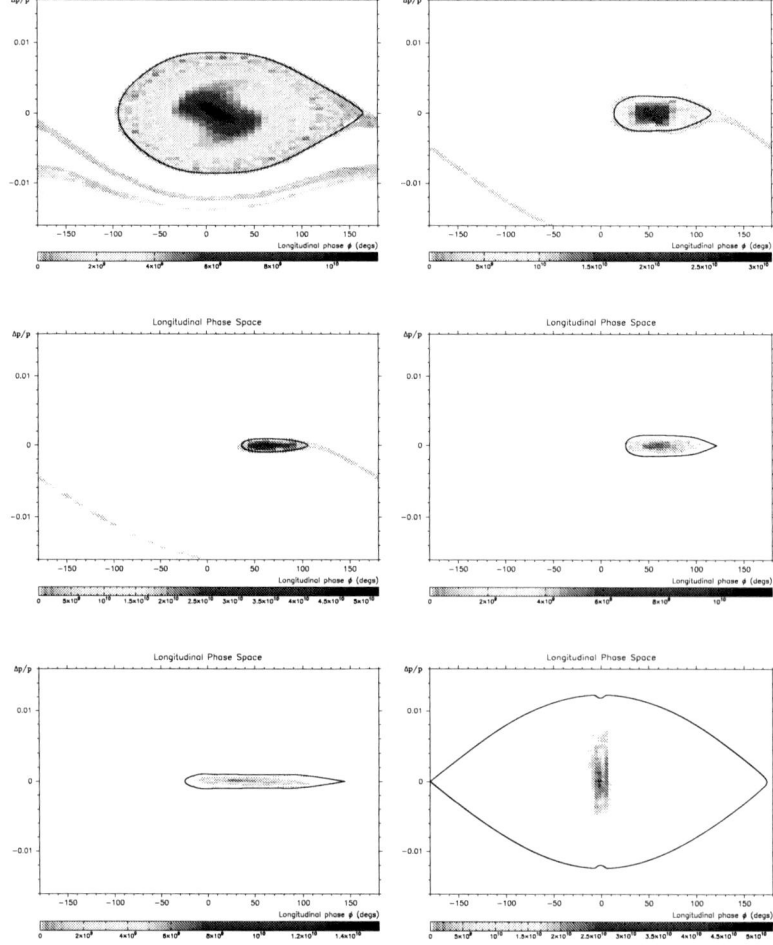

FIGURE 2. ISIS longitudinal bunch compression experiment

the bunch duration will be ∼ 20 ns (full), ∼ 2.5 ns (rms). Figure 1 shows the RF voltage programme needed to achieve this state, superimposed on the ISIS parameters when under normal operation on the neutron target.

ISIS MEGAWATT UPGRADES

Development of ISIS into a versatile, high power, high intensity proton driver is an attractive, economical option, and would provide a range of opportunities for fixed-target studies. A megawatt upgrade scheme has been proposed at RAL that could be implemented in three phases, with relatively little disruption to normal operation while each stage is built.

Phase 1: Addition of a new synchrotron

Installation of a dual harmonic ($h = 2/h = 4$) RF system in 2004 will enable ISIS to accelerate 3.75×10^{13} protons to 800 MeV at a repetition rate of 50 Hz, giving 240 kW of mean beam power. Beyond this, the machine is space-charge limited and the simplest way to increase the beam power to megawatt levels is through an increase in energy by means of an additional synchrotron. The proposed lattice is based on a racetrack design with a mean radius of 78 m (three times that of the present ring) and is described in [7]. The two bunches that make up a normal ISIS pulse will be transferred bucket to bucket to the new ring, and accelerated to an energy in the 3-8 GeV range, depending on the specific application of the proton beam. The overall dimensions of the ring are 184 m × 107 m and a possible layout on the RAL site is shown in Figure 3. Note the new 10 Hz spallation neutron target presently under construction following recent funding approval. In 3 GeV mode of operation, the machine would be used as a source of spallation neutrons. However, the possibility exists of operating at one third of the frequency, $f = 16.7$ Hz, directing two pulses in three to a beam dump, and accelerating the remainder to 8 GeV over 30 ms. No additional RF would be needed. Opportunities would thereby be provided for further experimental tests of bunch compression, for pion target tests and investigations of a prototype pion decay/muon capture channel.

Phase 2: A new injector

In Phase 2, a new booster injector would be built, replacing the ISIS accelerator complex as the feed to the 78 m racetrack synchrotron, which would be upgraded to be capable of delivering up to 2.5 MW of beam power. The dual role of a neutron facility with applications to neutrino factory development would be maintained. The proposed injector is based on a 180 MeV H$^-$ linac feeding two 1.2 GeV, 50 Hz booster synchrotrons. The linac comprises a 60 mA H$^-$ ion source, a 280 MHz RFQ, a 2.5 MeV fast beam chopper and a drift tube linac. Beam chopping is necessary for low levels of uncontrolled particle loss at ring injection and extraction, and is achieved by deflecting 30% of the train of micro-bunches to a water-cooled collector. Since the inter-bunch gap, during which the deflecting electric field has to rise for clean chopping, is of the order of only 2 ns, severe technological difficulties are presented. The RAL solution is a novel two-stage chopper in which two or three microbunches are pre-chopped with voltages $\sim \pm 2$ kV to create a larger gap and ease the rise-time requirements for the much stronger deflection of the remaining beam. A prototype is in the early stages of construction and the design is described in [8].

After the linac, momentum collimation is carried out using the high normalised disersion $D_h/\sqrt{\beta_h} \approx 6 \text{ m}^{1/2}$ created in an 180° achromatic arc. This has both positive and negative bends to meet the optical requirements within a relatively confined space.

FIGURE 3. New ISIS main synchrotron on the Rutherford Appleton Laboratory site

Horizontal and vertical betatron collimation are included in the transfer line, along with momentum ramping cavities for injection painting [2]. Two 1.2 GeV synchrotrons are used for the booster to maintain manageable space charge levels. They have mean radius 39 m (half the radius of the main synchrotron) and their optical parameters are designed for low loss beam accumulation. H^- injection is by charge exchange through a carbon, or possibly aluminium oxide, stripping foil[1] at a point in the latttice where the normalised dispersion is $\sim 1.6\,\mathrm{m}^{1/2}$. Simultaneous longitudinal and horizontal transverse phase space painting helps reduce space charge effects and minimise particle loss. The booster rings would be filled one after the other, each with three bunches of protons, and all six bunches would be transferred together to fill the main ring for final acceleration and, in the case of NF, bunch compression.

---

[1] The choice of stripping foil material depends on the predicted temperature rise. Although carbon is anticipated for most proton drivers, simulations suggest that Al_2O_3 ISIS-type foils may be acceptable here.

Phase 3: Addition of a Second Synchrotron

The proton driver would be completed in Phase III with the addition of a second main racetrack synchrotron, stacked vertically above the ring built in Phase I. An optimal layout would site the booster rings inside the main rings and a possible schematic diagram, omitting all but one of the transfer lines between the rings, is shown in Figure 4. The main synchrotrons would run at 25 Hz and would be filled alternately from the 50 Hz boosters, so that on extraction an operating frequency of 50 Hz would be recovered. Several options have been proposed: a top energy of $\sim$ 3 GeV could be used for a spallation neutron source and at 6 GeV, bunch compression could be carried out for a 5 MW NF proton driver. One could envisage a combination of applications such as one ring for neutrino studies and the other for neutron production.

At the time of writing, a limited sum of money has been awarded by the European Union under its Framework 6 Programme for fast beam chopper prototyping, the 2.5-180 MeV DTL design and some beam dynamics studies. A bid has also been submitted to the UK's Particle Physics and Astronomy Research Council (PPARC) for proton driver development and completion of the UK NF design. Full beam tests of the proton driver front end (ion source, RFQ and chopper) are regarded as vital for a feasibility demonstration of any future NF accelerator complex.

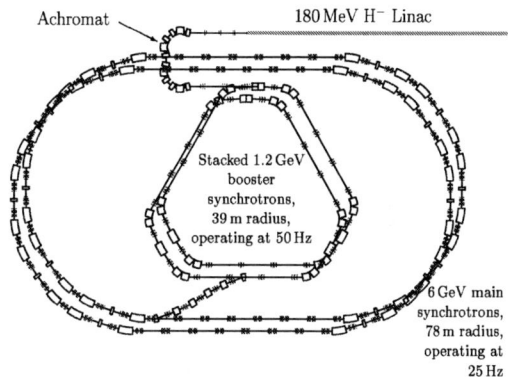

FIGURE 4. Schematic layout of the ISIS proton driver

REFERENCES

1. Prior, C.R. and Rees, G.H., *RAL Proton Driver Studies for a Neutrino Factory*. Proceedings of the Neutrino Factory Workshop, NUFACT'00, Monterey, California, June 2000.
2. Prior, C.R. and Rees, G.H., *Synchrotron-based Proton Drivers for a Neutrino Factory*. Proceedings of the 7th European Particle Accelerator Conference, EPAC'00, Vienna, June 2000.
3. Brooks, S.J., *Simulation of Neutrino Factory Muon Capture with Chicane*, in preparation.
4. Rees, G.H., Brooks, S.J., Harold, M.R. and Prior, C.R., *Muon Front-End Chicane and Acceleration*, J. Phys. G: Nucl. Part.Phys **29** (2003), 1673-1677.
5. Brooks, S.J., Harold, M.R., Prior, C.R. and Rees, G.H., *Features of a Muon Cooling Ring for a Neutrino Factory*, these proceedings.
6. Prior, C.R. et al, *ISIS Megawatt Upgrade Plans*, Proceedings of the US Particle Accelerator Conference PAC'03, Portland,Oregon, May 2003.
7. Rees, G.H., *Lattices for 8 and 30 GeV Proton Drivers*, Proceedings of the Neutrino Factory Workshop, NUFACT'02, London, June 2002.
8. Clarke-Gayther, M.C., *A Fast Chopper for the ESS 2.5 MeV Beam Transport Line*, Proceedings of the 8th European Particle Accelerator Conference EPAC'02, Paris, France, June 2002.

Bunch Production for a Muon Collider

V. Balbekov

Fermi National Accelerator Laboratory, Batavia, IL, 60510

Abstract. Production of a short high-intensity bunch for a $\mu^+\mu^-$ collider is considered. The system designed and simulated includes: 24 GeV proton driver; jet mercury target; 30 m phase rotation channel; bunch compressor based on a ring cooler with an intensive emittance exchange; RFOFO ring cooler for 6D cooling; linear Li lens channel for final transverse cooling. The cooled muon bunch contains about 0.046 muons per incident proton at transverse and longitudinal emittance 0.48 mm and 12 mm. Total merit factor of the system is of order 10^4.

INTRODUCTION

High luminosity $\mu^+\mu^-$ collider can only be realized if all its components achieve the required high performance [1]. In particular, muons wanted to be collected in a single bunch because at given number of muons the luminosity is inversely proportional to the number of bunches. The bunch should be enough short to be accepted by a high frequency and voltage cooling channel. For example, 201 MHz and 15 MV/m channel of the ν-factory can accept a bunch of ± 30 cm $\times \pm 35$ MeV with a total longitudinal emittance of about 10 cm [2, 3]. Simple RF capture in this channel could provide only of order 0.01 muons per proton per bunch. A phase rotation is not very effective in this case because it does not change phase density of the beam, and provides more muons by increase of number of bunches. Probably, a more promising way is a shaping of a short bunch soon after the particle production, using strong longitudinal cooling. Special ring cooler with intensive emittance exchange ("bunch compressor") was proposed for this in Ref. [4, 5]. The idea is developed in this paper where production of the "compressed" bunch is considered by front-end simulation according to scheme presented in Fig. 1.

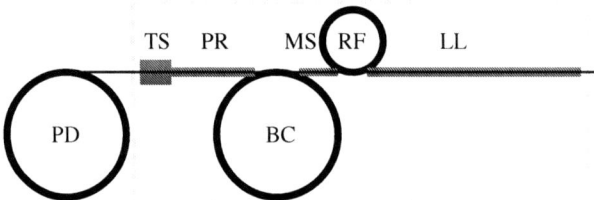

FIGURE 1. General layout of the bunch production system: PD – 24 GeV Proton Driver; TS – Target Station; PR – 30 m Phase Rotation channel; BC – 72.3 m Bunch Compressor; MS – 14 m Matching Section; RF – 33 m RFOFO ring cooler; LL – 117 m Li Lens cooling channel.

TARGET STATION AND PHASE ROTATION - DECAY CHANNEL

The system starts with 24 GeV proton bunch of r.m.s sizes $\sigma_x = \sigma_y = 1.5$ mm, $\sigma_z = 3$ ns. It impinges on a mercury jet target of radius 5 mm sitting in 20 T solenoid followed by 4.1 m long section where the field decays to 4.4 T. The phase rotating part is a solenoid of length 30 m with FR cavities installed on the first 20 m. Its axial field is 1.75 T everywhere except the first 5 m where it smoothly decreases from 4.4 to 1.75 T. Frequency and gradient of the cavities are 36.37 MHz and 6.4 MV/m that is the same as at the bunch compressor. More detailed description can be found in Ref. [5].

Longitudinal phase space of pions and muons is shown in Fig. 2 (left and center). Number of negative muons per incident proton in the window of 5 m × 140 MeV is plotted in the right figure for cases when the phase rotating RF field is on or off. It is seen that the phase rotation increase the yield by factor about 1.6.

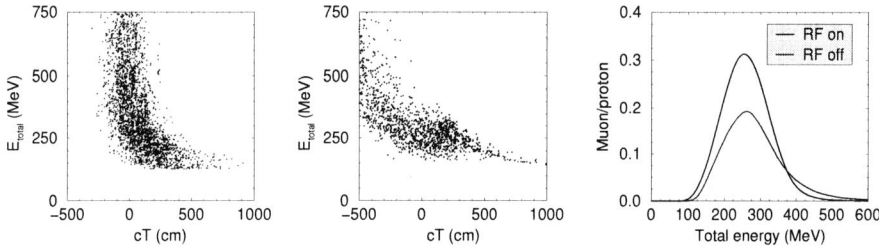

FIGURE 2. Longitudinal phase space of pions and muons after the target (left) and at the end of phase rotation channel (center). Right: Number of negative muons per proton vs central energy of the window.

BUNCH COMPRESSOR

The ring cooler proposed in Ref. [4, 5] and shown in Fig.3 is used as the bunch compressor. The bend of trajectories is carried out by combine function dipole magnets with field index 0.5. RF cavities and liquid hydrogen absorbers are placed in the long SS, where axial field ±1.75 T is applied. The field flip is performed in the short SS where there is non-zero dispersion, and LiH wedge absorbers are placed. The assumption $V'(\text{MV/m}) = 15\sqrt{F(\text{MHz})/200}$ is used at the optimization of frequency F providing

Nominal energy	220 MeV
Circumference	72.291 m
Bending radius	52 cm
Bending field	1.238 T
Normalized field gradient	0.5
Length of short SS	1.948 m
Length of long SS	6.68 m
Short solenoid max field	2.35 T
Long solenoid field	1.75 T
Revolution frequency	3.637 MHz
Accelerating frequency	36.37 MHz
Accelerating gradient	6.4 MeV/m
Synchronous phase	30 deg
LH main absorber, length	54.5 cm
LiH wedge absorber, dE/dy	1 MeV/m

FIGURE 3. Layout and parameters of the bunch compressor

maximal density of muons in longitudinal phase space. Injection/extraction kicker is not discussed in this paper and shown in Fig. 3 schematically. Evolution of the beam parameters is shown in left Fig. 4. The bunch shaping takes place during ∼ 2 turns when significant non-decay loss occurs. Therefore the longitudinal emittance and yield are not quite distinct there depending on the window used. It is illustrated by central and right Figs. 4 where longitudinal phase space is shown at injection and after 13 turns.

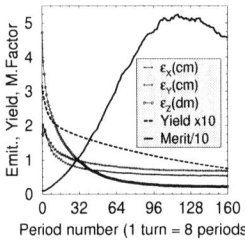
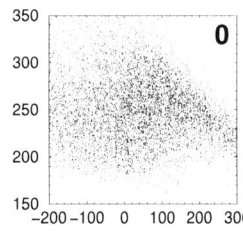
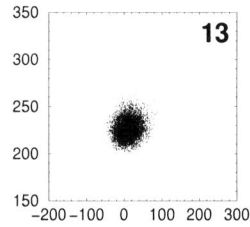

FIGURE 4. Left: evolution of the beam parameters at the bunch compression. Center and right: longitudinal phase space at the injection and after 13 turns.

MATCHING SECTION AND RFOFO RING COOLER

Next step is 6D cooling performed by RFOFO ring consisting of 12 cells shown in left Fig.6. [6, 7]. The cell includes 2 solenoid coils with opposite currents, 5 cavities, and liquid hydrogen wedge absorber. The coils have inner/outer radii 77/88 cm, length 50 cm, current density ±95.27 A/mm^2, and are tilted on ±52 mrad in vertical plane. Period length is 275 cm, and distance between the coil centers is 165 cm along the centerline which is a circle of radius 525.2 cm. Liquid hydrogen absorber has angle 125° and length 42.8 cm along the centerline. Each 203.4 MHz cavity develops 4.8 MV voltage. Seven similar cavities are used in the foregoing matching section which is 14 m long solenoid with axial magnetic field increasing from 1.75 to 3.5 T. Evolution of the beam parameters is plotted in Fig.6 (right). Considerable non-decay particle loss is observed at the first turn because longitudinal emittance of the beam is nevertheless rather large.

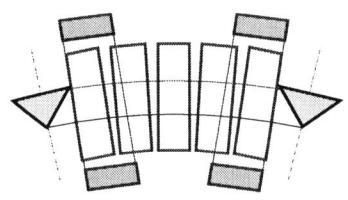
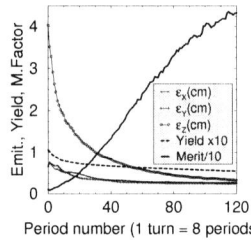

FIGURE 5. Left: cell (1/12) of the RFOFO ring. Right: Evolution of beam parameters at the cooling.

LITHIUM LENS COOLER

Li lens cooler proposed in Ref. [8] is used for the finale transverse cooling. It consist of 44 cells each including Li lens, 3 solenoids, and 6 RF cavities as shown in left Fig.6. Li lens of radius 3 cm carries 355 A/mm² providing field gradient 2.23 T/cm. The coils of inner/outer radius 6/14 and 69/71 cm with current density 84.6 and 81.7 A/mm² produce axial field plotted in central Fig.6. RF parameters are the same as at the RFOFO ring.

Two ideas are realized in this lattice: (i) small betatron phase advance, and (ii) suppression of 2 linear resonances by special shape of the field [8]. The phase advances per cell are 3π and 2π at momenta 164 and 238 MeV/c. However, corresponding linear resonances are suppressed and resonance free region extends from 145 to 300 MeV/c. Owing to this, high transmission is achieved (94% without decay) as it shown in Fig. 6, right. However, merit factor is modest because of strong longitudinal heating.

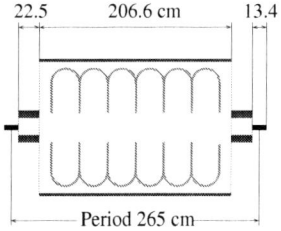 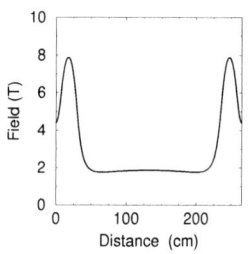 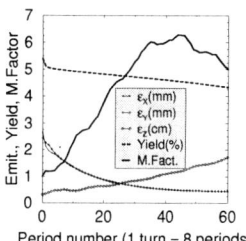

FIGURE 6. Left: Cell of the Li lens channel. Center: Axial field in the cell. Right: evolution of the beam parameters at the cooling.

SUMMARY AND CONCLUSION

Data obtained are summarized in the Table starting from the end of the phase rotation channel. Final transverse emittance is still large for a muon collider [1]. Including of emittance exchange to the Li lens cooler looks as the most important problem because its efficiency is restricted now by the longitudinal heating. Another serious problem is high field matching solenoids. A using of low-gradient part of the Li lens for the matching is possible only at long (at least ~ 1 m) lenses; however it again put in the forefront the problem of chromaticity. The bunch intensity decreases in the system by factor about 7. It is mostly decay loss, and it is hard thing to improve it considerably.

Step	Distance (m)	Trans.emit.(cm)	Long.emit.(cm)	Yield (μ/p)	Merit factor
PR	30	1.8	~ 47	~ 0.32	1
BC	30+940 = 970	.63	2.5	0.11	$1 \times 51 = 51$
MS	970+14 = 984	.66	4.0	0.11	$51 \times .60 \simeq 31$
RF	984+330 = 1314	.25	.32	0.054	$31 \times 43 \simeq 1300$
LL	1314+117 = 1431	.048	1.2	0.046	$1300 \times 6.3 \simeq 8200$

REFERENCES

1. C.M. Ankenbrandt et al. *Phys. Rev. ST. Accel. Beams*, **2**, 081001 (1999).
2. N. Holtkamp and D. Finley, eds. *Fermilab-Pub*-00/108-E (2000).
3. S.Ozaki, R.Palmer, M.Zisman, J.Gallardo, eds. *BNL*-52623 (2001).
4. V. Balbekov. *Fermilab MCNote*-245 (2002).
5. V. Balbekov and N. Mokhov. *Fermilab MCNote*-272 (2003).
6. J. Berg, R. Fernow, R. Palmer. *Fermilab MCNote*-239 (2002).
7. V. Balbekov. *Fermilab MCNote*-264 (2002).
8. V. Balbekov. *Fermilab-Conf*-03-091-T (2003).

Novel Ideas for Beam Profiling in a Muon Cooling Channel

K. D. Hoffman

The University of Chicago, Enrico Fermi Institute
5640 S Ellis Ave, Chicago, IL 60637, U.S.A.

Abstract. A muon cooling channel poses particular challenges to beam profiling. Two methods of beam measurement are being pursued: one senses the heat deposited by the beam using a thin film bolometer, and the other detects electrical pulses from the ionization of a thin CVD diamond substrate. Here we report on the detector designs and the status of prototype testing.

INTRODUCTION

Beam diagnostics are essential for the optimization and operation of any particle accelerator. Such instrumentation must provide an accurate profile of the shape, position, and intensity of the beam, while disturbing it as little as possible. In addition, the detection medium must operate for long periods of time in very high radiation environments with limited access. These requirements make the development of new instrumentation challenging, and at present, the tools available to the beam operator are limited, and include scanning wires, multiwires, optical radiation monitors, and, more recently, ionization profile monitors. The development of beam profilers for use in the front end of a muon storage ring is made all the more difficult by the production mechanism: muons are produced by pions decaying in flight which are in turn produced by slamming an intense proton beam into a target. This process yields a very large muon beam whose phase-space must be reduced by orders of magnitude within the short muon lifetime. Profilers must be developed that can operate in a harsh environment over a wide range of intensities.

BOLOMETRY

The challenge of producing a sufficiently coalesced muon beam may be accomplished through ionization cooling, a novel technique which rapidly cools a beam by alternately passing it through an energy absorbing material and high gradient accelerating cavities. Low Z materials provide the maximum ionization loss with the minimum heating from multiple Coulomb scattering [1]. Current designs employ liquid hydrogen contained within an aluminum vessel, whose thickness is minimized along the beam axis, as the energy absorbing medium [2]. Much effort within the muon collaboration has gone into studying the flow of heat deposited by the beam in the absorber to ensure that there is adequate heat exchange to keep the system in equilibrium. This suggests a way to monitor the beam: a measurement of this heat deposition may be used as a beam profile. This technique, known as bolometry, is widely used in astrophysics. A typical bolometer consists of an absorber, a thermometer, and a heat sink. Our version of the bolometer is a thin thermometric film which is encapsulated in an electrical insulator and applied directly to the surface of the liquid hydrogen absorber which acts as the heat sink.

Two prototype bolometers have been tested. One is a commercially manufactured thin film Nickel thermometer encased in polyimide, the other consists of a thin graphite foil laminated in kapton tape. The choice of these two materials complements one another because their change in resisitivity in response to temperature is oppositely signed. This provides an important control that can be used to confirm that any signal is thermometric and not the result of electromagnetic induction or electrons knocked by the beam from the surface of the sensitive material. In order to simulate the conditions on the surface of the absorber, the bolometers were applied to an aluminum backing of thickness comparable to the absorber vessel at its thinnest point, and placed inside of a Janis ST-100 cryostat which features an evacuated sample chamber. The cryostat's temperature could be varied with a resistive heating element connected to a Lakeshore 331 temperature controller. While most tests were done at liquid hydrogen temperatures, the ability to vary the temperature provided another important method of verifying a thermometric signal, since the slope of the resistivity vs. temperature curve for most materials is a strong function of the temperature.

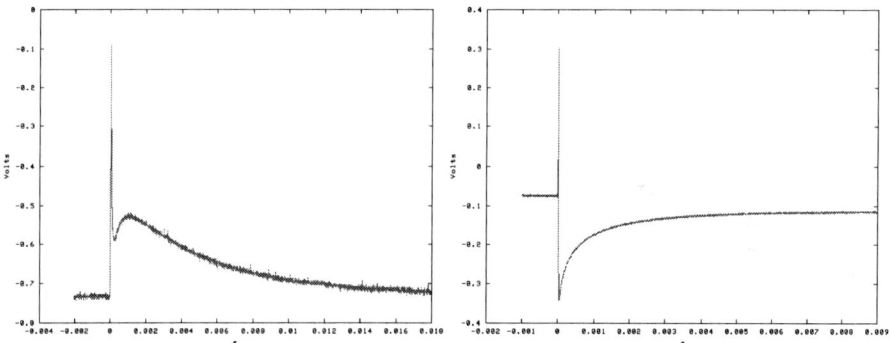

FIGURE 1. The time evolution of a signal from a Xe flashlamp in nickel (right) and graphite (left) bolometers. The noise spike is induced by the discharging flashlamp.

The bolometers were studied in the laboratory using the signal from first a Xenon flashlamp, and later a pulsed laser. Figure 1 shows the signals in nickel and carbon at liquid hydrogen temperatures. The opposite polarity of the signals confirms that the signals were thermally induced. Once proof of principle was established, the apparatus was tested in an electron beam of 840 nA pulses of 10 ns duration at 30 Hz. The signal proved more difficult to distinguish in beam conditions, since the current induced by the charged particles masked the thermometric signal. This might be remedied using an electronic filter, since the thermal time constant is much longer than the pulse duration, however, this is not a viable solution for the front end of a muon cooling channel since the muons will be produced in a long decay line which will introduce a large time spread. Another possible solution is to use a more thermally sensitive medium such as a superconducting edge thermometer. The challenge is to develop a material that retains its thermometric properties in an extreme radiation environment and is flexible enough to apply to the LH2 absorber window.

DIAMOND

Diamond has a number of extreme properties that make it an excellent choice for a particle detector. With an electrical resistivity of more than $10^{13}\Omega$ cm, it can hold a sufficient voltage to operate as a solid state ionization chamber without the need for a reverse-biased p-n junction. Diamond's bandgap, at 5.4 eV, is large compared to traditional crystalline detector media such as silicon and germanium. Although this means a weaker signal, it also results in much smaller dark current, thus maintaining a reasonable signal to noise ratio. The high binding energy of the carbon atoms within the crystal which is responsible for the material's notable strength also protects it from radiation damage. With a binding energy of 80 GeV, diamond has been shown to retain its electrical properties even after irradiation to fluences in excess of $10^{15}/cm^2$. The RD42 collaboration at CERN has been developing diamond as a detector medium, with particular emphasis on its potential use as a single particle tracking device for use at small radii in the LHC detectors [3]. It has also been used as a diagnostic in heavy ion transfer lines where its fast intrinsic response can be exploited to count individual ions as they are transferred at rates in excess of 1 GHz [4]. Here we propose that diamond may be suited as a beam profiler, if its signal yield scales with the beam intensity. If this proves successful, it may find beam profiling applications outside of the Muon Collaboration.

As with a tracking device, diamond grown for a beam profiler must have a high enough resistivity to maintain an adequate potential difference for electron hole pair separation, but this is where the specifications begin to diverge. Diamond tracking devices must have single particle efficiency, thus diamond grown for this purpose has fewer grain boundaries and defects where charge can be reabsorbed and must be thick to provide a long enough track intersection to induce a reasonable pulse. The operating voltage for such a device is chosen to maximize the carrier velocity, typically 1V/μm of thickness. For a bunched beam profiler, the amplification is more than adequate, and we are more concerned about the potential for nonlinear space charge effects. We propose that the ideal bunched beam profile would have a large number of trapping sites and would be as thin as possible. We are preparing to test this hypothesis in the Argonne electron beam using diamond of varying quality and thickness. A prototype detector is shown in Figure 2. Another approach to limiting the space charge effects might be to decrease the electric field, and thus the separation between electron-hole pairs, although, this will have the unwanted consequence of degrading the time resolution.

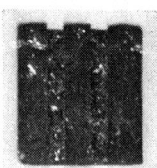

FIGURE 2. Prototype CVD diamond strip detector for beam profiling. The detector shown measures 11mm x 11mm x 0.5mm. The sputtered contacts are titanium gold, with three strips on the front, and a solid backplane.

OUTLOOK

Of the two detector media we have studied, diamond looks more promising due to its higher signal to noise ratio. In fact, some development may be necessary to damp the signal to reduce space charge effects and design electronics that prevents a voltage lag in the wake of a bunch crossing. Although diamond is radiation hard, it almost certainly cannot be operated in a beam indefinitely, however, most direct beam profilers are used in a mode where they are only placed in the path of the beam during tuning and beam studies. Beam tests will be needed to determine the intensity range and lifetime over which diamond profilers may be operated, and well as the optimal material characteristics operating parameters.

ACKNOWLEDGMENTS

This work was supported by the Illinois Board of Higher Education.

REFERENCES

1. Bross, A. D., *these proceedings*.
2. Cummings, M. A. C., *these proceedings*.
3. Adam, W., et. al., *Nucl. Instrum. Meth. A* **453**, 141-148 (2000).
4. Berdermann, E., et. al., *Nucl. Phys. Proc. Suppl.* **61B**, 399-403 (1998).

The MuCool/MICE LH$_2$ Absorber Program

Mary Anne Cummings

Northern Illinois University
for the MuCool Collaboration

Abstract. Hydrogen absorber R&D for the MuCool Collaboration is actively pushing ahead on two parallel and complementary fronts. The continuing LH$_2$ engineering and technical developments by the MuCool group, conducted by ICAR institutions (NIU, IIT and UIUC), the University of Mississippi and Oxford University in cooperation with Fermilab, are summarized here, including plans for the first tests of an absorber prototype from Osaka University and KEK cooled by internal convection at the newly constructed FNAL MuCool Test Area (MTA). Designs for the high-power test of another absorber prototype (employing external heat exchange) are complete and the system will be installed by summer 2004. A convection-cooled absorber design is being developed for the approved MICE cooling demonstration at Rutherford Appleton Laboratory.

The development of high-power liquid-hydrogen (LH$_2$) absorbers has been a critical goal in the MuCool program. The driving issues are to minimize the multiple scattering (beam heating) and to handle large heat loads while maintaining uniform temperature, and hence density, within the absorber volume. To meet the former requirement, we have developed new shapes for the ends of the hydrogen flasks that allow for significant reduction in their thickness, particularly near the center where the beam intensity is at maximum [1] (Fig. 1). We have successfully fabricated tapered, curved windows out of aluminum alloy (6061-T6) disks using a numerically controlled lathe. We have also devised novel means to test these nonstandard windows and demonstrate that they meet design specifications and satisfy applicable safety requirements. By optimizing the maximum stress as a function of pressure, our best design, the inflected "thinned bellows" window, was able to achieve a minimum central thickness about one fourth that of the ASME-standard "torispherical" window of equivalent strength and diameter [1].

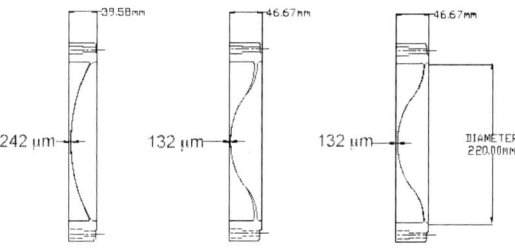

Fgure 1: Evolution of non-standard thin window profiles

Verifying that the as-built geometry of such windows meets the design specification is a challenging task. We have developed a shape-measurement technique employing photogrammetry which is also applicable to monitoring window performance in pressure tests. In photogrammetry, a pattern of points of light projected onto the window's surfaces allows three-dimensional shape measurement, using images captured with a digital camera and parallax calculations to determine the location of each point. The advantages of photogrammetry are the non-contact nature of the measurement and the large number (~10^3) of points that can be measured simultaneously (to be compared with the ~10 points that were practical to sample with a coordinate-measurement machine and strain gages) [1].

We have completed manufacture and pressure tests (see Fig. 2) of our first series of nonstandard windows, of the "tapered torispherical" design, establishing the testing and certification procedures, including finite element analysis (FEA) predictions for window performance [1]. Fabrication of our next window series, of the "thinned bellows" design, is now underway. We are considering the use of lithium-aluminum alloys, such as the 2195 alloy used in the Space Shuttle (45% stronger than 6061-T6); the thinness of the resulting window could challenge our current fabrication technique, and any new alloy will have to be certified for machinability and high-radiation application.

The power to be dissipated in these absorbers in the cooling-channel designs considered so far (~10^2 W) is within the limits achieved in LH_2 targets for current high-beam-power experiments. However, the highly turbulent fluid dynamics involved in the heat-exchange process necessitates R&D for each new configuration. We are pursuing two approaches to heat extraction: a conventional flow-through design with an external heat exchanger, similar to that used for high-powered LH_2 targets, and a convection-cooled design with an internal heat exchanger built into the absorber vessel. The convection design has desirable mechanical simplicity and minimizes the total hydrogen volume in the cooling channel (a significant safety concern), but is expected to be limited in the amount of power it can handle compared to the flow-through design. Convection-cooled absorbers are planned for the MICE experiment, but, depending on the results of upcoming tests, a full-intensity muon-cooling channel may require flow-through absorbers.

The approach adopted for heat transfer within and heat extraction from the absorber volume is one of the main design concerns for the absorber manifold, with the goals being the minimization of temperature and density fluctuations and the avoidance of boiling. For both absorber cooling approaches, the maximum heat load that can be dissipated is still an open question. Two- and three-dimensional fluid-flow simulations are being developed (by IIT's K. Cassel and Oxford's W. Lau) for both the forced-flow and internal-convection designs. To study the fluid mixing and heat transfer properties of these designs, we have been exploring ways to visualize the flow patterns and temperature distributions within the fluid and calibrate the simulations. We have conducted tests using the Schlieren optical method with water in cylindrical test cells; these were done at Argonne National Lab using a 20 MeV electron beam [2]. The Schlieren method determines density fluctuations in a fluid via changes in the refraction of light. Fig. 3 shows a picture and a simplified schematic of one

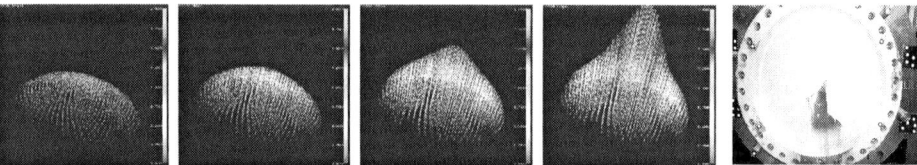

Figure 2: Successive photogrammetry measurements for a window pressure test to bursting (the far right-hand picture). Hundreds of points can be measured without coming into contact with delicate window surfaces.

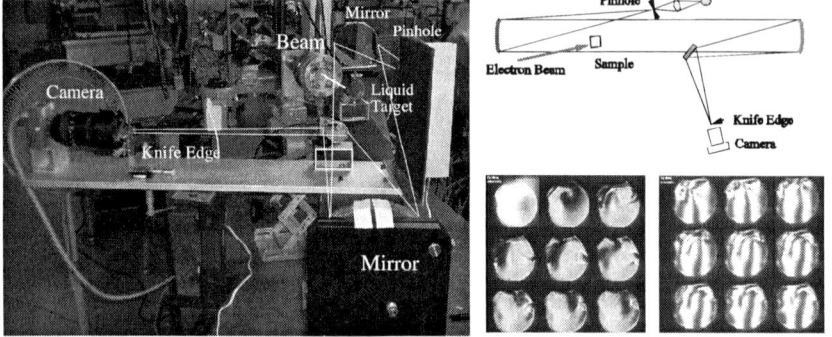

Figure 3. Setup for Schlieren optical test at Argonne's 20 MeV electron beam. At the lower right are examples of data taken at successive time intervals after 50 pulses (30 mC/pulse) of electrons were deposited.

experimental setup, along with a sample of the data. In addition to these convection tests, future plans include Schlieren tests on cells containing liquid circulating through nozzles. So far, tests have been done with water only, but tests are planned with liquids whose viscosity is more similar to that of LH_2.

The cryogenic operation of absorbers will be monitored by temperature probes inside the manifold, the initial placement of which will be determined based on the results of flow tests and simulations. The control systems for refrigeration and hydrogen-system monitoring will be developed and implemented in collaboration with the FNAL cryogenics group. Plans include a first hydrogen test of an absorber in late fall 2003 with a convection-cooled absorber prototype built by KEK; the setup is shown in Fig. 4. To simulate a beam, heat will be deposited by a flow of helium gas through tubes within the absorber. The heat exchange is accomplished at the finned interior surface of the manifold: colder helium gas flowing within the manifold cools this surface, which encloses the hydrogen volume. This test will exercise the controls and data acquisition systems and instrumentation, which will then be adapted for the subsequent test of a forced-flow absorber, operating inside a 5-tesla superconducting magnet, scheduled for Fall 2004. The goal is to have at least these two absorber prototypes operational for a high-power test in late 2005 with a planned MTA beam, to be extracted from the FNAL Linac. The full MTA test program will include a complete cooling-cell test, beam instrumentation tests, and other R&D programs involving alterative cooling-channel technologies [3].

Fig. 5 shows the forced-flow setup and the layout of a complete cooling-cell test at the MTA. This program will be running concurrently with the construction and operation of the MICE experiment, now approved at Rutherford Appleton Laboratories. These two programs comprise a complete scientific and engineering test of the feasibility of hydrogen-absorber-based cooling channels for both neutrino factories and muon colliders.

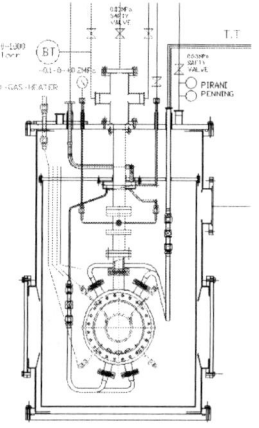

Figure 4. The KEK LH_2 absorber shown with inlets for warmer and colder helium gas and a central pipe for hydrogen (left), the system inside a vacuum vessel for MTA tests (center) and system schematic (right).

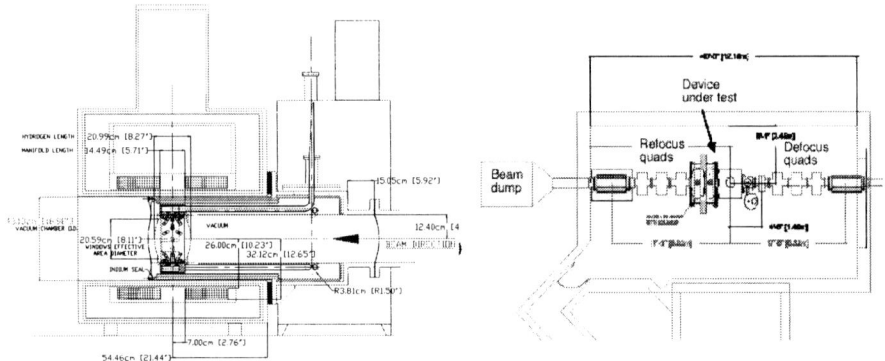

Figure 5. The forced-flow absorber prototype, shown inside its vacuum vessel, inserted into the FNAL Lab G

REFERENCES

[1] M.A.C. Cummings et al, "Current LH_2 Absorber Research in MuCool", NuFact '02, Imperial College, London, UK, 2002.
[2] J. Norem et al, "Measurement of Beam Driven Hydrodynamics", PAC2003 proceedings.
[3] A. Bross, "Mucool Results and Plans", these proceedings.

Plans for MICE at RAL

Paul Drumm, for the MICE Collaboration

CCLRC Rutherford Appleton Laboratory, Chilton, Didcot, OXON OX11 0QX, UK

Abstract. MICE is the Muon Ionisation Cooling Experiment; an engineering demonstration of ionisation cooling. The experiment is based on using the ISIS synchrotron and upgrading an existing beam line to provide a 300-400 MeV/c muon beam. Since the MICE proposal was submitted to the Rutherford Appleton Laboratory, the design of the beam line and of the infrastructure needed to make MICE a reality has been progressed. This paper describes some of the plans for implementing MICE at RAL.

INTRODUCTION

The Muon Ionisation Cooling Experiment (MICE) [1] is a demonstration of a key Neutrino Factory technology proposed [2] to be implemented on ISIS at the Rutherford Appleton Laboratory, UK. In MICE, it is expected that a muon beam of large transverse emittance (6π mm.rad) will be cooled by around 10% in one full cell of the cooling channel lattice, Fig. 1. The measurement of the cooling itself ($\varepsilon_{out}/\varepsilon_{in}$) must have a precision which is much better than this. The cooling cell consists of a cryogenic absorber (hydrogen is the material of choice) where the momentum of the muon beam is reduced both longitudinally and transversely. Subsequently, the beam is re-accelerated in a module of four RF cavities, replacing the lost momentum in the forward direction and thus reducing the transverse emittance of the beam. This is repeated in an identical section which provides a field flip to control the emittance increase that would otherwise occur. The real key to the success of MICE is in balancing the cooling against multiple scattering in the beam windows because these heat the beam.

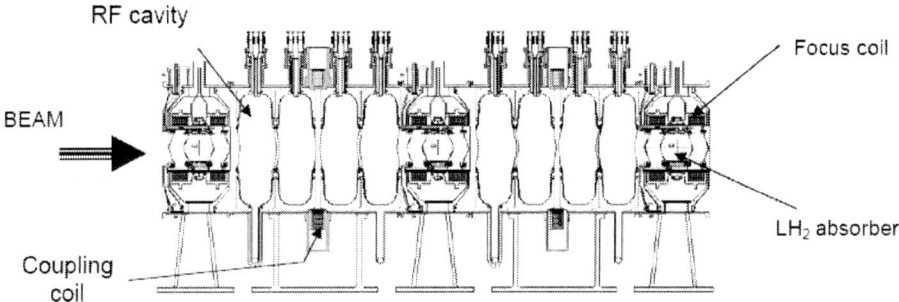

Figure 1. Layout of the MICE cooling channel.

It is proposed to provide a muon beam into a vacant hall by upgrading an existing beam line attached to the ISIS synchrotron at RAL. The work involved in mounting such an experiment is described below.

ISIS

ISIS is a powerful pulsed neutron source. It is driven by an 800 MeV proton synchrotron with a current averaging 200 µA. The ISIS proton pulse has a duration of ~400 ns and a repetition rate of 50 Hz. During the 20ms cycle time, a H– beam is injected from the ISIS (70MeV) linac, captured and accelerated to 800 MeV. As the beam approaches an energy of 800 MeV, it is circulating in the synchrotron as two bunches (dual harmonic) at ~1.5 MHz.

THE MUON BEAM

The synchrotron proton beam interacts with a thin blade-like target which intercepts the outer edge of the beam. The target is designed to dip into the beam 1ms prior to extraction. This minimises unnecessary activation of ISIS, gives the highest production energy and matches the pulsing of the MICE RF system. Particles from the target enter a beam line set at 25 degrees to the proton beam direction. Momentum selection takes place in the first of two dipole magnets. The muon beam is formed from the decay of pions in a 5T superconducting solenoid (from the muE4 beam line in collaboration with PSI). A second dipole is used to momentum select the muon beam and to separate the muons from contaminating pions and protons from the original beam.

The muon beam rate is, on average, relatively low (a few thousand per second). However, because the beam is pulsed for 1ms at 1 Hz, the instantaneous rates are rather high (MHz). The muon/proton/pion ratio is dependent on the relative π/μ momentum selection in the second dipole and on the deflection angles chosen for the dipoles.

THE MICE HALL

The MICE experiment benefits from the use of a large experimental hall previously occupied by the nimrod injector, and more recently home to a beam test area used for particle detector testing.

Infrastructure

The MICE apparatus itself will be built outside of the laboratory, with RAL playing a major role in the design of the RF system, the design and manufacture of sets of SC focus coils for the absorber unit in collaboration with Oxford University and in the safety aspects of the design and implementation of the absorber unit as a whole. The infrastructure needed for MICE covers the cryogenic system and its distribution, the hydrogen system for the absorbers and cooling plant to remove the 500 kW of electrical power MICE is expected to consume. The initial plan for the cryogenic and cooling plant requires new buildings to house compressors and cooling water plant.

Layout of the Hall

The Hall benefits from some 560 m^2 of floor space (12m wide by 8m high), is serviced by two 8 tonne cranes and has a 1.25 MW electrical installation available to MICE. Both the beam line and MICE fit comfortably in the hall. Figure. 2 shows the schematic layout of the ISIS synchrotron, the muon beam line and MICE. Considerable space is taken up by shielding around the muon beam line to protect from neutrons from proton beam loses in the synchrotron, and around MICE to protect from X-rays emitted by the cavities.

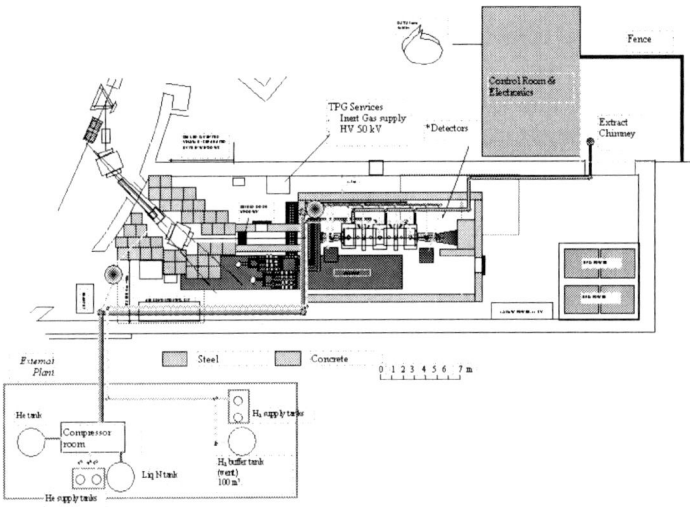

FIGURE 2. Schematic Layout of the ISIS synchrotron (left top), the muon beam line, MICE and the infrastructure needed to operate.

The RF System

The RF system for MICE operates at 201.25 MHz, similar to the frequency of the ISIS linac amplifiers (202.5 MHz, 4MW). The design of the MICE RF system anticipates re-using similar amplifier circuits from a number of laboratories (possibly 2 from CERN and 2 from LBL). Each circuit should be capable of delivering ~2 MW or RF power and so 4 amplifiers could be used to drive the 8 cavities of MICE (1 MW into each cavity with a flat top of 1 ms). Used power tubes from ISIS, where they are retired when their output reaches ~2MW, will be used in the power amplifiers. If more circuits can be obtained, then it is planed to drive each cavity with its own, separately phased and amplitude controlled amplifier system. Alternatively, the output from each circuit can be split between two cavities.

Safety

Many safety issues are raised by MICE and have to be addressed and dealt with in a competent manner leading to low risk operations for MICE. The most significant of which are
- Ionising Radiation: X-rays from the cavities, neutrons from ISIS,
- Strong peripheral magnetic fields: pacemaker issues and interference with the ISIS linac,
- Explosive risk from the hydrogen absorber system.

Many of these risks are routinely dealt with at high-energy accelerators. The particular issue for MICE is the use of some 60l of liquid hydrogen. While the ISIS neutron moderators use cryogenic hydrogen, the issues of containment are different. In ISIS the liquid hydrogen is contained in a thick, welded aluminium vessel that is surrounded by vacuum and He gas layers that proves no hindrance to the neutrons, Fig. 3. Such a solution would seriously degrade the performance of MICE. The MICE collaboration has invested considerable engineering effort into solve these problem.

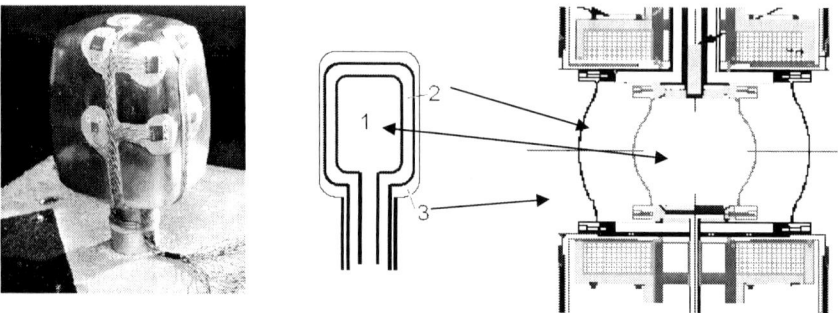

Figure 3. Comparison between the ISIS liquid hydrogen moderator (left and centre) and the MICE absorber (right). The numbered regions are 1: the liquid hydrogen volume; 2: vacuum inter-space and 3: a tertiary He/vacuum layer.

The particular issue is to prevent the possibility that hydrogen gas could come into contact with oxygen. The situation is solved in ISIS by surrounding the LH_2 by vacuum and He layers. In MICE, careful control of the temperatures of various surfaces (where oxygen could accumulate in an undetectable manner), and a separate vacuum layer around the absorber protected by Ar-jacketing of the external pipe-work allows active (and redundant) monitoring of the system state.

Timescales

The implementation of MICE is likely to culminate in detector testing in late 2006. This will also be an opportunity to characterise the muon beam. The first major milestones will occur at the end of 2003 when the hall will be cleared and in 2004 when, during an ISIS shutdown, the ISIS and MICE halls will be joined together in preparation for installing the µE4/PSI solenoid. It is a significant requirement of the installation planning that MICE does not impact on ISIS operations.

CONCLUSION

The implementation of MICE at RAL is a major task supported by an active and enthusiastic community. Plans are in place for all of the major infrastructure needs. Safety considerations have been given priority and solutions are being actively addressed. For the UK, an opportunity is being presented to host a significant R&D project which will prove to be a significant milestone on the path towards the Neutrino Factory.

REFERENCES

1. Y Torun, these proceedings.
2. The MICE proposal can be viewed on-line at
http://hep04.phys.iit.edu/cooldemo/micenotes/public/pdf/MICE0021/MICE0021.pdf

200MHz superconducting RF cavity development for RLAs[1]

R.L. Geng*, H. Padamsee*, D. Hartill*, P. Barnes*, J. Sears*,
R. Losito†, E. Chiaveri†, H. Preis† and S. Calatroni†

*LEPP, Cornell University, Ithaca, NY 14853-5001, USA
†CERN, Geneva, Switzerland

Abstract. A 200MHz single-cell superconducting Nb-Cu cavity has been fabricated and tested at 4.2 K and at 2.5 K. The low field Q_0 reached 1.5×10^{10} at 4.2 K. The accelerating gradient (E_{acc}) reached 11 MV/m at a Q_0 of 6×10^8. Two multipacting barriers show up at E_{acc} = 3 and 1 MV/m. Helium processing is effective in reducing field emission and improving accelerating gradients. The Q-drop is observed with a much stronger field dependence as compared to what is expected. Results on the radiation background near the cavity and the external magnetic field effect on the Q of a cold cavity are also reported.

INTRODUCTION

The proposed neutrino factory and muon collider ask for RF cavities operating at a frequency near 200 MHz for rapid acceleration of muons [1]. One scenario is to use superconducting RF cavities [2]. The desired accelerating gradient is at least 15 MV/m at a Q_0 of 6×10^9. Since there is no superconducting RF experience at 200 MHz, R&D in this regime should be started early. Cornell and CERN collaborated to fabricate and test a 200 MHz single-cell elliptical Nb-Cu cavity.

DESIGN AND FABRICATION

The cavity is fabricated with the standard film niobium sputtering technique that has been used for LEP2 cavities. The cell has a length of 750 mm and a major diameter of 1370 mm. The entire cavity, including the 400 mm diameter beam tubes, is nearly 2000 mm long. RF parameters are listed in Table 1.

The two half cells were formed by spinning two 8 mm thick OFE copper sheets. The RF surface was polished electrolytically for a 400 μm surface removal to reduce the imperfections induced by the mechanical process. The cavity was welded with an electron beam from the inside. Further mechanical smoothing has been done by grinding locally all the sharp points of the RF surface. Chemical polishing (SUBU) was performed twice on the whole cavity to remove 20 μm + 20 μm. The deposition was

---

[1] Work supported by National Science Foundation.

TABLE 1. Cavity RF parameters

Parameter	Value	Unit
G	250	Ω
R/Q	121	Ω
E_{pk}/E_{acc}	1.69	-
B_{pk}/E_{acc}	4.34	mT/(MV/m)
$E_{acc}/\sqrt{U}$	0.518	(MV/m)/$\sqrt{J}$

made by using the existing infrastructure developed for LEP2 cavities. The cavity was rinsed at 100 bars with ultra pure water.

PERFORMANCE

The cavity was tested in a vertical dewar fitted in a radiation shielded pit. The pit was lined with sheets of low carbon steel, attenuating the earth magnetic field to 0.2 G in the cavity space. Two multipacting barriers were identified at 3 and 1 MV/m [3], corresponding to first and second order two-point multipacting in the equator region respectively. These barriers were observed consistently during each cavity test and could be surmounted after a few hours of RF processing. Field emission was found to be rather strong during initial cavity tests. Additional high pressure water rinsing was applied and a significant improvement was achieved. Gas helium processing was found to be very effective in attacking field emitters and the accelerating gradient was improved by a factor of two in some cases. The achieved cavity performance is shown in Fig. 1. At 2.5 K, the low field Q reached 2.5×10^{10} and E_{acc} reached 11 MV/m. The cavity itself did not limit us from reaching higher gradients, but the RF coupler did.

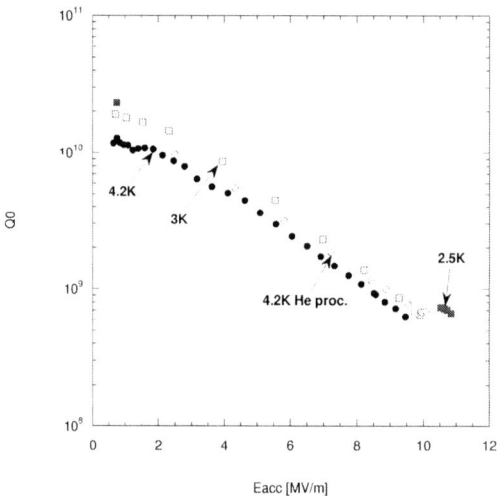

FIGURE 1. Performance of the 200 MHz superconducting Nb-Cu cavity.

The Q-slope (a characteristic of sputtered Nb-Cu cavities) of the 200 MHz Nb-Cu cavity is found to be about 10 times steeper than what is projected from the previous results at other frequencies. The fact that the cavity Q changes as the bath temperature changes (see Fig. 1) indicates that the Q-slope is resulted mainly from the intrinsic property of the niobium film. However, it should be pointed out that field emission remains to be a contributing factor to the Q-slope, even after helium processing.

RADIATION BACKGROUND

The radiation in the test pit is generated in the form of Bremsstrahlung x-rays when field emission or multipacting electrons hit the cavity wall. Measurements were made by a γ probe and a NaI detector on the top plate of the dewar. The detectors are roughly aligned with the cavity axis to allow the best detection of high energy γ's generated by electrons accelerated by the full cavity voltage. Fig. 2 shows the gradient dependence of the dose rate. The peak near 3 MV/m is a result of multipacting which generates a large number of low energy electrons. Above 4 MV/m, the radiation is mainly due to field emission electrons originated from the iris of the cavity and its field dependence fits very well into the modified Fowler-Nordheim theory. The field enhancement factor β is found to be in the range of 600 - 900.

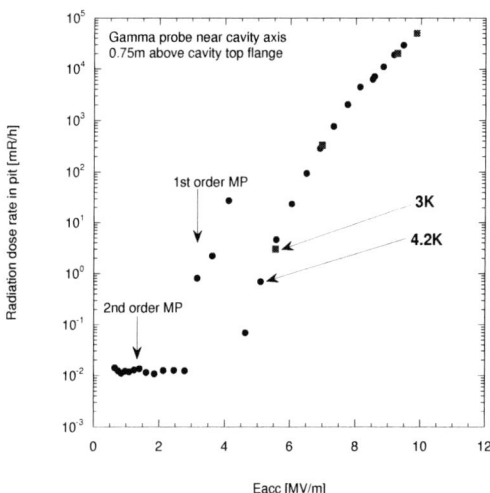

FIGURE 2. γ radiation near the 200 MHz superconducting cavity.

The energy of γ rays shows a low energy peak below 1 MeV. This can be attributed to the Compton effect when photons go through the cavity wall and the top plate of the dewar. The end point energy of the spectrum corresponds reasonably well to the full voltage of the cavity. No neutron radiation was detected at the highest achieved accelerating gradient.

EFFECT OF EXTERNAL MAGNETIC FIELDS

The envision of fitting a strong solenoid and a cavity into one cryostat motivated the study of the external magnetic field effect on a cold cavity. A rather simple configuration was adopted, in which a superconducting coil (the end face diameter being 40 mm) was installed against the cavity equator. The cavity was cooled down prior to turning on the magnet. As shown in Fig. 3, the cavity is not affected by an external magnetic field of $\leq$ 1200 Oe. Above 1200 Oe, a non-reversible power loss due to the external field is observed. These results are consistent with the fact that niobium is a type-II superconductor with its H_{c1} being close to 1200 Oe at 4.2 K.

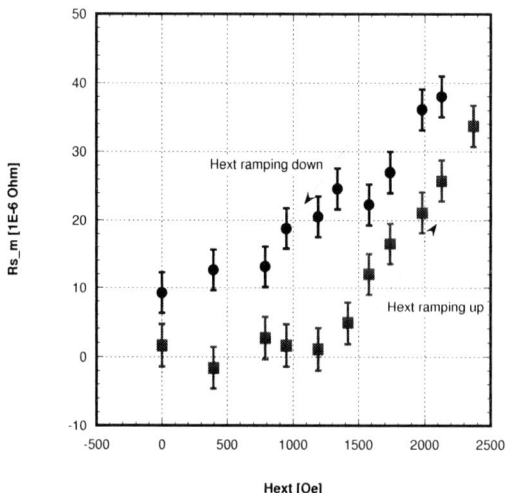

FIGURE 3. External magnetic field effect on the additional surface resistance of the 200 MHz cavity. The cavity is cooled down to superconducting prior to turning on the magnetic field.

CONCLUSION

The first 200 MHz superconducting elliptical cavity was successfully fabricated and tested. The accelerating gradient reached 11 MV/m. The low field Q reached 2×10^{10} at 2.5 K. The film is not responsible for the gradient limitation, but is responsible for a strong Q-drop, for which further work is needed.

REFERENCES

1. Feasibility Study-II of a Muon-Based Neutrino Source, ed., S. Ozaki, R. Palmer, M. Zisman, and J. Gallardo, BNL-52623(2001).
2. H. Padamsee, Proc. of the 9th workshop on RF superconductivity, 1999, p.587.
3. R.L. Geng et al., Proc. of the 2003 Particle Accelerator Conference, Portland, OR, May 12-16, 2003.

Optimized Beam Optics for Muon Acceleration

S.A. Bogacz

Center for Advanced Studies of Accelerators
Thomas Jefferson National Accelerator Facility
Newport News, VA 23606

Abstract. A conceptual design of a Ring Cooler driven muon acceleration based on recirculating superconducting linacs is proposed. In the presented scenario, acceleration starts after ionization cooling at 250 MeV/c and proceeds to 20 GeV, where the beam is injected into a neutrino factory storage ring. The key technical issues are addressed; such as: the choice of acceleration technology (superconducting versus normal conducting) and the choice of RF frequency, and finally, implementation of the overall acceleration scheme: capture, acceleration, transport and preservation of large phase space of fast decaying species. Beam transport issues for large-momentum-spread beams are accommodated by appropriate lattice design choices. The proposed arc optics is further optimized with a sextupole correction to suppress chromatic effects contributing to emittance dilution.

MUON ACCELERATION SCHEME

A neutrino factory [1] is aimed to produce narrow neutrino beams via decay of muons in long straight sections of a storage ring. Recent development of Ring Coolers promises much larger longitudinal phase space compression compared to previously considered straight Cooling Channels. Here we propose an optimized muon accelerator complex based on a Ring Cooler as a front end. As illustrated schematically in Figure 1, a proposed muon accelerator features a 250 MeV-to-2.8 GeV straight pre-accelerator linac and a 2.8-to-20 GeV four-pass recirculating linac accelerator (RLA).

The pre-accelerator captures a large muon phase space coming from the Ring Cooler and accelerates muons to relativistic energies of about 2.8 GeV. It makes the beam sufficiently relativistic and adiabatically decreases the phase-space volume, so that effective acceleration in recirculating linacs is possible. The RLA further compresses and shapes the longitudinal and transverse phase spaces, while increasing the energy.

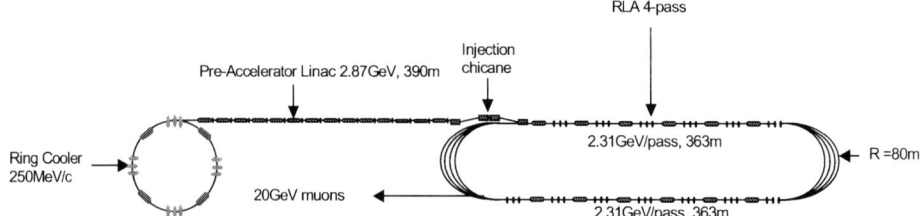

FIGURE 1. Ring Cooler driven design of a 20 GeV muon accelerator complex based on RLA – machine layout..

Accelerating Technology

To ensure adequate survival rates of short-lived muons, acceleration must occur at high average gradient. Initial estimate [2] shows that a "real estate average" RF gradient of 15 MV/m will allow survival of about 80% of source muons throughout the RLA. Since muons are generated as a secondary beam they occupy large phase-space volume. The accelerator must provide high average gradient, while maintaining very large transverse and longitudinal accelerator acceptances. The above requirement drives the design to low RF frequency, e.g. 200 MHz. If normal-conducting cavities at that frequency were used, the required high gradients would demand unachievably high peak RF sources. The RF power can then be delivered to the cavities over an extended time, and thus RF source peak power can be reduced.

Machine Architecture

In a recirculating linac accelerator one needs to separate different energy beams coming out of a linac and to direct them into appropriate arcs for recirculation. For multiple practical reasons horizontal rather than vertical beam separation was chosen [2]. Rather than suppressing horizontal dispersion created by the spreaders and recombiners it is smoothly matched to the horizontal dispersion of the arc. Finally, to assure compact arc architecture very short matching sections in spreaders and recombiners are desired. One also needs to maintain manageable beam sizes in the arcs [3]. This calls for short cells and for putting stringent limits on dispersion and beta functions (beam envelope). Since spreaders and recombiners were chosen in the horizontal plane, the uniform focusing and lattice regularity was broken in that plane and the horizontal beam envelope requires special attention. On the other hand, the vertical beam size remains small due to maintaining uniform focusing and small beta functions in that plane.

BEAM DYNAMICS CHOICES

The initial longitudinal acceptance of the linear accelerator (2.5σ) is set by the best performance promised by Ring Coolers. Initial transverse and longitudinal emittaces for Ring Coolers (as they compare to the Straight Cooling Channel) are summarized in Table 1.

TABLE 1. Beam emittance after cooling at 250 MeV/c.		
rms values	Cooling Channel	Ring Cooler
Transverse emittance : ε_x, ε_y (normalized)	2.4 mm·nrad	2.0 mm·rad
longitudinal emittance: $\varepsilon_\ell = \sigma_{\Delta p} \sigma_z/m_\mu c$	27 mm	3 mm
momentum spread: $\sigma_{\Delta p/p}$	0.08	0.025
bunch length: σ_z	163 mm	50mm

Longitudinal Dynamics

In the 'Ring Cooler scenario' to perform adiabatic bunching, the RF phase of the cavities needs to be shifted by only 30 deg at the beginning of the preaccelerator and gradually changed to zero by the linac end. In the first half of the linac, when the beam is still not sufficiently relativistic, the offset causes synchrotron motion, allowing bunch compression in both length and momentum spread to $\Delta p/p=\pm 6.5\%$ and $\Delta\phi=\pm 31$ deg. The synchrotron motion also suppresses the sag in acceleration for the bunch head and tail. The first frame in Figure 2 shows how the initially elliptical boundary of the bunch longitudinal phase space will be transformed by the end of the linac.

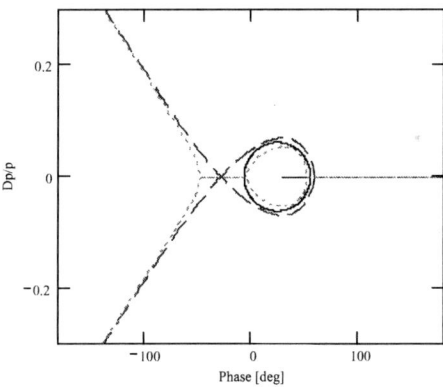

FIGURE 2. Longitudinal beam boundary (solid line) inside a separatrix (dashed line) shown at the beginning of the linac ($\Delta p/p=\pm 0.065$ or $\Delta\phi =\pm 31$deg.)

Longitudinal Bunch Compression

Bunch length and energy spread are still too large at the RLA input and their further compression is required in the course of acceleration. To achieve this, the beam is accelerated off-crest with non zero M_{56}. That causes synchrotron motion, which suppresses the longitudinal emittance growth related to non-linearity of accelerating voltage. Without synchrotron motion the minimum beam energy spread would be determined by non-linearity of RF voltage at bunch length and would be equal to $(1-\cos\phi) \approx 6\%$ for bunch length $\phi=20$ deg. The synchrotron motion causes particle motion within the bunch and averages the total energy gain of tail's particle to the energy gain of particles in the core. The optimum value is about 90cm, while optimal detuning of RF phase from on-crest position is different for different arcs. To perform a further bunch compression in 4-pass RLA (see the remaining frames on Figure 3) the beam is again accelerated off-crest with phase offsets in the range of 5 – 15 deg for different passes and M_{56} of 90 cm for all seven arcs.

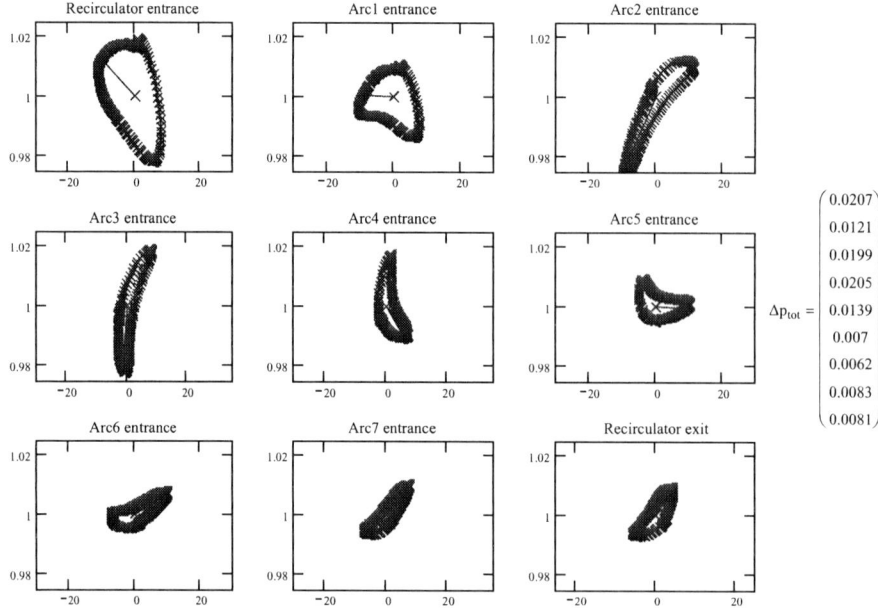

FIGURE 3. Boundaries of the beam longitudinal phase space at different locations in the recirculator; $M_{56}=90$ cm.

Optimized RLA Optics

To maintain uniform focusing periodicity between linacs and arcs one needs to minimize Twiss functions for multi-pass optics of both linacs. This can be achieved through designing a semi-periodic linac optics (introducing additional large scale lattice period). As illustrated in Figures 4 and 5, resulting Twiss functions (at linac ends) are significantly lowered, which facilitates uniform matching between different types of optics to alleviate emittance dilution due to chromatic aberrations. Finally, there is a need for suppression of chromatic effects via sextupole corrections in spreaders and recombiners. This was implemented via three families of sextupoles [2] to control the horizontal emittance blow-up.

Pass 1

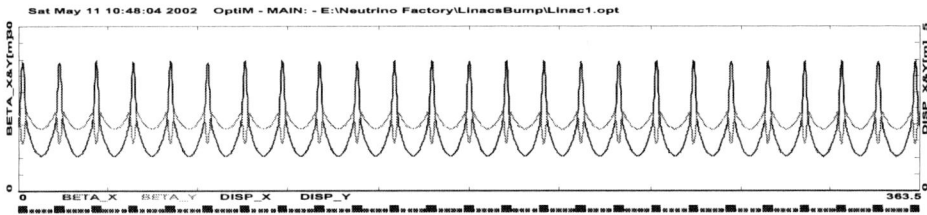

Pass 2

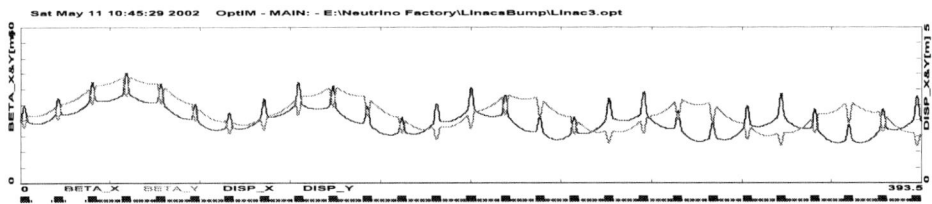

FIGURE 4. Optimized linac optics for multi pass beams – smooth transition Arc-linac.

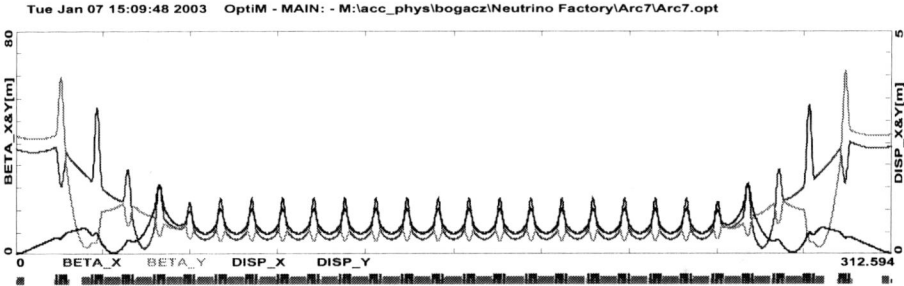

FIGURE 5. Arc 7 Optics – beta-functions and the horizontal dispersion matched to both adjacent linacs.

CONCLUSIONS

Results of this study suggest that there are no obvious physical or technical limitations precluding construction of an RLA for acceleration of muons to 20 GeV. The proposed acceleration and beam transport scheme is well suited for handling large phase space beams with no emittance dilution [4].

REFERENCES

1. M.M. Alsharo's et al., Physical Review Special Topics – Accelerators and Beams, Vol **6**, 081001 (2003)
2. S.A. Bogacz and V.A. Lebedev, Nuclear Instruments and Methods in Physics Research A, Vol **472/3**, 528, (2001)
3. S.A. Bogacz and V.A. Lebedev, Nuclear Instruments and Methods in Physics Research A, Vol **503**, 306, (2003)
4. S.A. Bogacz, Journal of Physics G: Nuclear and Particle Physics, Vol **29**, 1723 (2003)

Time-energy densities in $\pi \to \mu$ decay

B. Autin* and F. Méot†

*CERN AB, Geneva
†CEA DAPNIA, Saclay

Abstract. An analytical model is developed to describe the longitudinal phase space of a hybrid beam of pions and decay muons.

INTRODUCTION

The properties of the muon beam created by decay of pions are usually obtained by simulation. Here, an analytical model is developed to get insight into the properties of the beam and guidelines in the adjustment of the magnets of the decay channel and of the RF system that collect the muon beam. It is assumed that there is no coupling between transverse and longitudinal spaces. This treatment is thus more appropriate in the present stage for a quadrupolar than for a solenoidal decay channel. All the pions are supposed to travel along the axis of the channel. It is thus essentially the kinematic effects of the pion motion, muon creation and muon motion that are investigated. These effects are described in terms of energy or momentum density, of time density and of longitudinal phase space portraits. Densities are manipulated using random variable methods.

MOMENTUM AND ENERGY SPECTRA

Calculations that follow are based on the kinematic relations of the decay process $\pi \to \mu + \nu$. The following notations are used : pion mass m_π, its lifetime at rest τ_π^*, energy E_π, muon mass m_μ, center of mass decay angle θ_μ^* with respect to pion velocity. The basic kinematics ingredients needed are the pion
 - laboratory frame lifetime $\tau_\pi = \gamma_\pi \tau_\pi^*$,
 - decay law $N(s) = N_0 e^{-\eta s/p_\pi}$, wherein $\eta = m_\pi / c\tau_\pi^*$,
and the muon
 - center of mass energy $E_\mu^* = (m_\pi^2 + m_\mu^2)/2m_\pi$ and momentum $p_\mu^* = (m_\pi^2 - m_\mu^2)/2m_\pi$,
 - laboratory frame energy $E_\mu = \gamma_\pi (E_\mu^* + \beta_\pi p_\mu^* \cos \theta_\mu^*)$,

The technique used to calculate a density in some variable x as a function of a density in another variable y relies on the relation $g_x = g_y |dy/dx|$.

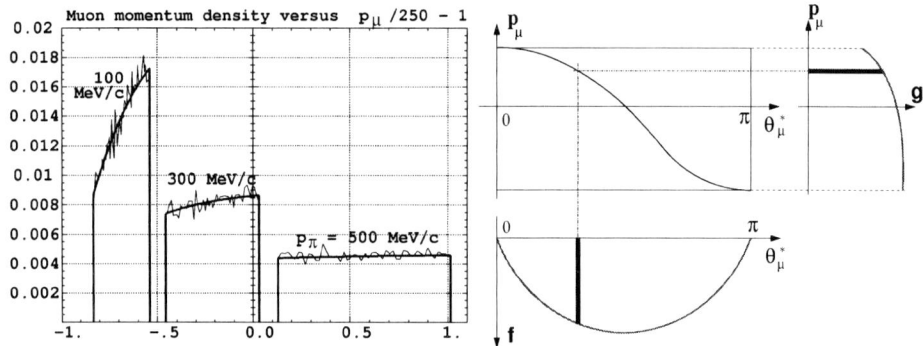

FIGURE 1. Density $g_{p_\mu|p_\pi}(p_\mu)$ for $p_\pi = 100$, 300 and 500 MeV/c (left graph), and geometrical understanding of its build-up from $g_{\theta^*}(\theta_\mu^*)$ in the change of variable $\theta_\mu^* \to p_\mu$ (right).

Muon spectra at fixed pion momentum

Muon momentum. Given fixed pion momentum p_π, the decay muon momentum satisfies a p_π-conditional density that writes

$$g_{p_\mu|p_\pi}(p_\mu) = g_{\theta^*}(\theta_\mu^*) \left| d\theta_\mu^*/dp_\mu \right| \qquad (1)$$

$$= \frac{m_\pi}{2p_\pi p_\mu^*} \frac{p_\mu}{\sqrt{p_\mu^2 + m_\mu^2}} = \frac{m_\pi}{2p_\pi p_\mu^*} \beta_\mu, \qquad p_\mu \in [\gamma_\pi(\beta_\pi E_\mu^* - p_\mu^*), \gamma_\pi(\beta_\pi E_\mu^* + p_\mu^*)]$$

wherein $g_{\theta^*}(\theta_\mu^*) = \sin\theta_\mu^*/2$ ($\theta_\mu^* \in [0,\pi]$) is the decay angle density. $g_{p_\mu|p_\pi}(p_\mu) \equiv 0$ outside the specified p_μ interval. FIG. 1 shows typical shapes of $g_{p_\mu|p_\pi}(p_\mu)$. Monte Carlo histograms $\Delta N_{p_\mu|p_\pi}/N_0\Delta p_\mu$ are superimposed for comparison.

Muon energy. Similar calculations in the case of a change of variable $\theta_\mu^* \to E_\mu$, or as well using $d/dE = (1/\beta)d/dp$ in Eq. 1, yield the energy density at fixed p_π

$$g_{E_\mu|p_\pi}(E_\mu) = \frac{m_\pi}{2p_\pi p_\mu^*}, \qquad E_\mu \in [\gamma_\pi(E_\mu^* - \beta_\pi p_\mu^*), \gamma_\pi(E_\mu^* + \beta_\pi p_\mu^*)] \qquad (2)$$

Pion and muon spectra versus flight distance

Pions. Pion densities properties that intervene in the sequel are as follows. The decay density as a function of flight distance s, given p_π, writes

$$g_{s|p_\pi}(s, p_\pi) = (\eta/p_\pi)\exp(-\eta s/p_\pi) \qquad (3)$$

Given parent pions with initial momentum density $g_{p_\pi}(p_\pi)$ (say, at $s = 0$), one gets the 2-D density at arbitrary $s > 0$

$$g_{s,p_\pi}(s,p_\pi) = g_{s|p_\pi} \times g_{p_\pi} \qquad \text{(and } \int_{s=0}^{\infty}\int g_{s,p_\pi}(s,p_\pi)dsdp_\pi = 1\text{)} \qquad (4)$$

In the following, for the sake of simplification, we will illustrate things using a uniform initial pion momentum density

$$g_{p_\pi}(p_\pi) = \mathbf{1}_{\Delta p_\pi}(p_\pi) = 1/(p_{\pi_2} - p_{\pi_1}) \qquad (p_\pi \in [p_{\pi_1}, p_{\pi_2}]) \qquad (5)$$

The ensuing form of $g_{s,p_\pi}(s,p_\pi)$ is shown in FIG. 2, given a pion bunch launched at $s = 0$ with zero size and $p_\pi \in [100, 500]$ MeV/c. Integrating Eq. 4 with respect to s yields the p_π-density of the decayed parent pions at distance s,

$$g_{p_\pi}(p_\pi)\big|_s = \int_0^s g_{s,p_\pi}(s,p_\pi)\,ds = \mathbf{1}_{\Delta p_\pi}(p_\pi)\,(1 - \exp(-\eta s/p_\pi)) \qquad (6)$$

The p_π-density of the *non-decayed* pion population ensues,

$$\bar{g}_{p_\pi}(p_\pi)\big|_s = (g_{p_\pi}(p_\pi) - g_{p_\pi}(p_\pi)\big|_s) = \mathbf{1}_{\Delta p_\pi}(p_\pi)\exp(-\eta s/p_\pi) \qquad (7)$$

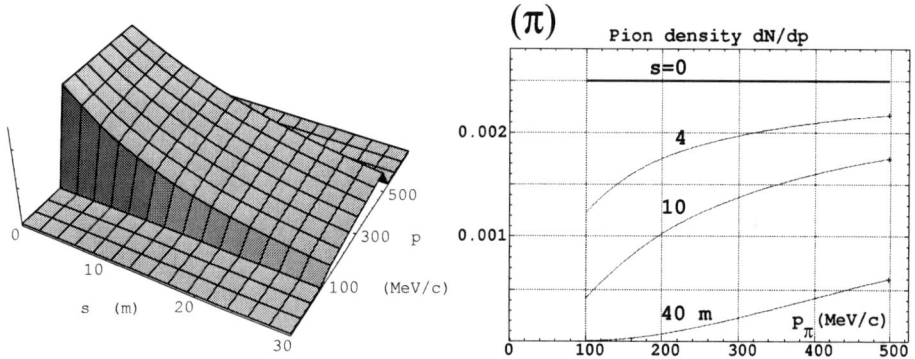

FIGURE 2. Pion momentum density along the decay channel (left graph) and s-sections (right).

Muons. A non-zero pion momentum byte is accounted for by multiplying the p_π-conditional density $g_{p_\mu|p_\pi}(p_\mu)$ (Eq. 1) by the muon density at s at given p_π, $g_{s,p_\pi}(s,p_\pi)$ (Eq. 4). (The muon decay is not taken into account in the following for simplicity, doing so would mean accounting for an s-dependent muon survival additional factor.) This yields the muon momentum spectrum at s under the integral form

$$g_{p_\mu}(p_\mu)\big|_s = \int_{\Delta p_\pi} g_{p_\mu|p_\pi}\,dp_\pi \int_0^s g_{s,p_\pi}(s,p_\pi)\,ds \qquad \text{(and } \lim_{s\to\infty} g_{p_\mu}(p_\mu)\big|_s = 1\text{)} \qquad (8)$$

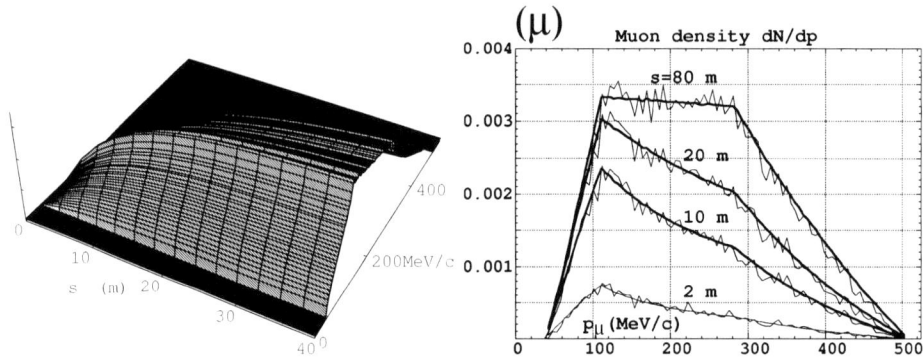

FIGURE 3. *Left :* muon momentum density along the decay channel. *Right :* s-sections.

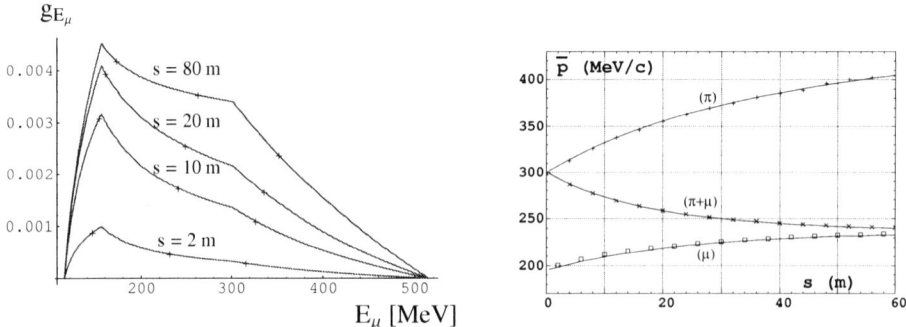

FIGURE 4. *Left :* Muon energy density at various distances s ; the crosses are at $\overline{E}_\mu$ and at $\overline{E}_\mu \pm \sigma_{E_\mu}$. *Right :* Average momentum of beams over 60 m flight distance (markers are from Monte Carlo simulations).

The Δp_π integration interval is a function of p_μ following the dependence given in Eq. 1 (not all pions can produce a muon of momentum p_μ). A similar integral holds for the energy spectrum $g_{E_\mu}(E_\mu)|_s$, given $g_{E_\mu|p_\pi}$ (Eq. 2).

The summation in Eq. 8 can be viewed as a superposition of the fixed-p_π muon spectra of FIG. 1, this is the way the muon spectra shown in FIG. 3 has been numerically calculated (Monte Carlo histograms $\Delta N_{p_\mu}|_s/N_0 \Delta p_\mu$ have been superimposed for comparison). However the calculations can been completed analytically, as performed for obtaining the energy spectra in FIG. 4-left, which we do not detail for shortness.

Mean value and standard deviation

Parent pion beam. The pion beam average momentum as a function of s is obtained from Eq. 7 that yields

$$\bar{p}_\pi(s) = \int p\, \bar{g}_{p_\pi}(p)|_s\, dp \Big/ \int \bar{g}_{p_\pi}(p)|_s\, dp = \frac{\sum_{i=1,2}(-)^i \dfrac{p_{\pi_i}^2 - \eta s p_{\pi_i} - \eta^2 s^2 e^{\frac{\eta s}{p_{\pi_i}}} \operatorname{Ei}(-\frac{\eta s}{p_{\pi_i}})}{2 e^{\eta s/p_{\pi_i}}}}{\sum_{i=1,2}(-)^i \dfrac{p_{\pi_i} + \eta s e^{\frac{\eta s}{p_{\pi_i}}} \operatorname{Ei}(-\frac{\eta s}{p_{\pi_i}})}{e^{\eta s/p_{\pi_i}}}}$$

Muon beam. Similar calculations apply to the determination of the mean momentum $\bar{p}_\mu(s)$ and energy $\bar{E}_\mu(s)$ of the muons and to the momentum of the center of gravity of the hybrid beam.

Average momenta of both pion and muon beams are increasing functions of the distance, because the lower energy parent pions decay faster, whereas the average momentum of the $\pi + \mu$ beam decreases monotonically here (FIG. 4-right), a behavior that can be accounted for to maintain constant focusing strength in tuning the decay channel [1].

Another parameter, relevant to the voltage of the RF system, is the second moment of the energy density:

$$\sigma_{E_\mu}(l) = \left(\int_{E_{\mu 1}}^{E_{\mu 2}} (E - \bar{E}_\mu)^2 g_{E_\mu}\, dE \Big/ \int_{E_{\mu 1}}^{E_{\mu 2}} g_{E_\mu}\, dE \right)^{1/2} \quad (9)$$

Both mean energy and ends of the standard energy interval are displayed in FIG. 4-right. The capture efficiency can be defined as

$$y_{E_\mu}(s) = \int_{\bar{E}_\mu - \sigma_{E_\mu}}^{\bar{E}_\mu + \sigma_{E_\mu}} g_{E_\mu}\, dE \Big/ \int_{E_{\mu 1}}^{E_{\mu 2}} g_{E_\mu}\, dE \quad (10)$$

TIME SPECTRA

The approach followed for the energy distribution can be resumed for time distribution. The p_π-conditional time density $g_{t_\mu|p_\pi}(p_\mu)$ of the muons at arbitrary s can be derived from $g_{\theta^*}(\theta_\mu^*)$ through a change of variable

$$\theta_\mu^* \to t_\mu = s_d/c\beta_\pi + (s - s_d)/c\beta_\mu$$

On the other hand $g_{t_\mu|p_\pi}(p_\mu)$ can be derived from $g_{p_\mu|p_\pi}(p_\mu)$ (Eq. 1) using a change of variable $p_\mu \to t_\mu$.

A non-zero pion momentum byte Δp_π is accounted for in the muon density calculation, by introducing the decayed pion density at s_d, under the form of a $g_{s,p_\pi}(s, p_\pi)$ factor (Eq. 4). This yields the muon time density under the integral form

$$g_{t_\mu}(t_\mu)|_s = \int_{s_d=0}^{s} ds_d \int_{\Delta p_\pi} g_{t_\mu|p_\pi}(t_\mu)\, g_{s,p_\pi}(s_d, p_\pi)\, dp_\pi \quad (11)$$

with still a p_μ dependence, and in addition a s_d dependence, of the integration domain Δp_π. The calculation cannot be performed analytically because of the presence of β_π together with p_π in the integrand. Moreover, integrating over s cannot be done separately.

The numerical method of histograms superimposition followed for calculating the momentum and energy densities remains however valid. The boundary times correspond to the fastest and slowest muon emitted by the fastest and slowest pion. The typical shape of $g_{t_\mu}(t_\mu)|_s$ is displayed as a projected density in FIG. 6. The first two moments of $g_{t_\mu}(t_\mu)|_s$ can be calculated so as to derive capture efficiencies as was done for the energy spectra.

Proton bunch length. The time distribution is affected by the length τ of the proton bunch which generates the pions by interaction with the target (which in turn affects the muon yield [2,Tab. 5.2]). This is a matter of re-writing the density function under the form of a convolution product $g_{t_\mu}(t_\mu)|_s = \frac{1}{\tau}\int_0^\tau S(t_\mu - \tau)g_{t_\mu}(\tau)|_s \, d\tau$, wherein S characterizes the density of pions inside the bunch at the time of production.

LONGITUDINAL PHASE-SPACE

The muon time density $g_{t_\mu}(t_\mu)|_s$ (Eq. 11) has an explicit dependence on s. Note that, given the pion energies in concern here, the flight distance s can be considered in good approximation as the position along the channel length. $g_{t_\mu}(t_\mu)|_s$ also has an implicit

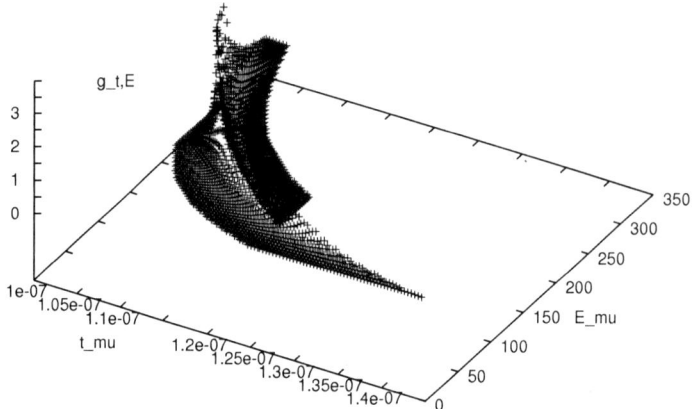

FIGURE 5. Muon density $g_{t_\mu,E_\mu}(t_\mu,E_\mu)|_s$ (arbitrary units) at $s = 30$ m.

dependence on p_μ or E_μ (from the $p_\mu \leftrightarrow t_\mu$ correlation). As a consequence $g_{t_\mu}(t_\mu)|_s$ can be considered as a 2-D density $g_{t_\mu,E_\mu}(t_\mu,E_\mu)|_s$ in the longitudinal phase-space with parameter s. This is illustrated in FIG. 5 in the case $p_\pi \in [200,400]$ MeV/c and $s = 30$ m.

The muon population at arbitrary s can also be reconstructed from Eq. 11 : the (t_μ,E_μ) space can be meshed, $N_0 g_{t_\mu,E_\mu}(t_\mu,E_\mu)|_s \Delta p_\mu \Delta E_\mu$ gives the local number of points on the mesh. This is illustrated in FIG. 6-left, taking $p_\pi \in [200,400]$ MeV/c and $s = 40$ m.

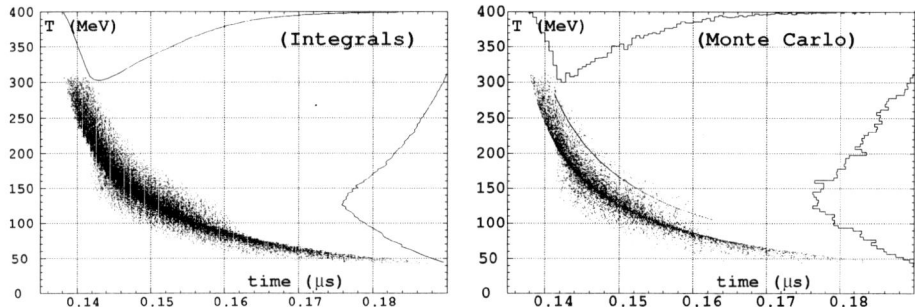

FIGURE 6. Muon longitudinal phase-space (time-Kinetic energy) at $s = 40$ m (the thin arc on the right plot is the survived pion bunch).

Monte Carlo simulations of longitudinal phase-space at distance s along a drift axis are displayed in Fig. 6-right for comparison, showing excellent agreement, in particular it is seen that time and energy projected densities superimpose fairly well.

RF parameters

The first two moments of the marginal densities g_{t_μ} and g_{E_μ} can be calculated as described earlier, in the conditions illustrated in FIG. 5, which yields the following results. The bunch distribution in time and in energy satisfy

$$\overline{ct}_\mu - \sigma_{ct_\mu} = 33.69 \pm 1.69 \text{ m}, \qquad \overline{E}_\mu \pm \sigma_{E_\mu} = 148 \pm 54.4 \text{ MV}$$

The *rms* time extent determines the choice of 44 MHz RF frequency for a half-wave extent, whereas σ_{E_μ} determines a ≈ 54 MV total RF voltage, for bunch rotation (consistent with the CERN design parameters [2,Sec.5.2]).

The *rms* bunch emittance is $\varepsilon/\pi = 1.23$ eV.s, yielding a

capture efficiency of 64%

namely, the ratio of the number of muons contained in the *rms* bunch to the total number of muons. The ct_μ to E_μ correlation coefficient is -0.87. The proton bunch length upper limit in order to avoid excessive muon bunch lengthening is about 5 ns (in quadratic mean).

CONCLUSION

The model described in this paper explains the shape of the density functions of a muon beam and allows calculating the 2-D longitudinal phase-space density. The calculations can be applied to a realistic pion spectrum once HARP will have provided its results [3], yet the hypothesis of uniform spectrum is fairly well fulfilled for a $\pi^\pm$ momentum interval of 150 − 500 MeV/c [4], allowing an estimate of various parameters entering muon capture dynamics.

The energy and time densities can be given an simple integral form and thus calculated numerically almost instantaneously. Moreover several of the integral expressions involved have an analytical primitive, this has not been detailed for the sake of shortness.

Various quantities relevant to beam dynamics can be derived : average beam momentum applies to the adjustment of the focusing strength of the quadrupoles in the decay channel. The mean energy and the energy spread affect the RF voltage. The mean arrival time and the muon bunch length are relevant to the RF phase and the choice of the RF frequency. The predictions of the model have been compared with results from Monte Carlo simulations for validation.

Calculation and transport of the *transverse* densities and phase-space portraits have been undertaken in a similar way [5], and will be subject to further publication.

Further developments and applications can be foreseen, including fast propagation of densities by methods of second order transport, using techniques of random variables and their combination.

Acknowledgments

We thank A. Verdier (CERN) for fruitful discussions, O. Napoly (CEA) for pointing out an error on $g_{\theta^*}(\theta_\mu^*)$ in an earlier stage of the work and J. Doornbos (TRIUMF) for Monte Carlo cross checks using GEANT.

References

1. B. Autin, F. Lemuet, F. Méot, A. Verdier, Optimization Of A Quadrupole Funnel/Decay Channel, these proceedings, and
B. Autin, F. Méot, A. Verdier, Efficiency Of An Alternating Gradient Muon Collection Channel, CERN NUFACT Note 128 (2003)
2. Study Of A European Neutrino Factory Complex, CERN NUFACT Note 122 (2002)
3. Expected for July 2003, cf. A. Tonazzo, "Status of HARP", Muon week 17-19 Nov. 2003, CERN, http://muonstoragerings.web.cern.ch/muonstoragerings/Events/200311
4. B. Autin, Technical challenges of neutrino factories, NIM A Vol. 451, Issue 1, 21 Aug. 2000, Pages 244-254, and
Johann Collot, Harold G. Kirk and Nikolai V. Mokhov, Pion production models and neutrino factories, *ibidem*, Pages 327-330
5. F. Méot, Proc. FFAG03 workshop, KEK, Tsukuba (July 2003)

A Pulsed Synchrotron for Muon Acceleration at a Neutrino Factory

D. J. Summers*, A. A. Garren[†], J. S. Berg[¶] and R. B. Palmer[¶]

*Dept. of Physics and Astronomy, University of Mississippi–Oxford, University, MS 38677
[†]Dept. of Physics, University of California, Los Angeles, CA 90095
[¶]Brookhaven National Laboratory, Upton, NY 11973

Abstract. A 4600 Hz pulsed synchrotron is considered as a means of accelerating cool muons with superconducting RF cavities from 4 to 20 GeV/c for a neutrino factory. Eddy current losses are held to less than a megawatt by the low machine duty cycle plus 100 micron thick grain oriented silicon steel laminations and 250 micron diameter copper wires. Combined function magnets with 20 T/m gradients alternating within single magnets form the lattice. Muon survival is 83%.

Historically synchrotrons have provided economical particle acceleration. Here we consider a pulsed muon synchrotron [1] for a neutrino factory [2]. The accelerated muons are stored in a racetrack to produce neutrino beams ($\mu^- \to e^- \bar{\nu}_e \nu_\mu$ and $\mu^+ \to e^+ \nu_e \bar{\nu}_\mu$). Neutrino oscillations have been observed at experiments [3] such as Homestake, Super–Kamiokande, SNO, and KamLAND. Further exploration using a neutrino factory could reveal CP violation in the lepton sector [4].

This synchrotron must accelerate muons from 4 to 20 GeV/c with moderate decay loss ($\tau_{\mu^\pm} = 2.2\ \mu$S), using magnet power supplies with reasonable voltages. To reduce voltage, magnet gaps are minimized to store less magnetic energy. Cool muons [5] with low beam emittance allow this. Acceleration to 4 GeV/c might feature fixed field dogbone arcs [6, 7] to minimize muon decay loss. Fast ramping synchrotrons [6, 8] might also accelerate very cool muons to higher energies for a $\mu^+\mu^-$ collider [9].

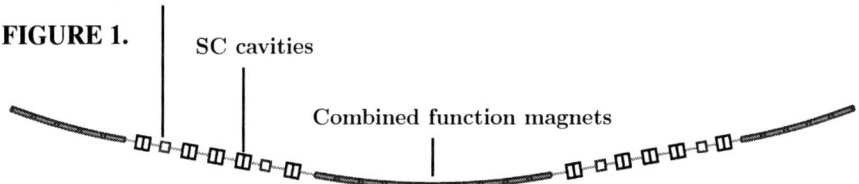

FIGURE 1.

We form arcs with sequences of combined function cells within continuous long magnets, whose poles are alternately shaped to give focusing gradients of each sign. A cell has been simulated using SYNCH [10]. Gradients alternate from positive 20 T/m gradient (2.24 m long), to zero gradient (.4 m long) to negative 20 T/m gradient (2.24 m) to zero gradient (0.4 m), etc. See Fig. 1 and Table 1. It is proposed to use 5 such arc cells to form an arc segment. These segments are alternated with straight sections containing RF. The phase advance through one arc segment is $5 \times 72^0 = 360^0$. This being so, dispersion suppression between straights and arcs can be omitted. There are 18 arc segments and 18 straight sections, forming 18 superperiods in the ring. Straight sections (22 m) without dispersion are used for superconducting RF, and, in two longer straights (44 m), the injection and extraction. To assure sufficiently low magnetic fields at the cavities, relatively long field free regions are desirable. A straight consisting of two half cells would allow a central gap of 10 m between quadrupoles, and two smaller gaps at the ends. Details are given in Table 3. Matching between the arcs and straights is not yet designed. The total circumference of the ring including combined functions magnets and straight sections adds up to 917 m ($18 \times 26.5 + 16 \times 22 + 2 \times 44$).

TABLE 1. Combined function magnet cell parameters. 5 cells/arc. 18 arcs form the ring.

Cell length	5.28 m
Combined Dipole length	2.24 m
Combined Dipole $B_{central}$	0.9 T
Combined Dipole Gradient	20.2 T/m
Pure Dipole Length	0.4 m
Pure Dipole B	1.8 T
Momentum	20 GeV/c
Phase advance/cell	72^0
beta max	8.1 m
Dispersion max	0.392 m
Norm. Trans. Acceptance	4π mm rad

TABLE 2. Superconducting RF.

Frequency	201 MHz
Gap	.75 m
Gradient	15 MV/m
Stored Energy	900 J
Muons per train	5×10^{12}
Orbits (4 to 20 GeV/c)	12
No. of RF Cavities	160
RF Total	1800 MV
ΔU_{beam}	110 J
Energy Loading	.082
Voltage Drop	.041
Acceleration Time	37 μS
Muon Survival	.83

The superconducting RF (see Fig. 1 and Table 2 and note that 11 MV/m has been achieved so far [11]) must be distributed around the ring to avoid large differences between the beam momentum (which rises in steps at each RF section) and the magnetic field (which rises continuously). The amount of RF is a tradeoff between cost and muon survival. Time dilation permits extra orbits with little muon decay if the RF sags.

TABLE 3. Straight section lattice parameters.

ϕ	77^0
$L_{cell}/2$	11 m
L_{quad}	1 m
dB/dx	7.54 T/m
a	5.8 cm
β_{max}	36.6 m
σ_{max}	1.95 cm
B_{pole}	0.44 T
U_{mag}/quad	$\approx$ 3000 J

TABLE 4. Permeability ($B/\mu_0 H$). Grain oriented silicon (3% Si) steel has a far higher permeability parallel ($\parallel$) to than perpendicular ($\perp$) to its rolling direction [12]. Grain oriented silicon steel permits high fields with little energy ($B^2/2\mu$) stored in the yoke.

Material	1.0 T	1.5 T	1.8 T
1008 Steel	3000	2000	200
Grain Oriented ($\parallel$)	40000	30000	3000
Grain Oriented ($\perp$)	4000	1000	

The muons accelerate from 4 to 20 GeV. If they are extracted at 95% of full field they will be injected at 19% of full field. For acceleration with a plain sine wave, injection occurs at 11^0 and extraction occurs at 72^0. So the phase must change by 61^0 in 37 μS. Thus the sine wave goes through 360^0 in 218 μsec, giving 4600 Hz.

Estimate the energy stored in each 26.5 m long combined function magnet. The gap is about .14 m wide and has an average height of h = .06 m. Assume an average field of

1.1 Tesla. The permeability constant, μ_0, is $4\pi \times 10^{-7}$. $W = B^2/2\mu_0$[Volume] = 110 000 Joules. Next given one turn (N = 1), an LC circuit capacitor, and a 4600 Hz frequency; estimate current, inductance, capacitance, and voltage.

$$B = \frac{\mu_0 NI}{h} \to I = \frac{Bh}{\mu_0 N} = 52\,\text{kA}; \qquad W = \tfrac{1}{2}LI^2 \to L = \frac{2W}{I^2} = 80\,\mu\text{H} \qquad (1)$$

$$f = \frac{1}{2\pi}\sqrt{\frac{1}{LC}} \to C = \frac{1}{L(2\pi f)^2} = 15\,\mu\text{F}; \quad W = \tfrac{1}{2}CV^2 \to V = \sqrt{\frac{2W}{C}} = 120\,\text{kV} \qquad (2)$$

The stack of SCRs driving each coil might be center tapped to halve the 120 kV. Nine equally spaced 6 cm coil slots could be created in the top and bottom of each yoke using 6 cm of taller laminations to cut the voltage by ten, while leaving the pole faces continuous. 6 kV is easier to insulate than 120 kV. It will be useful to shield [1] and/or chamfer [13] magnet ends to avoid large eddy currents where the field lines typically do not follow laminations. Neutrino horn power supplies are of interest.

Calculate the resistive energy loss in the copper coils. There are two 5 cm square copper conductors each 5300 cm long. R = 5300 $(1.8\,\mu\Omega\text{-cm})/(2)(5^2) = 190\,\mu\Omega$. So, $P = I^2 R \int_0^{2\pi} \cos^2(\theta)\,d\theta = 260\,000$ w/magnet. Eighteen magnets give a total loss of 4680 kW. But the neutrino factory runs at 30 Hz. Thirty half cycles of 109 μsec per second gives a duty factor of 300 and a total I^2R loss of 16 kW. Muons are orbited in opposite directions on alternate cycles. If this proves too cumbersome, the duty cycle factor could be lowered to 150. See if .25 mm (30 gauge) wire is usable. The skin depth [14], δ, of copper at 4600 Hz is $(\rho/\pi f \mu_0)^{1/2} = (1.8 \times 10^{-8}/\pi 4600 \mu_0)^{1/2} = 0.97$ mm.

Now calculate the dissipation due to eddy currents [15] in a w = .25 mm wide conductor, which consists of transposed strands to reduce this loss [13, 15]. To get an idea, take the maximum B-field during a cycle to be that generated by a 0.025m radius conductor carrying 26 kA. The eddy current loss in a conductor made of square wires .25 mm wide (Litz wire [16]) with a perpendicular magnetic field is as follows. $B = \mu_0 I/2\pi r = 0.2$ Tesla.

$$P = [\text{Volume}]\frac{(2\pi f B w)^2}{24\rho} = [2\,.05^2\,53]\frac{(2\pi\,4600\,.2\,.00025)^2}{(24)1.8 \times 10^{-8}} = 1400\,\text{kW} \qquad (3)$$

Multiply by 18 magnets and divide by a duty factor of 300 to get an eddy current loss in the copper of 85 kW. Stainless steel water cooling tubes will dissipate a similar amount of power [6]. Alloy titanium cooling tubes would dissipate half as much.

Grain oriented silicon steel is chosen for the yoke due to its high permeability at high field at noted in Table 4. The skin depth [14], δ, of a lamination is $(\rho/\pi f \mu)^{1/2}$ = $(47 \times 10^{-8}/\pi 4600\,1000\,\mu_0)^{1/2}$ = 160 μm. ρ is resistivity. Take $\mu = 1000\mu_0$ as a limit on magnetic saturation and hence energy storage in the yoke. Next estimate the fraction of the yoke inductance that remains after eddy currents shield the laminations [17]. The lamination thickness, t, is 100 μm [18]. $L/L_0 = (\delta/t)(\sinh(t/\delta) + \sin(t/\delta))/(\cosh(t/\delta) + \cos(t/\delta)) = 0.995$. So it appears that magnetic fields can penetrate 100 μm thick laminations at 4600 Hz. Thicker 175 μm laminations [12] would be half as costly and can achieve a bit higher packing fraction. $L/L_0(t = 175\,\mu\text{m}) = 0.956$.

Do the eddy current losses [15] in the 100 μm thick iron laminations. Use equation 3 with a quarter meter square area, a 26.5 m length, and an average field of 1.1 Tesla. P = $[(26.5)(.5^2)](2\pi\,2600\,1.1\,.0001)^2/[(24)47 \times 10^{-8}]$ = 5900 kW. Multiply by 18 magnets and divide by a duty factor of 300 to get an eddy current loss in the iron laminations of 350 kW or 700 watts/m of magnet. So the iron will need some cooling. The ring only ramps 30 times per second, so the $\int \mathbf{H} \cdot d\mathbf{B}$ hysteresis losses will be low, even more so because of the low coercive force (H_c = 0.1 Oersteds) of grain oriented silicon steel. This value of H_c is eight times less than 1008 low carbon steel.

The low duty cycle of the neutrino factory leads to eddy current losses of less than a megawatt in a 4600 Hz, 917 m circumference ring. Gradients are switched within dipoles to minimize eddy current losses in ends. Muon survival is 83%.

This work was supported by the U. S. DOE and NSF. Many thanks to K. Bourkland, S. Bracker [19], C. Jensen, S. Kahn, H. Pfeffer, G. Rees, Y. Zhao, and M. Zisman.

1. D. Summers *et al.*, J. Phys. **G29** (2003) 1727; PAC, hep-ex/0305070; S. Berg, A. Garren, R. Palmer, G. Rees, D. Summers, Y. Zhao, www-mucool.fnal.gov/mcnotes/public/pdf/muc0259/muc0259.pdf.
2. A. Blondel *et al.*, Nucl. Instrum. Meth. **A451** (2000) 102;   R. Palmer *et al., ibid.,* 265; D. Neuffer, IEEE Trans. NS–28 (1981) 2034; D. Cline, D. Neuffer, AIP Conf. Proc. **68** (1980) 846; D. Ayres *et al*, physics/9911009;   N. Holtkamp, D. Finley, *et al*, "A feasibility study of a neutrino source based on a muon storage ring," Ferrmilab-Pub-00-108-E;   S. Ozaki, R. Palmer, M. Zisman, J. Gallardo, *et al*, "Feasibility study II of a muon based neutrino source," (2001) BNL-52623.
3. R. Davis *et al.* (Homestake), PRL **20** (1968) 1205; B. Cleveland *et al.*, Astrophys. J. **496** (1998) 505; Y. Fukuda *et al.* (Super-K), Phys. Rev. Lett. **81** (1998) 1562; Q. Ahmad *et al.* (SNO), Phys. Rev. Lett. **89** (2002) 011301;   011302;   **87** (2001) 071301;   S. Ahmed *et al.* (SNO), nucl-ex/0309004; H. H. Chen, Phys. Rev. Lett. **55** (1985) 1534; K. Eguchi *et al.* (KamLAND), *ibid.* **90** (2003) 021802.
4. V. Barger *et al.*, Phys. Rev. Lett. **45** (1980) 2084;   Phys. Rev. **D62** (2000) 073002;   013004; S. Geer, Phys. Rev. **D57** (1998) 6989; S. Bilenky *et al., ibid.* **D58** (1998) 033001; J. Burguet-Castell *et al.*, hep-ph/0207080; C. Albright *et al.*, hep-ex/0008064; K. Kodama *et al.*, hep-ex/0012035; A. De Rujula *et al.*, Nucl. Phys. **B547** (1999) 21;   A. Romanino, Nucl. Phys. **B574** (2000) 675; A. Cervera *et al.*, Nucl. Phys. **B579** (2000) 17; M. Koike and J. Sato, Phys. Rev. **D61** (2000) 073012.
5. A. Skrinsky, V. Parkhomchuk, Sov. J. Part. Nucl. **12** (1981) 223; D. Neuffer, Part. Accel. **14** (1983) 75; R. Fernow, J. Gallardo, Phys. Rev. **E52** (1995) 1039; M. Alsharo'a *et al.*, PRSTAB **6** (2003) 081001; G. Penn and J. Wurtele, Phys. Rev. Lett. **85** (2000) 764; C. Wang and K. Kim, *ibid.* **88** (2002) 184801; J. Norem *et al.*, PRSTAB **6** (2003) 072001;   D. Li *et al.,* J. Phys. **G29** (2003) 1683; R. Palmer, "Ring coolers," J. Phys. **G29** (2003) 1577; S. Berg, R. Fernow and R. Palmer, *ibid.,* 1657; D. M. Kaplan *et al.,* Nucl. Instrum. Meth. **A503** (2003) 392;   D. M. Kaplan, physics/0306135; R. P. Johnson *et al.*, AIP Conf. Proc. **671** (2003) 328; Y. Derbenev and R. Johnson, "Six-dimensional muon beam cooling in a continuous, homogeneous, hydrogen absorber," COOL '03, Mt. Fuji, Japan.
6. D. J. Summers, Snowmass 2001, hep-ex/0208010.
7. S. Berg *et al.,*, PAC '01, http://accelconf.web.cern.ch/AccelConf/p01/PAPERS/RPPH044.PDF.
8. D. Summers, D. Neuffer, Q. S. Shu and E. Willen, PAC 97, Vancouver, physics/0109002; D. Summers, Snowmass '96, physics/0108001;   Bull. Am. Phys. Soc. **39** (1994) 1818.
9. D. Cline, Nucl. Instr. Meth. **A350** (1994) 24; D. Neuffer, *ibid.,* 27; AIP Conf. Proc. **156** (1987) 201; V. Barger *et al.,* Phys. Rev. Lett. **75** (1995) 1462; hep-ph/9803480; S. Choi, J. Lee, hep-ph/9909315; C. M. Ankenbrandt *et al.,* Phys. Rev. ST Accel. Beams **2** (1999) 081001;   R. Palmer *et al.,* Nucl. Phys. Proc. Suppl. **51A** (1996) 61;   R. Raja and A. Tollestrup, Phys. Rev. **D58** (1998) 013005.
10. A. Garren *et al.,* AIP Conf. Proc. **297** (1994) 403.
11. R. L. Geng, P. Barnes, D. Hartill, H. Padamse, J. Sears, S. Calatroni, E. Chiaveri, R. Losito and H. Preis, "First RF test at 4.2K of a 200MHz superconducting Nb–Cu cavity," TPAB049, PAC '03.
12. http://www.aksteel.com/markets/electrical_steels.asp; R. Bozorth, "Ferromagnetism," (1950) 90–1.
13. N. Marks, "Conventional Magnets – I and II," Jyväskylä Accel. School, CERN 94-01, **II**, 867–911.
14. P. Lorrain, D. Corson, and F. Lorrain, "Electromagnetic fields and waves," 3rd ed. (1988) 537–42.
15. H. Sasaki, "Magnets for fast–cycling synchrotrons," Indore Conf. Synchrotron Rad., KEK 91-216.
16. MWS Wire Industries, Westlake Village, CA 91362, http://www.mwswire.com/litzmain.htm.
17. K. L. Scott, Proc. Inst. Radio Eng. **18** (1930) 1750–64.
18. Arnold Engineering, 300 North West St., Marengo, IL 60152, http://www.grouparnold.com.
19. B. M. Lasker, S. B. Bracker and W. E. Kunkel, Publ. Astron. Soc. Pac. **85** (1973) 109.

Nonlinear Acceleration Modes in FFAGs with Fixed RF [1]

Shane Koscielniak* and C. J. Johnstone[†]

*TRIUMF, 4004 Wesbrook Mall, Vancouver, B.C., V6T 2A3, Canada
[†]Fermi National Accelerator Laboratory, Batavia, IL 60510, USA

Abstract. A signature of fixed-field acceleration is that the orbit of the beam centroid unavoidably changes with energy. resulting in a phase slip of the particle beam relative to a fixed-frequency accelerating waveform. The present work explores the influence of the fixed points and of rf manipulations on the longitudinal dynamics in FFAGs with path-length dependencies on kinetic energy; emphasis is given to quadratic dependence as occurs in a type of accelerator currently proposed for rapid acceleration of muons.

INTRODUCTION

A signature of fixed-field acceleration is that the orbit of the beam centroid unavoidably changes with energy. The corresponding, often nonlinear, change in path length results in phase slip of the particle beam relative to a fixed-frequency accelerating waveform. Nonetheless, depending on the number and location of the fixed-points of the motion, acceleration is possible for a limited number of turns by allowing the beam to cross back and forth across the crest. For multiple fixed points, this asynchronous acceleration is facilitated by a libration path that extends from injection to extraction energy when a thresold value of accelerating voltage is exceeded. Successful acceleration is accomplished when the radio frequency is made synchronous with the revolution period at a specific machine central orbit, or when the rf is offset in a manner that leads to staggering of phase traces on consecutive turns. This work explores the influence of the fixed points and of rf manipulations on the longitudinal dynamics in FFAGs with emphasis on a quadratic dependence as occurs in a type of accelerator currently proposed for rapid acceleration of muons.

NON-SCALING FFAGS

One very important feature allowed by fast acceleration is the freedom to cross betatron resonances. In a Fixed-Field Alternating Gradient (FFAG) accelerator lattice of the non-

---

[1] TRIUMF receives federal funding via a contribution agreement through the National Research Council of Canada. Work also supported by the Universities Research Association, Inc. under contract DE-AC02-76CH03000

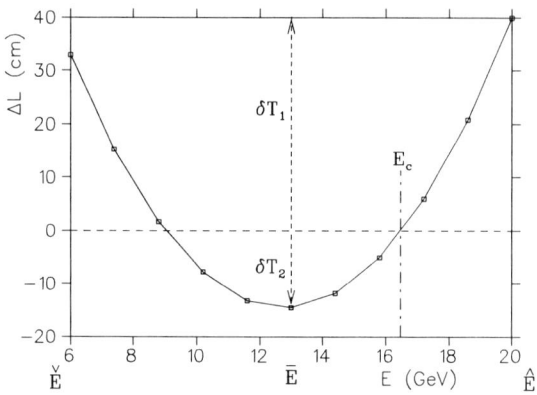

FIGURE 1. Path-length increment versus energy for a proposed muon accelerator.

scaling type considered here[1, 2], the optics change slowly with energy, crossing resonance tunes. With variable optics the machine lattice can be built from linear magnetic elements only (quadrupoles and dipoles) with a corresponding large dynamic aperture. In such a machine, the orbital path length is (almost) parabolic as a function of energy, figure 1, and the sign of the phase-slip reverses (possibly twice) during acceleration for a single, fixed rf frequency. A variable-optics, or *nonscaling*, type of FFAG is of particular interest because it provides an opportunity to consider a strong nonlinear oscillator as the model for the longitudinal dynamics. Here we demonstrate that the lack of isochronism can be overcome and bucket-less acceleration supported by utilising a curious form of libration that links injection to extraction energy when the rf voltage exceeds a threshold value. The insight gained using different variants of FFAG acceleration (scaling versus non-scaling, for example) has lead to a general prescription for acceleration between multiple fixed points each of which represents a condition of perfect synchronism with the rf. As with conventional acceleration, the limiting performance criterion remains an acceptable extraction phase space. In the first part of this work we introduce a theoretical background for interpreting the results of particle tracking studies that are reported in the second part.

Acceleration in a non-scaling FFAG

Although the reference orbit may vary with energy (figure 1), it is customary to define some "central orbit" and corresponding "central energy" E_c. This orbit is not necessarily at the centre of the magnet aperture, and nor is the energy necessarily the mean energy of the acceleration range; rather E_c it the location of a zero of the path-length dependence. For the purpose of simple calculations, accelerating cavities are assumed equally spaced a distance L_0 apart on the central orbit, and are driven by a single frequency ω sinusoid with peak voltage V. Initially we allow L_0 and ω to be independent variables. Consider the ultra-relativistic limit $\beta \to 1$ and $d\beta/d\gamma \to 0$. In this case non-isochronism results only from path-length dependence on energy $\Delta L(E)$. We define $T_0 = L_0/c$ and $\Delta T = \Delta L/c$ with c the speed of light.

When the means of acceleration is discrete rather than distributed, the motion is governed by difference equations. Let the iteration index be n. The energy E and arrival times t are

$$\begin{aligned} E_{n+1} &= E_n + eV\cos(\omega t_n) \\ t_{n+1} &= t_n + T_0 + \Delta T(E_{n+1}) \,. \end{aligned} \quad (1)$$

We introduce a relative time coordinate $T_n = t_n - n\tau_s$ where $\tau_s \equiv 2h\pi/\omega$ and the harmonic number $h = \text{Int}[\omega\tau_0/2\pi]$ is an integer. This constitutes a transformation to a moving frame in which we anticipate the physics will look simpler.

$$\begin{aligned} E_{n+1} &= E_n + eV\cos(\omega T_n) \\ T_{n+1} &= T_n + \Delta T(E_{n+1}) + (\tau_0 - \tau_s) \,. \end{aligned} \quad (2)$$

Because τ_s is directly linked to ω there is a physically observable overall slip-rate unless ω is made to depend on τ_0. We now make the distinction between *normal* ($\tau_0 = \tau_s$) and *slip* ($\tau_0 \neq \tau_s$) modes of operation. In the former case, the rf is synchronous with the iteration period on the central orbit. In the latter case, the rf is synchronous with orbits at some other energy $E \neq E_c$. In the normal mode, L_0 is the only free variable; apart from the choice of a harmonic number, ω (and hence τ_s) is determined uniquely from L_0. In the slip mode, both L_0 and ω retain the status of free variables (to be fixed after developing a rationale for optimization).

One may wonder if the equations (1) or (2) are useful for acceleration; this depends on the nature and location of the sets of first-order fixed points which are the simultaneous solutions of $E_{n+1} = E_n$ and $T_{n+1} = T_n$. The latter condition is that of *perfect synchronism*. At a more prosaic level, what is important is whether the phase slip during energy traversal between adjacent fixed points of $T_{n+1} = T_n$ is limited to a value less than π. The question is essentially whether there is enough rf voltage to slide the beam in energy from one condition of synchronism to another before the accumulated phase slip modulus exceeds π. A similar argument applies to the permissable energy range of a manifold of periodic orbits, such as those contained within an rf bucket. Although determining conditions may be found for the difference equations in terms of discrete sums, it is far simpler to obtain them from the hamiltonian of the approximating differential equations; typically one finds relations between energy excursions y and the restoring constant λ which is itself proportional to voltage.

Acceleration with a quadratic path-length dependence on energy

Whereas orthodox particle accelerator designs are predicated on the use of systems which are oscillators for excursions about some reference orbit, we shall consider reference orbits which are themselves nonlinear oscillations. For mathematical symplicity we shall replace the difference equations (2) by their continuous approximations. In doing so, we appeal to the following property. The fixed points of first-order symplectic difference equations and of their differential equation counterparts are identical.

In a fixed/varying field machine, reference energy varies with radius/time. In the case of a scaling, radial-sector FFAG[3, 5, 6, 7], orbits are staggered radially outward as a function of momentum (or, equivalently, energy for a relativistic machine) and the path-length change can be approximately linear with energy. The use, in a scaling FFAG, of gradient magnets with large field index to encompass orbits with a large energy change typically results in lattice nonlinearity and compromises the transverse aperture; and it is for that reason that we consider here the non-scaling type.

The case of quadratic path-length dependence on energy is therefore quite important since it corresponds to that of the non-scaling FFAG accelerator, which is considered for rapid acceleration of muons[1, 2, 8, 9, 10, 11, 12]. Figure 1 shows an example of the variation in path length encountered in these machines; evidently the approximation by a parabola is well justified. Consider the model equations:

$$dy/ds = \lambda(\pi/2)\cos(x\pi/2), \quad dx/ds = y^2 - 1, \quad y \propto (E - \bar{E}). \qquad (3)$$

The motion is synchronous at the two energies $y = \pm 1$; and the direction of phase slip reverses above/below these values. In each interval $\Delta x = 4$ (i.e. $\omega \Delta T = 2\pi$) there are two elliptic $\mathbf{x}_{1,2} = \pm(1,1)$ and two hyperbolic $\mathbf{x}_{3,4} = \pm(1,-1)$ fixed points[2]. For $\lambda \neq 0$ there is always rotational motion about $\mathbf{x}_{1,2}$ bounded by a separatrix; we shall refer to each of these two sets of periodic motions as a *manifold*. Although there is no linear term y in x', there is some similarity to behaviour of buckets at transition crossing[4, 13] in a synchrotron.

Figure 2 show contours of constant hamiltonian, and how the topography changes in response to varying λ. The changes are discontinuous at $\lambda = 2/3$. For smaller λ values there are three bands of libration separated by manifolds; whereas for larger values a striking new feature of the phase space is a serpentine libration that meanders between $y = +2, -2, +2, -2, \ldots$ while x increases without limit. This 'gutter' feature[3], emphasised in figure 2, can be used to augment the range of acceleration over and above the scaling or linear case.

A beamlet may be accelerated in either of the manifolds over a range of 3 units in y via a nonlinear oscillation; or the beam may accelerated in the gutter over the range $\mathbf{x} = \pm(1,2)$ for a energy gain of 4 units in y. In the former case, the crest of the waveform is crossed twice; whereas in the latter it is crossed three times giving a greater energy gain. Thus two reversals of phase slip is actually an advantage; because it allows the beam to cross the crest three times before the phase slips to values where the beamlet is decelerated.

The waveform may be locally flattened at crest ($x = 0$) by the addition of a higher harmonic; however, this is decelerating and must be compensated by boosting the fundamental. Hence the restoring force becomes

$$y'(x,n) \propto [n^2 \cos(x\pi/2) - \cos(nx\pi/2)]/(n^2 - 1). \qquad (4)$$

---

[2] In assigning coordinates to respective indices, we take the positive sign first.
[3] We have in mind a street gutter.

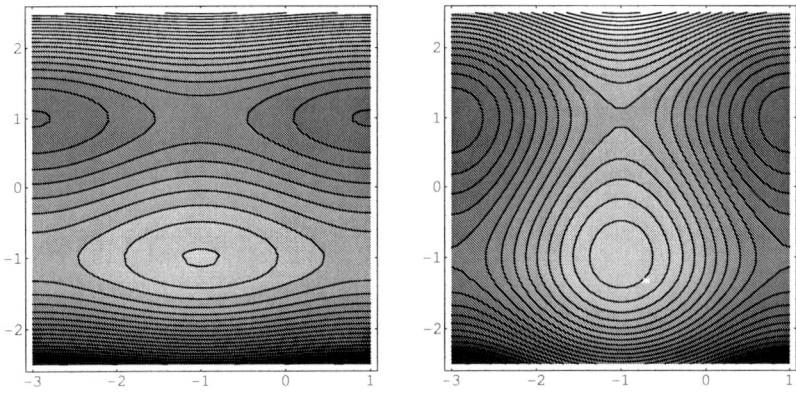

FIGURE 2. Quadratic pendulum. At left $\lambda = 1/4$; at right $\lambda = 4/3$.

There are two options for selecting the voltage. An *aggressive* approach in which one determines the number of turns N_t for acceleration, $\delta E = \Delta E/N_t/N_c$, resulting possibly in a high value for λ. A *conservative* approach starting with the selection of λ. Using the critical value results in miniscule acceptance and paths which waste time near the waveform zero-crossing. Adopting $\lambda \geq 1$ typically yields a healthy acceptance and allows to inject and extract closer to crest. In this case the energy increment is estimated to be

$$\delta E \simeq \psi/\alpha = \omega \delta T_2 \Delta E/4 = \omega(\delta T_1 + \delta T_2)\Delta E/16 . \qquad (5)$$

This is the fundamental relation between energy increment and range and phase slip for the non-scaling FFAG. Evidently it is advantageous to lower the radio-frequency and reduce the total change in path length.

Particle Tracking

As a particular example, we take a non-scaling FFAG intended for muon acceleration between 6 and 20 GeV. The machine[2] of 2 km circumference is comprised of 300 FODO cells and equipped with 200 MHz rf cavities. The ring has path-length variation $\Delta L(E)$ up to 50 cm (120° of rf phase) as shown in figure 1. The fixed rf is roughly the 1334th harmonic of the revolution frequency. If the ring were operated with synchronous rf, a frequency sweep of $\simeq 44$ kHz would be needed.

Longitiudinal particle tracking was performed for synchronous on-crest acceleration and for both styles (normal and slip rf)[14] of fixed-frequency, asynchronous cross-crest acceleration. In this study, acceleration was completed in a number of turns ranging from two to twenty. The tracking includes the variation of kinematic β with γ and interpolates the path-length dependence from a table since this is not precisely parabolic. Some details of the computer program are given in references [15, 16]. Initially the longitudinal phase plane is uniformly flooded with trial particles on a rectangular grid; and one attempts to accelerate them. Particles which survive the complete acceleration to $\simeq 20$ GeV are recorded and used to map out the input acceptance and corresponding output emittance. We follow the convention that these quantities are the actual occupied areas; there is no need to multiply by π. The tracking results (beam transmission and quality) are very similar whether 300 or 20 cells are employed; and for computational speed 100 cells were used.

A prediction of the previous section is that transmission be related to the ratio of path lengths $\delta T_1/\delta T_2$ which parametrises the parabola; and that the optimum ratio is roughly 3. This is confirmed in figure 3 which was generated by sweeping δT_2 while holding $(\delta T_1 + \delta T_2)$ fixed.

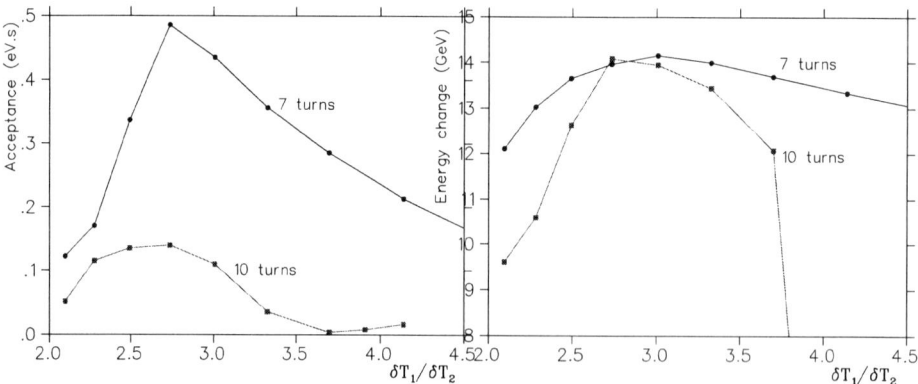

FIGURE 3. Acceptance and energy versus $\delta T_1/\delta T_2$ for acceleration completed in 7 and 10 turns

In a synchronous rf scheme, the phase of the reference particle is fixed. In the normal and slip rf schemes, the reference particle crosses (repeatedly) the crest of the accelerating waveform and so both are asynchronous rf schemes. Slip rf results in a phase profile (versus energy) that is discontinuous between turns but that has a smller r.m.s. phase variation than that generated by the normal rf scheme, which exhibits a continuous variation of phase. Whereas the slip mode minimises the quadratic sum $\sum \Delta\phi_{\text{ref}}^2$, the normal mode ideally minimises the linear sum $\sum \Delta\phi_{\text{ref}}$ provided that $\Delta L(E)$ and the injection/extraction conditions are perfectly symmetric.

For six or less turns, the output emittance for the slip scheme is superior to either normal or synchronous (i.e. rephased) rf schemes. For seven or more turns, the normal rf scheme appears somewhat preferable, but in either asynchronous scheme acceleration cannot be extended beyond ten turns.

When anti-phased second harmonic voltage is added to the fundamental waveform, several benefits are derived: (i) the acceptance is increased by 80% on average because a wider range of rf phases are transmitted; (ii) the efficiency factor increases and the over-voltage may be reduced slightly; and (iii) the 'gutter' paths do not close off until 14-15 turn acceleration.

Third harmonic has a smaller impact; on average the acceptance is increased by only 22%. Though the gains in transmission are smaller than for second harmonic, the third harmonic voltage requirement and increment in fundamental are a factor of three smaller; and this is a strong argument in its favour.

CONCLUSION

The 'fast regime' in a fixed-field accelerator opens a new frontier of beam dynamics in which path-length variations with energy and fixed radio-frequency systems combine to give a longitudinal phase space that is both useful for acceleration and rich in new physics. Although conventional synchronous acceleration is not possible in these machines, an asynchronous mode of acceleration has been discovered in which the reference particle performs a nonlinear oscillation involving repetitive crossing of the crest of the rf waveform. In order to understand this mode, a new, encompassing concept of acceleration has been developed in terms of the general equations which describe nonlinear oscillators. Based on this model, a comprehensive treatment of acceleration has been developed which accurately describes the evolution of machine dynamics beyond the simple linear description as applied to a synchrotron (for example). The beauty of the concept is that it appears to apply to any magnetic-rf based acceleration scheme in which a reference orbit can be accurately parameterized in the energy coordinate.

Also compared in this work is a practical rephasing of rf cavities at injection. In the simple scenario, termed "normal rf", the radio-frequency is matched to the revolution period on an optimally-chosen central orbit. In this and previous work, it was proposed and found to be advantageous to introduce a slight rf mismatch that results in inter-turn phase jumps; under certain conditions this "slip rf" results in improved transmission.

In the 6-20 GeV FFAG example, the magnet lattice and 200 MHz rf conspire to give a miniumum $120°$ phase-slip and, under such conditions, acceleration from injection to extraction is not, in general, achievable. With the sophisticated approach elaborated here, both rf schemes lead to successful acceleration with a viable phase space acceptance of about $\simeq 0.25$ eV.s. In conclusion, the authors can definitively state the viability of nonlinear acceleration and predict successful application to rapid acceleration.

REFERENCES

1. C. Johnstone and S. Koscielniak: *FFAGs for Rapid Acceleration*, NIM-A, Vol. 503, issue 3, May 2003.
2. C. Johnstone and S. Koscielniak: *Recent Progress on FFAGs For Rapid Acceleration*, Proc. Summer Study on Future of Particle Physics, Snowmass CO, 30 Jun-Jul 2001 (SLAC eConf).
3. D. Kerst et al: *FFAG Particle Accelerators*, CERN Symposium on High Energy Accelerators, Geneva Swtizerland, 1956, pg.32
4. K. Symon and A. Sessler: *Methods of RF acceleration in FFAGs*, ibid, pp.44-58.
5. S. Machida et al: *Beam Optics Design of an FFAG Synchrotron*, Proc. 7th Proc. European Particle Accelerator Conf. June 2000, Vienna Austria, pg.557.
6. M. Yoshimoto et al: *Recent Beam Studies of the PoP FFAG Proton Synchrotron*, Proc. 2001 Particle Accelerator Conf. June 2001, Chicago IL, pg.51
7. M. Aiba et al: *Study of Acceptance of FFAG Accelerator*, Proc. 8th European Particle Accelerator Conf. June 2002, Paris France, pg.1226.
8. J.S. Berg: *Longitudinal Reference Particle Motion in Nearly Isochronous FFAG Recirculating Accelerators*, ibid.
9. C. Johnstone and S. Koscielniak: *Rapid Acceleration in a FFAG Using High-Frequency RF*, 20th ICFA Advanced Beam Dynamics Workshop on High Intensity and High Brightness Hadron Beams, Batavia Illinois, 2002, AIP Conf. Proc. 642.
10. J.S. Berg: *Longitudinal dynamics on high-frequency FFAG recirculating accelerators*, ibid.

11. C. Johnstone and S. Koscielniak: *Recent Progress on FFAGs for Rapid Acceleration*, Proceedings of EPAC 2002, Paris, France.
12. J.S. Berg: *Dynamics in Imperfectly-Isochronous FFAG Accelerators*, ibid.
13. C. Limborg: *Review of Difficulties in Achieving Short Bunches*, Proc. 6th European Particle Accelerator Conf., June 1998, Stockholm Sweden, pg.151
14. S. Koscielniak: *Mechanisms for Nonlinear Acceleration in FFAGs with Fixed RF*; to be published in NIM-A.
15. S. Koscielniak: *Preliminary longitudinal dynamics studies of a muon beam in a 6-20 GeV non-scaling FFAG*; TRIUMF laboratory report TRI-DN-01-15.
16. S. Koscielniak: *Multiple radio-frequency schemes for acceleration of muons in a non-scaling FFAG*; TRIUMF laboratory report TRI-DN-02-06.

Lattice Design and Particle Tracking of FFAG

S.Machida

KEK, 1-1 Oho, Tsukuba-shi, 305-0801, Japan

Abstract. Recent update of lattice design and beam tracking results of Fixed Field Alternating Gradient (FFAG) synchrotrons for muon acceleration is described. Doublet lattice works well as a 10 to 20 GeV/c final ring. Tracking results show there is not much coupling between horizontal and vertical planes.

INTRODUCTION

Japanese design of neutrino factory is based on a cascade of FFAG synchrotrons from 0.3 GeV/c to 20 GeV/c. Both in size and in cost, the final ring design from 10 GeV/c to 20 GeV/c is crutial. In the first proposal of FFAG based Japanese Neutrino Factory design, we chose triplet focusing structure. We have further studied other focusing schemes and performed particle tracking to estimate dynamic aperture.

1. LATTICE DESIGN

Let us first summarize constraints on the lattice for a final muon acceleration ring in Table 1. The FFAG lattice we first proposed is triplet radial sector type shown in Fig. 1. That is similar to the Proof of Princeple (PoP) proton FFAG synchrotron that has been commissioned in 2000 [1]. To satisfy the listed constraints, triplet is not the only solution. If a singlet focusing is employed, lattice functions become the one shown in Fig. 2 and a doublet focusing is shown in Fig. 3.

Table 1: Constraints on the final FFAG acceleration ring.

Momentum range	10 to 20 GeV/c
Orbit excursion	0.25 m or less
Vertical beam size	0.15 m (full size)
Physical emittance	300 π mm-mrad
Average machine radius	120 m or less
Maximum bend field	8 T (6 T is preferred)

Although triplet focusing is proven and works well in a small-scaled model such as the PoP FFAG, it may not be an optimum choice for high momentum ring. Edge focusing becomes small because bending angle per cell is small. That results in small phase advance in vertical plane. In that respect, singlet (FODO) lattice is better since there is not much difference in horizontal and vertical phase advance. One obvious drawback, however, is a shorter straight length. If the number of focusing cell is the same, the straight length is a half of triplet or doublet. Therefore, doublet focusing is a compromise. It has similar lattice functions in both transverse planes like singlet. The straight length is similar to that of triplet focusing. In those considerations, now we choose a doublet focusing lattice as a baseline. Table 2 shows main parameters.

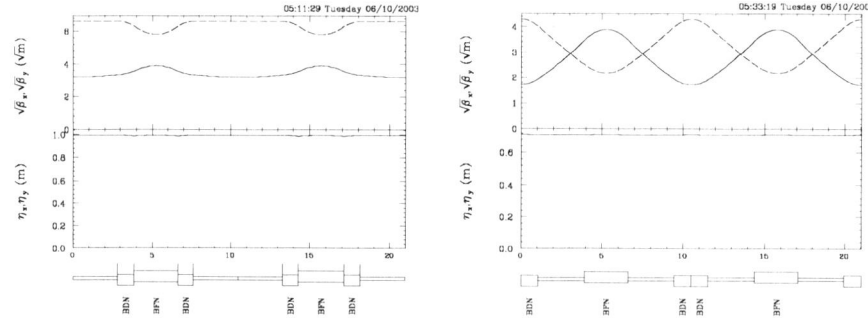

FIGURE 1. Lattice functions of 10 to 20 GeV/c final muon accelerating ring with triplet focusing.   **FIGURE 2.** Lattice functions of 10 to 20 GeV/c final muon accelerating ring with singlet focusing.

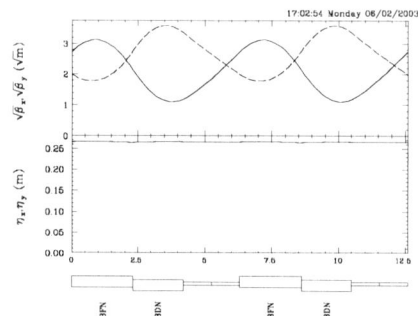

FIGURE 3. Lattice functions of 10 to 20 GeV/c final muon accelerating ring with doublet focusing.

Table 2: Main parameters of doublet focusing lattice.

	Low field version	High field version
Average radius	120 m	120 m
Number of cell	120	180
k-value	450	800
Orbit excursion	0.18 m	0.10 m
Maximum bend field	6 T	8 T
Average energy gain	0.83 MV/m	0.83 MV/m
Turn number	22	22
Length of straight	2 m	1.4 m

2. PARTICLE TRACKING

The FFAG based design does not assume cooling so that muon emittance is large. The ring should have an enough dynamic aperture although turn number is only a few 10 turns. Multi-particle tracking has been done to confirm the enough aperture. The following tracking has been done on the triplet lattice with a soft-edge magnet.

Polymorphic tracking code (PTC) developed by Forest [2] is employed. First, realistic magnetic field in 3-D is calculated with TOSCA. PTC reads the field map and then advances particle coordinates by kick-drift-kick algorism. Since it is a straightforward tracking, there is no assumption of center momentum and results should be exact for any momentum. That is important for tracking simulation of muon ring because the momentum changes a lot, from −33% to 33%.

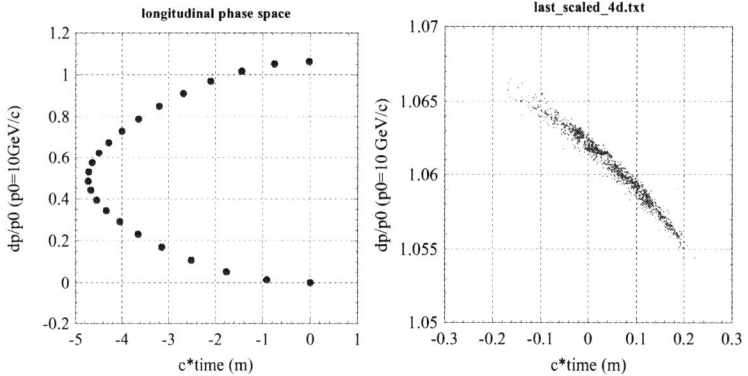
FIGURE 4. Each point (left) corresponds to each turn. Right is distribution at the final turn.

Figure 4 shows longitudinal phase plot during the accelerating process and particle distribution at the final momentum. One huge RF bucket has momentum height covering from 10 to 20 GeV/c. A half period of a synchrotron oscillation brings a beam to the final momentum.

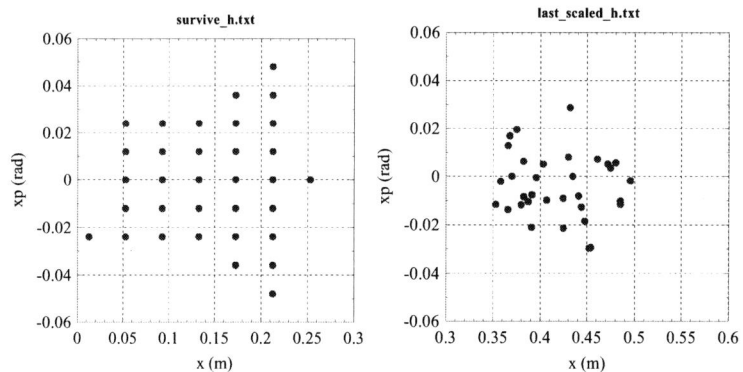
FIGURE 5. Survived particles at the beginning and its final distribution in horizontal plane.

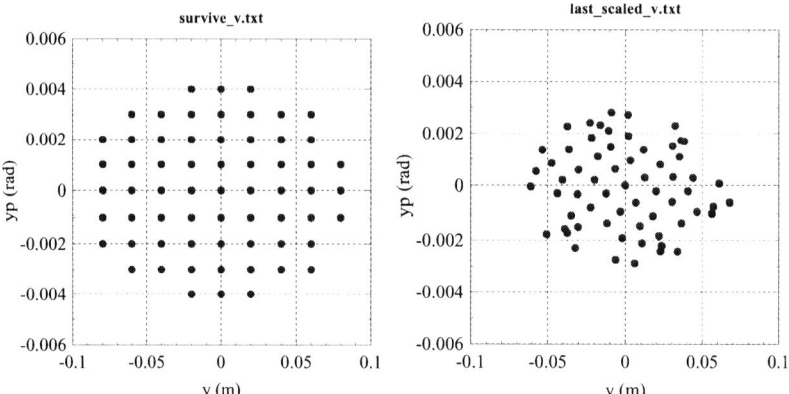

FIGURE 6. Survived particles at the beginning and its final distribution in vertical plane.

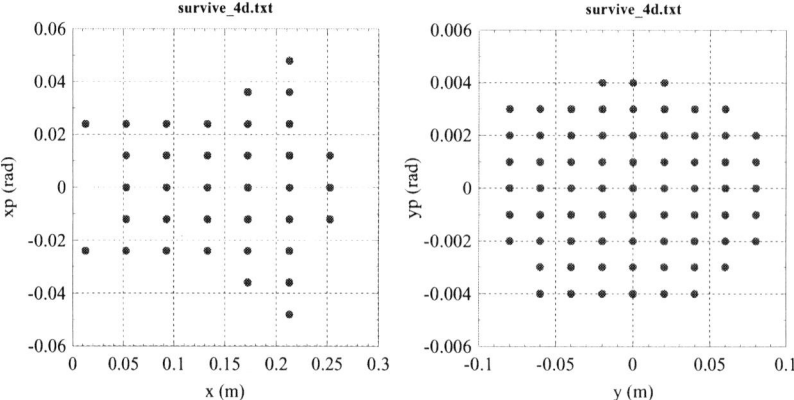

FIGURE 7. Survived particles in H (left) and in V (right) when we start with all 4-D distribution.

We have first looked only horizontal dynamic aperture during those accelerating process. Grid points in the horizontal phase space show particles survived as in Fig. 5. Horizontal particle distribution at the final momentum is also depicted. Similarly, Fig. 6 shows particles survived in vertical plane and final particle distribution. Then, we start with all 4-D transverse particle distribution. Figure 7 shows horizontal and vertical grid points which survive for a whole acceleration period. In that tracking, 1682 particles out of 11 x 11 x 11 x 11 grid points survives. In comparison, either only horizontal or vertical tracking shows that 33 particles out of 11 x 11 grids in horizontal plane or 63 particles out of 11 x 11 in vertical plane survives. Therefore, the reduction of dynamic aperture due to horizontal and vertical coupling is about 20%, namely 1682 vs. 33 x 63. It is not negligible, but not significant either.

REFERENCES

1. Y. Mori, et. al., "Development of a FFAG proton synchrotron", p. 581, Proc. of EPAC2000.
2. E. Forest, et. al., "Polymorphic Tracking Code", KEK Report 2002-3, also CERN-SL-2002-044 (AP).

FFAG Construction for PRISM

Akira Sato (for the PRISM-Working Group)

Department of Physics, Osaka University, 1-1 Machikane-yama, Toyonaka, Osaka 560-0043, Japan

Abstract. The construction of the PRISM-FFAG ring has been started. It would be completed by the end of JFY 2005. Then phase-rotation, muon acceleration and muon ionization cooling could be studied using the ring. This paper describes the present designs of the RF system and the magnet system of PRISM-FFAG and its construction schedule.

INTRODUCTION

PRISM [1] is a project to realize a super muon beam which combines high-intensity, low-energy, narrow energy-spread and high purity. Its aimed intensity is about $10^{11} - 10^{12}$ $\mu^{\pm}$/sec. The muon beam would be provided with a low kinetic energy of 20MeV±6% to optimize for the stopped muon experiments such as a $\mu^- - e^-$ conversion search experiment [2]. PRISM consists of mainly three sections. They are a large solid-angle pion capture section with a solenoid magnet field of about 10 T, a $\pi - \mu$ decay section consisting of a 10-m long super-conducting solenoid magnet, and a phase rotation section to make the beam energy spread narrower. To achieve such superb beam characteristics, a Fixed Field Alternating Gradient synchrotron (FFAG) [3] will be used as a phase rotator. The advantages of FFAG are that it has a large momentum (longitudinal) acceptance, and it also has a wide transverse acceptance with strong focusing, and it has synchrotron oscillation which is needed to perform phase rotation. According to simulations, initial energy spread of 20MeV±40% is reduce down to ±6% after 5 turns of muons in the FFAG ring. In the FFAG ring almost all pions decay into muon, hence extracted beam has extremely low pion contamination. We plan to locate the PRISM at the Tokai-site in Japan to utilize the J-PARC 50GeV proton synchrotron as a proton driver.

PRISM-FFAG CONSTRUCTION

Among the all PRISM components, the phase rotator section can be constructed from japanese fiscal year (JFY) of 2003 for five years. The PRISM-FFAG ring with eight triplet magnets, one RF system, and one kicker magnet will be constructed as shown in Fig.1, although originally the PRISM-FFAG needs twelve RF systems and two kickers in order to accomplish the aimed full performance. Other components would be constructed in near future. Using this prototype PRISM-FFAG ring, phase rotation, muon acceleration and muon ionization cooling will be studied. PRISM-FFAG has two chal-

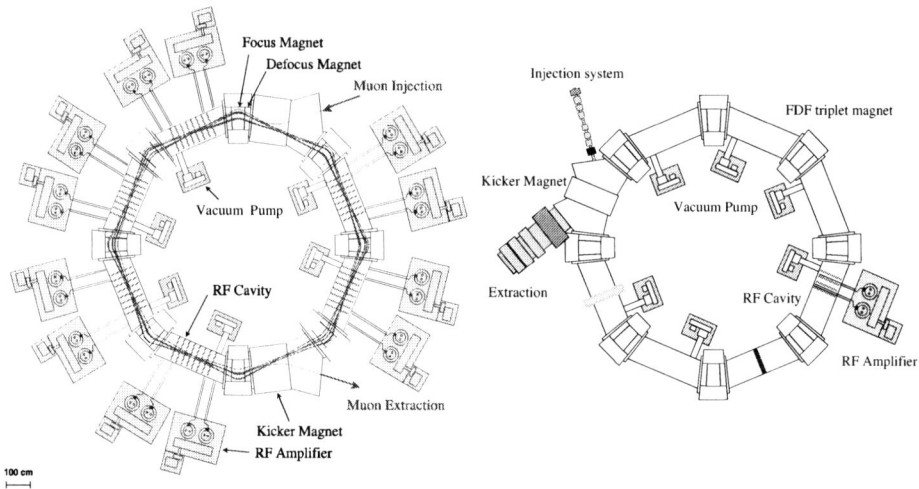

FIGURE 1. Layout of the PRISM-FFAG ring for full performance (left) and for the prototype (right). A red line indicate simulated muon track.

lenging R&D components. They are a high field gradient RF system and a large aperture FFAG-magnets. The present designs of these items and their construction schedule are described in the following sections.

High Field Gradient RF system

To increase a number of surviving muons, phase rotation has to be completed as short time as possible. Simulations showed that 5 turns (1μsec) in the FFAG ring of 5m radius is sufficient to complete phase rotation, keeping a survival rate of 60%. In present design, the PRISM-FFAG ring has eight straight sections of about 2.5m long. Among these, two of them are used for injection and extraction muons. RF systems can be installed into the other sections. Hence PRISM requires very high field gradient of 200-300kV/m at the low frequency (5 MHz). As compared with usual cavities, PRISM has to operate its cavities at a remarkably outstanding condition as shown in Fig.2.

Such a operation can be achieved by a low duty factor and ultra-thin magnetic alloy (MA) cavities [4]. MA core [6] has stable impedance at a required magnetic field for PRISM (320-490 Gauss), while the impedance of ferrite cores change at a high RF field, as shown in Fig.2. To optimize phase rotation, not only a high field gradient but also the shape of RF voltage are important. According to our simulations, a saw-tooth RF voltage makes final energy spread narrower than that by a sinusoidal one. Therefore, adding higher frequency harmonics to form a saw-tooth voltage is being considered. By using the cut core configuration [7], a wide band RF system with μQf @ 5MHz = 5.5×10^9 can be designed. The first and second harmonics could be applied on a RF

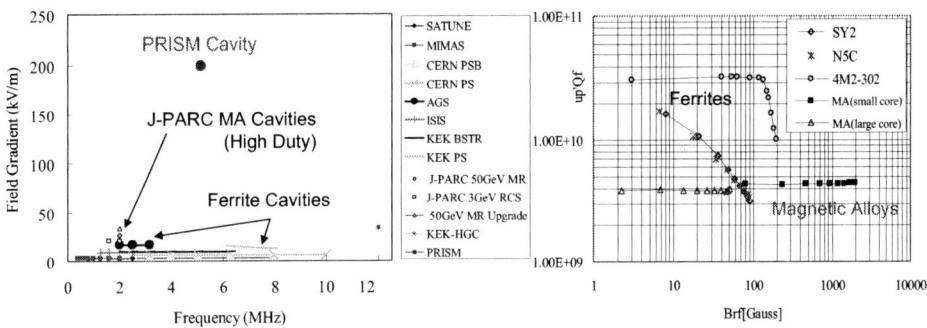

FIGURE 2. (Left) Field Gradient vs. Frequency for usual cavities [5] and PRISM cavity. (Right) Characteristics of magnetic cores.

TABLE 1. Parameters of PRISM-FFAG RF system.

Number of gap per cavity	4	RF frequency	5MHz
Length of cavity	1 m	Field gradient	200kV/m
Number of core per cavity	16	Flux density in core	320 Gauss
Core material	Magnetic Alloy	Tetrode	4CW150,000E
Core shape	Racetrack	Duty	<0.1%
Core size	1.4m × 1.0m × 3.5cm		
Expected shunt impedance	>180Ω/core		

simultaneously with sufficient efficiency. Side and front views of the two cavities are shown in Fig. 3. Two sets of cavities are installed in one straight section. The 1 m cavity consists of 4 gaps, and each gap has 4 MA cores. Each gap generates the RF voltage of 25-37.5 kV and is driven by two bus bars which are connected to a RF amplifier. To drive 4 gaps in a cavity, 4 push-pull amplifiers with 4CW150,000E tetrodes are used. A low duty factor (<0.1%=15μsec × 100/3.4sec) enables the tubes to generate the maximum RF power of 1.8 MW. Parameters of the RF system are summarized in Table1.

Large Aperture Magnet

The FFAG magnet has to produce a magnetic field of $B(r) = B_0(r/r_0)^k$, where r is a radial position, and B_0 is a magnitude of a magnetic field at the radius of r_0. k is a constant. For previous FFAGs build at KEK, PoP FFAG [8] and 150MeV proton FFAG [9], such a magnetic field gradient was created by changing a magnet gap size by $h(r) = h_0(r_0/r)^k$, where h_0 is a gap size at r_0. Hence the smaller gap size at outer radius makes the acceptance smaller. In order to improve the acceptance to larger, we are developing a new type of FFAG magnets, which create field gradient by thin coils so that the magnet gap can be flat and large. A principle and a preliminary design of this magnet are described in Ref.[10].

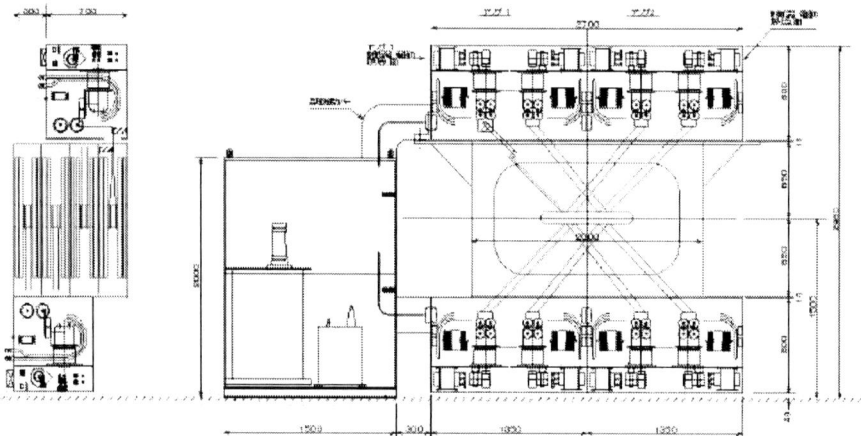

FIGURE 3. Side(left) and front(right) views of PRISM RF system. Two amplifiers are located above the racetrack-shape cavities and others are set below it.

Schedule

The PRISM-FFAG magnet design will be finalized by the end of JFY 2003. The design of RF system had almost finished, and its construction has been started. RF tests, magnets construction and its field measurements will be carried out in JFY 2004 to JFY 2005. The FFAG-ring will be completed by the end of JFY 2005. After commissioning, phase rotation, muon acceleration, and muon ionization will be studied.

REFERENCES

1. "The PRISM Project – A Muon Source of the World-Highest Brightness by Phase Rotation –", LOI for Nuclear and Particle Physics Experiments at the J-PARC (2003)
2. "An Experimental Search for the $\mu^- - e^-$ Conversion Process at an Ultimate Sensitivity of the Order of 10^{-18} with PRISM", LOI for Nuclear and Particle Physics Experiments at the J-PARC (2003)
3. C. Ohkawa, Proceedings of annual meeting of JPS (1953) , K.R.Symon et. al., Phys. Rev. **103** (1956) 1837-1859
4. C.Ohmori et al., to be published as a proceedings of SAST03
5. M. Fujieda et al., gStudies of Magnetic Cores for JHF Synchrotrons hProceedings of PAC97.
6. C.Ohmori et al., gA Wide Band RF Cavity for JHF Synchrotrons hProceedings of PAC97.
7. C. Ohmori et al., gHigh Field Gradient Cavity loadedwith MA for Synchrotrons hPAC99
8. M.Aiba et al., "Development of a FFAG Proton Synchrotron", Proceedings of EPAC2000 , M.Yoshimoto et al., "Recent Beam Studies of the POP FFAG Proton Synchrotron", PAC2001 , M.Yoshimoto et al., "Dynamic Aperture of the PoP-FFAG Proton Synchrotron", EPAC 2002
9. T.Adachi et al., "A 150MeV FFAG Synchrotron with "Return-Yoke Free" Magnet", Proceedings of PAC2001 , M.Aiba et al., "150 MeV Fixed Field Alternating Gradient Accelerator and Return-yoke Free Magnet", EPAC 2002
10. M.Yoshimoto, in this proceedings

Author Index

A

Abazajian, K., 256
Ajima, Y., 325
Alberico, W. M., 350
Aoki, M., 136, 325
Appelquist, T., 261
Autin, B., 48, 325, 338, 455

B

Bakule, P., 313
Balbekov, V., 387, 428
Baldini, A., 68, 136, 289
Barger, V., 3
Barnes, P., 445
Bennett, J. R. J., 48, 317
Berg, J. S., 391, 463
Berz, M., 413, 418
Bilenky, S. M., 350
Bodek, A., 358
Bodek, K., 305
Bogacz, S. A., 449
Bouchez, J., 37
Brooks, S. J., 399
Bross, A. D., 97
Budzanowski, A., 305
Burguet-Castell, J., 235

C

Calatroni, S., 445
Campanelli, M., 202
Casper, D., 235
Chen, M.-C., 269
Chiaveri, E., 445
Choubey, S., 198
Cline, D., 395
Cummings, M. A., 436

D

Danneberg, N., 305
de Gouvêa, A., 175

Donini, A., 219
Drumm, P., 441

E

Edgecock, T. R., 114, 144

F

Fernow, R. C., 90, 387, 391
Fetscher, W., 305
Fleming, B. T., 379
Formaggio, J. A., 183
Fukasawa, N., 325
Fukui, Y., 395

G

Galea, R., 403
Gallardo, J. C., 391
Garen, A. A., 463
Garren, A., 395
Geng, R. L., 445
Gilardoni, S., 48, 334
Giunti, C., 170
Grawer, G., 334

H

Harold, M. R., 399
Hartill, D., 445
Haseroth, H. D., 48
Hilbes, C., 305
Hoffman, K. D., 432
Huber, P., 194

I

Ikedo, Y., 342
Ishibashi, K., 325
Ishida, K., 77, 293, 309, 313, 342, 346

Itahashi, K., 309
Iwasaki, M., 293, 309, 313

J

Jarczyk, L., 305
Johnstone, C. J., 329, 413, 467

K

Kahn, S., 239, 387
Kirch, K., 305
Kirk, H., 395
Kirstryn, S., 305
Kitching, P., 301
Kodama, K., 231
Köhler, K., 305
Koscielniak, S., 467
Kozela, A., 305
Kumano, S., 29
Kuno, Y., 325

L

Lang, J., 305
Lemuet, F., 338
Lindroos, M., 37
Link, J. M., 83
Losito, R., 445
Louis, W. C., 20

M

Machida, S., 144, 475
Mahanthappa, K. T., 269
Maieron, C., 350
Maire, G., 334
Makimura, S., 313
Makino, K., 413, 418
Maltoni, M., 165
Mason, G., 293
Matsuda, Y., 293, 309, 313
Matsuzaki, T., 293, 309, 313
Maugain, J.-M., 334
McConnel, K., 247
Méot, F., 48, 338, 455

Mezzetto, M., 37
Migliozzi, P., 223
Minakata, H., 206
Miura, T., 325
Miyachi, Y., 354
Miyadera, H., 342, 346
Miyake, Y., 313
Morse, W., 285
Murayama, H., 122

N

Nagamine, K., 293, 309, 313, 342, 346
Nakahara, K., 325
Nakamoto, T., 325
Nakamura, K., 12
Nakamura, S. N., 293
Neuffer, D., 407
Nishiyama, K., 342
Nosaka, N., 325
Numajiri, M., 325

O

Ogitsu, T., 325
Ohlsson, T., 265
Ohnishi, H., 325
Onderwater, C. J. G., 297
Ota, T., 252

P

Padamsee, H., 445
Palmer, R. B., 391, 463
Paul, K., 329
Popp, J. L., 321
Preis, H., 445
Prior, C. R., 48, 399, 422

R

Raffelt, G. G., 130
Raja, R., 387
Rajasekaran, G., 243

Rangod, S., 334
Rees, G. H., 399
Reimer, P. E., 371
Rimmer, R. A., 144
Roberts, B. L., 281

S

Sakamoto, S., 293
Sakuda, M., 247
Sato, A., 325, 479
Sato, J., 252
Sears, J., 445
Sergiampietri, F., 363
Shimomura, K., 313, 342, 346
Shrock, R., 261
Sievers, P., 48, 325
Smyrski, J., 305
Stephan, E., 305
Strasser, P., 309, 313
Strzałkowski, A., 305
Sugiyama, H., 211
Summers, D. J., 463

T

Tanaka, H., 342
Tomono, D., 293
Torun, Y., 106
Tzanakos, G. S., 179

U

Usubov, Z., 387

V

Verdier, A., 48, 338
Voelker, F., 334
Von Allmen, A., 305

W

Walter, C. W., 247
Watanabe, I., 293
Whisnant, K., 215
Winter, W., 227

Y

Yamamoto, A., 325
Yamanoi, A., 325
Yang, U.-k., 358
Yasuda, O., 190
Yoshimura, K., 325
Younus, I., 367

Z

Zejma, J., 305
Zeller, G. P., 375
Zisman, M. S., 60

Over the years there has been much concern as to just what constitutes a "Learning Disability." Because of so much conflicting data one may question whether such a disability exists. One thing is certain, if a child is not learning, we as teachers must question the validity of our teaching strategy. We must keep changing our strategies until we get the desired performance from the child. If we do not try alternative strategies we must be concerned with a teaching disability rather than a learning disability.

Traditionally the approach to learning disabilities has been one of causal orientation. This approach has grown out of the interests of medical research. For educators, however, another approach is necessary; what can *we* do for a brain damaged child in our classroom? Educators need a model which looks at the variables of learning in order to enhance learning. The model must be flexible enough to take into account the individual differences in learning among children. Such a model is presented.

In this chapter we will discuss the present definition of learning disabilities and the characteristics that are typically associated with the term learning disabilities.

Let's take a brief look at some of the labels typically applied to children with learning problems. Some you will notice grew out of the causal approach while others grew out of an effectual approach.

chapter 1

what is the concept "learning disability?"

contents

1. what is the concept "learning disability?"

2. learning disabilities vs. teaching disabilities

3. controversy regarding educational programming

4. how to teach children with learning problems

Mainstreaming requires that regular classroom teachers accept greater responsibility for children who are not succeeding. Many special education programs are full of children who do not belong there.

The majority of learning disabled children spend most of their time in the regular class program anyway, so what we are concerned with is providing regular classroom teachers with information and suggestions that will help them work more effectively with children who are experiencing learning difficulties. After reading this material, we hope you will have a better understanding of the learning disabled child's uniqueness, and some of the causal factors involved. We hope you will understand the issues surrounding the concept of learning disabilities, and recognize the advantages and disadvantages of the two main approaches of providing services to learning disabled children. Most importantly, we hope that the chapter on teaching children with learning problems provides you with some useful suggestions for working with this child. Remember, the suggestions can be employed even if a child does not have the categorical label "specific learning disabilities" attached to him/or her.

THOMAS N. FAIRCHILD & FERRIS O. HENSON, II

introduction

The concept of learning disabilities is relatively new in special education. It has only been around since the late 50's and early 60's. In the 70's we see a rapid proliferation of programs for children identified as having specific learning disabilities.

Many persons raise the question "weren't there learning disabled children in our schools prior to the late 50's?" The answer is an unequivocal "YES." There were many children who were failing in our schools as a direct result of an inherent learning problem, or as a result of situational factors which the child could not control. As we began to learn more and more about how children learn—their uniqueness, and their styles, we began to recognize a special group of children who were having academic difficulties that were not related to mental retardation. The failing learner is not a novelty of our time. He/she has been around for all time. Fortunately, our educational technology has advanced to the point where children with learning disabilities can be identified and provided with educational experiences that will enable them to achieve academic success.

Learning disabilities are probably the most controversial area in the field of special education. There is considerable disagreement regarding the definition of learning disabilities, the causes of learning disabilities, and the remediational techniques which should be employed. Special education programs and services for learning disabled children have mushroomed as the prevalence figures have continued to rise. The current state of affairs in the area of learning disabilities is that programs for learning disabled children are serving an overwhelmingly large number of nonlearning disabled children. Because the definition is so general and vague it has been easy to "fit" many children into the definition in order to provide them with needed individual attention.

Unfortunately, this has reinforced and perpetuated the attitude that if a child is not learning at grade level using grade level materials, he or she should be receiving special educational services. Frequently when a child does not succeed in the regular classroom situation it is assumed that he or she has a "problem." It is believed that there is something "wrong" with the child and he or she is sent to a special education program. Some refer to this as the "disordered child" viewpoint. Obviously, there are children with disorders, but there are also large numbers of "nondisordered" children receiving special services. If we always assume that a child is having a learning problem becaue of an inherent disorder, we forget to look for other possible explanations for the child's learning problem, e.g., our teaching methodology, inappropriate curriculum, or other factors that are operating in the child's environment.

preface

In the past, the educational needs of exceptional children were met by removing them from the "mainstream" of the regular classrooms, and serving them in a variety of segregated self-contained special classes. The trend in the '70's is educating exceptional children in the least restrictive educational setting; that is, as close as possible to their normal peers. This concept of "mainstreaming" exceptional children has received considerable support from within and outside the educational community. Although self-contained special classes will always be a meaningful alternative for some children, the personal and educational needs of many exceptional children can better be served in the regular class program with the supportive services of ancillary personnel and/or resource room help.

With the emphasis on "mainstreaming," the regular classroom teacher is now expected to meet the needs of exceptional children in his or her classroom along with all the other children in the class. The problem is that most regular classroom teachers have little or no preparation in the area of educating exceptional children. Regular classroom teachers need basic information regarding the various exceptionalities, and more specifically, practical suggestions which they can employ to enhance the "mainstreamed" exceptional child's personal and educational development.

The MAINSTREAMING SERIES was written to fill this need. Each book in the SERIES addresses itself to one area of exceptionality allowing teachers to select from the SERIES according to their interest or need. Each text provides information designed to correct misconceptions and stereotypes, and to improve the teacher's understanding of the exceptional child's uniqueness. Numerous practical suggestions are offered which will help the teacher work more effectively with the exceptional child in the "mainstream" of the regular classroom.

Currently, there is a great deal of controversy surrounding the use of categories and labels. The books in the SERIES are organized according to categories of exceptionality because the content within each book is only relevant for a child with a specific handicapping condition. The intent is not to propagate labeling; in fact, labeling children is inconsistent with the philosophy of the SERIES. The books address themselves to behaviors, and how teachers can work with these behaviors in exceptional children. The books in the SERIES are categorized—not the children. The books are categorized in order to cue teachers to the particular content for which they might be looking.

There is much truth in the old saying, "A picture is worth a thousand words." A cartoon format was used for each book in the MAINSTREAMING SERIES as a means of sustaining interest and emphasizing important concepts. The cartoon format also allows for easy, relaxed reading. We felt that teachers, being on the firing line all day, would be more likely to read and refer to our material, than to a lengthy text filled with theory and jargon. Typically cartoons exaggerate, stereotype, and focus on weaknesses. I sincerely hope that these cartoons do not offend any children, parents, or professionals, because that is not the purpose for which they were intended. They are intended to make you think.

I hope you find this book helpful in your work with mainstreamed exceptional children, or with any other children, since they are all special.

THOMAS N. FAIRCHILD
SERIES EDITOR

acknowledgments

As I sit at my desk looking at the manuscript which consumed so much time and energy, I would like to sincerely express my appreciation to Ferris Henson for being my friend and dropping everything to assist me; my wife, Carolyn, for her constant encouragement and support; and last *but* not least Marcy Taylor, for typing the rough, the final, and proofreading it all. She has worked as hard on the manuscript as anyone.

THOMAS N. FAIRCHILD

To
Carolyn and Leatha

371.9
H398m

Library of Congress Cataloging in Publication Data

Henson, Ferris O
   Mainstreaming children with learning disabilities.

   (Mainstreaming series)
   1. Learning disabilities. I. Fairchild, Thomas N., joint author. II. Title.
LC4704.H46   371.9   76-50641
ISBN 0-89384-009-2

Learning Concepts
2501 N. Lamar
Austin, Texas 78705
(512) 474-6911

Copyright© 1976 by Ferris O. Henson, II, Ph.D., Thomas N. Fairchild, Ph.D. and Danial B. Fairchild. All Rights Reserved. Printed in the United States of America. No part of this publication may be reproduced, stored in a retrieval system or transmitted, in any form or by any means, electronic, mechanical, photocopying, recording, or otherwsie, without written permission of the publisher.

mainstreaming children with learning disabilities

by
ferris o. henson, II
university of idaho

thomas n. fairchild
university of idaho

thomas n. fairchild
series editor

danial b. fairchild
thomas n. fairchild
illustrators

LEARNING CONCEPTS
2501 N. Lamar, Austin, Texas 78705 (512) 474-6911